DESIGN OF
LAND TREATMENT SYSTEMS
FOR INDUSTRIAL WASTES—
Theory and Practice

DESIGN OF LAND TREATMENT SYSTEMS FOR INDUSTRIAL WASTES—
Theory and Practice

by

MICHAEL RAY OVERCASH

Associate Professor
Department of Biological and Agricultural
    Engineering
Department of Chemical Engineering
North Carolina State University
Raleigh, North Carolina

Soil Systems, Inc.
Marietta, Georgia

DHIRAJ PAL

Post Doctorate Research Associate
Department of Biological and Argicultural
    Engineering
North Carolina State University
Raleigh, North Carolina

ANN ARBOR SCIENCE
PUBLISHERS INC / THE BUTTERWORTH GROUP

Second Printing, 1981

Copyright © 1979 by Ann Arbor Science Publishers, Inc.
230 Collingwood, P. O. Box 1425, Ann Arbor, Michigan 48106

Library of Congress Catalog Card No. 79-88908
ISBN 0-250-40291-2

Butterworths, Ltd., Borough Green, Sevenoaks, Kent TN15 8PH, England

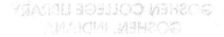

PREFACE

Land treatment is a complete technology representing a major innovative and future approach to industrial waste management. Design of these systems is absolutely critical to achieving environmentally acceptable performance. This book is an effort to unify and develop a total engineering approach for the complete evaluation or design of a land treatment system. We feel this information is timely since within virtually every industrial category there has been consideration of land treatment.

A fundamental concept of the entire land treatment design process as developed in this book is that of nondegradation or maintenance of usable conditions associated with the terrestrial receiver system. That is, careful consideration of the soil, vegetation and groundwater is made throughout the design. This concept predates by two to three years the regulatory efforts of the Resource Conservation and Recovery Act and thus is essential for future regulatory compliance.

The complete methodology or design procedures in this book were developed to include all wastes, regardless of industrial origin or physical properties, such as sludges, effluents, slurries, solid waste. Thus municipal effluents or sludges are one *subset* of possible wastes and can be readily designed. The objective of this work is to establish clearly the technical and engineering basis for design of land treatment systems by means of a unified approach.

Topics covered in the book are: (1) the fundamental design basis; (2) a total design procedure, (3) all categories of waste constituents and the respective information on assimilation in soils, (4) full-scale system design, (5) pretreatment or in-plant source control rationale and options, (6) economic analyses of land treatment and pretreatment, and (7) example systems.

A large body of information collected from a wide variety of experiments not generally related to land treatment is presented in this book. We have interpreted and extrapolated this information for direct use in the technology of land treatment. Often this information is quantitatively tenuous, but serves very effectively as a starting point for design calculations coupled with monitoring of full-scale operation or with pilot-scale testing. Such

v

extrapolation and interpretation rather than the normal review of literature information is strongly needed in this evolutionary stage of land treatment.

Within the book we have used metric units for simplicity. We accept responsibility for the material in the book and sincerely hope that we have held mistakes to a minimum. The natural review and comment that will accompany the use of this book will be accepted by the authors as professional input.

Michael R. Overcash is an Associate Professor in the Department of Biological and Agricultural Engineering and Associate Professor in the Department of Chemical Engineering at North Carolina State University. He has conducted extensive research and undertaken design of a variety of full-scale systems for the land treatment of industrial waste. Dr. Overcash also teaches, in industry and professional meetings, a special intensive course covering the complete design and implementation of industrial pretreatment-land application systems. He received a PhD in Chemical Engineering in 1972 from the University of Minnesota after earning a MS degree as a Fulbright Scholar to the University of New South Wales, Sydney, Australia. In professional societies he has been an active member of a number of environmental and international committees of the American Society of Agricultural Engineers. He is the vice chairman of the Environmental Division of the American Institute of Chemical Engineers. Dr. Overcash has authored over 75 professional papers and research reports in his field of research and development.

Dhiraj Pal is an Agronomist and Soil Scientist presently working as a Research Associate in the Department of Biological and Agricultural Engineering at North Carolina State University. Dr. Pal is internationally known for his scholarly contributions in soil microbiology and biochemistry, agricultural waste and soil management, characterization and disposal of industrial wastewaters, and agricultural chemistry.

The author received a PhD in Soil Science in 1973 from the University of California and has earned a number of awards and honors throughout his academic career. Dr. Pal has visited and worked with numerous research organizations in Canada, India, New Zealand and the United States. He is an active member of several professional societies in agronomy, environment and soil science, and has participated in many workshops, conferences and professional meetings organized to teach and extend research information on pretreatment-land application technology. He has presented more than 15 technical papers before regional, national and international audiences. Dr. Pal is author of over 20 publications.

Dedicated to

Mary
Surekha

Our Parents

ACKNOWLEDGMENTS

The written documentation of those persons responsible for this book is as difficult as the complete characterization of waste from a complex organic chemical manufacturing plant.

To Dr. F. J. (Pat) Hassler and Dr. Frank J. Humenik, I most gratefully return the silver spoon. Dr. Philip W. Westerman has always been counted on for his support as a colleague and hence was instrumental in this work. The scientific encouragement of Mr. David H. Howells, Dr. George J. Kriz and Dr. J. K. Ferrell, and the diverse inputs of Dr. Jim C. Barker, Dr. A. R. Rubin, Dr. G. W. Gilliam, Dr. K. R. Reddy and Dr. R. Khaleel deserve full appreciation. Special thanks to Dr. K. H. Keller for transmitting his approach to science and chemical engineering as the basis for this technological effort. The chief reviewers of this book, Dr. James A. Moore and Dr. Jim Davidson, are sincerely thanked for their efforts. This book was written with the full awareness that as we proceed to newer areas of terrestrial, environmental, and land treatment understanding *au pays des aveugles les borgnes sont rois* (in the land of the blind, the one-eyed men are kings).

The contributions of valuable time, knowledge and full-scale system economics has been substantial from various industries and designer/consultants. Of major importance has been the full cooperation and substantial input from Soil Systems, Inc. Marietta, Georgia, including, Dr. N. S. Fox, Dr. Kirk W. Brown, Mr. Glen Taylor, Mr. Roy Wilson and Mr. Bud Northcutt. Of special note is Dr. Wade L. Nutter for his chapter on hydrologic assimilative capacity. The staff of Barry's Greenhouse has thus had a unique input to this book. Finally the long term professional support of Mr. E. C. Ladd (FMC) is continually appreciated. We also acknowledge our sincere thanks to the authors and publishers of the scientific works that represent the structure and foundation of this book.

Table of Contents

CHAPTER 1
PERSPECTIVE OF LAND-BASED TREATMENT SYSTEMS

GOALS OF INDUSTRIAL WASTE MANAGEMENT

Introduction

Federal Water Pollution Control Act Amendments (FWPCA) of 1972 have established a detailed program and timetable to achieve the renovation of our natural resources through the complete treatment of the wastes and wastewaters of municipalities and industries. The Congressional program defined certain stages or incremental goals to gradually approach the national objective of zero discharge of pollutants into our air and water resources. The incremental stages defined were: (1) evaluation of the United States industrial base and establishment of industrial categories, definition of significantly different manufacturing processes from the waste production standpoint within each industrial category and characterization of these waste streams for potential pollutants; (2) review of all waste treatment practices; (3) selection and implementation by 1977 the Best Practical Control Technology Currently Available (BPCTCA); and (4) the review and upgrading by 1983 to achieve the Best Available Technology Economically Achievable (BATEA).

A development document for Interim Final Effluent Limitations Guidelines and Proposed New Source Performance Standards for each industrial category, beginning with those deemed as the greatest environmental hazard, was prepared as the first stage requirement. Environmental hazard was assessed either for the size and magnitude of the industry waste effluents or for toxicity of the waste streams.

In each development document, the significantly different manufacturing processes were characterized with verbal descriptions of the industry. Included were products manufactured; the basic technology of the processes (chemical, biological or physical) used to transform raw materials to finished products; geographical distribution; and assumptions used in subdividing the industry. Each process was subjected to a mass balance approach considering

1

all inputs, the product outputs and the waste effluents. Further elaboration on the waste streams included constituent concentrations, amounts and variability or ranges for species present in this industry composite.

Controversy exists concerning the precision of the development document effluent characterization, especially when applied to a particular operating plant. However, as an assessment of the major production processes and a comparsion of different industries, the development documents present an excellent perspective of industrial waste management. The framework of significantly different processes will be used in this book to provide a methodical basis for considering subsystems of each total industry category.

The portion of the development documents that addressed the waste treatment alternatives and, subsequently, the selection of BPCTCA and BATEA, is also a subject of considerable debate. The quality and conclusions of the development documents, when compared across industrial categories, demonstrate a wide variability. As a consequence, environmentalists oppose the conclusions in one industry; while in another manufacturing category, those in industry oppose the conclusions.

Viewed from another perspective, there is another basis on which both public environmentalists and industrial environmental specialists may be critical of the development documents. This basis is the absence of any substantial consideration of land-based treatment systems for industrial waste. This important alternative was included in the description of waste treatment alternatives only if it already existed in the industry. Thus, no extrapolation of land treatment technology from one industry to similar industries was made, nor within an industry from one waste stream to another. As a consequence, few industry development documents concluded that zero-discharge goals could be approached utilizing land-based treatment. The reasons for exclusion of land systems were numerous, including the existing state-of-the-art; however, at present, considerably more should and can be written concerning the potential of land-based treatment. In essence that is the goal of this book.

The impact of pollutants from industry, agriculture and municipalities involves both short-term, dramatic effects (eutrophication, fish kills, recreational degradation) and long-term, low-level effects (certain toxic substances). Whether short- or long-term effects, the waste constituents interact with the air and water environments and are basically dispersed with time over an increasing range. The dispersion phenomena are difficult to control or correct, hence, the use of treatment systems prior to reaching these receiver systems is highly desirable.

Initial regulatory constraints were centered on the industrial waste constituents that produced an immediate impact, usually associated with water receiver eutrophication. The waste characterization in the development documents was also concerned with eutrophic species (biochemical oxygen demand, nitrogen, phosphorus), with public nuisance (color, odor, turbidity, solids), and with certain highly prevalent and toxic species (cyanide, acids, etc.). From these environmental constraints were developed effluent quality standards and recommended waste treatment practices for the various industrial processes.

The concepts of BPCTCA and BATEA were developed to control and upgrade stream quality with respect to eutrophic, nuisance and highly toxic parameters. As evidence of this, effluent standards for most industries are specified for a certain level of biochemical oxygen demand (BOD_5), nitrogen, phosphorus and solids. Such an approach was necessarily directed at an immediate problem of water quality in the conventional context.

Expanded research and improved data collection have shown that a severe problem exists outside the above described water quality parameters (CRS 1975). The problem is the introduction of pollutants with low-level but long-term toxicity. Mutagenicity, carcinogenicity and teratogenicity are included in the concept of toxic. As a result of this regulatory gap, several successful lawsuits against the U.S. Environmental Protection Agency (EPA) have initiated a revision of effluent guidelines to include toxic substances. The process is just beginning and will have a tremendous impact on industrial waste management.

As impetus for promulgating regulations for toxic substances, the 307a Section of PL 92-500 and the Toxic Substances Control Act (1976) are relevant. Section 307a was envisioned by Congress as a stringent, technology-forcing regulatory system, which emphasizes concern for public health and is not expressly constrained by cost or the capability of technology (Ward 1977). Difficulties of short time frames, lack of data base, absence of control technology, etc. have limited Section 307a. Therefore, only a few substances are controlled under this section (aldrin/dieldrin, DDT/DDE/DDD, endrin, toxaphene, benzidene and PCE). Other toxic substances will be part of achievement of BAT, thus allowing future development of solutions. The Toxic Substances Control Act takes a long-range approach by attempting to define toxicity of all future compounds manufactured or sold, so that a better base for source and pollution control will be available.

A thorough review of toxic substances is not warranted here, but the reader should consult one of several sources (CRS 1975, Staples 1974, Leland *et al.* 1974) to appraise the impact of the Toxic Substances Control Act on a particular process or industry of concern. Briefly, the waste constituents of concern are (1) natural toxicants and (2) synthetic material with toxic properties.

The first category is primarily heavy metals such as cadmium, arsenic, lead and mercury, which can be viewed as conservative species. These constituents are moved from concentrated locations where they are mined to industrial processes and then are dispersed throughout the environment as the waste effluents. A wide variety of oxidation state changes, bioconcentration and relocation phenomena occur to present a very complex environmental effect. Also in this category of toxicants are naturally radioactive species or certain parameters such as fluoride, which also appear in the effluents from raw materials processing.

The second, category includes man made compounds that can have toxic low-level effects, but which usually exhibit some susceptibility to degradation and detoxification in the environment. The rate of breakdown varies from hundreds of years to several hours. In this category are chlorinated hydrocarbons, polybrominated biphenyls (PBB), aromatic amines, vinyl chloride, nitrosamines, etc.

It is important to evaluate what precisely is meant by toxicity and, in particular, what system is being considered. Species that enter drinking water supplies would then be of concern as "toxic substances." However, a toxic substance that degrades or is satisfactorily assimilated in a plant-soil system should not remain in the same category of public health concern. As the list of toxic substances expands, there are more examples of species that can be assimilated by a plant-soil system in an environmentally acceptable manner, thus providing an industrial waste treatment alternative.

In summary, the constituents in industrial wastes that will be subject to control regulations include the eutrophic and nuisance parameters. In addition, many species even present at low levels will be included; therefore, waste management practices will move to reflect these new constraints. Basically, the toxic substances will have the greatest impact on industrial wastes and few, if any, constituents arising from industrial use can be viewed as acceptable for stream discharge.

Industrial Waste Generation

Considering a large number of industrial processes, plant facilities and modes of production operation leads to some commonalities in the definition of waste effluents. The reason for the similarities is that each industry (from swine production to petroleum refining to electroplating) is converting certain raw materials and inputs through a variety of processes into product. Many of the unit processes are similar and, as always, the raw material-product conversion is not 100% efficient. Four categories of effluents reflect industrial water use as well as sources of pollution:

1. noncontact cooling water,
2. process-generated effluents,
3. auxiliary processes water, and
4. rainfall and groundwater

Noncontact water is generally used for cooling purposes and does not come in direct contact with any raw material, intermediate product or by-product, or with the final product. These waters are either used on a once-through or a recycle basis and discharged accordingly. Although predominantly noncontact, the water interaction with pumps, compressors, etc. introduces a certain amount of grease and oil, metals, salts, etc. to this effluent. The level of contamination is very dependent on the condition of cooling equipment. However, by the present standards, most industrial, noncontact water effluents need no further treatment. Future approaches toward zero discharge may result in reexamination of these noncontact effluents.

Process-generated wates (liquids and solids) are the major source of industrial waste constituents. Included are washwater, transport water, process failures, unreacted or unseparated species, etc. The initial focus of regulatory criteria was on this portion of the industrial plant effluent. Much of the waste generated by an industrial process can be quantified in a reliable manner. However, variability is provided by process failures, in which large quantities of material are included in the raw waste. These pulse inputs can greatly affect certain waste treatment processes and create a sudden adverse environmental impact. Therefore, this segment of industrial wastes is characterized by relatively high concentrations and subject to considerable variation.

In the auxiliary water category are the waste inputs not directly involved in the production process. Cooling tower blowdown, the waste streams from water purification treatment and plant domestic waste, are examples. In general, these streams are a concentrated effluent, but are extremely low in volume compared to the process-generated wastes.

The fourth category is relatively new in many industries as a source of regulatory concern. Basically, rainfall or, for mining operations, groundwater, come in contact with industrial raw materials, manufactured products, industrial facilities or stockpiled wastes and become contaminated. For example, rainfall leaching from iron ore piles, rainfall contacting surfaces of a sulfuric acid plant on which air pollutants are deposited and sludge pits through which rainfall leaches, are sources of rainfall-generated effluents. The randomness of rain occurrence and data concerning the magnitude of this category of waste preclude a major control effort until the previous industrial waste categories are controlled.

Each of these four industrial effluent categories must be considered in the design of industrial plants and the consideration of waste management alternatives. The basic differences in these streams due to pollutant content, magnitude of effluents and degree of controllability generally dictate a different control-treatment strategy for each. All four should also be considered in relation to land application as a treatment alternative.

Generalized Industrial Waste Management Alternatives

The environmental manager responsible for industrial wastes has certain alternative procedures available when considering the overall concept of waste management. These alternatives are depicted in Figure 1.1 and are fairly universal for industrial processes. The selection among these approaches is based on the relative economics of each, subject to satisfying necessary environmental regulations, over the life of the manufacturing plant being considered. Obviously, somewhat less flexibility in existing facilities exists than for the planning of a new one.

It is also important to remember that the economic comparison of alternatives should be done after each is specified at or near the respective cost-effective optimum. That is, an alternative designed and engineered for a particular waste should not be compared to another evaluated using general design estimates, since considerable savings result when a treatment process is designed and improved for a particular waste. A brief description of each of these alternatives is presented below.

In-Plant Changes

Reduction of the amount or nature of material that appears in the various waste streams is one of the most cost-effective alternatives. For example, conversion from wet to dry scrubbing processes removes the need for liquid treatment and yields a dry waste that can be treated more easily. Recycle of waste streams to allow further reaction or separation can reduce the material present in the effluent. Specific recommendations for in-plant changes can be done most effectively by the process engineers, since most factors to be

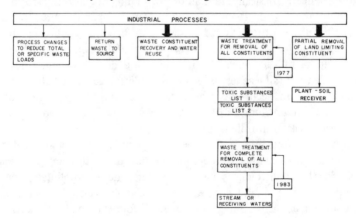

Figure 1.1 Alternatives available for industrial compliance with federal environmental regulations.

included are very process specific; however, the full benefit of this alternative is generally not utilized because the total system is not considered. That is, the usual extent of in-plant changes is based on the cost justification of the product manufactured, not on the combined process of product manufacture plus waste treatment. Under the constraints of total system costs meeting the increasingly stringent streams regulations, greater progress will be achieved by using in-plant changes as an industrial waste alternative.

Reuse and Pretreatment-Recycle

A large amount of the liquid in effluent streams can be recycled for process use after varying degrees of treatment. Associated with this water recycling is the recovery of constituents in the waste streams for reuse or sale. The recovery of materials from waste effluents represents a direct savings in the further cost of waste treatment and potential economic return from the recovered materials. As with in-plant changes, the decisions for materials recovery too often require that the recycle process be self-supporting. A more cost-effective analysis would evaluate recycle processes in terms of both market value of recovered constituents and reduced waste treatment costs. Many regulatory and industry efforts are currently directed at partial or total recycle of industrial effluents, resulting in numerous articles and proceedings (Cecil 1976, Cecil 1975).

Return Wastes to Source Location

At present, two major industry categories employ this waste treatment concept. The first is the seafood processing industry, which by the Ocean Dumping Act is permitted to return fish parts to the sea as part of the natural ocean food cycle. The mining and processing industry, when in proximity to the actual mining operation, often recycles material back to the mine. The localized high concentration of a specific metal or other ore is considered a part of that environment and, therefore, does not constitute a pollutant needing control. Further consideration of returning waste such as heavy metals to locations containing high natural concentrations should be undertaken. This recycling reduces the dispersion of these constituents throughout the environment and returns them to the logical place within it—the places of high concentration.

Pretreatment-Water Receiver Systems

Most industrial waste systems utilize streams and waters as the terminal receiver system. As previously described, the regulatory constraints for water discharge systems are expanding the focus of control parameters and becoming increasingly stringent. Therefore, the economics of pretreatment prior to stream discharge are characterized by very large cost increases associated with incremental removals as the goal of zero discharge is approached. Technology surrounding waste treatment prior to stream discharge is, however, well developed, and within many existing systems will remain the predominant waste management alternative.

Pretreatment-Land Application Systems

As a major alternative, industrial wastes can be pretreated as needed (usually at moderate to low removal levels) and recycled to a plant-soil system. Pretreatment processes are necessary to reduce toxic or pollutant species as they affect land requirements and to improve the overall economics of the total system. The important difference is that because the land-based and water-based receivers have substantially different waste assimilative capabilities, the cost-effective pretreatment processes are usually not the same. Therefore, conventional treatment systems are not optimal, but rather various unit processes must be selected and rearranged when land is the terminal receiver system.

These five alternative approaches must be evaluated to provide an adequate, least-cost waste management system. With the goal of zero discharge, the long-term selections will evolve from the present state-of-the-art. The use of in-plant changes, reuse and recycle, and return to source location, offer the best approach to zero discharge because they do not introduce effluents into the environment, except to the originating location. However, it appears that some industrial effluents would not be recycled, so a receiver system must be selected. The choice of land receiver has certain inherent advantages over stream receivers (as delineated in a later section) and will gain substantial acceptance in the long term because it is essentially a recycling opportunity to a plant-soil system.

Economic Incentive for Consideration of Land-Based Alternative

Land treatment of industrial wastes and the technology for design represent a relatively new alternative for the industrial manager. Within the realm of industrial waste management there is only limited experience with pretreatment-land application systems. Thus, a certain inertia exists toward evaluation of the land receiver alternative. The incentive for overcoming this is the economic advantages of land treatment when compared to conventional systems with stream discharge.

It is impossible to generalize with respect to costs of industrial waste management because of considerable variation; however, there appears to be a substantial and recognizable difference in the economics of land

Table 1.1 Comparative Industrial Waste Management Alternatives

System	Land Treatment Alternative		Conventional Treatment Alternative
	Size (ha)	Investment Cost ($)	Investment Cost ($)
Pharmaceutical	49	490,000	745,000
Poultry Processing		220,000	720,000
Potato Processing	67	350,000	780,000
Nylon and Polyester	12	140,000	1,300,000
Refinery	10	6.50/wet ton	11.50/wet ton (Incineration

treatment vs treatment for BPT and BAT stream standards. These differences are based on the limited data of actual systems and designs where alternatives are compared (Table 1.1).

Land treatment, as 11-66% of the conventional alternative at these five sites, was a substantial economic incentive for considering the land-based alternative. As will be developed in later chapters, the net operational costs are often lower for land treatment. Based on preliminary data on land treatment systems, it appears that there may be considerable economic incentives for land treatment of industrial wastes.

Land Treatment Technology—Present Status

In the review of all industrial categories, the myriad of process waste streams in each category and the massive amount of industrial waste research directed at BPCTA, BATEA and toxic substances, it is evident that pretreatment-land application systems are not widely accepted. Even where wastes are similar to those successfully applied to the plant-soil, there has been little evaluation of the potential for land treatment. At this stage of development there certainly has been no optimization or adaptation of land-based treatment for specific industrial waste objectives.

In addition to the relatively infrequent use of pretreatment-land application systems, there are a number of biases and misconceptions regarding the potential of plant-soil receiver systems. Often heard is the statement "land application is only possible for food or domestic wastes." This book is dedicated to removing at least some of these misconceptions and directing attention to land-based receivers.

The literature available on land treatment systems consists of a series of studies of individual systems. An industrial plant has a waste that is applied over a certain land area and monitored. Preliminary tests often include greenhouse and field plots. The missing factor is transferability of information, *i.e.*, what basic principles and assimilatory data can be distilled from the information to allow better design of future systems involving different waste and site conditions. Without a basic structure for design, engineers and scientists are forced to repeat experiments for each new waste or geoclimatic condition. Because of the magnitude of such studies, they often risk not obtaining the critical data with enough reliability to guarantee the resulting system to function successfully.

In other words, land application is not viewed as a technology by industrial waste engineers, but rather as a nondescript system of irrigation or spreading. The lack of firm design criteria has made land treatment the equivalent of shovelling smoke. These factors have contributed directly to the nonacceptance and misconceptions regarding land-based treatment. The large influence of site characteristics, the diversity of industrial wastes, the concept of pretreatment and the assimilatory capacity of a plant-soil system cause land-based treatment to be complex, but not intractable. The diverse inputs require a procedure or methodology, not a standard, off-the-shelf design. This book is designed to develop the technological principles of land-based treatment. The land being used for waste treatment must, with minor management inputs, be convertible to other useful purposes. If there are limits on concentration or total accumulation, then the industrial use must remain within these critical level.

A second important concept is that an environmentally acceptable rate of application to a plant-soil system can be determined *for any and all industrial waste constituents,* with the possible exception of radioactive species. Therefore, the wastes from all industrial categories can be assimilated satisfactorily in a land-based treatment system. Whether land application is chosen as the treatment alternative then hinges on economic rather than technical consideration and should be judged accordingly. A third concept is that the intimate mixing and dispersion of wastes with the plant-soil system is the objective of land application. This contrasts with the containment in concentrated form associated with landfill and deep well injection. Dispersion or spreading of industrial wastes promotes microbial stabilization, reaction, sorption and reduced toxicity. Finally, the methodology developed in this book holds for all waste streams (raw waste, treated effluents, sludges and solids), while the mechanics and equipment used in the actual land application depend on the particular wastes.

Why then, has land application not been widely practiced, even for the most amenable industrial wastes? Part of the answer lies with the structure of the regulations in PL 92-500, which were established to approach zero discharge of industrial wastes by a stagewise approach. Each stage was more restrictive; but by sequencing, the industries could more economically achieve the ultimate goals. This was logical and necessary. A typical cost curve reflecting the appropriate regulatory stages is shown in Figure 1.2. At any stage the environmental engineer considers the alternatives to reach the next required level of treatment with land application being one choice. Figure 1.2 shows that the land-based system is typically more expensive than the first stage. Hence, land application is not selected. Then the analysis is repeated at the second stage, with considerable investment already in place. Again land application may not be less expensive than the next incremental step, so it is not selected. However, if compared to the selection of land application at stage one, a real savings would have resulted as compared to the step one and two processes chosen. That is, land application reaches very high levels of environmental protection in one process, but these substantial, long-term benefits are not included in the economic analysis. After more repetitions (possibly oriented toward toxic substances) the pattern is clear. However, if the long-range sequence were considered, land-based treatment may be the most cost effective (Figure 1.2). Therefore *while land application may appear at any point in environmental decision-making to be a safari gun approach to the mosquito-sized problem of waste treatment; in the end the problem of total pollution control is actually a tiger.*

FUNDAMENTAL PRINCIPLES AND DESIGN CONSTRAINTS FOR LAND TREATMENT

A number of reports on land application systems exist with more municipal effluent and sludge systems than those for industry. Furthermore it is obvious that:

1. land treatment systems involve a complex mixture of components;
2. while reading 20 literature articles is interesting, it does not allow one to design the 21st system; and
3. there are only the most approximate criteria for determining whether a system is a success or failure.

The objective of this book is to directly remedy this situation through developing a basic design objective or criterion for all land treatment systems. This book contains a complete methodology or procedure for the design of an entire combined pretreatment-land application system for all waste types, thus allowing for wide transferability of information.

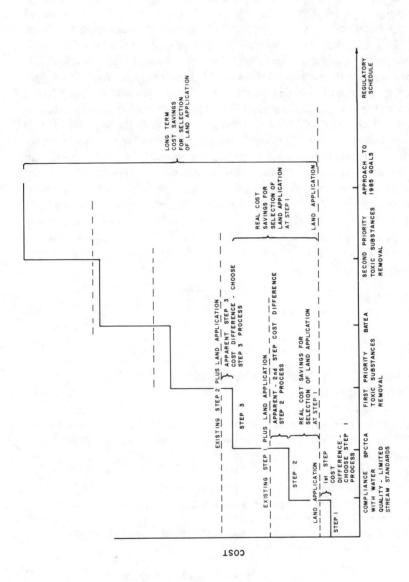

Figure 1.2 Economic consequences of stage-wise federal approach to environmental goals–penalty of major innovative approaches.

Basic Nondegradation Constraint

Land treatment technology represents a major, different approach to industrial waste treatment—"a philosophical change." The central idea of this *alternative* approach is that the plant-soil system is the ultimate receiver, rather than the conventional receiver of a stream or water body. This difference necessitates a new design method, reconsideration of pretreatment requirements and new monitoring approaches as the technology of land-based treatment evolves. This book presents the technology of industrial waste management systems in which land treatment is the ultimate receiver.

The technology of industrial waste land-based treatment centers on one primary design objective or philosophical constraint—*the industrial waste, when considered on a constituent-by-constituent basis, shall be applied to the plant-soil system at such rates or over such limited time spans that no land is irreversibly removed from some other potential societal usage (agriculture, development, forestation).* Such a constraint is basically one of nondegradation, which seems to be the central lesson learned with other uncontrolled pollution in aquatic systems or with pesticide application. Nondegradation is a severe constraint and, given the state-of-the-art, cannot always be guaranteed before initiating a project; thus, monitoring is used to gauge progress. However, approaching land application with this philosophy appears to (1) reflect most accurately the thrust of all environmental regulations, and (2) improve the probability that the design criteria used will provide long-term, reliable performance by land application. This primary design objective is translated into three calculational constraints or approaches, depending on the waste constituent being considered (these are described in the later methodology section).

When the design of land application systems attempts to incorporate the nondegradation constraint within the limits of existing data, the potential for land-based treatment is greatly expanded. Designed correctly, land application rates can be specified for *any and all industrial wastes*, with the possible exception of radioactive wastes. This means that for all industrial categories and subcategories delineated by the Environmental Protection Agency, it is theoretically possible to specify or develop the specifications for land-based treatment. However, as resulting land areas may be extremely large compared to the other industrial waste alternatives (Figure 1.1), land application would not be the most cost-effective. The decision regarding the use of land-based treatment for any industry is one of economic considerations and not one of technical limitations. Note: In a steadily increasing number of industrial situations, land treatment is becoming the most cost effective alternative.

Land Treatment—A Definition

This definition is qualitative and operational because it describes an overall technology or approach. *Land application is the intimate mixing or dispersion of wastes into the upper zone of the soil-plant system with the objective of microbial stabilization, adsorption, immobilization, selective dispersion or crop recovery, leading to an environmentally acceptable assimilation of the waste.* Industrial wastes are applied in relatively thin layers over land areas that permit a large degree of waste constituent interaction with the soil and there exist substantial soil zones between waste and relevant surface and groundwaters. The ratio of waste to soil is very low over the impacted area by contrast with other land-based approaches such as land fills and deep well injection. The direct contrast is described later in this chapter. Several terms will be used interchangeably in this book as equivalents to land treatment. These are land-based treatment, plant-soil receiver system, sludge or waste farming, land as the ultimate receiver or pretreatment-land application systems.

Site-Specific Importance

A fundamental principle emerges as extremely important in all future discussions of land-based treatment—that the use of a plant-soil receiver system for industrial waste is absolutely site-dependent or site-specific. The specific site or potential sites must first be identified and the data be used from these sites to determine the rate of application of a waste or constituents within the waste. This was learned from experiences with successful and unsuccessful systems, in which the critical difference was in accounting for site-specific properties. Land-based treatment design criteria are generally *not* transferable from location to location, only the method of data collection and design calculations are general. To some, this limitation is severe and may have contributed to the slower acceptance of land application by engineers and regulating bodies. However, this design requirement is the essence of the need for engineering in this field and is more than offset by the economic and technical advantages of land-based treatment.

UNIFIED METHODOLOGY FOR PRETREATMENT– LAND APPLICATION SYSTEM DESIGN

Introduction

The problems of acceptance and of understanding land-based treatment as a technology for industrial waste management have been discussed. Complexity is introduced immediately by the highly site-specific nature of waste constituent assimilation and by the diversity of industrial waste streams and characteristics. Under these conditions, the solution for the successful design of pretreatment-land application systems is a unified methodology or approach.

There are several reasons to favor a methodology or procedural approach for industrial waste pretreatment-land application systems. First, since design is very site-specific, values for the assimilation of constituents are not transferable. Only the method for arriving at those design criteria, based on agreed constraints and site data, is transferable. With basic design principles associated with this methodology, limited experiments are needed to obtain the most critical data. Redundant data or data for noncritical parameters can be avoided. The existence of a formal methodology with constraints and necessary data input serves as a basis for agreement between the designer and the permitting agency, thus alleviating unnecessary and undefined disagreement. The wide variation in industrial wastes favors a flexible design methodology. Finally, since this unified approach includes the costs of pretreatment alternatives and land application, a least-cost total waste management system can be evaluated. Only with this optimized cost estimate can the various industrial waste management alternatives be compared (Figure 1.1).

Total System Concept

In the design of a total industrial waste management system is the consideration of (1) the source or generation of wastes, (2) the terminal or ultimate receiver system, and (3) those intermediate pretreatment unit processes that alter the wastes prior to the terminal receiver (Figure 1.3).

The design methodology for the land application of industrial wastes deals predominantly with the plant-soil system as the ultimate receiver, although for certain conditions the interaction with receiving waters must also be evaluated. The constraints that determine the environmentally acceptable assimilation rates are based on the constituents within a waste and not on the total mass of the aggregate waste from any particular industrial process. Therefore, the design methodology is based on constituent parameters or classes of constituents. It is equally valid as an approach for wastewaters, sludges, slurries and solid wastes. Furthermore, any industrial process, category or subcategory (from organic chemicals to metal plating, from petroleum to poultry processing, *ad libitum*) can be utilized in the procedure for evaluating a pretreatment-land application for that particular waste and its characteristics. This discussion has focused on industrial waste because of the wide diversity of waste characteristics—hence the need for a unified methodology. The procedural approach will be particularly useful in the area of municipal wastes, especially when greater recognition is made of the industrial waste input to municipal systems.

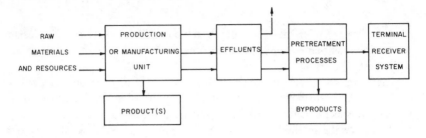

Figure 1.3 Representative environmental flow sheet for any processing or manufacturing facility.

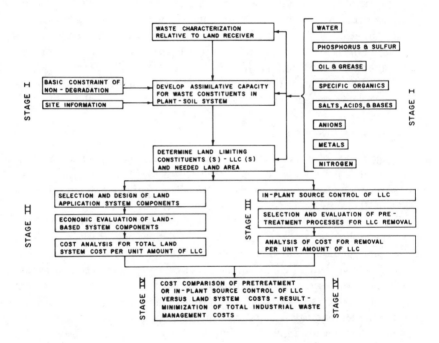

Figure 1.4 Four-stage procedure or methodology for complete pretreatment-land application system design.

Design Stages in Unified Methodology

There are four major stages involved in the design of a total waste management system for an industrial waste (Figure 1.4):

I.   the determination, on a constituent-by-constituent basis, of plant-soil assimilation characteristics, the assessment of waste generation and the comparison of these two values to select the parameter or class of constituents that requires the greatest land area. This is then the controlling parameter(s) for required land areas or the land-limiting constituent (LLC);

II. the design evaluation of all required components for the land application system and the cost analysis based on different amounts of the LLC, expressed in investment costs or average annual costs per unit amount of LLC;

III. the selection and cost analysis of pretreatment or in-plant alternatives for reducing the total level of the LLC, with investment or average annual costs expressed per unit of LLC; and

IV. the economic balance between the cost of the total land receiver and the cost of pretreatment processes so that the sum total system cost is at a minimum; thus, the pretreatment-land application alternative can be compared to other industrial waste management alternatives to allow selection of the most cost-effective alternative over the lifetime of the manufacturing facility.

These four stages, briefly discussed here, will be dealt with more fully in succeeding chapters.

Stage I

Determination of assimilative capacity is often specific to the plant-soil system at the site(s) selected for land application. In comparison to the other stages of the design methodology, the assimilative capacity is one of the most difficult yet most critical facets. The constraints regarding irreversible damage to a site must be included, and where priority pollutants or other specific compounds of concern are present in a waste, it must be recognized that quantitative data are not always available. These data are generally not available for other pretreatment-stream discharge or industrial waste alternatives, so this deficiency is universal. For many organic compounds, more is known about the degradation and interaction in soils than in conventional biological treatment processes. The designer may have to generate needed assimilative capacity data from greenhouse or field experiments. The factors and constraints used in overall design methodology can reduce the amount of experimentation by identifying the critical components for testing.

To facilitate consideration of the many constituents present in industrial wastes, effluents, sludges and solids, eight broad categories of parameters were established:

1. Water or hydraulic loading
2. Phosphorus and Sulfur
3. Oil and grease
4. Specific organics
5. Salts, acids and bases
6. Anionic species
7. Heavy metals
8. Nitrogen

Within each category are many compounds and species for which the soil-plant assimilative capacity is established; the broad category indicates the general response of those compounds.

The primary design constraint regarding the long-term effect or assimilative capacity of the soil-plant system is translated into land application rates by one or more of three calculational procedures. That is, from experience with determining the assimilative capacity of many industrial waste constituents, the calculation methods are of three broad types:

1. those that degrade or require plant uptake for assimilation in the plant-soil system, *e.g.,* oils or organics;
2. those that are relatively immobile and nondegradative, thus are permitted to accumulate in soils to predetermined critical levels, *e.g.,* heavy metals; and
3. those that are mobile and nondegradative and must be assimilated over land areas so that receiving waters are not altered to a degree requiring further drinking water treatment, *e.g.,* anionic species.

Results of these calculations for the assimilative capacity at a specific plant-soil system site are usually expressed as mass of a constituent in the waste per unit land area per unit time (kg/ha/yr), or as the concentration in the soil (ppm or % of soil weight).

The characterization of the waste, whether solid residue, sludges or wastewater, must be performed for *parameters that are important in the plant-soil system.* Many of the conventional tests in the industrial waste field are oriented toward an aquatic receiver. Color, turbidity, filterable soilds, biochemical oxygen demand, etc. are relatively unimportant in relation to a land-based receiver. The nature of waste constituent assimilative capacity is such that waste constituent generation must be known on a basis of mass per unit time, kg/yr, rather than simply as a concentration in the waste stream. Often this imposes greater requirements on evaluation of waste generation from an industrial plant.

As the final portion of stage I the ratio of waste generation (kg/yr) to plant-soil assimilative capacity (kg/ha/yr) is calculated for each waste constituent. This ratio is the land area (ha) required to assimilate a waste constituent present in that particular waste stream. The calculational process is shown in Table 1.2 where broad categories of waste constituents are used. Usually each category would consist of a number of specific compounds, each

Table 1.2 Land-Limiting Constituent (LLC) Analysis

Parameter (grouped by category)	Generation Rate in Waste Stream (kg/yr)	Plant-Soil Assimilative Capacity at Land Treatment Site(kg/ha/yr)	Land Area Required (ha)	
Water	-	-	-	
Phosphorus and Sulfur	-	-	-	
Oil and Grease	-	-	-	
Specific Organics	-	-	-	
Salts, Acids, Bases	-	-	-	
Anions	-	-	-	Largest land area, hence this constituent is defined as LLC
Metals	-	-	-	
Nitrogen	-	-	-	

with a generation and assimilative rate. That parameter requiring the largest land area is identified as the LLC. If the total waste is then applied to the land area established by the LLC, all other parameters will be assimilated satisfactorily. Changes in the amount of LLC would alter land area, so the costs of various size land application sites can be related to the amount of this LLC. Also pretreatment must be directed at the LLC. Thus the main result from stage I is to document and identify the LLC and needed land area for the particular site(s) selected and for the waste stream of interest.

Stage II

The specific possible site(s), the LLC and required land area for assimilation of a particular waste stream have been identified in stage I of the overall design methodology. With these data, the design engineer can evaluate the cost of the entire land application system (stage II). Land system costs are those necessary to transport the waste from the plant or manufacturing facility and assure it is applied according to the best practices available, meeting all environmental regulations. The overall methodology described here is for any waste stream, so the land application material may be a liquid, slurry, sludge or solid. The principal difference in stage II would be the equipment used in actually spreading the waste onto the plant-soil system.

A land application system consists of a large number of components and must be selected and designed for the specific application site conditions, then the projected costs summed to determine the total cost of the land system. Typical components are:

- transmission or conveyance
- storage
- application system (spreader, irrigation, subsurface incorporation, etc.)
- land purchase and preparation
- buffer zone
- monitoring
- operational control systems
- diversions and land management practices
- agricultural equipment for vegetative cover
- operation and management manual

Not all these components are needed for each land application system, but design and economic factors for each are presented in Chapter 12 related to stage II of the overall design process. From the list of possible components, the designer selects and develops the costs for those needed at the proposed industrial waste land application site. Total cost is that needed for the environmentally acceptable assimilation of the initial waste stream considered. Based on so many kg/yr of the identified LLC this represents the maximum cost for the complete land application of the entire raw waste stream under consideration.

Since the overall design methodology considers the total system of pretreatment-land application, the land system costs must be adjustable for less than the total raw waste amount of LLC. If pretreatment or in-plant changes are employed, the land system would be scaled down for some percent (*e.g.,* 90%, 75%, 50%, 30%, 10%) of the total LLC generated. The costs for a smaller land system should next be calculated by the designer. Some of the components are insensitive to varying levels of the LLC, while others (land area) are directly proportional to the amount of LLC. Thus, the dependence of total land system costs on the amount or percentage of LLC is usually nonlinear and often quite site specific. Costs values for several levels of the LLC define a land system cost curve that can be plotted vs the amount or percentage of LLC. Such a cost-dependence relationship is the principal result of the stage II analysis.

Stage III

Pretreatment or in-plant source control must aim to reduce the amount of the LLC identified in stage I. If a pretreatment process cannot reduce the LLC at a cost less than the savings in reduced land system requirements, then in terms of the total industrial waste management system, the pretreatment is not justifiable. The level of pretreatment and land system requirements must be balanced to minimize total system costs.

The concepts in stage III will be discussed using pretreatment, but the reader should remember that in-plant source control can and should be considered in a parallel manner to pretreatment. The principal question in stage

III is, "how costly is the removal of different levels or percentages of the LLC from a particular industrial stream?" This is quite different from the situation for conventional pretreatment prior to stream discharge in which the economic questions are "how costly is it to remove a very high and increasingly larger percentage ($95^+\%$, $99^+\%$, etc) of virtually all constituents from an industrial stream?"

Especially where there is selective pretreatment removal, the LLC analysis must be continually reconsidered. As the amount of the LLC derived on the basis of the untreated waste is reduced, one of the other parameters may become the limiting constituent. At that point all further pretreatment must result in the simultaneous reduction of two or more constituents. For example, at a location in North Carolina, the soil assimilative capacity and the waste generated from a poultry processing plant indicated that nitrogen content was the LLC. However, after the removal of 30% of this nitrogen, the hydraulic loading became a limiting constituent. Therefore, investment to reduce nitrogen beyond 30% was not justified without also reducing the overall waste volume or water loading. Thus, the evaluation of effective pretreatment is an iterative process, to assure that the LLC(s) are being removed and that there is an effect on the land requirements.

Pretreatment alternatives prior to land application are typically selected from two broad categories of engineering processes. The first category is based on a given unit process to remove an LLC (*e.g.*, ammonia stripping for nitrogen removal) and the changes in design or operating variables that lead to different efficiencies of removal. That is, for a unit process, the size, flowrate, temperature of operation, nature of gas-liquid contact etc. can be changed to vary the concentration of the chemical species in the pretreatment effluent, *e.g.*, increasing the air flow per unit of effluent volume decreases the percent removal of ammonia from a stripping device. Usually, the higher percentage reductions in a waste of the LLC the greater the cost, since larger investments are associated with greater efficiency of removals.

The second general route for evaluting pretreatment alternatives is to select and rank a number of unit process with regard to the respective percent removal of a given LLC. In this portion of the overall design methodology, the design engineer must consider the typical removal characteristics of all available pretreatment processes and establish the estimated costs for each. Again, those processes that remove high levels of an LLC are typically those requiring the greatest costs for a given waste stream. A significant problem with the evaluation of pretreatment alternatives is that the majority of available technical literature is oriented toward very high levels of treatment or removal. This is required for tertiary or advanced wastewater treatment prior to stream discharge. If a given unit process or a certain operating range of a unit process are not highly efficient in removing waste constituents, then these are often not reported and, hence, are unavailable to the designer as pretreatment prior to land application. Also, to selectively remove one or a few constituents without complete removal of all constituents is not a well-developed part of the pretreatment technology.

Within the limits of available pretreatment information there is an advantage for partial or limited pretreatment. The costs of removing the last 10%, 5%, or 1% of waste constituents are increasing almost exponentially, particularly where regulations are focusing on low-level removal of priority pollutants. Therefore, the cost reductions in using a modest amount of removal (30-60%) are orders of magnitude lower than those required by BCPTCA, BATEA and other regulatory statutes.

From the engineering analysis of pretreatment to remove the LLC identified for a given waste stream, a cost curve is generated. This pretreatment economic curve relates the amount or percentage of LLC removed to the cost of achieving such reductions by means of one or both of the general categories of pretreatment design discussed above. The pretreatment cost curve is the principal result of stage III and represents the critical evaluation of the economics of reducing the land limiting constituent prior to the land system.

Stage IV

In the previous stages of the design analysis for land treatment, the economic relationship is established between (1) the total components of the land system vs the amount or percentage of LLC for a given waste stream, and (2) the pretreatment or in-plant alternatives vs the amount or percentage of LLC removed from a given waste stream. These two cost curves are plotted together and the sum is the cost of the total pretreatment-land application (Figure 1.5). This is a simplified cost optimization or balance between the level of pretreatment or in-plant alteration and the resultant land system cost requirements. Stage IV is thus the selection of the least-cost total system.

From utilizing this overall methodology for a number of industrial wastes, the relationship in Figure 1.5 has been shown as representative. Neither complete land application of the process generated waste nor complete pretreatment or in-plant control of waste constituents is the most cost-effective. Rather some intermediate combination of pretreatment and the use of the plant-soil assimilative capacity results in a least-cost, total industrial waste management system.

The four-stage design process described in this chapter is sensitive to site conditions which is a primary requirement for any land treatment technology. The variations due to specific site conditions and waste types can be very large. Therefore, economic and technical evaluations for industrial wastes

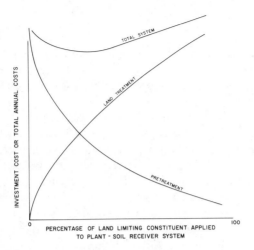

Figure 1.5 Representative economic balance between pretreatment and land treatment to minimize total system costs.

must consider these factors at some level. With the overall design methodology described, a sufficient level of site- and waste-specific information is included, so that the components of the pretreatment-land application system are reasonably estimated and the resultant economics are realistic. The cost-effective combination of pretreatment and land application is a necessary starting point for the industrial manager to compare the various waste management alternatives in Figure 1.1.

OTHER USES OF SOIL SYSTEMS FOR WASTE DISPOSAL

Besides the spreading assimilation of wastes in the surface zone of a plant-soil system implied in the concept of land treatment, there are a number of other uses of soil or soil-plant systems for waste treatment or disposal. These include (1) landfill, (2) deep well injection, (3) overland flow, and (4) soil reactors. A number of detailed sources can be studied to obtain a more thorough understanding of these alternative uses of land in industrial waste management. A brief discussion of each is presented so the reader can differentiate such systems from the dominant focus of this book—the use of land application for industrial wastes.

Landfill

Landfill or waste burial is a common industrial waste practice. There are also a large number of landfills for municipal sludge and refuse. Two well-developed methods for design, construction and operation of a landfill are depicted in Figure 1.6.

The two contaminants of environmental concern from refuse disposal are gas and leachate. Gas is always generated as the organic matter decomposes, whereas leachate will only be generated when there is an excess of water infiltrating through the stored waste. Gas production normally begins as the refuse is being placed in a landfill. The principal gases produced from decomposing refuse are carbon dioxide and methane. Carbon dioxide is the gas of consequence for water quality, since the other gases of decomposition are relatively insoluble in water. When the carbon dioxide is dissolved in water, it causes a lowering in pH, resulting in a corrosive environment and probably an increase in water hardness. Carbon dioxide predominates in the early decomposition of the refuse and reaches a peak concentration within several months to years after placement. Thereafter, there is a steady decrease in the percentage of carbon dioxide in the gas mixture.

The typical water content for municipal refuse is between 30% and 60% by weight, the wide range being attributed to seasonal and geographic variances. The optimal water content for gas production lies closer to 80-90%, a condition where excess water is available.

Leachate from a landfill results when the amount of water input exceeds that which can be retained by the wastes. Either infiltrating surface or groundwaters can create this excess condition. Landfills located in areas receiving more than 75 cm of rainfall annually will undoubtedly have a net infiltration of water into the refuse and, therefore, eventually discharge leachate. Where there is less than 50 cm of annual rainfall, significant leachate may not develop, since the water balance of runoff, evaporation and evapotranspiration probably will exceed the infiltration of water.

Leachate, as an effluent from a waste containment site is a liquid of complex nature. Many variables can interact to change the quality and quantity of the leachate. Organic and inorganic salts, heavy metals, pesticides, toxic chemicals, pH, virus and pathogens are among the parameters of concern in leachate composition.

The potential for leachate degradation of groundwater from a disposal site is directly related to the geological and hydrological characteristics of the site. The geological factors include the soil and bedrock type and condition of their respective abilities to attenuate or restrict movement of leachate and gas emanating from the disposal site. Hydrological factors include the groundwater location and movement, the amount and intensity of rainfall and the ability to control surface water or drainage. Occasionally, a site is situated

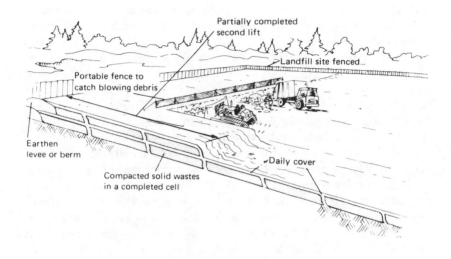

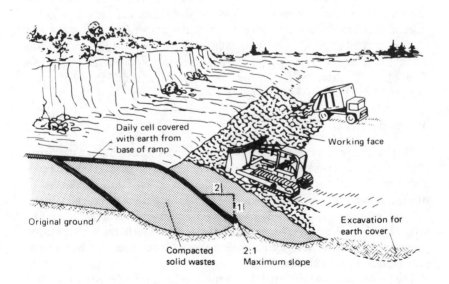

Figure 1.6 Schematic of landfill techniques (Tchobanoglaus *et. al.* 1977).

above the water table, is in an area of low rainfall, and is underlain by impervious clay soils which completely contain leachate and gases. Such ideal conditions are not normally available; therefore, special design features and/or operational methods must be provided.

The desired short–term containment of leachate and the gases of decomposition can be achieved by lining disposal sites in combination with leachate collection/extraction systems. Gas venting may be required even where linings are not required but where carbon dioxide may be of concern. Leachate recirculation may help reduce leachate containment loading.

The essential difference between land application and landfill is that land application leads to treatment or assimilation, while landfill leads to containment and only for an unspecified time. Land application aims at dispersing industrial waste constituents to provide intimate contact with the soil system over large areas for environmentally acceptable assimilation. The full microbial and chemical treatment potential of the plant-soil mantle is utilized in land application systems. In this manner, long-term systems are obtained with the end result being agricultural land. By contrast, landfill technology attempts to concentrate industrial waste constituents at one spot and to contain them to prevent their entry into the environment. There is little or no contact with soil; hence, little treatment is obtained in a landfill with the end result being the same concentrated waste source that was originally placed in the landfill. The period of containment is about 5-30 years, at which time this concentrated source may pose a significant environmental hazard. Another problem that seems endemic with landfills is the propensity to forget where they are located after 50 years (especially smaller industrial fills), and then to utilize the land for other purposes. The resultant problems can be severe (Deutsch 1961, Lazar 1976). So concern exists that a landfill does not represent an ultimate disposal but a temporary storage.

Deep Well Injection

This has as the underlying concept the permanent isolation of waste materials from the usable environment by means of confining them to selected geological strata, although there is some evidence of biological treatment in selected situations (Cox and Walker 1973). The injection technique (Figure 1.7) can provide a final solution to certain liquid waste problems provided that: (1) disposal strata of low utility exist into which the waste can be injected without undesirable consequences, and (2) the waste can be permanently confined to the chosen strata.

The problems associated with long-term confinement of waste to the disposal zone are perhaps more difficult to assess than the question of damages occurring within the disposal zone. Confinement refers primarily to prevention of the vertical movement of the waste and is normally provided by an

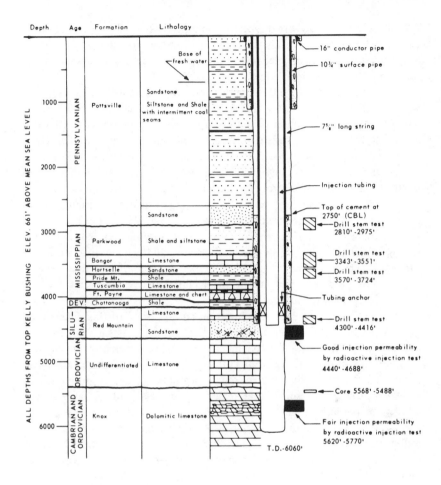

Figure 1.7 Schematic cross section of deep well injection facility (Braunstein 1973).

impermeable stratum over the disposal zone. Vertical migration of the wastes may take place by means of natural features such as faults; abandoned wells which penetrate the confining layer; or through the injection well itself. Without complete knowledge of underground conditions, confinement of the waste can never be certain. The confining stratum may be destroyed by the injection process if excessive injection pressures are used. In addition, loss of confinement may result from collapse of the confining stratum, if the injected waste produces large solution cavities in underlying support materials.

For more than 50 years the petroleum industry has reinjected brines both for secondary recovery of oil and for waste disposal. Today, nearly 1,200 ha-m of brine wastes are injected yearly through many thousands of wells in the oil-producing states (Piper 1969).

Deep-well disposal is, not proposed as a cure-all for problems related to waste-liquid disposal (Barlow 1973). It is relatively limited, considering the wide divergence in chemical composition of wastes; however, if installations are properly conceived, constructed, operated and installed, they can fulfill a need without creating other problems such as can occur with waste-retention basins, incineration or even sludge disposal. The deep-well method does remove the waste initially from the biosphere.

Because the capacity of potential receiving formations, although enormous, is still finite, unrestricted deep well disposal should not be allowed. The use of a formation for this purpose in any specific area should be controlled, which includes the awarding of permits and the delineation of factors such as acceptable injection rates and pressures, types of materials used for construction, and tests and monitoring facilities that can ensure the utility and safety of the installation.

The legal aspects of deep well injection are in a state of flux. There is no general philosophical agreement as to whether injection is a valid form of ultimate waste disposal. The full impact of implementation of the FWPCA of 1972 and subsequent environmental bills is not yet known. This uncertainty and the possibility of new federal legislation, have produced a wait-and-see attitude with some state regulatory agencies and probably with a number of potential users of the technique. The case law that may provide a precedent for resolution of legal disputes between private parties is not well-developed in most areas and non-existent in others. It, appears therefore, that legal uncertainty will hamper the injection well concept, at least in the immediate future (McJunkin and Vesilind 1973, Braunstein 1973).

Overland Flow (OLF)

OLF is often described as one of three methods for putting waste on land (along with spray irrigation and infiltration-percolation) with the potential advantages being greatest on less permeable soils. That is, in areas with moderate to low permeability of the upper soil horizons, OLF is the method of choice. This has led to some confusion between OLF and irrigation with traditional gated pipe distribution of wastewater with subsequent infiltration since both have some liquid movement at the soil surface. It is also uncertain what OLF actually accomplishes.

OLF must be viewed as a pretreatment process, not as a method for terminal land application (Figure 1.8), which will separate OLF from ground-level application (GLA) of wastewater. Ground-level application can be used to distribute wastewater to an OLF system, to an irrigation infiltration system or to a high-rate, infiltration-percolation process. GLA is basically gated pipe or similar distributors, either pumped under pressure or utilizing gravity, in which liquid wastewater discharges 15-30 cm above the ground. The technology of GLA is extremely important in reducing certain costs associated with land application of wastewater (Overcash 1978).

The OLF process is a pretreatment system in which wastewater is applied in a manner conducive to developing controlled sheet flow over the soil surface. More than half the applied wastewater is available as runoff for reuse, terminal land application, artificial recharge or return to surface waters (Trax 1973). Waste constituent separation and conversions occur by physical, chemical or biological processes, which take place during flow and during resting periods between overland flow applications. Activitiy is at the soil microbial mat-air-plant interface. Several aspects of this operational definition are critical. The first is that there is an effluent from OLF, often 50-60% of the incoming wastewater volume; therefore, another pretreatment or a terminal system must be selected to deal with this effluent. Thus, OLF is a pretreatment process. Second, wastewater conversions are occurring at a complex interface, so OLF has some capacity to reduce wastewater constituents. Depending on the waste source, these constituents may be controlling factors to the needed land area or stream flow, so, OLF could serve an effective pretreatment role in a total system.

With this information, OLF can be evaluated as part of an overall waste management system. Data are available primarily for municipal and food processing and canning wastes since these account for 90% of all existing systems. A summary of performance of OLF systems is given in Table 1.3 (Overcash 1978). Many of the treatment processes occuring in OLF are similar to those of land application systems.

Soil Reactors

These use the soil microbial populations under more controlled and isolated conditions than land application. The most recognized example is the barriered landscape water renovation system (BLWRS) (Erickson *et al.* 1974). In this system a soil block is isolated with plastic sheets in a box shape, with the applied waste collected in drains (Figure 1.9). The vertical soil profile is maintained with two parts—an aerobic upper zone and an anaerobic zone.

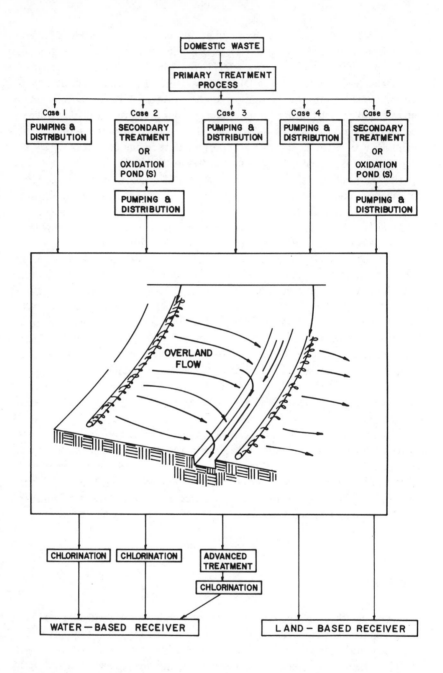

Figure 1.8 Alternative uses of overland flow for waste pretreatment (Overcash 1978).

Table 1.3 Literature Data Summary for Overland Flow (OLF) Systems

Percent reduction, %, with effluent concentration from OLF system, mg/l, in parenthesis

Input to OLF System	Reference	Length of OLF System (m)	Application Rate (cm/wk)	H_2O	BOD_5	COD	TSS	TN	NH_3-N	O-N	NO_3-N (Effluent)	TP	TS	TVS	Na	Al	Ca	Cl	Total Coliform	Fecal Coliform
Raw or minimally treated domestic waste	Kirby (1970)	365	13 (cool period)	–	96 (25)	–	95 (25)	45	–	–	–	35	–	–	–	–	–	–	–	–
	Thomas et al. (1974)	36	7.4-9.8 (warm per.)	50	93 (11)	77 (73)	95 (8)	89 (2.6)	94 (1)	86 (0.8)	(0.4)	60 (4)	20 (826)	53 (140)	–	–	–	–	–	–
		36	7.4-9.8 (cool per.)	50	92 (12)	83 (53)	92 (12)	77 (5.4)	97 (0.5)	67 (1.9)	(2.8)	56 (4.4)	31 (720)	42 (170)	–	–	–	–	–	–
	Thomas (1975)	36	9.8 (alum addition)	50	96 (7)	85 (41)	93 (16)	88 (2.5)	94 (0.8)	86 (1.2)	0.6	84 (1.6)	27 (810)	61 (120)	–	60 (0.28)	–	–	97.3 (0.2×10^6)	97.5 (0.025×10^6)
		36	9.8 (no alum addition)	50	94 (9)	84 (54)	93 (16)	86 (2.9)	95 (0.8)	82 (1.6)	(0.5)	62 (3.7)	31 (770)	62 (120)	–	80 (0.15)	–	–	95.9 (0.3×10^6)	91 (0.09×10^6)
Secondary Treated Domestic Waste	Locke (1972)	62	330 (warm period)	–	62 (12)	–	75 (8)	–	0 (16)	–	(0.2)	–	–	–	–	–	–	–	–	–
	Walker (1972)	39	400	–	59 (9.7)	–	73 (15)	–	20 (8.4)	–	10 (21.9)	–	–	–	–	–	–	–	–	–
	Meyer (1974)	46	6-25 (based on area of first terrace)	45-55	(0.5-7)	–	–	15 (15)	0 (12)	–	10 (5-6)	15 (5)	–	–	0 (21)	–	0 (38)	0	–	–
	Carlson et al. (1974)	6	5	–	0 (15)	–	–	–	100 (0)	45 (2.2)	99 (0.1)	40 (8)	–	–	–	–	–	–	–	–
Cannery	Stevens and Dunn (1969)	150	3.8-8 (warm period)	40	96 (80)	–	65 (210)	80 (6)	–	–	–	–	–	–	–	–	–	–	–	–
	Bendixen et al. (1969)	30-60	16	60	85 (185)	81 (100)	89 (27)	73 (16)	–	74 (7)	50 (0.2)	65 (0.3)	–	–	–	–	–	–	–	–
	Mather (1969)	100	8	40-60	99 (9)	–	94 (16)	84 (3)	–	81 (3)	–	42 (4.3)	–	–	–	–	–	–	–	–
Feedlot Runoff	Thomas (1972)	9	6	60	–	23 (230)	–	47 (40)	–	–	–	26 (7.4)	–	–	–	–	–	–	–	–
Meatpacking Anaerobic Lagoon	Witherow (1973)	40	4.2	45	83	70	80	80	84	–	–	64	–	–	–	–	–	–	–	–
Swine Lagoon	Stevers et al. (1975)	260	Infrequent	–	90 (25)	81 (130)	–	85 (12)	95 (7)	75 (5)	–	78 (8)	27 (700)	51 (210)	65 (20)	–	60 (42)	–	–	–
	Humenik et al. (1975)	30	~5	–	–	10 (690)	–	35 (130)	–	–	–	22 (17)	–	–	–	–	–	–	–	–
	Wiltrich and Boda (1976)	30	5-27	–	–	45-72	–	–	–	–	38-75	40-80	–	–	–	–	–	–	–	–
Poultry Manure	Overcash (1978) (mass basis)	30 m	3.8-7.6	60-80	–	94 (220)	–	93 (30)	97 (5)	90 (20)	(3)	90 (9)	–	–	–	–	–	–	–	–
	(concentration basis)			–	–	77	–	70	87	56	–	59	–	–	–	–	–	–	–	–

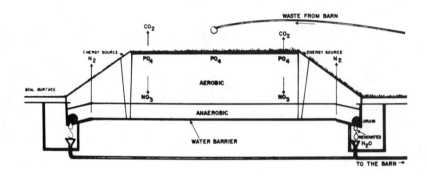

Figure 1.9 Schematic of soil reactor for nitrogen removal (Erickson *et al.* 1974).

Applied waste nitrogen is nitrified, then denitrified with waste application rates greatly exceeding crop uptake data. The resultant nitrogen losses are large. These systems are totally confined because of the high waste applications, the unexpected variablility of effluent, temperature effects, etc. Similarly, soil decomposition processes could be utilized for organics in which any migration can be trapped and recycled or treated elsewhere. As plant-soil treatment becomes more developed, greater use of the high levels of constituent decomposition for soil reactors will be undertaken.

MAGNITUDE OF CURRENT USE OF LAND-BASED WASTE TREATMENT

Industry and Waste Type

A list of land application systems or laboratory-scale test data which would substantiate the consideration of land treatment was compiled (Pal and Overcash 1976) according to the industrial point source categories established by the Environmental Protection Agency (Table 1.4). The listing is not total, since new projects are continually being reported. Presence of a particular waste in Table 1.4 often does not assure an optimal system with respect to economics of pretreatment and land application. All the listed systems are not assured of long term success since plant-soil design criteria are often exceeded. Even short term success has not always been achieved because of underdesign.

The list of Table 1.4 primarily demonstrates the wide spread interest in using land application and, more importantly, that in a wide variety of industries an economical alternative can be obtained with land systems. To understand the fundamental pathways for treatment, the broad diversity of waste types listed strengthens the land treatment methodology.

Industrial Systems with Available Detailed Design Information

Table 1.5 is a partial list of land application systems for which the overall design methodology developed in this book has been employed with some level of detailed design criteria and cost analyses. It demonstrates the extent of evaluations of land treatment technology. In the long term, more industrial wastes will have both fundamental design bases as well as field demonstration information.

For the textiles, petroleum refining and seafood processing industrial categories, a computer-based design data system is being developed (Overcash 1978). Variables such as the particular production process (as delineated in Environmental Protection Agency Development Documents for each industry), plant size, available pretreatment processes and cost information are evaluated to yield a least-cost estimate of the pretreatment-land application system. These designs must be refined for each location because of the strong site-specific dependence of land treatment. These refinements proceed in a fundamental manner from the initial design.

IMPACT OF LAND APPLICATION TECHNOLOGY

Improvement of Water Quality

The most immediate effect of using the plant-soil receiver is that it stops the discharge of industrial waste with toxic substances to a country's receiving waters, a main drinking water supply. This would have substantial benefit to water quality and public health. Placing wastes on land increases the probability of satisfactory assimilation and reduces the widespread dispersion of pollutants to streams, rivers, impoundments, oceans and other aquatic systems.

Table 1.4 Investigations of Land Treatment of Industrial Wastes

Industry Source	Waste Category or Constituents	Remarks	Reference
Petroleum	Oily wastes (0.4-40% oil)	Field studies	Huddleston and Cresswell (1976)
Refining	Oils		Raymond *et al.* (1976)
	Simulated oily waste		Kincannon (1972)
	Oil spill and contamination		Udo and Fayemi (1975)
			Udo (1972)
	Oil, fertilizer and bacteria		Jobson *et al.* (1974)
	Oil-contaminated field		Gudin and Syratt (1975)
	Kerosine contamination		Jones *et al.* (1970)
	Oil-contaminated soil		Schwendinger (1968)
	Crude petroleum		Plice (1948)
	Spent motor oil	Greenhouse	Giddens (1976)
	Oil pollution of salt marsh		Stebbings (1970)
	Oil degradation		Schwendinger (1968)
	Oil pollutants	Lab studies (degradation by bacteria and fungi in aquatic systems)	Ahearn and Meyers (1973)
		Low-temp. degradation	Walker and Colwell (1974)
		Hydrocarbon decay	Gibson (1974)
	Crude petroleum		Baldwin (1922)
Potato Processing	Vegetable oil—soybean and palm oils		Smith (1976, 1974) Anderson and Wallace (1971)
Petroleum	Hydrocarbon in soils		Dobson and Wilson (1964)
Chemicals	Crude petroleum-NO_3 production		Murphy (1929)
	Polynuclear aromatic hydrocarbons		McKenna and Heath (1976)
Aluminum Casting	Dilute oily wastewater		Ongerth (1975)
Synthetic Fiber	Solvents—difficult organics	Lab	Dolgova (1973)
Organic	Phenolic wastes	Lab	Dolgova (1973)
Chemicals	Hydroquinones and benzoquinones	Lab	Medvedev and Davidov (1972, 1974)
	Alkyl ethers	Lab	Valoras *et al.* (1976)

Table 1-4, continued

Industry Source	Waste Category or Constituents	Remarks	Reference
Coke Industry	Phenolic compound wood distillate waste	Lab	Haider and Martin (1975)
Resin and Plant Manufacture Wood Distillation, etc.	Benzene, toluene, cyclohexane, chloroform, acetone phenols, quinones, hydroquinones, pyridine, etc. land application of phenolic wastes	Lab 400 ppm application rate	Hickerson and McMahon (1960) Buddin (1914) Miller-Neuhaus (1957)
Pharmaceutical Organic Chemicals	Organic solvents		Colovos and Tinklenberg (1962), Woodley (1968)
Textile Industries	Acids and bases		
Kerosine Refinery	Acid sludge		Jones et al. (1970)
Mines	Acid wash Sulfuric acid on soils Sulfuric acid in irrigation water		Blevins et al. (1970) Skinner (1973) Miyamoto and Ryan (1976) Miyamoto et al. (1975)
Cannery	Alkali waste		Webber et al. (1973)
Winery	Acid stillage		Schroeder et al. (1973)
Paper and Pulp	Sulfite, sulfate		Wisniewski et al. (1955), Billings (1958), Gellman and Blosser (1959)
Strawboard, Hardboard Boxboard, Paper Board, Groundwood and Insulation			Blosser and Owens (1964), Crawford (1958), Wallace et al. (1975) Voights (1955) Meighan (1958)
Fruit and Vegetable Canning	Alkali and salts		Bendixen et al. (1969), Glide et al. (1971), Hands et al. (1968), Ludwig and Stone (1962), Anderson et al. (1966), Molloy (1964),

Table 1-4, continued

Industry Source	Waste Category or Constituents	Remarks	Reference
			Sandborn (1953), Lane (1955), Canham (1958), Dietz and Frodey (1960), Monson (1958), Luley (1963)
Municipal Sludge	Nitrogen		Carlile and Phillips (1976)
Dynamite Explosive Factory			Lever (1966) Kirkham (1975) Gracia *et al.* (1974)
Potato Starch Processing			deHaan and Zwerman (1973)
Dairy and Cheese Industry	Nitrogen, salts, acids		Breska *et al.* (1957), Lane (1955), Lawton *et al.* (1959), McKee (1955), Scott (1962)
Meat Industry	Wastes with high organic content		Scott (1962) Schraufnagel (1962)
Detergents, Fireproofing Cosmetics Rocket Fuel Photography Nuclear Device	Borates and boron	Toxicity limit >1 ppm	Bradford (1966)
Hydrochloric Manufacture, Chlorination Process, Metal Cleaning Process	Chloride	Toxicity limit >21 ppm	Eaton (1966)
Coke Ovens Acrylics Manufacture, Ferromanganese Blast Furnaces, Electroplating, Mining and Ore Processing,	Cyanide Cyanate Thiocyanates	Toxicity limit >0.5 ppm	Jarnea and Salay (1969)

Table 1-4, continued

Industry Source	Waste Category or Constituents	Remarks	Reference
Color Photography			
Nuclear Fuels (spent)	CH$_3$I, HI		Sittig (1973)
Glass Manufacture Phosphate Fertilizer Manufacture	Fluoride		Sittig (1973)
Metallic and Sulfur Works and High Se Trash Incineration	Selenate Selenite		Sittig (1973)
Fruit, Canning and Seafood Processing Municipal Sewage Wastes	Salts		Webber *et al.* (1973), Chawla (1973), Carlile and Phillips (1976) Parsons (1967) Philipp (1971) Koch and Bloodgood (1959)
Municipal and Industrial Sludges, Electroplating and Metal Industries Wastes	Heavy Metals in Sludges	Field Studies	Carlile and Phillips (1976) Halderson and Petersen (1976) Illinois EPA
Petroleum Refinery Wastes	Pb, Zn, Cu	Lab Studies	Raymond *et al.* (1976) Lagerwerff *et al.* (1976)
	Hg, Cu, Zn, Cd		Vanderpost and Corke (1975)
	Hg		Huisingh and Huisingh (1974)
Tanning	Cr		Varadarajan *et al.* (1970) Parker (1967)

Table 1.5. Land Application Systems Receiving Detailed Pretreatment-Land Analysis

Textile Process Effluents
Nylon Plant
Petroleum Process Effluents
Poultry Processing
Seafood Processing
Poultry Hatchery
Municipal Sludge
Municipal Effluent
Enzyme Manufacturing
Electroplating
Complex Organic Chemicals Manufacturing

Industrial Development

For an industry planning a new facility, the land application alternative greatly increases the opportunities and flexibility of site selection. A potential site no longer must be located on a major receiving body of water or on an existing municipal sewer system. Instead, large areas without such facilities can be considered for industrial sites based on the availability of water and amenability for land application. Industry can locate in rural areas, where land costs are much lower, tax considerations more favorable, expansion is easily accomplished and advantage can be taken of the trend in lifestyles toward more nonurban areas.

Low-Level Pollutant Control

As the focus of water quality regulations turns to the next increments of pollution control (99+% removal) and to the complete control of certain low concentration priority pollutants, land-based treatment may offer an important industrial waste management alternative. When designed using the previous constraints for an environmentally acceptable system, the plant-soil system provides extended periods for microbial breakdown of organic priority pollutants or safe storage of many inorganic compounds of concern. This treatment potential is operational where applied wastewaters have very low concentrations of pollutants and even with sludges which contain very high concentrations of pollutants. The key to the successful performance of land application for a particular wastewater or sludge is that the constraints of nondegradation be included in the design for all constituents or classes of constituents in the waste.

Economic and Energy Factors

A number of other characteristics of pretreatment-land application systems for industrial wastes have evolved through studies and experience with existing systems: (1) agriculturally oriented, (2) low investment and operating costs, (3) energy-efficient, and (4) not technology-intensive.

The essence of land-based treatment is the application of waste to a plant-soil system so that the land remains productive and the waste is satisfactorily assimilated, which is very similar to the operation of a crop production system. The equipment for applying solid waste or sludges or for applying liquids includes spreaders and irrigation equipment, which evolved from similar farm equipment. The monitoring and assimilation criteria are largely related to the behavior of the soils and plants, which is also agronomically related. While these systems are agriculturally related, the agricultural aspects are secondary to the goal of providing waste treatment and environmental protection.

A vegetative cover on a land application site will be the general recommendation throughout this book. A vegetative cover provides enhanced rainfall and irrigation infiltration over noncropped areas due to plant roots and senescent plant matter. Increased infiltration decreases runoff, which because of this characteristic nonpoint source impact, is the recommended method for area wide pollution control, the so-called "best management practice." The vegetative cover is one of the most effective monitoring systems for satisfactory waste assimilation by the soil-plant system. The ability to maintain viable crop or forest production is roughly equated to acceptable application rates of waste.

Under certain conditions the vegetative cover can be used to recover plant nutrients in an industrial waste, providing an economic return. The value of such crops can often equal the annual operation and maintenance costs, providing a strong incentive for land application. In terms of management and operation of a land application system, the requirement to maintain agricultural yields stimulates the responsible operation of the waste management system, another advantage of a vegetative cover. Responsible operation is a large part of a successful waste management system; hence, factors that encourage such operation should be recommended. Finally, an important contribution of a vegetative cover is to promote an aesthetic system. With a crop and more natural conditions, neighbors, passersby, manufacturing personnel and customers are more favorably impressed.

The low investment and operating costs characteristic of pretreatment-land application systems is an implied comparison to advanced wastewater treatment and stream discharge. Generally, pretreatment is limited to moderate percent removals (25-75%) with land as the ultimate receiver. Such pretreatment is not on the portion of the cost curves where cost increases are exponentially increasing (95-99+%); hence, savings result. Further, the capital investment in the components of a land application is not high when compared to conventional units providing the same high levels of pollution control. Much of the discussion of economics is included in Chapters 12 and 13.

Energy efficiency is becoming important in all areas of society, including pollution control. An unrefined, but effective, comparison was made of several secondary treatment units (tricking filter and activated sludge) followed by sludge landfill or incineration at existing municipalities versus pond pretreatment followed by effluent land application (Williams 1973). When expressed on a basis of energy used per unit amount of BOD_5 removed the conventional systems with stream discharge were two to three times more energy intensive than pretreatment-land application. It is anticipated that a similar or possibly wider energy use spread exists for industrial waste treatment.

The technology characteristic of pretreatment-land application systems is not as sophisticated as those advanced wastewater treatment processes which even approach the same level of waste constituent and priority pollutant removals. A plant-soil receiver involves such agricultural operations as pumping and irrigation, hauling and spreading, crop harvesting, storage ponds and lagoons, or conservation practices to prevent runoff. These can be accomplished using less formally trained personnel than are required for the successful operation of advanced wastewater treatment facilities. Therefore, pretreatment-land application is an appropriate technology for many areas of the world, as well as areas of the United States.

SUMMARY

This chapter sets the regulatory and industry use stage for the detailed development of the technology of pretreatment-land application systems for industrial wastes. The basic design objective of nondegradation of the land resource has been presented along with the overall design methodology for pretreatment-land application systems. These concepts are used repeatedly throughout the book.

Chapter 2 thoroughly examines stage I of the design methodology. The reader is introduced to basic soil and agronomy concepts, which are important to land treatment design in Chapter 3. The critical assimilative capacities for waste constituents are developed in Chapters 4-11. Present state-of-the-art is used. After Chapter 11, the LLC analysis can be completed with the constituent controlling needed land area and the actual area requirements being determined.

Design and economics of the land system are covered in detail in Chapter 12, and pretreatment design and economics are the subjects of Chapter 13.

Finally, in Chapter 14, the balance between pretreatment and land application for a total industrial system is discussed, as are land treatment systems.

REFERENCES

Ahearn, D. G., and S. P. Meyers, Eds. "The Microbial Degradation of Oil Pollutants," Center for Wetland Resources, Louisiana State University, Baton Rouge, LA (1973).

Anderson, D. G., and A. T. Wallace. "Innovations in Terrestrial Disposal of Steam Peel Potato Wastewater," *Proc. Pacific North West Ind. Waste Managem. Conf.* (1971), p. 45.

Anderson, D. R., *et al.* "Percolation of Citrus Wastes Through Soil," *Proc. 21st Ind. Waste Conf.*, Purdue University 121:892 (1966).

Baldwin, I. L. "Modification of the Soil Flora Induced by Application of Crude Petroleum," *Soil Sci.* 14:465-477 (1922).

Barlow, A. C. "Philosophy of Deep-Well Disposal," In *Underground Waste Management and Artificial Recharge,* Vol. 2, J. Braunstein, Ed. American Association of Petroleum Geologists, George Bonta Co., Menasha, WI (1973), p. 667.

Bendixen, T. W., R. D. Hill, F. T. DuByne and G. G. Robeck. "Cannery Waste Treatment by Spray Irrigation Runoff," *J. Water Poll. Control Fed.* 41(3):385-391 (1969).

Billings, R. M. "Stream Improvement Through Spray Disposal of Sulfite Liquor at Kimberly-Clark Corporation, Niagara, Wisconsin, Mill, *Proc. 13th Ind. Waste Conf.*, Purdue University 96:71 (1958).

Blevins, R. L., H. H. Bailey and G. E. Ballard. "The Effect of Acid Mine Water on Flood Plain Soils in the Western Kentucky Coal Fields," *Soil Sci.* 110:191-196 (1970).

Blosser, R. O., and E. L. Owens. "Irrigation and Disposal of Pulp Mill Effluents," *Water Sew. Works* 111:424 (1964).

Bradford, G. R. "Boron," In *Diagnostic Criteria for Plants and Soils,* H. D. Chapman, Ed., University of California, Division of Agricultural Science (1966), p. 33.

Braunstein, J., Ed. *Underground Waste Management and Artificial Recharge,* Vols 1 and 2, American Association of Petroleum Geologists, George Banta Co., Menasha, WI (1973).

Breska, G. V., *et al.* "Objectives and Procedures for a Study of Spray Irrigation of Dairy Wastes," *Proc. 12th Ind. Waste Conf.*, Purdue University 94:636 (1957).

Buddin, W. "Partial Sterilization of Soil by Volatile and Non-volatile Antiseptics," *J. Agric. Sci.* 6:417-451 (1914).

Canham, R. A. "Comminuted Solids Inclusion with Spray Irrigated Canning Wastes," *Sew. Ind. Wastes* 30:1028 (1958).

Carlile, B. L., and J. A. Phillips. "Evaluation of Soil Systems for Land Disposal of Industrial and Municipal Effluents," WRRI of UNC. Report No. 118 (1976).

Carlson, C. A., P. G. Hung and T. B. Delaney, Jr. "Overland Flow Treatment of Wastewater," U.S. Army Engr. Waterways Expt. Sta., Vicksburg, MS Miscellaneous Paper Y-74-3, U.S. Army Corps of Engineers (1974).

Chawla, V. K. "Treatment of Fish and Vegetable Processing Waste-Lagoon Effluent," In *Food Processing Waste Management, Proc. 1973 Cornell Agric. Waste Managememt Conf.*, R. C. Loehr, Ed. Syracuse, NY (1973), pp. 74-85.

Cecil, L. K., Ed. *Proc. 2nd Nat. Conf. Complete Wateruse*, AIChE, New York (1975).

Cecil, L. K., Ed. *Proc. 3rd Nat. Conf. Complete Water Resuse-Symbiosis as a Means of Abatement for Multimedia Pollution*, AIChE and EPA, New York (1976).

Colovos, G. C., and N. Tinklenberg. "Land Disposal of Pharmaceutical Manufacturing Wastes," *Biotech. Bioeng.* 4:153 (1962).

CRS (Congressional Research Service). "Effects of Chronic Exposure to Low-Level Pollutants in the Environment," prepared for subcommittee on the Environment and the Atmpshere of the Committee on Science and Technology, U.S. House of Representatives (1975), p. 402.

Cox, W. E., and W. R. Walker. "Legal and Institutional Consideration of Deep-Well Waste Disposal," in *Underground Waste Management and Artificial Recharge*, Vol. 1, American Association Petroleum Geologists, George Banta Co. Menasho, WI (1973), p. 3.

Crawford, S. C. "Spray Irrigation of Certain Sulfate Pulp Mill Wastes," *Sew, Ind. Wastes* 30:1266 (1958).

de Haan, F. A. M., and P. J. Zwerman. "Land Disposal of Potato Starch Processing Wastewater in the Netherlands," in *Food Processing Waste Management, Proc. 1973 Cornell Agric. Waste Managemt. Conft.*, R. C. Loehr, Ed., Syracuse, NY (1973), pp. 222-228.

Deutsch, M. "Incidents of Chromium Contamination of Groundwater in Michigan," *Proc. 1961 Symp. Groundwater Contam.*, Tech. Rep. W 61-5, USDHEW PHS. Robert A. Taft Center, Cincinnati, OH (1961), pp. 98-104.

Dietz, M. R., and R. C. Frodey. "Cannery Waste Disposal by Gerber Products," *Compost Sci.* 1:22 (1960).

Dobson, A. L., and H. A. Wilson. "Respiration Studies on Soil Treated with some Hydrocarbons," *Soil Sci. Soc. Am. Proc.* 28:536-538 (1964).

Dolgova, L. G. "Phenoloxidase Activity of Soil Under Conditions of Industrial Pollution," *Pochvovedenie No.* (Russian) 9:64-69 (1973).

Eaton, F. M. "Total Salt and Water Quality Appraisal," *Diagnostic Criteria for Plants and Soils*, H. D. Chapman, Ed., University of California, Division of Agricultural Science (1966).

Erickson, A. E., *et al.* "Soil Modification for Denitrification and Phosphate Reduction of Feedlot Waste," EPA-660/2-74-057 (1974).

Gellman, I., and R. O. Blosser. "Disposal of Pulp and Paper Mill Waste by Land Application and Irrigation Use," *Proc. 14th Ind. Waste Conf.*, Purdue University 104:479 (1959).

Gibson, D. T. "Microbial Degradation of Hydrocarbons," in *The Nature of*

Seawater, E. D. Goldberg, Ed., Physical and Chemical Sci. Report 1 (1974).

Giddens, J. "Spent Motor Oil Effects on Soil and Crops," *J. Environ. Qual.* 5(2):179-181 (1976).

Gilde, L. C., *et al.* "A Spray Irrigation System for Treatment of Cannery Wastes," *J. Water Poll. Control Fed.* 43:2011 (1971).

Gracia, W. J., C. W. Blessin, G. E. Inglett and R. O. Carlson. "Physical Chemical Characteristics and Heavy Metal Content of Corn Grown on Sludge-Treated Strip-Mined Soil," *J. Agric. Food Chem.* 22(5):810-815 (1974).

Gudin, C., and W. J. Syratt. "Biological Aspects of Land Rehabilitation Following Hydrocarbon Contamination," *Environ. Poll.* 8(2):107-112 (1975).

Haider, K., and J. P. Martin. "Decomposition of Specifically Carbon-14 Labelled Benzoic and Cinnamic Acid Derivatives in Soil," *Soil Sci. Soc. Am. Proc.* 39(4):657-662 (1975).

Halderson, J. L., and J. R. Peterson. "Environmental Monitoring at Agricultural Sites for Sludge Utilization," American Society Agricultural Engineers, St. Joseph, MI, Paper No. 76-2064 (1976).

Hands, F. J., J. R. Lambert and P. S. Opliger. "Hydrologic and Quality Effects of Disposal of Peach Cannery Waste," *Trans. Am. Soc. Agric. Eng.* 11:90 (1968).

Hickerson, R. D., and E. K. McMahon. "Spray Irrigation of Wood Distillation Wastes," *J. Water Poll. Control Fed.* 32:55 (1960).

Huddleston, R. L., and L. W. Cresswell. "The Disposal of Oily Wastes by Land Farming," paper presented to the Management of Petroleum Refinery Wastewaters Forum at Tulsa, OK, Sponsored by EPA, API, WPRA and UT (1976).

Huisingh, D., and V. Huisingh. "Factors Influencing the Toxicity of Heavy Metals in Food," *Ecol. Food Nutr.* 3:263-272 (1974).

Humenik, F. J., *et al.* "Transformations of Swine Wastewater in Laboratory Soil Profiles," *Trans. Am. Soc. Agric. Eng.* 18:1130-1135 (1975).

Illinois Environmental Protection Agency "Design Criteria for Municipal Sludge Utilization of Agricultural Land," Technical Policy WPC-3, Draft copy, pp. 1-24.

Jarnea, S., and G. Salay. "Irrigation with Water Contaminated with Compounds of Cyanic Acid," *Annov. Inst. Circ. Imb. Func. Ser. Imb. Func.* 2:117-127 (1969).

Jobson, A., M. McLaughlin, F. D. Cook and D. W. S. Westlake. "Effects of Amendments on the Microbial Utilization of Oil Applied to Soil," *J. Appl. Microbiol.* 27(1):166-171 (1974).

Jones, J. G., M. Knight and J. A. Byron. "Effect of Grass Pollution by Kerosene Hydrocarbons on the Microflora of Moorland Soil," *Nature* 227: 116-118 (1970).

Kincannon, C. B. "Oily Waste Disposal by Soil Cultivation Process, EPA Report EPA-R2-72-110, Washington, DC (1972).

Kirby, C. F. "Sewage Treatment Farms," Dept. Civil Eng., Melborne University (1970), p. 16.

Kirkham, M. B. "Trace Elements in Corn Grown on Long-Term Sludge Disposal Site," *Environ. Sci. Technol.* 9(8):765-768 (1975).

Koch, H. C., and D. E. Bloodgood. "Experimental Spray Irrigation of Paper Board Mill Wastes," *Sew. Ind. Wastes* 31:827 (1959).

Lagerwerff, J. V. "Heavy Metal Contamination of Soils," in *Agriculture and Quality of Our Environment,* N. C. Brady, Ed., American Association for the Advancement of Science Publ. 85:343-364 (1976).

Lane, L. C. "Disposal of Liquid and Solid Wastes by Means of Spray Irrigation in the Canning and Dairy Industries," *Proc. 10th Ind. Waste Conf.,* Purdue University 89:508 (1955).

Lawton, G. W., et al. "Spray Irrigation of Dairy Wastes," *Sew. Ind. Wastes* 31:923 (1959).

Lazar, E. C. "Damage Incidents from Improper Land Disposal," *J. Hazard. Mat.* 1:157-164 (1976).

Leland, H. V., E. D. Copenhaver and L. S. Carrill. "Heavy Metals and Other Trace Elements," *J. Water Poll. Control Fed.* 46(6):1452-1476 (1974).

Lever, N. A. "Disposal of Nitrogenous Liquid Effluent from Modderfontein Dynamite Factory," *Proc. 21st Ind. Waste Conf.,* Purdue University 121:902 (1966).

Locke, E. T. "Lakelands Grass Filter Removes Most BOD, Suspended Solids - but not Enough Nutrients," *Overflow* (May 8-10, 1972).

Ludwig, R. G., and R. V. Stone. "Disposal Effects of Citrus By-Products Wastes," *Water Sew. Works* 109:410 (1962).

Luley, H. G. "Spray Irrigation of Vegetable and Fruit Processing Wastes," *J. Water Poll. Control Fed.* 35:1252 (1963).

Mather, J. R. "An Evaluation of Cannery Waste Disposal by Overland Flow Spray Irrigation," Campbell Soup Co., Paris Plant, *Publ. Climatol.* SSII (2):1-73 (1969).

McJunkin, F. E., and P. A. Vesilind. "Ultimate Disposal of Wastewaters and Their Residues," *Proc. WRRI of UNC,* North Carolina State University Raleigh, NC (1973).

McKee, F. J. "Spray Irrigation of Dairy Wastes." *Proc. 10th Ind. Waste Conf.,* Purdue University 89:514 (1955).

McKenna, E. J., and R. D. Heath. "Biodegradation of Polynuclear Aromatic Hydrocarbon Pollutants by Soil and Water Microorganisms," Res. Report No. 113, University of Illinois at Urbana, Water Resources Center (1976).

Medvedev, V. A., and V. D. Davidov. "Transformation of Individual Organic Products of the Coke Industry in Chernozemic Soil," *Pochvovedenie No.* (Russian) 11:22-28 (1972).

Medvedev, V. A., and V. D. Davidov. "The Rate of Degradation of Phenols and Quinones in Chernozemic Soil According to Data Relating to Oxidation-Reduction Potential and Infra-red Spectroscopy," *Pochvovedenie No.* (Russian) 1:133-137 (1974).

Meighan, A. D. "Experimental Spray Irrigation of Strawboard Wastes," *Proc. 13th Ind. Waste Conf.,* Purdue University 96:456 (1958).

Meyer, E. A., and R. M. Butler. "Effects of Hydrologic Regime on Nutrient

Removal from Wastewater Using Grass Filtration for Final Treatment," Inst. for Res. on Land & Water Resources, Pennsylvania State University, University Park, PA, Res. Publ. 88 (1974).

Miller-Neuhaus, G. "Industrielle abwasse unter Berücksichtigung besonderer Verhältnisse im Rheinisch-westfalischen Kohlen," *Gluckauf* 93:684 (1954).

Miyamoto, S., and J. Ryan. "Sulfuric Acid for Treatment of Ammoniated Irrigation Water: II Reducing Calcium Precipitation and Sodium Hazard," *SSSA J.* 40(2):305-308 (1976).

Miyamoto, S., J. Ryan and J. L. Strochlein. "Potentially Beneficial Uses of Sulfuric Acid in Southwestern Agriculture," *J. Environ. Qual.* 4:431-437 (1975).

Molloy, D. J. "Instant Waste Treatment," *Water Works Wastes Eng.* 1:68 (1964).

Monson, H. "Cannery Waste Disposal by Spray Irrigation - After 10 Years," *Proc. 13th Ind. Waste Conf.,* Purdue University 96:449 (1958).

Murphy, J. F. "Some Effects of Crude Petroleum on Nitrate Production, Seed Germination and Growth," *Soil Sci.* 27:117-120 (1929).

Odu, C. T. I. "Microbiology of Soils Contaminated with Petroleum Hydrocarbons. I. Extent of Contamination and Some Soil and Microbiological Properties After Contamination," *J. Inst. Pet.* 58:201-208 (1972).

Ongerth, J. E. "Feasibility Studies for Land Disposal of a Dilute Oily Wastewater," paper presented at 30th Industrial Waste Conf., Purdue University, West Lafayette, IA (1975).

Overcash, M. R. "Implications of Overland Flow for Municipal Waste Management," *J. Water Poll. Control Fed.* 2337-2347 (1978).

Pal, D., and M. R. Overcash. "List of Investigations on the Fate of Various Constituents in Soil-Waste Systems," Department of Biological and Agricultural Engineering, North Carolina State University, Raleigh, NC (1976).

Pal, D., and M. R. Overcash. "Appraisal of Pretreatment-Land Application Technology for the Management of Textile Industry Wastes," WRRI of UNC Report (1979).

Parker, R. R. "Disposal of Tannery Wastes," *Proc. 22nd Ind. Waste Conf.,* Purdue University 129:36 (1967).

Parsons, W. C. "Spray Irrigation of Wastes from the Manufacture of Hardboard," *Proc. 22nd Ind. Waste Conf.,* Purdue University 129:602 (1967).

Philip, A. H. "Disposal of Insulation Board Mill Effluent by Land Irrigation," *J. Water Poll. Control Fed.* 43:1749 (1971).

Piper, A. M. "Disposal of Liquid Wastes by Injection Underground- Neither Myth or Millennium" *U. S. Geol. Surv. Circ.* 631 (1969).

Plice, M. J. "Some Effects of Crude Petroleum on Soil Fertility," *Soil Sci. Soc. Am. Proc.* 13:413-416 (1948).

Raymond, R. L., J. P. Hudson and V. W. Jamison. "Oil Degradation in Soil," *Appl. Environ. Microbiol.* 31(4):522-535 (1976).

Sandborn, N. H. "Disposal of Food Processing Wastes by Irrigation," *Sew. Ind. Wastes* 25:1034 (1953).

Schraufnagel, F. H. "Ridge-and-Furrow Irrigation for Industrial Waste Disposal," *J. Water Poll. Control Fed.* 34:1117 (1962).

Schroeder, E. D., D. J. Reardon, F. Matteoli and W. H. Movey. "Biological Treatment of Winery Stillage," in *Food Processing Waste Management, Proc. 1973 Cornell Agric. Waste Managemt. Conf.*, R. C. Loehr, Ed., Syracuse, NY (1973).

Schwendinger, R. B. "Reclamation of Soil Contaminated with oil," *J. Inst. Pet.* 54:182-197 (1968).

Scott, R. H. "Disposal of High Organic Content Wastes on Land," *J. Water Poll. Control Fed.* 34:932 (1962).

Sievers, D. M., G. B. Garner and E. E. Pickett. "A Lagoon-Grass Terrace System to Treat Swine Wastes," in *Proc. 3rd Int. Symp Livestock Wastes*, St. Joseph, MI (1975), pp. 541-543, 548.

Sittig, M. *Pollutant Removal Handbook*, Noyes Data Corp., Park Ridge, NJ (1973).

Skinner, R. J. "Contamination of Soil by Oil and Other Chemicals," in *Aspects of Current Use and Misuse of Soil Resources Report*, Welsh, Soils Discussion Group No. 14:131-141 (1973).

Smith, J. H. "Decomposition in Soil of Waste Cooking Oils Used in Potato Processing," *J. Environ. Qual.* 3(3):279-281 (1974).

Smith, J. H. "Treatment of Potato Processing Wastewater on Agricultural Land," *J. Environ. Qual.* 5(1):113-116 (1976).

Staples, R. E. "Teratogens and the Delaney Clause," *Environ. Health Pers.* VII (August 1974).

Stebbings, R. V. E. "Recovery of Salt-Marsh in Brittany Sixteen Months after Heavy Pollution by Oil," *Environ. Poll.* 1:163-167 (1970).

Stevens, T. G., and G. G. Dunn. "Grass Filtration-Pond Stabilization of Canning Waste, a Two Step Process," in, *16th Ont. Ind. Waste Conf. Proc.* Toronto, Ontario, Ontario Water Resources Commission (1969), pp. 161-175.

Thomas, R. E., "Spray-Runoff Treatment of Feedlot Runoff," EPA, R. S. Kerr Water Research Center, Ada, OK (1972).

Thomas, R. E. "Overland Flow Treatment of Raw Domestic Wastewater," R. S. Kerr Environ. Res. Lab., Ada, OK, EPA, presented at 2nd Nat. Conf. on Water Reuse, AIChE and EPA, Chicago, IL, May (1975).

Thomas, R. E., K. Jackson and L. Penrod. "Feasibility of Overland Flow for Treatment of Raw Domestic Wastewater," R. S. Kerr Environ. Res. Lab., Ada, OK, EPA-660/2-74-087 (1974).

Trax, J. R. "EPA Viewpoint on Land Application of Liquid Effluents," in *Proc. Land Disposal of Municipal Effl. Sludges*, Rutgers University (1973), pp. 733-742.

Udo, E. J., and A. A. A. Fayemi. "The Effect of Oil Pollution of Soil on Germination, Growth and Nutrient Uptake of Corn," *J. Environ. Qual.* 4(4):537-540 (1975).

Valoras, N., J. Letey, J. P. Martin and J. Osborn. "Degradation of a Nonionic Surfactant in Soils and Peat," *Soil Sci. Soc. Am. J.* 40(1):60-63 (1976).

Vanderpost, J., and C. T. Corke. "Interactions of Metals with Microbial

Populations in Soils," in *Metals in the Biosphere,* Proc. of a Symp. by Dept. of Land Resource Sci., University of Guelph, Guelph, Ontario (1975).

Varadarajan, S., T. A. Govinda and A. Gopalswamy. "Influence of Tannery Effluents on Soils and Crops and Proper Disposal of Effluents," *Madras Agric. J.* 57:353-360 (1970).

Voights, D. "Lagooning and Spray Disposal of Neutral Sulfite Semichemical Pulp Mill Liquors," *Proc. 10th Ind. Waste Conf.,* Purdue University 89:497 (1955).

Walker, J. D., and R. R. Colwell. "Mercury Resistant Bacteria and Petroleum Degradation," *Appl. Microbiol.* 27(1):285-287 (1974).

Walker, R. G. "Tertiary Treatment of Effluent from Small Sewage Works," *Water Poll. Control* 198-201 (1972).

Wallace, A. T. "Land Disposal of Liquid Industrial Wastes," in *Land Treatment and Disposal of Municipal and Industrial Wastewater,* R. L. Sanks and T. A. Asano, Eds. (Ann Arbor, MI: Ann Arbor Science Publishers, Inc., 1976), pp. 147-162.

Ward, P. S. "Toxic Pollutants - Control: Progress at Last," *J. Water Poll. Control Fed.* 49(1):6-9 (1977).

Webber, L. R., T. G. Stevens and D. A. Tel. "Exchange Properties of a Soil Used for the Disposal of Alkali Cannery Wastes," in *Food Processing Waste Management, Proc. 1973 Cornell Agric. Waste Managemt. Conf.,* Syracuse, NY (1973).

Williams, T. C. "Spray Irrigation for Wastewater Disposal in Small Residential Subdivisions," *Proc. Nat. Symp. on Ultimate Disposal of Wastewaters and their Residuals,* Water Resources Research Institute North Carolina State University, Raleigh, NC (1973), pp. 301-311.

Willrich, T. L., and J. O. Boda. "Overland Flow Teatment of Lagoon Effluent," Oregon State University, presented at ASAE Meeting, (December 1976).

Wisniewski, T. F., A. J. Wiley and B. F. Lueck. "Ponding and Soil Filtration for Disposal of Spent Sulphite Liquor in Wisconsin," *Proc. 10th Ind. Waste Conf.,* Purdue University 89:480 (1955).

Witherow, J. L. "Small Meat Packers Wastes Treatment Systems," Pacific Northwest Environmental Research Lab., EPA, presented at 4th Nat. Symp. on Food Processing Wastes, Syracuse, NY (1973).

Woodley, R. A. "Spray Irrigation of Organic Chemical Wastes," *Proc. 23rd Ind. Waste Conf.,* Purdue University 132:251 (1968).

CHAPTER 2

LAND LIMITING CONSTITUENT ANALYSIS—STAGE I

INTRODUCTION

The design and evaluation of industrial waste pretreatment-land application systems involves a four-stage methodology. This chapter describes the first stage of that design procedure. The perspective of this chapter to the total methodology is presented in Figure 2.1. A basic part of stage I is the plant-soil assimilative capacity, which is built on the fundamental nondegradation constraint described in Chapter 1. It is important to remember the nature of that constraint since all subsequent site design criteria (assimilative capacities) have already been developed in an environmentally acceptable manner. Furthermore, the development of the assimilative capacities (kg constituent/ha/yr) allows use of variable-size land areas (of the same soil characteristics) as needed based on the waste generation. Thus, stage I is critical to successful industrial waste land application design.

The concept of land application described here is the incorporation of wastes in the surface zone of a plant-soil system to ensure environmental assimilations. This does not imply landfills, overland flow pretreatment, deep well injection or soil reactor units.

Stage I is probably the most complex and difficult phase involved in the overall land application design, particularly for industrial wastes for which the first stage is typically complex. Experience with a substantial number of actual industrial systems indicates that expertise from several disciplines is frequently needed.* Land application specialists, soil specialists, soil chemists and microbiologists, hydrologists and system design engineers are needed in the average industrial system. In addition, soil drilling and

---

*B. Greenhouse, SSI, 525 Industrial Dr., Marietta, GA. Personal communication.

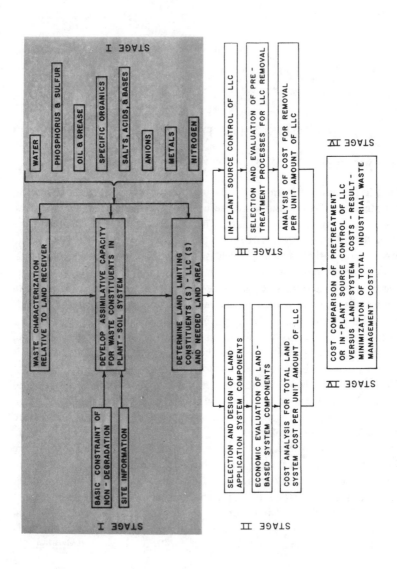

Figure 2.1 Relationship of Chapter 2 to overall design methodology for industrial pretreatment-land application systems.

undisturbed sampling are essential field equipment. Chapter 3 enables the industrial manager or environmental specialist utilizing this book to converse knowledgeably about or use such expertise. It describes basic soil science concepts. For the reader already acquainted with such material, Chapter 3 can be bypassed.

Examination of stage I (Figure 2.1), reveals a number of primary facts that throughout the book are reviewed briefly here. An extremely important part of stage I is the use of date from the site(s) under consideration. This is referred to as site-specific inputs, which are absolutely essential to any LLC analysis, even a preliminary one. Land application design has a highly site-specific character, which reinforces the concept of a design procedure rather than arbitrary design criteria (kg/ha/yr) for all sites. Whether an industrial system is on owned land or on a number of potential future sites, the design cannot be realistic without such site data. Of the land application systems that have required troubleshooting or redesign because of potential failure circumstances, 80-90% are the result of not considering the site-specific characteristics.*

A wide variation in the assimilative capacity among sites has substantial economic implications for new industrial facilities. If a preliminary study of several sites, such an Industrial Terrain Analysis,®** is performed, then factors such as groundwater, drainage, topography, vegetation and assimilative capacity can be determined. Other industrial siting factors are also included. This analysis is performed in an anonymous manner to avoid influencing land costs and acceptance. Then the site requiring the least land for a pretreatment-land application system can be selected at considerable financial savings.

WASTE CHARACTERIZATION

Individual waste constituents or classes of parameters are the basis for the overall design methodology, so an industrial waste must be quantified on an individual parameter basis. Obviously, regardless of concentration, all compounds cannot be analyzed in an industrial waste. A compromise is based on a thorough understanding of (1) which parameters are important with respect to land as a receiver (as opposed to stream discharge), (2) which constituents are of particular environmental concern, and (3) the chemical and manufacturing process inputs and products. The second category is based on a list of designated or priority pollutants likely to be in the discharge of a certain industry. Table 2.1 should be consulted for the present version of these

---

*Greenhouse, B., SSI, 525 Industrial Dr., Marietta, GA. Personal communication.

**Fox, N., SSI, 525 Industrial Dr., Marietta, GA. Personal communication.

Table 2.1 Recommended List of Consent Decree Priority Pollutants (EPA 1976)

Compound Name	Compound Name
1. Acenaphthene[a]	D Dichloroethylenes (1,1-dichloro-
2. Acrolein[a]	ethylene and 1,2-dichloro-
3. Acrylonitrile[a]	ethylene)[a]
4. Benzene[a]	29. 1,1-dichloroethylene
5. Benzidine[a]	30. 1,2-trans-dichloropene
6. Carbon Tetrachloride[a]	31. 2,4-dichloropropane and dichloro-
(tetrachloromethane)	pene[a]
Chlorinated Benezenes (other than	Dichloropropane and dichloropene[a]
Dichlorobenzenes)[a]	32. 1,2-dichloropropane
7. Chlorobenzene	33. 1,2-dichloropropylene (1,3-
8. 1,2,4-trichlorobenzene	dichloropropene)
9. Hexachlorobenzene	34. 2,4-dimethylphenol[a]
Chlorinated Ethanes (including	Dinitrotoluene[a]
1,2-dichloroethane, 1,1,1-	35. 2,4-dinitrotoluene
trichloroethane and hexa-	36. 2,6-dinitrotoluene
chloroethane)[a]	37. 1,2-dinitrotoluene[a]
10. 1,2-dichloroethane	38. Ethylbenzene[a]
11. 1,1,1-trichloroethane	39. Fluoranthene[a]
12. Hexachloroethane	Haloethers (other than those listed
13. 1,1-dichloroethane	elsewhere)[a]
14. 1,1,2-trichloroethane	40. 4-chlorophenyl phenyl ether
15. 1,1,2-thrichloroethane	41. 4-bromophenyl phenyl ether
16. chloroethane	42. bis (2-chloroisopropyl) ether
Chloroalkyl Ethers (chloromethyl,	43. bis (2-chloroethoxy) methane
chloroethyl and mixed ethers)[a]	Halomethanes (other than those
17. bis (chloromethyl) ether	listed elsewhere)[a]
18. bis/2 chloroethyl/ether	44. Methylene chloride (chloro-
19. 2-chloroethyl vinyl ether (mixed)	methane)
Chlorinated Naphthalene[a]	45. Methylchloride (chloromethane)
20. 2-chloronaphthalene	46. Methyl bromide (bromomethane)
Chlorinated phenols (other than	47. Bromoform (tribromomethane)
those listed elsewhere; includes	48. Dichlorobromomethane
trichlorophenols and chlorinated	49. Trichlorofluoromethane
cresols)[a]	50. Dichlorodifluoromethane
21. 2,4,6-trichlorophenol	51. Chlorodibromomethane
22. parachlorometa cresol	52. Hexachlorobutadiene[a]
23. Chloroform (trichloromethane)[a]	53. Hexachlorocyclopentadiene[a]
24. 2-chlorophenol[a]	54. Isophorone[a]
Dichlorobenzenes[a]	55. Naphthalene[a]
25. 1,2-dichlorobenzene	56. Nitrobenzene[a]
26. 1,3-dichlorobenzene	Nitrophenols (including 2,4-dinitro-
27. 1,4-dichlorobenzene	phenol and dinitrocresol)[a]
Dichlorobenzidine[a]	57. 2-nitrophenol
28. 3,3'-dichlorobenzidine	58. 4-nitrophenol
	59. 2,4-dinitrophenol[a]
	60. 4-6-dinitro-o-cresol

Table 2.1 Continued

Compound Name		Compound Name
Nitrosamines[a]		DDT and Metabolities[a]
61. N-nitrosodimethylamine	92.	4,4'-DDT
62. N-nitrosodiphenylamine	93.	4,4'-DDE (p,p'-DDX)
63. N-nitrosodi-n-propylamine	94.	4,4'-DDD (p,p'-TDE)
64. Pentachlorophenol[a]		Endosulfan and metabolities[a]
65. Phenol[a]	95.	a-endosulfan-Alpha
Phthalate Esters	96.	b-endosulfan-Deta
66. bis (2-ethylhexyl) phthalate	97.	endosulfan sulfate
67. Butyl benzyl phthalate		Endrin and Metabolities[a]
68. Di-n-butyl phthalate	98.	Endrin
69. Di-n-octyl phthalate	99.	Endrin aldehyde
70. Diethyl phthalate		Heptachlor and Metabolities[a]
71. Dimethyl phthalate	100.	Heptachlor
Polynuclear Aromatic Hydro-	101.	Heptachlor epoxide
carbons [a]		Hexachlorocyclohexane (all isomers)[a]
72. Benzo (a) anthracene (1,2-ben-	102.	a-BHC-Appha
zanthracene)	103.	b-BHC-Beta
73. Benzo (a) pyrene (3,4-benzo-	104.	r-BHC (lindene)-Gamma
pyrene)	105.	g-BHC-Delta
74. 3.4-benzofluoranthene		Polychlorinated biphenyls (PCB)[a]
75. Benzo (k) fluoranthane (11,12-	106.	PCB-1242 (Arochlor 1242)
benzofluoranthene)	107.	PCB-1254 (Arochlor 1254)
76. Chrysene	108.	PCB-1221 (Arochlor 1221)
77. Acenaphthylene	109.	PCB-1232 (Arochlor 1232)
78. Anthracene	110.	PCB-1248 (Arochlor 1248)
79. Benzo (ghi) perylene (1,12-benzo-	111.	PCB-1260 (Arochlor 1260)
perylene)	112.	PCB-1016 (Arochlor 1016)
80. Fluoroene	113.	Toxaphene[a]
81. Phenathrene	114.	Antimony (total)[a]
82. Dibenzo (a,h) anthracene (1,2,5,6-	115.	Arsenic (total)[a]
dibenzathracene)	116.	Asbestos (fibrous)[a]
83. Indeno (1,2,3-cd) pyrene (2,3-0-	117.	Beryllium (total)[a]
phylenepyrene)	118.	Cadmium (total)[a]
84. Pyrene	119.	Chromium (total)[a]
85. Tetrachloroethylene[a]	120.	Copper (total)[a]
86. Toluene[a]	121.	Cyanide (total)[a]
87. Trichloroethylene[a]	122.	Lead (total)[a]
88. Vinyl chloride (chloroethylene)[a]	123.	Mercury (total)[a]
Pesticides and Metabolities	124.	Nickel (total)[a]
89. Aldrin[a]	125.	Selenium (total)[a]
90. Dieldrin[a]	126.	Silver (total)[a]
91. Chlordane (technical mixture and	127.	Thallium (total)[a]
metabolites)[a]	128.	Zinc (total)[a]

[a]Specific compounds and chemical classes as listed in the consent degree.

priority pollutants. The third category is often best described by in-plant manufacturing personnel or chemical engineers closely associated with likely materials in the waste stream.

With the diverse inputs described above, an in-depth waste characterization can be assembled. Because of the basic pathways involved in land-based treatment, mass generation of waste constituents (kg/yr) is generally more important than constituent concentration (mg/l). Hence, waste characterization involves obtaining a true composite sample, volume average, which, when multiplied times an average flow, will yield the weekly, monthly or annual rate of waste constituent generation from a particular plant. For certain parameters, the concentration of material applied to land is also important. Equalization and less-frequent-than daily land application are often used, so the range of concentrations often is not required, which differs substantially from pretreatment-stream discharge systems, in which brief surges can be deleterious.

The design engineer is often faced with the question of which industrial waste stream can be land applied. The same waste constituents are often present in the raw waste from a plant (either as solids, slurries or liquids) as well as existing pretreatment effluent or sludge. Thus, waste characterization differs only in the magnitude of these parameters. Likewise, the assimilative capacity must be developed for each waste constituent, so these capacities are valid if used in the design for land application of raw waste, effluent, sludge, etc. This common information needed, based on constituents and not necessarily on waste stream, means that all waste streams can be designed for land treatment. Thus waste characterization on a mass basis is advantageous for all waste streams.

A typical industrial waste characterization used in preliminary constituent evaluation involves a number of parameters (Table 2.2), which is necessary for a variety of wastes from food processing to textiles to complex organics manufacture. This list of 30 constituents is usually only a starting point, since specific priority pollutants or other organics are often added. Other parameters must be added based on knowledge of likely chemicals present in substantial quantities from the particular industrial process. Furthermore, with the wide variation in industrial processes, such characterization is usually needed on each different plant.

As much information as possible about the industrial raw waste or other plant waste streams must be obtained to project the assimilatory capacity and a knowledgeable design. The use of land application for long time periods means that lower concentration levels of parameters may be important and must be addressed.

Table 2.2 Partial List of Parameters to be Characterized in Industrial Waste as a
Part of Pretreatment-Land Application Design

Parameters	
Volume	Boron
Total Phoshorus	Arsenate
Total Sulfur	Copper
Sulfate	Zinc
Conductivity or Total Dissolved	Lead
Solids (TDS)	Nickel
Chloride	Cadmium
Sodium	Other Relevant Metals
Potassuim	Chemical Oxygen Demand (COD)
Calcium	Total Nitrogen (TN)
Magnesium	Ammonium
Carbonate	Total Solids (TS)
pH	

PLANT-SOIL ASSIMILATIVE CAPACITY

This phase of stage I is often the most complex and requires the greatest amount of technological and site-specific input. Assimilative capacity is usually expressed in terms of the mass of a given constituent, which is accommodated in an environmentally acceptable manner per unit area and per unit time (*e.g.,* kg parameter/ha/yr). An environmentally acceptable manner refers to the basic design constraint of nondegradation (see Chapter 1). In addition, the mass of constituent per unit volume or area of soil (concentration) present in any short time period can also be critical in the assimilative capacity.

Procedural Tools

Determining assimilative capacity based on an environmental objective of nondegradation can be accomplished by one or more of three broad categories of calculated tools (Table 2.3). For any constituents of industrial waste (excepting radioactive materials) there are only three types of design methods or calculations needed to arrive at land loading rates.

Parameters that Decompose

The first of these procedural tools is for those parameters that are degraded at some finite rate or are substantially removed by plants as the major mechanism of assimilation. If an industrial waste constituent is taken up by the vegetative cover, but only in small amounts relative to the amount applied, removal by crop harvesting is not the principal assimilative pathway; thus, the first calculational tool would be inappropriate.

Table 2.3   Calculational Categories for Achieving Plant-Soil Assmiliative Capacity for Constituents of Industrial Wastes

1. Those compounds that degrade or require plant uptake for assimilation in the plant-soil system, *e.g.,* oils or specific organics.
2. Those compounds that are relatively immobile and nondegradative and, thus, are permitted to accumulate in soils to predetermined critical levels, *e.g.,* heavy metals.
3. Those mobile and nendegradative compounds which must be assimilated over land areas so that receiving waters are not altered to a degree that would require further drinking water treatment, *e.g.,* anionic species.

For those parameters that undergo decomposition or substantial crop uptake, the assimilative capacity is related directly to the rate of compound transformation or crop removal. As examples, nitrogenous, oil and grease, and specific organic compounds are assigned assimilative capacities (kg parameter/ha/yr) based on this first calculational tool.

As illustrative of the considerations involved in this decomposition pathway, design criteria for organic compounds can be developed. When an organic species is applied to a plant-soil system in a dose or application event, the soil concentration of that material is raised immediately. The magnitude of the resultant soil concentration depends on the amount applied per unit area and the depth of contact. As the amount applied is increased, different plant responses occur. For a growing crop, no effect or a stimulatory effect on yield is often observed at low rates of compound addition (Dekock and Vaughan 1976, Buddin 1914, Overcash, *et al.* 1978). If the rate of application is increased to a medium or high rate, the crop will evidence decreased yield response (Figure 2.2). Thus, the single dose amount of a particular organic can be controlled to minimize adverse plant response by remaining below some critical soil level. In addition to adverse yield, the uptake level or crop concentration of a particular compound can also be used to establish the critical dose level in a specific plant-soil system.

Once applied at the critical single dose amount, the organic compound would degrade at some finite, nonzero rate expressed in kg/ha/unit time. No more compound would be applied until soil concentrations had decreased to 10-20% above background levels. More of this constituent then could be applied to the plant-soil system up to the critical soil level. The resultant pattern of applications and soil concentrations is shown in Figure 2.3 This saw-toothed management of a particular organic represents a pattern and depicts the critical design data needed: (1) the critical single dose levels, and (2) the rate of decomposition.

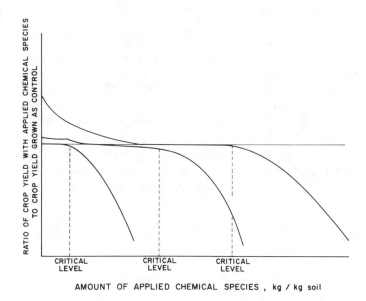

Figure 2.2 Representative crop yield response curves for three vegetative species to which application of a chemical substance has occurred.

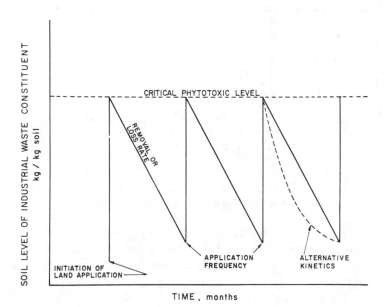

Figure 2.3 Typical soil concentration response pattern for land application of organic constituents.

Parameters that Accumulate

The second procedural tool is to be used for those industrial waste constituents that do not decompose (*i.e.,* are conservative) or migrate substantially
in the plant-soil systems. As an example, elements in the metals category and
some cations have assimilative capacities based on this second calculational
procedure. These constituents may undergo transformations that alter the
effect on land, so must be accounted for in design. However, the element
remains in the plant-soil system, and so the assimilative capacity is based on
establishment of an upper acceptable limit for accumulation. That is, as
industrial waste constituents in this category are continually land applied, the
level in the soil increases until a critical concentration is reached (Figure 2.4).

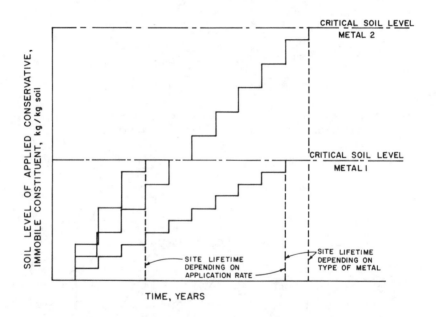

Figure 2.4 Accumulation pattern and critical soil level for various metal application
rates and metal types in industrial waste applied to land.

An adverse pressure on the food chain through vegetation concentration
occurs above this level, or soil conditions exist that substantially reduce plant
yields. Since crop removal from the soil is not large, exceeding this critical
level would render the land unusable for future agricultural use, thus violating
the basic design constraint of nondegradation (see Chapter 1). The assimilative capacity is the critical soil concentration above which the plant-soil
system evidences the adverse plant responses described above.

Parameters that Migrate

The third calculational method used is for those industrial waste constituents that do not decompose, yet migrate in the plant-soil system. Migration is with the water movement whether from rain and snow sources or from the waste-applied liquid. Therefore, the water movement, total amount proceeding beyond the root zone, fraction in lateral direction, fraction to deep percolation, etc. must be established along with the movement of the individual industrial constituent. Inorganic cations and anions fit into this category.

Assimilative capacity for this category is based on utilizing sufficient land area so that the concentration of the industrial constituent when reaching the receiving water does not require further water treatment for drinking purposes. That is, greater expense would not have to be incurred to use that receiving water for societal purposes, whether the groundwater or surface waters are being considered. The application design criterion is the amount of industrial waste parameter/ha/unit time that allows the maintenance of established drinking water concentrations in the soil water reaching receiving waters.

Once developed for a site, the assimilative capacity can be used regardless of the mass of constituent generated or land applied. This implies that any waste stream (raw effluent, sludge, etc.) can be sized for land application, once the assimilative capacities of individual constituents are determined. However it is then almost as difficult to design for a small as for a large waste stream, since all the same parameters must be evaluated for plant-soil assimilative capacity in each size system.

The assimilative capacities for a wide variety of industrial constituents are presented in Chapters 4-11. Each chapter represents one of these broad categories of waste constituents (Table 2.4). Within each chapter or category many individual compounds are described with respect to the assimilation in the plant-soil receiver system. Again, these constituent assimilation rates for industrial waste are based on the fundamental constraint of nondegradation

---

Table 2.4   Inclusive Categories for Industrial Waste Constituents

---

Water
Phosphorus and Sulfur
Oil and Grease
Specifie Organics
Salts, Acids and Bases
Anions
Metals
Nitrogen

described in Chapter 1. Therefore, the use of these data implies that land is not dedicated, but rather remains viable as a societal resource.

The industrial land application designer would first determine the waste characteristics as described above, then delineate the potential site(s). Data necessary to specify the assimilative capacity are collected from these sites. Thus, the designer must consult each chapter for the nature of the assimilative capacity calculations for each constituent, specific data on rates, concentrations, etc., and, if necessary, the type of experiment required to obtain design data.

Industrial waste constituents for which no assimilative capacity data exist must be examined with experiments. Laboratory or pilot experiments can be conducted. It is important to consult the chapter (4-11) with similar waste constituents for a firm understanding of basic assimilative mechanism and the needed design data. This stage of land treatment design is analogous to bench- or pilot-scale treatability tests conducted for conventional or advanced waste treatment processes. These experimental tests must be directed at measuring the basic pathways such as soil-waste constituents exchange capacity, degradation rates, immediate phytotoxicity and soil-water transport capacities. It is important to use conditions similar to actual or likely field application rates. Experience with very high rates has proved that long-term effects are not portrayed accurately, and under overloaded conditions, the nature of the assimilative pathway is changed. One should obtain basic mechanistic data, accepting the short-term validity of the experiments, and attempting to predict long-term successful assimilation. From the respective chapters most closely related to the unknown constituent assimilative capacity, the citations and discussion can be used to actually formulate needed experimental protocols.

Site-specific data are essential to determine the assimilative capacities using the above three calculational tools. Until a specific piece of land, or better yet four or five alternative land sites are chosen, then the assimilative capacities cannot be established. For the location of a new facility in which land-based treatment is to be used, approximately five sites should be selected, since the variation and potential cost savings can be high when optimal land is selected. Then a preliminary confidential screening analysis can be performed (such as an Industrial Terrain Analysis*) to narrow the selection to the two best sites. These two sites would then receive the indepth LLC analysis, the most cost-effective of which would be chosen.

For an existing industrial facility the land options are often less broad. Within the available owned land a preliminary analysis can select the best areas for land treatment. Alternatively, land may not be presently owned or

---

*Fox, N., SSI, 525 Industrial Dr., Marietta, GA. Personal communication.

available. One manufacturer hauls waste sludge up to 50 km and then applies it to the land as the most cost-effective waste management program. Thus, options to use land-based treatment exist even where pumping or hauling 1-50 km is necessary. These options must be examined before any decision is made regarding land application feasibility.

LAND LIMITING CONSTITUENTS (LLC) ANALYSIS

With the assimilative capacity data, based on site-specific characteristics, and the detailed waste characterization, the designer is ready to complete the LLC analysis and stage I. The waste characterization for each constituent is expressed as kg/unit time, while the assimilative capacity is usually expressed as kg of parameter/unit area/unit time (Table 1.2, page 19). The ratio of waste generation to assimilative capacity is the area (hectares or acres) required for the environmentally acceptable application of each constituent to the plant-soil system. A comparison of the areas for the parameters will show one or more requiring the greatest land size, and these are designated the LLC. It is this constituent(s) that controls the land area required. Using the LLC area, the other constituents are thus put on the land at a lower rate than would be required to meet the environmental constraints of Chapter 1.

The LLC is not defined in the context of prohibiting the use of a plant-soil system in the sense that there is no assimilative capacity or that the constituent is too toxic. Instead, it is simply a design tool that focuses on the particular species which, if removed, would directly reduce necessary land area and hence receiver system costs. Conversely, removal of species other than the LLC will not result in lowering land area needs, so is economically counterproductive.

Example (Solution Given on Page 65)

The following two waste streams are to be land applied. From the waste generation rate and the waste assimilative capacity, determine the LLC for each effluent:

Waste Stream Constituent	Effluent (kg/yr)		Assimilative Capacity kg/ha/yr
	Waste 1	Waste 2	
Water	58×10^6	0.63×10^9	5.2×10^6
N	200	36,000	400
P	2,600	5,000	500
Cu	1	18	2
Oil	24,500	48,000	500
Phenol	4,000	-	100
COD	70,000	120,000	120,000

The manipulations required to determine the LLC are straightforward as evidenced by these example calculations from the petroleum and the poultry hatchery industries. The major effort is involved in determining the constants for waste assimilation and the analytical mass balance work involved in the waste generation.

Some industrial effluents involve more than one LLC, however. That is, two or more species are shown to require the largest and nearly the same land areas. Under these circumstances, pretreatment processes to remove both species are needed to reduce the land area. As an example, secondary treated domestic waste has both nitrogen and water as well as LLCs; hence, removal of nitrogen alone does not reduce the land area. Land areas are usually designed for hydraulic considerations because the water reduction appears unachievable.

The LLC analysis may be used iteratively within the overall four-stage design process (Figure 2.1) for pretreatment-land application systems. The pretreatment of the waste stream for the LLC results in a substantial reduction in the amount to be land spread. Recalculation of the land area requirements for each species in the pretreated waste stream reveals that the LLC has changed. In fact, pretreatment for the LLC beyond the point where another species becomes limiting leads to no further reduction in land area. At that level, pretreatment for both the new and the initial LLC are needed to achieve land system savings. These same arguments would hold for evaluating the economic level of in-plant changes with respect to anticipated savings in the land treatment system. Thus, the LLC concept must be iteratively calculated as pretreatment unit processes, and in-plant modifications are introduced to the existing plant raw waste streams.

Use of the LLC analysis and stage I has several advantages:

1. It provides a mechanism or standard methodology for including important site-specific data, basic environmental constraints and a direct relationship between waste characteristics and land capacity.
2. The LLC analysis is completely general, so any waste stream can be investigated.
3. This methodology is useful for both the designer and the regulatory agency responsible for review and approval, because, if performed correctly, all facets of the waste will have been addressed with specific site and waste data. This prevents the use of generalized design criteria that are unsuccessful because of wide natural variations in plant-soil and industrial manufacturing systems.
4. Identification of the LLC(s) focuses proper design attention on possible plant-soil site selection or management techniques to further improve the assimilative capacity for the critical parameter controlling land area.
5. Pretreatment or in-plant alternatives must be directed specifically at the LLC. It is economically unjustified to remove parameters other

than the LLC because the cost of the land and land system would remain unchanged.

6. Monitoring programs of vegetative cover, soils and receiving waters can be centered on the more critical parameters, the LLC. This is a logical, justifiable alternative to broad-brush, expensive requirements for monitoring all parameters.

7. A scientific design procedure on which research and development programs can focus is emphasized, so required data can be obtained efficiently. With a standard design structure, future research can be more easily justified based on impact for land application improvements.

Several important points relative to the stage I design process can be illustrated by examples of LLC analysis results. First, with a fixed industrial waste, a large range of land area requirements can result from the LLC analysis. This is the direct result of climatological or plant-soil characteristics, which are an essential part of the land application design. Table 2.5 gives an example of an industrial waste along with the assimilative capacities for three different sites. Water or hydraulic loading is the LLC in all cases, but the land area

Table 2.5 LLC Analysis Results from a Single Industrial Waste and Three Possible Land Treatment Sites

Parameter	Area Required Based on Site Assimilative Capacity and Waste Generation (ha)		
	Site		
	1	2	3
P	8	6	8
SO$_4$	3	3	3
CO D	1.5	1.5	1
H$_2$0 (LLC)	$\boxed{10}$	$\boxed{18}$	$\boxed{26.5}$
Sn	6	3	6
Sb	1	0.5	1
Cu	0.5	0.5	0.5
N	6	6	6

varies by a factor of 2.65 because of different site soil characteristics. In terms of total waste management costs, this industry realized· substantial savings by proper site selection. Similar examples also exist of land area variation in which other parameters were the LLC.

Site or climatological characteristics can also change the LLC for the same waste (Table 2.6). In this instance the entire pretreatment options were changed for waste when applied to site A vs site B. Again, broad design criteria would not have differentiated between sites, which demonstrates the advantage of LLC analysis.

Table 2.6   LLC Analysis Results from a Single Industrial Waste and Two Possible Land Treatment Sites

Parameter	Area Required Based on Site Assimilative Capacity and Waste Generation (ha)	
	Site	
	1	2
P	3	1
H_2O (LLC 1)	20	10
Ethylene Glycol	5	5
N (LLC 2)	15	15
Zn	0.7	3
Cl	4	4
K	7	7
CO D	5	5

The second concept evolving from the stage I analysis is that when a number of waste constituents are analyzed in industrial wastes, the variations between plants or facilities become obvious. For a given set of plant-soil conditions, two similar industrial facilities inevitably have different waste characteristics, so the LLC analysis may predict different land areas, particularly when the more refined waste analyses on constituent-by-constituent basis are conducted. In other words, land application design is also very waste specific. Table 2.7 shows that the differences in these two wastes from the same industry resulted in different parameters being identified as the LLC. The in-plant source control objectives were quite different for these two plants, since the LLC parameters were not the same.

These examples and the explanation of stage I LLC analysis preview the material in the remainder of this book. Chapters 4-11 give data on the important assimilative capacity as necessary inputs for land application design and the LLC analysis. Chapter 12 presents the design and economics of the land

Table 2.7   LLC Analysis for Two Manufacturing Plants Producing the Same Product with Similar Soils at Each Point

Parameter	Area Required Based on Site Assimilative Capacity and Waste Generation (ha)	
	Line	
	1	2
H_2O (LLC 2)	30	35
CO D	16	16
Chlorobenzene	8	11
Phenol	10	8
Oil (LLC 1)	40	30
Cr	4	4

system with all relevant components–stage II of the overall methodology. That is, once the LLC analysis is complete and the land area identified and sized, the transmission, application and monitoring components can be designed. Chapter 13 presents pretreatments or stage III concepts and economics and the economic balance between pretreatment and land application are discussed in Chapter 14 as the fourth stage. Examples of the full design methodology are also a part of Chapter 14.

Solution to Example Problem

1. Waste Stream Constituent	Area Required Based on Site Assimilative Capacity and Waste Generation (ha)	
	Waste 1	Waste 2
Water	11.1	121 LLC
N	0.5	90
P	5.2	10
Cu	0.5	9
Oil	49 LLC	96
Phenol	40	–
COD	0.6	1

REFERENCES

Buddin, W. "Partial Sterilization of Soil by Volatile and Nonvolatile Antiseptics," *J. Agric. Sci.* 6:417-451 (1914).
Dekock, P. C., and D. Vaughan. "Effects of Some Chelating and Phenolic Substances on the Growth of Abused Pea Root Segment," *Planta* 126:187-195 (1975).
Overcash, M. R., J. B. Weber, M. L. Miles and P. W. Westerman. "Land Application of Organic Priority Pollutants, North Carolina State University, U.S. Department of Interior Grant (1978).
U.S. Environmental Protection Agency. "Rationale for Recommended List of Priority Pollutants," Memo to director of Effluent Guidelines Division (November 19, 1976).

CHAPTER 3

FUNDAMENTALS OF PLANT-SOIL SYSTEMS

INTRODUCTION

This chapter introduces the agronomic nomenclature and presents a basic primer of the plant-soil system; as it is so essential to the pretreatment-land application design for industrial wastes. The introductory nature of these fundamentals will be useful to those outside the traditional agricultural field, while those having related training may wish to proceed to the next chapter.

A format for plant-soil fundamentals begins with a macroscopic view of soil systems. These concepts are most frequently employed in the design and description of land application systems. The following section deals with the microlevel of soils and is essential to understanding of the assimilative capacity of industrial waste constituents. Understanding the pathways for the interaction of waste in plant-soil systems is the basis for the utilization of the design methodology presented in Chapter 1 for a variety of industrial wastes. The relationship of understanding agronomic fundamentals to the total design process of this book is given in Figure 3.1.

Soil and Land

Soil is the three-dimensional unconsolidated mass on the surface of the earth that serves as a natural medium for the growth of terrestrial life, such as plants, earthworms, microorganisms and microflora. Soil consists of mineral particles, organic matter, water and air in varying proportions, as influenced by the genesis from parent material (rocks), climate (precipitation, temperature), macro- and microorganisms (flora, fauna, vegetation and higher animals), topography (relief, slope) and time. Many physical, chemical and biological characteristics are necessary to adequately understand and describe the plant-soil system.

Soil is a natural system with three phases—solid, liquid and gas—in a dynamic equilibrium capable of supporting plant growth and microbiological

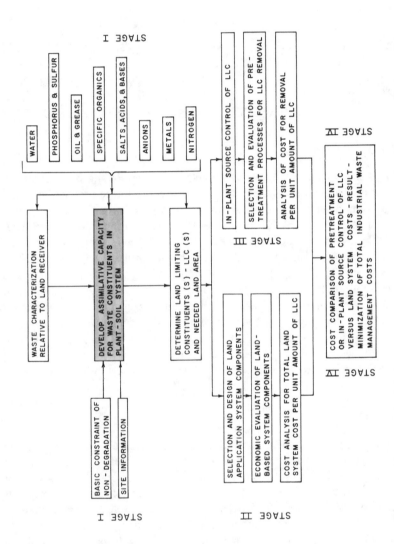

Figure 3.1 Relationship of Chapter 3 to overall design methodology for industrial pretreatment-land application systems.

activity. Land refers to the two-dimensional surface of the earth lithosphere, *i.e.*, terrestrial area under various uses such as cultivation, recreation, wild-life, forests, range or grasses, urban development, rural housing and waste treatment. Finally, to some, the entirety of soil and land will forever be referred to as dirt (á la Woody Allen).

SOIL SYSTEM–MACROSCALE

Solid and Fluid Composition

Natural soil is a heterogenous mixture composed of solids and fluids (Figure 3.2). The solid phase constitutes roughly one-half the total volume and includes both organic and inorganic or mineral components. The organic percentage of the dry solid portion of the soil is 5-12% by weight, with extreme organic or peat soils containing 20% organic matter.

Organics in soil are heterogenous in nature and occur as a micelle of decomposed, partially decomposed and undecomposed litter, as live and dead microbial cells, plant roots and/or faunal bodies. The organic or humus fraction is interwoven cohesively with the mineral fraction. In most soils,

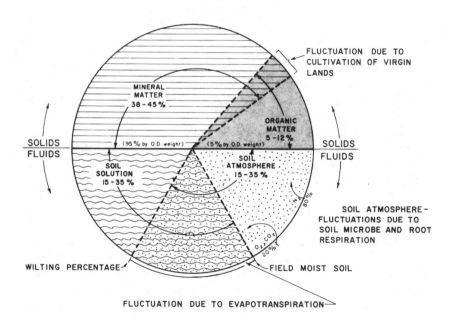

Figure 3.2 Solid and fluid fractions of soil.

the organic matter is highest near the soil surface and decreases with depth. Based on the organic constituents listed above, the gradation is logical, since the upper zone is the most biologically active, with microorganisms, microflora and plants. Although the organic fraction is small, it is the major source of nitrogen, an essential nutrient for all plant growth. The humus fraction is very dynamic in composition.

The predominant solid phase in soil is mineral and provides the mass and support in a soil system. In terms of the assimilation of industrial wastes, the properties of the solid phase of the soil (organic and inorganic) are very important.

The remaining 50% by volume of soil is air or water. The ratio of water to gas or atmosphere is altered considerably during natural events. Immediately after a large rainfall the soil is nearly saturated, so the voids are water-filled. As a practical matter, air is not completely displaced because of entrapment and surface bubble adhesion, so that only about 35-45% of the soil volume is liquid. As water drains from the soil and drying occurs, the shift between liquid and air in soil voids is nearly complete. The soil atmosphere is altered in chemical composition from the above-soil atmosphere. Soil gases are at nearly 100% relative humidity—high in carbon dioxide and low in oxygen. Again, because of the chemisorption of water to soil microsurfaces, not all the water is displaced under even extended drought field conditions. Thus, the cyclic fluid changes in soil voids is inevitable.

Field Soil Profile

Soil, consisting of the solid and void fractions described, is found in the earth crust not as a homogeneous mass, but as a distinctly layered system (Figure 3.3). This is termed the soil profile or horizon (layers) structure. Soil horizons in any profile vary in thickness and are identified visually based on morphological characteristics. The distribution of profiles varies over geological time; thus, a young soil will have a shallow surface layer above the parent material, while an old soil may be very deep. However, the depth of soil profile layers is also dictated by the water or wind transport of soil to a particular site—alluvial soils. Erosion reduces layer thickness, for example, in the southeastern United States, the upper horizon has been completely lost in many areas.

A cross-sectional view of a field soil down to the bedrock consists of distinct layers known as genetic or diagnostic horizons. The topmost layer, which contains high organic matter, constitutes the 0 horizons. The next upper horizons of the soil profile are collectively called solum (horizon A plus B). Between the solum and bedrock is the parent material, substratum, or horizon C, which is consolidated but much less affected by soil-forming factors. Each soil horizon or layer extends approximately parallel to the land

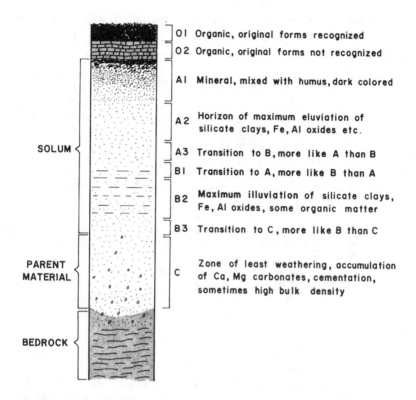

Figure 3.3 Profile features present in varying degrees with field soils.

surface and differs distinctly from the adjacent ones in physical, chemical and biological characteristics, such as color, texture, structure, degree of acidity or alkalinity, organic matter level, and number and type of soil population. Designations and properties of major soil horizons are listed below in Table 3.1. No soil is known to possess all of these horizons but each has some of them in different developmental stages.

Soil profile development is a result of rock weathering and genesis of soil material. Processes of physical disintegration, chemical decomposition and biological assimilation of rocks and minerals are:

1. physical, such as cracking, crumbling, crushing and mellowing to bring about a change in shape, size and form of rocks and minerals as glaciation, freezing, thawing, wetting, drying, expansion, contraction, shrinking, swelling, gravitational force and wind transport;
2. chemical, such as oxidation-reduction, hydration, hydrolysis, carbonation and solution; and

Table 3.1 Modern Nomenclature and Explanation of the Horizons in Most Soil Profiles

Horizon Designation	Description
0	Organic horizons of mineral soils. Horizons: (i) formed or forming in the upper part of mineral soils above the mineral part; (ii) dominated by fresh or partly decomposed organic material; and (iii) containing $> 30\%$ organic matter if the mineral fraction is $> 50\%$ clay, or $> 20\%$ organic matter if the mineral fraction has no clay. Intermediate clay content requires proportional organic matter content.
01	Organic Horizons, in which essentially the orginal form of most vegetative matter is visible to the naked eye. The 01 correspond to the L (litter) and some F (fermentation) layers in forest soils designations, and to the horizon formerly called Aoo.
02	Organic horizons in which the original form of most plant or animal matter cannot be recognized with the naked eye. The 02 corresponds to the H (humus) and some F (fermentation) layers in forest soils designations, and to the horizon formerly called Ao.
A	Mineral horizons consisting of: (i) horizons of organic matter accumulation formed or forming at, or adjacent to, the surface; (ii) elevated horizons that have lost clay, iron or aluminum with resultant concentration of quartz or other resistant minerals (i) or (ii) above, but transitional to an underlying B or C.
A1	Mineral horizons, formed or forming at or adjacent to the surface, in which the feature emphasized is an accumulation of humified organic matter intimately associated with the mineral fraction.
A2	Mineral horizons, in which the feature emphasized is loss of clay, iron or aluminum, with resultant concentration of quartz or other resistant minerals in sand and silt sizes.
A3	A transitional horizon between A and B and dominated by properties characteristic of an overlying A1 or A2, but having some subordinate properties of an underlying B.
AB	A horizon transitional between A and B, having an upper part dominated by properties of A and a lower part dominated by properties of B, and the two parts cannot be conveniently separated into A3 and B1.
A & B	Horizons that would qualify for A2 except for included parts constituting $< 50\%$ of the volume that would qualify as B.
AC	A horizon transitional between A and C, having subordinate properties of both A and C, but not dominated by properties characteristic of either A or C.
B & A	Any horizon qualifying as B in $> 50\%$ of its volume, including parts that qualify as A2.
B	Horizons in which the dominant feature or features is one or more of the following: (i) an illuvial concentration of silicate clay, iron, aluminum or humus, alone or in combination; (ii) a residual concentration of sesquioxides or silicate clays, alone or mixed, that has formed by means other than solution and removal of carbonates or more soluble salts; (iii) coatings of sesquioxides adequate to give conspicuously darker, stronger or redder

Table 3.1 Continued

Horizon Designation	Description
	colors than overlying and underlying horizons in the same sequum but without apparent illuviation of iron and not genetically related to B horizons that meet requirements of (i) or (ii) in the same sequum; or (iv) an alteration or material from its original condition in sequums lacking conditions defined in (i), (ii) and (iii) that obliterates original rock structure, that forms silicate clays, liberates oxides or both, and that forms granular, blocky or prismatic structure, if textures are such that volume changes accompany changes in moisture.
B1	A transitional horizon between B and A1 or between B and A2, in which the horizon is dominated by properties of an underlying B2 but has some subordinate properties of an overlying A1 or A2.
B2	That part of the B horizon where the properties on which the B is based are without clearly expressed subordinate characteristics, indicating that the horizon is transitional to an adjacent overlying A or an adjacent underlying C or R.
B3	A transitional horizon between B and C or R, in which the properties diagnostic of an overlying B2 are clearly expressed but are associated with clearly expressed properties characteristic of C or R.
C	A mineral horizon or layer, excluding bedrock, that is neither like or unlike the material from which the solum is presumed to have formed, relatively little affected by pedogenic processes, and lacking properties diagnostic of A or B but including materials modified by: (i) weathering outside the zone of major biological activity; (ii) reversible cementation, development of brittleness, development of high bulk density, and other properties characteristic of fragipans; (iii) gleying; (iv) accumulation of calcium or magnesium carbonate or more soluble salts; (v) cementation by accumulations such as calcium or magnesium carbonate or more soluble salts; or (vi) cementation by alkali-soluble siliceous material or by iron and silica.
R	Underlying consolidated bedrock, such as granite, sandstone or limestone. If presumed to be like the parent rock, from which the adjacent overlying layer or horizon was formed, the symbol R is used alone. If presumed to be unlike the overlying material, the R is preceded by a Roman numeral denoting lithologic discontinuity.

3. biological, such as plant roots intrusion, algal growth, invasion by mosses, lichens, fungi, actinomycetes and bacteria in addition to grinding by earthworms, rodents, higher animals and man.

Soil Classification

The concept of horizon definition is the basis of soil classification systems and is the most common description available for the design of industrial waste land treatment systems. Soils have been grouped systematically or arranged into categories based on their characteristics and intended use. Among the different soil classification schemes developed and adopted in different countries for various reasons, the latest is a comprehensive system of soil classification known as 7th Approximation. This was developed in the 1960s by the Soil Survey Staff of the USDA (Kirkham 1964) and has since been adopted as the official system in the United States.

Diagnostic horizons are used in the new system to differentiate soil classes or categories. The presence or absence of a specific diagnostic horizon in an individual soil (hereinafter called as pedon) is the basis of defining a soil class or category. The major diagnostic horizons are presented in Table 3.2. So far, only six epipedons or surface diagnostic horizons have been established, based on color or organic matter content, percent base saturation, thickness and

Table 3.2   Major Features of Diagnostic Horizons used to Differentiate at the Higher Levels of the Comprehensive Classification Scheme.

Diagnostic Horizon	Major Feature
Surface Horizons (Epipedons)	
Mollic	Thick, dark-colored, high-base saturation, strong structure
Umbric	Same as Mollic except low-base saturation
Ochric	Light colored, low organic content, may be hard and massive when dry
Histic	Very high in organic content, wet during some part of year
Anthropic	Cultivated soil layer rich in N, P, K and bases
Plaggen	Manmade surface layer more than 20 in. thick caused by continuous manuring
Subsurface Horizons	
Argillic	Silicate clay accumulation
Natric	Argillic, high in sodium, columnar or prismatic structure
Spodic	Organic matter, Fe and Al oxide accumulation
Cambic	Changed or altered by physical movements or by chemical reactions
Agric	Organic and clay accumulation just below plow layer
Oxic	Primarily mixture of Fe, Al oxides, and 1:1-type minerals

structure. The epipedons have formed at the surface, and as a rule, occur nowhere else. The remaining six subsurface horizons may have been exposed to atmosphere by the subsequent removal of surface layers. Properties that can be measured quantitatively are estimated for each diagnostic horizon in each pedon. The smallest volume that can be called "a soil," ranges in size from 1-10 m^2. Many of the names used for diagnostic horizons and orders are of Latin and Greek derivation. There are six categories of the new comprehensive system: (1) Order, (2), Suborder, (3) Great Group, (4) Subgroup, (5) Family, and (6) Series.

Order

This category is the highest. An order includes soils of similar genesis or soil-forming factors and processes as reflected in the morphology of the epipedons and subsurface diagnostic horizons. There are 10 orders (Table 3.3),

Table 3.3 Orders of the New Comprehensive System

Order	Formative Element	Descriptive Connotation
1. Entisol	ent	Recent soils without pedogenic horizon
2. Vertisol	ert	Inverted soil with high content of swelling clays
3. Inceptisol	ept	Young (inception) soils
4. Ardisol	id	Desert (arid) soils that are dry for more than six months every year
5. Mollisol	oll	Soft crumby (mollify) surface with black, organic-rich material, high in base saturation
6. Spodosol	od	Gray ash (spodos) horizon of free sesquioxides in subsurface
7. Alfisol	alf	Gray-to-brown color epipedon where A1, Fe are leached down and ashy gray subsurface with high clay content
8. Ultisol	ult	Very old soils in humid climates (ultimus) that possess low base saturation
9. Oxisol	ox	Strongly weathered with oxides of Si, A1 and Fe
10. Histosol	ist	Organic (tissue) soils

all names ending in "sol," derived from solum that means top soil, horizon A. and B. Formative elements of Greek or Latin origin connote the broad categories of soils included in respective orders. For example, Entisol is derived from "Ent" and "sol," representing recent soils without pedogenic horizons.

Suborder

The suborder category within each order emphasizes genetic homogeniety, *e.g.*, moisture, vegetation and temperature effects in the development and arrangement of pedons. The names of suborders are obtained by adding a prefix syllable to a formative element taken from that order name. The prefix syllables for the suborder names are given in Table 3.4. For illustrative purposes, Order Entisol theoretically can have as many suborders as the number formative elements presented in Table 3.4. Suborders within Entisol would be albent, andent, aquent, arent, argent, etc. Psamment is a suborder that refers to a sandy-entisol, which possesses coarse texture, and psammult would mean a sandy-ultisol.

Table 3.4   Formative Elements in Names of Suborders Used in Comprehensive Soil Classification System as the Prefix Syllables

Formative Elements (Prefix Syllables)	Derivation of Formative Element	Connotation
alb	L. albus, white	Albic horizon (a bleached eluvial horizon A)
and	Modified from Ando	Ando-like, rotcanic allophane
aqu	L. aqua, water	Associated with wetness
ar	L. arare to plow	Mixed horizons
arg	Argilla, white clay	Argillic horizon (a horizon B with illuvial clay)
bor	Gk. boreas, northern	Cool
ferr	L. ferrum, iron	Iron present
fibr	L. Fibra, fiber	Least decomposed stage
Fluv	L. fluvius, river	Floodplains
hem	Gk. hemi, half	Intermediate stage of decomposition
hum	L. humus, earth	Organic matter, fully decomposed
lept	Gk. leptos, thin	Thin horizons
ochr	Gk. base of ochros, pale	Ochric epipedon (a light pale surface)
orth	Gk. orthos, true	Common types
plag	Modified from Ger. plaggen, sod	Presence of plaggen epipedon
psamm	Gk. psammos, sand	Sand texture
rend	Rendzina	Rendzine-like
sapr	Gk. sapros, rotten	Most decomposed stage
torr	L. torridus, hot and dry	Usually dry
trop	Modified from Gk. tropikos, of the solstice	Continually warm
ud	L. udus, humid	Humid climates
umbr	L. umbra, shade	Umbric epipedon (a dark surface)
ust	L. ustus, burnt	Dry hot climates
xer	Gk. xeros, dry	Annual dry season, deserts

Great Group

Great groups in a suborder are differentiated based on kind and arrangement of diagnostic horizons or uses of the epipedons. The great groups are named by prefixing the suborder names by an additional descriptive syllable given in Table 3.5. For example, *Agraquent* belongs to that aquent suborder of the entisol order which is under cultivation for farming. *Cryaquent* refers to that aquent suborder which is found in cold and wet temperature regions with no development of genetic horizons.

Subgroup

Subgroups are subdivisions of the great groups. Subgroup names are put in front of that of the great group denoting the typics and deviates or intergrades. For example, *typic cryaquent* refers to the typical of the Great Group cryaquent, and *othic cryaquent* means the subgroup most nearly representing the central concept of the Great Group, cryaquent.

Theoretically, it is possible to classify the world soils into well over 150,000 subgroups, which are further divisible into lower categories of family, series and phase.

Family

This category is based on properties important to plant root growth. For example, textural classes (fine, silty, fine loamy, sandy and clayey), mineralogical classes (montmorillonitic, kaolinitic, siliceous and mixed), and temperature classes (frigid, mesic, and thermic) are used to denote a family preceding the suborder to which it belongs. *Silty, mixed, thermic orthic araquent* means a family of soils within subgroup orthic agraquent that possess silty texture, mixture of clay minerals and is located in hot tropical climate. Each family is subdivided into a number of soil series and, therefore, soil families are at times named after the best known series put under it.

Soil Series

A series is a group of soils developed from the same kind of parent material by the same genetic combination of processes and whose horizons are quite similar in their arrangement and general characteristics. Soils of any one series possess a unique characteristic profile distinctly different in some way from other series within the same family. However, the differences between various soils of the same series are in the texture of the surface layer. Usually city, place, region, county, river or other important name is used initially to identify the major location of a series. For instance, Yolo loam, Sacramento clay, Stockton clay and Columbia fine sandy loam are names of some soil series in California.

Table 3.5 The Formative Elements Used as Prefix Syllables to Name Great Groups Within Each Suborder

Formative Element	Connotation	Connotation
acr	Extreme weathering	–
agr	Agric horizon	Ager (field)
alb	Albic horizon	Albus (white)
and	Ando-like	
anthr	Anthropic epipedon	Anthropos (man)
aqu	Wetness	
arg	Argillic horizon	Argilla (white clay)
calc	Calcic horizon	Calx (chalk)
camb	Cambic horizon	Cambiare (exchange)
chrom	High chroma	
cry	Cold	Kryos (cold)
dur	Duripan	Durus (hard)
dystr, dys	Low base saturation	Dystrophic (infertile)
eutr, eu	High base saturation	Eutrophic (fertile)
ferr	Iron	Ferrum (iron)
frag	Fragipan	Fragilus (brittle)
fragloss	See frag and gloss	
gibbs	Gibbsite	
gloss	Tongued	Glossa (tongued)
hal	Salty	Hals (salt)
hapl	Minimum horizon	Haplon (simple)
hum	Humus	Humus (earth)
hydr	Water	Hydor (water)
hyp	Hypnum moss	
luo, lu	Illuvial	
moll	Mollic epipedon	
nadur	See natr and dur	
natr	Natric horizon	Natrium (sodium)
ochr	Ochric epipedon	Ochros (pale)
pale	Old development	Palso
pell	Low chroma	
plac	Thin pan	
plag	Plaggen horizon	Plaggen (sod)
plinth	Plinthite	
quartz	High quartz	Quartz, $(S_i0_2)_n$
rend	Rendzina-like	
rhod	Dark red colors	Rhodon (rose)
sal	Salic horizon	Salt
sider	Free iron oxides	
sphango	Spahagnum moss	
torr	Usually dry	

Table 3.5 Continued

Formative Element	Connotation	Connotation
trop	Continually warm	
ud	Humid climates	
umbr	Umbric epipedon	Umbra (shade)
ust	Dry climate, usually hot in summer	Ustrus (burnt)
verm	Wormy, or mixed by animals	Vermes (worm)
vitr	Glass	
xer	Annual dry season	
sombr	Dark horizon	

Soil Phase

Within a soil series is represented surface texture, slope, erosion, soluble salt content, presence of stones, deposits of limestones or Chilean nitrate, etc. Soil phase description follows the series name, *e.g.,* Salado loam, calcareous phase and Norfolk fine sandy loam, 5% slope.

Other soil classification systems developed and used for agricultural purposes are:

1. 1949 Pedological systems (Brady 1974);
2. land capability classification system used by U. S. Soil Conservation Service (Hockensmith and Steele 1949);
3. European and forest soils classification system (Wilde 1958);
4. British soil classification into Division of Automorphic and Hydromorphic soils (Wilde 1958); and
5. soil classifications based on climate and profile leaching (Franz 1960, Scheffer and Schachtschabel 1952).

Other soil classifications systems that have been developed for engineering purposes (Sprangler 1960, Woods and Levell 1960) include: (1) Engineering Unified Soil Classification System used by Federal Housing Administration, (2) American Association of State Highway Officials (AASHO) classification system for soils, and (3) U.S. Civil Aeronautics Administration classification system. The details of these classification systems can be found elsewhere (Sprangler 1960, Woods and Levell 1960).

From agronomic and waste management standpoints, information from the 7th Approximation system is practical because it provides knowledge of soil profile layers, their characteristics and arrangement, which are far more important than the pedological classification *per se.* To determine the water movement and the rate of waste loading, it is essential to know the depth, texture, structure and other properties of surface layers (O, A horizons), subsurface layers (A and B horizons) and the parent material (the C horizon).

Soil Texture

Within the generalized 7th Approximation system, the surface phase properties are included as a classification factor. However, the surface phase or soil texture is important as one of the most common designations for soils. Soil texture has a strong historical usage and, hence, is used in basic design calculations for industrial waste land application.

Soil consists of particles of varying sizes (diameter). There are no sharp, natural divisions between any particle size and, therefore, some arbitrary limits have been proposed based on particle diameter. Figure 3.4 illustrates the size limits and names of particle-size classes according to six systems used by engineers, geologists and soil scientists. In general, there are three broad groups of primary particle sizes. The percentage of a given soil in each fraction is determined by sedimentation and sieving tests. There are: clay-size fraction ($<$ 2 μ diameter), silt-size fraction (2-50 μ diameter) and sand-size fraction (50 μ-2 mm). The percentage size of gravel, pebbles, cobbles or boulders is usually not measured. The soil separates are expressed as weight percentages of the total mineral matter.

Sands are the weathering products formed from the parent rock material. The irregular shape and larger size indicate that sand is an early stage in soil development. As the weathering or breakdown process continues, smaller particle sizes, termed silt, are formed. The weathering process continues to the molecular level, at which point restructuring and recrystallization occur. The particles formed in this building process are the clays. Major differences in physical and chemical properties between clays and sands/silts are attributable to this difference in formation stages.

Within the clay category are two principal groups:

1. aluminosilicates,
2. iron and aluminum hydrous oxide clays.

The first category is a predominant clay in temperate zones of high agricultural importance and hence has been thoroughly investigated. The second category is more typical of tropical zones. On a molecular level the structure of aluminosilicates is crystalline, with silicon tetrahedrons (often modified with an Al proxy for Si) and the aluminum octahedron (with oxygen or hydroxyl surrounding the Al). The hydrous oxides are Fe or Al oxides which are highly hydrated. Hydrous oxides are more amorphous and have generally better soil physical conditions than silicates.

The relative proportions of clay, sand and silt particles in a soil is referred to as the mechanical or particle-size composition of soil. This is used to determine the soil textural class, again based on certain arbitrary division, as shown in Figure 3.5. Each arbitrary division within the triangle represents a soil class or texture. The terms sand, silt, and clay denote not only primary

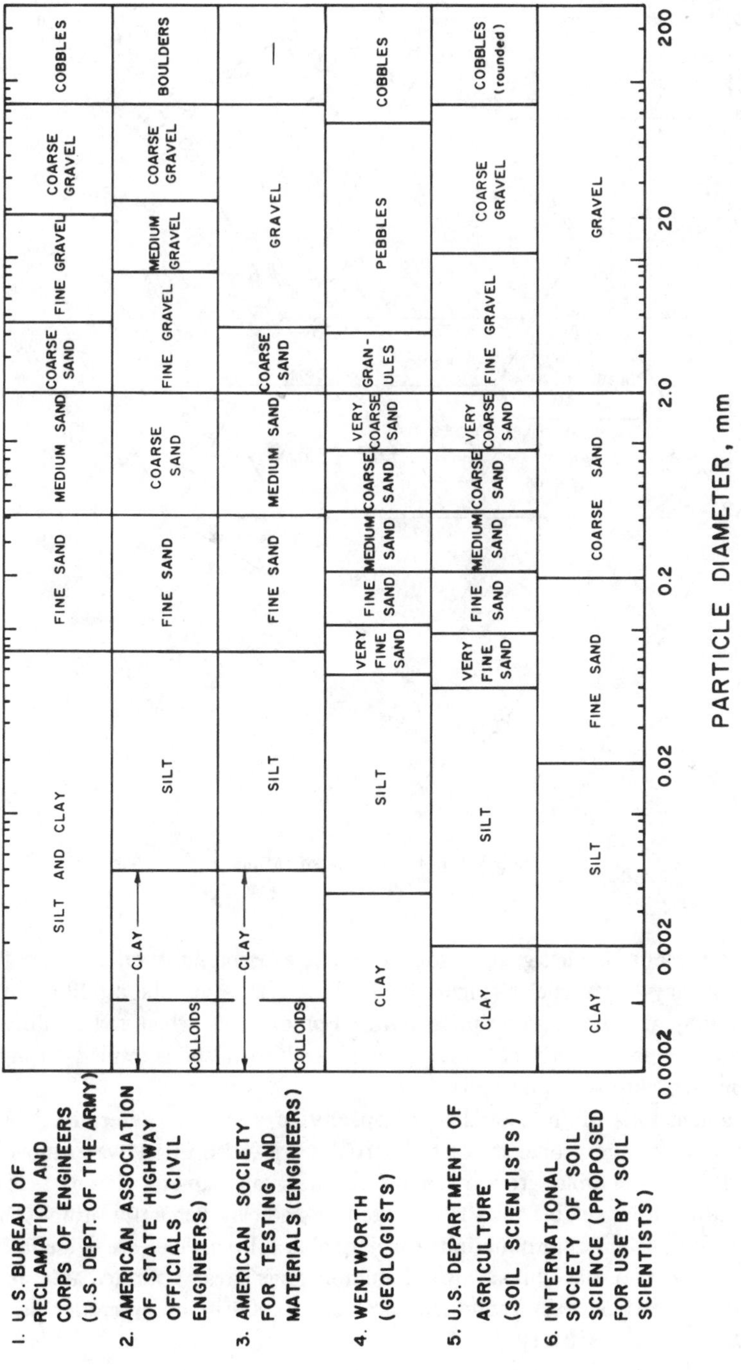

Figure 3.4 Particle size classifications used for soils.

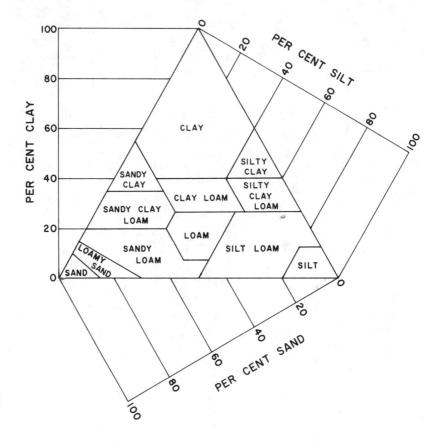

Figure 3.5 Soil texture classification.

soil particles but also designate a soil type. For example, clay refers to a soil with 50% clay size particles (Figure 3.5). These classes may be modified by the addition of suitable descriptive terms. For example when coarse stone fragments are present in substantial amounts a silt loam soil is termed "stony silt loam" or "silt loam, stony phase."

Certain aspects of the assimilative capacity of a soil are determined by the surface area and properties of soil particles, *e.g.*, adsorption, water movement, sites for microbial growth. Therefore, the small-diameter clay particles assume a much larger role than is indicated by the clay percent of total soil volume. Surface are determinations for definable soil particles are presented in Table 3.6. Clay has almost a 0.5-1 million times greater surface area per unit mass of particle than a sand and almost a 20,000 times larger area than the next particle size, the silt.

Table 3.6 Size and Surface Area of Soil Particles (Foth and Turk 1972)

Particle Type	Diameter	No. of Particles/g[a]	Surface Area
	mm		cm^2/g
Very Coarse Sand	2.00-1.00	90	11
Coarse Sand	1.00-0.50	722	23
Medium Sand	0.50-0.25	5,780	45
Fine Sand	0.25-0.10	46,200	91
Very Fine Sand	0.10-0.05	722,000	227
Silt	0.05-0.002	5,780,000	454
Clay	<0.002	90,300,000,000	8,000,000

[a]Assumed to have spherical shape (maximum particle diameter).

The soil texture is one of several factors that control water movement in soils. In an unrestricted situation such as a laboratory sample the conductivity of water is related to the soil textural class (Table 3.7). As the particle sizes become smaller, the physical dimensions of the void spaces between particles becomes smaller and hence water flow is slower. However, the magnitude of water that can be held when a soil is saturated is greatly increased with a clay versus a sandy soil (Table 3.8). The moisture retention capacity is an important factor in improved plant growth and microbial activity; thus, a balance is needed between the ability to move water into a soil and the

Table 3.7 Approximate Saturated Hydraulic Conductivities (Ross *et al.* 1977)

Soil Texture	SHC (cm/hr)
Coarse Sandy Loam, CSL	38
Sand, S	63
Loamy Sand, LS	57
Sandy Loam, SL	12.5
Fine Sandy Loam, fSL	7.5
Loam, L	2.5
Silt Loam, SiL	2.5
Sandy Clay Loam, SCL	2.3
Clay Loam, CL	0.88
Silty Clay Loam, SiCL	0.6
Sandy Clay, SC	0.8
Silty Clay, SiC	0.4
Silt, Si	6.7
Clay, C	0.5

Table 3.8   Available Water Capacity (water held between 0.34 and 15.2 bars matrix suction) of Soils of Different Textural Classes in Tennessee (Longwell 1963)

Textural Class	Available Water Capacity (cm of water/cm of soil depth)
Sand	0.015
Loamy Sand	0.074
Sandy Loam	0.121
Fine Sandy Loam	0.171
Very Fine Sandy Loam	0.257
Loam	0.191
Silt Loam	0.234
Silt	0.256
Sandy Clay Loam	0.209
Silty Clay Loam	0.204
Sandy Clay	0.085
Silty Clay	0.180
Clay	0.156

ability to retain materials in the upper soil zone for waste decomposition. The emphasis is on which waste constituents are the LLC rather than a single soil type as superior for all waste land application situations.

Soil Structure

Soils are composed of structural units obtained by combining primary particles (clay, sand, and silt) into secondary particles, units, peds, granules, or aggregates. Structure formation involves: (1) the development of inter-particle bonds that provide stability, and (2) the separation of structural units from one another that confers the identity (size and shape) characteristic to each individual unit. These secondary or structural units are characterized and grouped on the basis of size, to give soil structure classes and shape, and to give soil structure types (Table 3.9).

Knowledge of soil structure is particularly useful in understanding water and waste constituent movement. Platelike structure would inhibit flow perpendicular to the layers. Granular structure would permit water flow in any direction (Black 1968).

Another important characterization of soil structure is the presence and stability of aggregates as described by Kemper (1965). Aggregate-analysis is performed by a wet sieving procedure or simulated raindrop impact method. Aggregate-size fractions include diameters of < 0.2 mm, 0.2-1.0, 1-2 and 2-5 mm. Water-stable aggregates influence and modify plant growth, as

Table 3.9 Types and Classes of Soil Structure (SCS 1951)

Type (shape and arrangement of peds)			Class
Platelike with one dimension (the vertical) limited and greatly less than the other two; arranged around a horizontal plane; faces mostly horizontal			Platy
Prismlike with two dimensions (the horizontal) limited and considerably less than the vertical; arranged around a vertical line; vertical faces well defined; vertices angular	Without rounded caps		Prismatic
	With rounded caps		Columnar
Blocklike; polyhedronlike or spheroidal, with three dimensions of the same order of magnitude, arranged around a point	Blocklike; blocks or polyhedrons having plane or curved surfaces that are casts of the molds formed by the faces of the surrounding peds	Faces flattened; most vertices sharply angular	(Angular) Blocky[a]
		Mixed rounded and flattened faces with many rounded vertices	Subangular Blocky[b]
	Spheroids of polyhedrons having plane or curved surfaces that have slight or no accommodation to the faces of surrounding peds	Relatively nonporous peds	Granular
		Porous peds	Crumb

[a] Sometimes called nut. The word "angular" in the name ordinarily can be omitted.

[b] Sometimes called nuciform, nut or subangular nut. Since the size connotation of these terms is a source of great confusion to many, they are not recommended.

demonstrated for the yield of corn (Figure 3.6). In the design of a land treatment system, factors that improve water-stable aggregates should be used and represent a benefit from the usage of wastes on land.

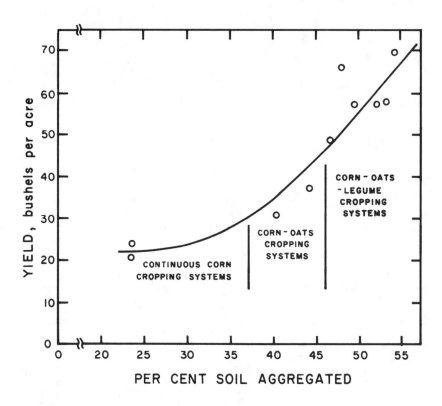

Figure 3.6 Relationship of soil aggregation and corn yield (Page and Willard 1946).

Field-Scale Factors

The description of soil composition and structure is not sufficient to fully describe and design for the site-specific assimilative capacity of industrial waste. Several field-scale parameters must also be determined, including vegetation, field size, topography and climatic inputs. A brief description of the importance of these factors follows.

Field size and topography govern the lateral and deep movement of water. If water must flow long distances (300+ m) to an outlet with relatively flat topography, then applications of liquids will be restricted. Under such conditions, a sandy soil actually may be poorly drained, thus emphasizing that neither soil texture nor drainage class are sufficient to predict hydraulic applications.

The Soil Conservation Service performs detailed soil mapping in many areas and soil classification is in series designation. Associated with each series is descriptive information on saturated conductivities and drainage classes (excessively, well, moderate, somewhat poor and poorly drained). However, these data should only be used as a preliminary tool to evaluate suitability for land application. This classification system should not be used for design purposes. The SCS designations:

1. do not represent actual field measurements and, hence, hydraulic conductivities can be substantially in error;
2. do not reflect any soil properties below 2 m, thus not establishing an adequate boundary condition for water flow, and
3. do not take into account the hydraulic behavior of a soil when wastewater is applied year round at levels equaling rainfall input (*i.e.,* soil drainage classes are for natural conditions without accounting for waste management inputs). Thus, for wastes in which hydraulic loading is a land limiting factor, a detailed geotechnical analysis is required.

Vegetation at a site is important in selecting least-cost areas for waste application and in achieving natural buffer zones and aesthetic characteristics in land treatment. The pattern of vegetation can also be interpreted in regard to natural water movement, past usage of site and soil types. Climatic factors are important because of the uncontrollable nature of these inputs. Rainfall, temperature, frozen soil and runoff characteristics must be obtained by the design engineer for land application sites under consideration.

These field-scale factors should be obtained in a study of a site(s) being considered for land application use. The first stage is a site analysis based on available data. As an example, the Industrial Terrain Analysis®* technique is described here as the necessary approach to land treatment design. Soils, water flow, percent usable area, drainage basin, vegetation, slope, and topographical data are assembled from published and unpublished sources. Confidentiality can be maintained so that site analysis can proceed without public notification. These field factors are essential to assess land treatment suitability and to avoid a common problem of land purchase only to find that a site is insufficient or predominantly unusable. The cost savings from such prestudies are substantial.

---

*Soil Systems, Inc., 525 Webb Industrial Dr., Marietta, GA 30062.

SOIL SYSTEM–MICROSCALE

Physical Properties

Many of the phenomena for the decomposition, adsorption and accumulation in soils depend on the concentration of a waste constituent, kg constituent/kg soil. To translate this information into waste application per unit land area, the bulk density of a soil is needed. Bulk density is determined by drying a known volume of soil to constant weight at 105° C.

For sandy and sandy loam soils and for silt and clay soils, a common range of bulk densities are shown in Figure 3.7. Bulk densities are known

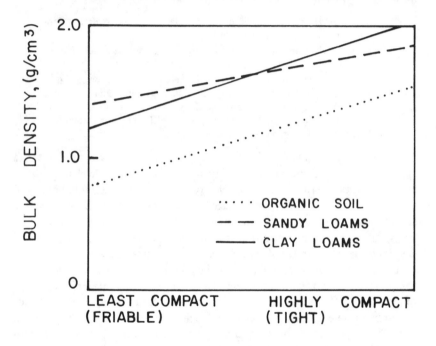

Figure 3.7 Representative bulk densities of soil types as a function of soil compaction.

to increase with depth from the soil surface (Figure 3.8), which is attributed to a decrease in organic matter and aggregation and an increase in compaction. Organic peat and muck soils have bulk densities of about 1.0. The actual mineral particle density of a soil is usually 2.6-2.75 g/cm^3; thus, bulk density accounts for the air-filled void space.

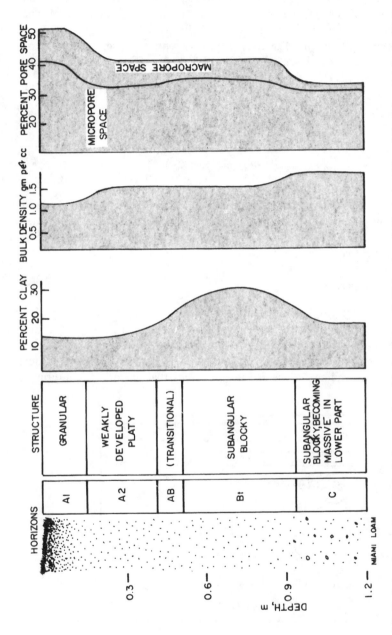

Figure 3.8 Variation in certain soil physical properties with depth in soil (University of Illinois Agricultural Experimental Station Bulletin 1960).

The most common expression of industrial waste constituent application is mass per unit area. This expression assumes the traditional agricultural zone of 15 cm (6 in.) or the plow layer. Thus kg/ha is, in actuality, kg/ha-15 cm or mass per unit volume. The conversion between soil concentration in parts per million (ppm) and soil concentration in kg/ha-15 cm is as follows (for bulk density of 1.4):

$$\frac{kg}{ha \cdot 15\,cm} = \frac{kg}{ha \cdot 15\,cm} \frac{ha}{1 \times 10^4 m^2} \frac{1m^2}{1 \times 10^4 cm^2} \frac{cm^3}{1.4\,g} \frac{1000\,g}{kg} = 2.2 \frac{kg}{1 \times 10^6 kg} \quad (3.1)$$

$$(kg/ha \cdot 15\,cm) = ppm \quad (2.2)$$

Conversion between waste constituent level is ppm, and application rate in kg/ha is used frequently in the design process for land application. If a greater soil depth than 15 cm is used, Equation 3.1 can be corrected accordingly to obtain the actual soil concentration.

Chemical Properties

The predominant soil elements are oxygen and silicon, Figure 3.9. The remaining 1.5% contains nearly all the other elements in periods 1-6 of the Periodic Table. Thus most metals in industrial waste would also be expected to be present already in the plant-soil system. The elemental composition may vary depending on the nature of parent rock materials, soil-forming

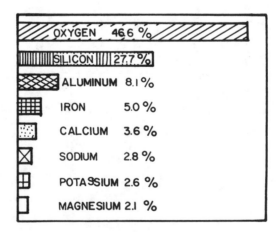

Figure 3.9 Eight elements comprising over 1% by weight of earth crust, remaining elements make up 1.5%.

factors and processes, and the management practices of the soils. Analyses of various size fractions of many soils show that Ca, Mg, K, and P are highest in the clay fraction and lowest in the sand. The difference in elemental composition among separates is most pronounced with P and Mg.

The organic solid phase of soil consists of chemical elements like carbon (44%), oxygen (40%), hydrogen (8%), nitrogen (1%), phosphorus (0.2%), sulfur (0.05%), and traces of chlorine, calcium, magnesium, potassium, sodium, zinc, iron, manganese, copper, molybdenum and cobalt. Organics in the soil occur in the form of

1. litter and debris of plant residues,
2. partially decomposed residues with molds and organisms,
3. fully decomposed organics and humus,
4. biomass of the living organisms, and
5. simple or complex organic chemicals from natural origin or by soil amendment.

Fresh plant residues contain 75-90% water and 10-25% dry matter, which depending on the maturity consists largely of carbohydrates, lignins, proteins, fats, waxes and tannins. The decomposition of these constituents leads to the synthesis of humus. The simple organic molecules that have been characterized in the hydrolysates of humus polymers are listed in Table 3.10. Chemical composition of humic substances has recently been reviewed by Flaig (1966). Native soil organics or humus materials are relatively resistant to decay, and only a small portion of the total is soluble in water, but much can be brought into solution by alkali or acid extraction. In the organic soil fraction, the C:N:P ratio is approximately 130:13:1.

The soil-water is essentially a solution in equilibrium with the solid phase. This liquid phase is of great importance to the organisms and plant roots as a source of both water and nutrients. Chemically, soil solution consists of the dissolved ions listed in Table 3.11. With the movement of soil-water, the dissolved solutes (ions, radicals and salts) move by mass flow.

The pore volume not filled by water is occupied by soil-air, whose chemical composition changes as a result of respiration by soil-inhabiting organisms and plant roots. The carbon dioxide partial pressure exceeds the atmospheric level by a factor of 10- to 100-fold and the oxygen concentration becomes depleted. The extent to which soil air differs in composition from atmospheric air is determined by:

1. the rate at which oxygen is consumed and other gases are produced by the living organisms and plant roots;
2. the rate of gaseous exchange between the soil and atmosphere; and
3. the special soil management practices such as fumigation, tillage, type of crop planted, copping techniques, and requirements.

Typical gaseous components of soil are given in Table 3.12.

Table 3.10   Several Simple Organic Molecules Detected in the Hydrolysates of Humus Polymer

I.	Aliphatic acids	VI.	Methyl sugars
	Acetic acid		Rhamnose
	Formic acid		Fucose
	Lactic acid		2-0-methyl-D-xylose
	Succinic acid		2-0-methyl-D-arabinose
II.	Amino acids	VII.	Pentose sugars
	Glutamic acid		Xylose
	Alanine		Arabinose
	Valine		Ribose
	Proline	VIII.	Purines
	Cystine		Guanine
	Phenylalanine		Adenine
III.	Amino sugars	IX.	Pyrimidines
	Glucosamine		Cytosine
	N-acetylglucosamine		Thymine
IV.	Aromatic molecules		Uracil
	Vanillic	X.	Sugar alcohols
	Syringic		Inositol
	Ferulic		Mannitol
	Coniferyl alcohol	XI.	Uronic acids
	Phenols		Glucuronic acid
	Quinones		Galacturonic acid
V.	Hexose sugars		
	Glucose		
	Galactose		
	Mannose		

Table 3.11 Approximate Composition of Soil Solution

Element	Extreme Range for Most Soils	Usual Range for Acid Soil (pH 4.8)	Usual Concentration for Calcareous Soil (pH 8.1)	Normal Soil Solution Concentration
			ppm	
Ca	20-1,520	40-136	560	750
Mg	17-2,400	46-100	168	250
K	8-390	28-100	40	100
Na	9-3,450	10-20	660	50
N	2-3,000	50-170	185	2,000
P	0.03-31	0.22	0.9	2-7
S	3-4,800	16-24	800	200
Cl	7-8,000	40	700	300
Mn	5-500	75-500	25	50
Al	0.0-1	0.1	0.0	0.0

Table 3.12 Composition of Soil Air and Atmospheric Gases by Volumes

Gaseous Components	Soil Air (% by volume)	Atmospheric Air (% by volume)
N_2	78.2	78
O_2	20.3[a]	20.9
CO_2	0.5[b]	0.03
Ar	1.0	0.95
Water Vapors	Almost saturated (except in air and oven dry soils)	Most times unsaturated (except during rain)
CH_4,[c] NH_3, N_2O	Nil to variable amounts (significant)	Nil to trace (insignificant)
NO,[c] H_2S, etc.	Amounts under flooded conditions for prolonged periods	Negligible amounts depending on accidental spills Aerial discharge of pollutant gases, or by volatilization form soils and dumps including sanitary land fills
SO_2,[c] CH_3Br	Nil to variable amounts depending on soil fumigation and other management practices	Same as above
Volatile organics,[c] e.g., alcohols, acids, esters, aldehydes, ketones, pesticides and other agricultural chemicals	Traces to few ppm in variable amounts depending on soil type, application methods, management practices, volatility of compound and soil temperature	Same as above

[a] Oxygen level in soil air may deplete to 12% or even lower under flood conditions.

[b] CO_2 level in soil air may increase to 6-8% or even higher under strongly reducing conditions, but not necessarily in flooded soils where the carbon dioxide initially forms carbonates and bicarbonates.

[c] Not normal constituents of soil or atmospheric air.

Cation Exchange

Clay and humus fractions of soil possess negative charges which are satisfied by cations of different charges and radii. These cations are near the surface of clay colloids and organic micelle are held with varying energy levels of attraction. The interchange between a cation in solution and another cation on the surface of the soil colloid or micelle is referred to as the cation exchange. The cations that are adsorbed on the surface of colloid particles and that can be replaced by those in the solution phase are designated as exchangeable cations. Principal exchangeable cations of a normal neutral

soil are calcium, magnesium, potassium, and sodium, and are usually designated as XCa, XMg, XK and XNa, respectively. Under extreme acid conditions, XH and XAl occupy most of the colloid surface, and in alkaline soils
XNa usually dominates the exchange surface cation exchange capacity (CEC)
of a soil material is defined as the sum total of exchangeable cations that a
soil can adsorb. Sometimes, CEC is called "total exchange capacity," "base
exchange capacity" or "cation adsorption capacity." The CEC is expressed
in units of meq/100 g of soil or other adsorbing material such as clay, humus,
etc. The simplest diagramatic expression of cation exchange reactions is
shown in Figure 3.10. It is in the form of an equation and can be written as:

$$2XCa + 2\,Na^+\,Cl^- \rightleftharpoons 2XNa + 2Ca^{+2} + (Cl^-)_2$$

where X is the negatively charged surface of the soil exchange colloid complex.
The clay size mineral fraction and the organic micelle exhibit a great deal
of external and internal surface with negative charges. The internal and
external surface of a silicate clay crystal is shown in Figure 3.10, where a
large number of cations are adsorbed. The cation adsorption and exchange
reaction have been systematically studied for the factors that influence the
kinetics and energetics of exchange. At an equal concentration of ions,
divalents are more strongly adsorbed than monovalent, so divalents can
readily displace monovalents from the exchange complex. Among the monovalents, Marshall (1964) quoted the series Li > Na > K > Rb > Cs with

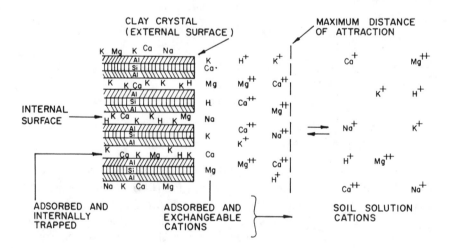

Figure 3.10 Relationship between solution and adsorbed cations for soil clay crystal.

increasing radii, decreasing hydration and increasing ease of displacement. For the divalents, the sequence $Mg > Ca > Sr > Ba$, is in an order of increasing radii, decreasing hydration of the ion and increased ease of displacement. The series for heterovalent cations is $Na > K > Mg > Ca$ where Na is released most readily in a fractional exchange.

The adsorption and exchange of cations is also related to the type and amount of clay and humus materials. The CEC of soil particles increases with fineness and surface area (Table 3.13). Various heavy metals ions such as

Table 3.13 Cation-Exchange Capacity and Specific Surface of Mineral Soil Particles Separated from a Clay Soil in Missouri (Whitt and Baver 1930)

Soil Separate	Equivalent Diameter of Particles (mm)	Calculated Surface Area per g[a] (cm^2)	Cation-Exchange Capacity (meq/100 g)
Silt	0.02 -0.005	1,800	3
	0.005 -0.002	6,200	7
Coarse Clay	0.002 -0.001	16,000	22
	0.001 -0.0005	30,000	35
	0.0005-0.0001	74,000	52
Fine Clay	0.0001-0.00005	320,000	56
	<0.00005	920,000	63

[a]Calculated on the basis of the average sixe of particles of each group.

Zn, Cu, Fe, Mn, Cu and Cr are also adsorbed and exchanged on the colloid matrix depending on the pH and presence of other salts. Each metal ion can have several cationic species such as Zn^{2+}, $Zn(OH)^+$, $ZnCl^+$, $Fe(OH)^{2+}$, $Fe(OH)_2^+$, $Mg(OH)^+$ and $Cu(OH)^+$, that occupy exchange sites. The level of these heavy metal ions is realtively low and insignificant as compared to the principal basic cations (Ca, Mg, K, and Na).

The CEC also varies with the type of cation used in the determination procedure and the pH at which the determination is made. The pH dependence arises from the fact that organic colloids as well as the mineral clays carry a pH dependent charge apart from the permanent negative charge as shown in Figure 3.11. Usually, CEC increases with soil pH because of the increasing ionization of the hydronium and phenolic groups to give a negative charge. In contrast to the clay, essentially all of the charges on the organic colloid are considered pH-dependent.

The extent to which the adsorption complex of a soil is occupied by a particular cation is referred to as exchangeable-cation percentage (ECP). For example, exchangeable sodium percentage (ESP) is the extent to which the

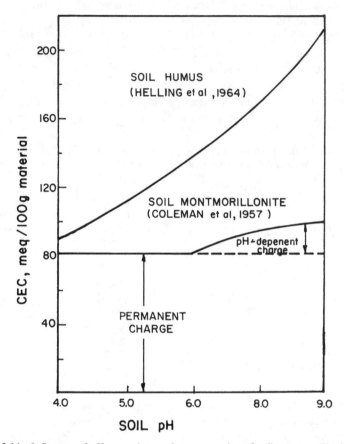

Figure 3.11 Influence of pH on cation exchange capacity of soil montmorillonite clay and soil humus.

adsorption complex of a soil is occupied by sodium. The sum total of all the four principal exchangeable bases is used to compute the percent base saturation, % BS:

$$\% \text{ BS} = \frac{(X\text{Ca} + X\text{Mg} + X\text{K} + X \text{ Na}) \, 100}{\text{CEC}} \qquad (3.2)$$

where the units of exchangeable cations and CEC are meq/100 g soil. The portion of the cation exchange capacity not occupied by the exchangeable bases can be measured and is usually attributed to exchange acidity, hydrogen, and aluminum. To maintain good soil fertility, % base saturation should exceed 80%, below which problems of metal toxicity and mobility may develop.

The reaction of a soil refers to the degree of acidity or alkalinity and is usually expressed in units of pH value which is defined as the negative logarithm of hydrogen-ion activity.

$$pH = -\log_{10}A_{H^+}$$

Soil pH is measured in aqueous suspensions or solutions of soils in 1:1 ratio for soil:water suspension; 1:2 ratio of soil:0.01 N $CaCl_2$ solution, or 1:2 ratio of soil:1N KCl solution.

Biological Properties

Soil environment provides the suitable habitat with space, substrate and other essential factors for the growth and activities of most organisms—microscopic as well as macroscopic (Figure 3.12). Microscopic organisms are called microorganisms, microbes or simply microflora and -fauna. Besides these organisms, certain biochemical agents such as enzymes, coenzymes, viruses or bacteriophages and hormones are found in soils. These biochemical agents may induce or carry out specific biochemical reactions by acting as

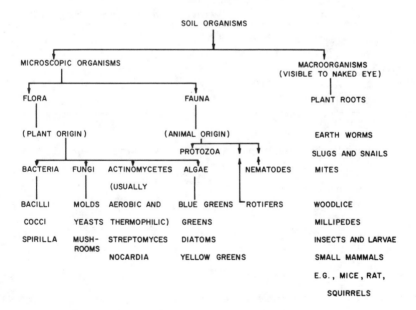

Figure 3.12 Representative soil organisms.

catalysts, synergists, or by being part of the reactions. Among these agents only viruses are known to reproduce and multiply in living cells. The other agents such as extracellular enzymes, are synthesized within cells and excreted into the soil for extracellular action.

Microogranisms are associated actively with the organic debris and mineral particles forming structural units. Microorganisms may be in the soil solution and/or attached to the specific surfaces of soil colloids while growing into the soil pores, extending into and around the soil structural units, and functioning to carry out or result in specific transformations of nutrient elements, inorganic minerals and organics.

The population of the soil microorganisms is highest in surface organic and Al horizons. The number of microorganisms declines rapidly with increasing depth of soil profile (Table 3.14) due to decreased organic matter, substrates, reduced oxygen supply, increased moisture supply, lack of sunlight, etc. The relative population and biomass of flora and fauna in surface soil layers are presented in Table 3.14 (Brady 1974).

Table 3.14 Distribution of Microorganisms in Various Horizons of the Soil Profile (Alexander 1977)

Horizon	Depth (cm)	Aerobic Bacteria	Anaerobic Bacteria	Actinomycetes	Fungi	Algae
A_1	3-8	7,800	1,950	2,080	119	25
A_2	20-25	1,800	379	245	50	5
A_2-B_1	35-40	472	98	49	14	0.5
B_1	65-75	10	1	5	6	0.1
B_2	135-145	1	0.4	-	3	-

Organisms/g of Soil X 10^3

The soil microbial population is a mixture of five major groups:

1. bacteria,
2. actinomycetes,
3. fungi,
4. algae, and
5. protozoa.

Bacteria

Bacteria are the single-celled organisms with nuclear material suspended freely in the protoplasm without a nuclear membrane. Most range in diameter from 0.5-2 μ. Bacteria are the most abundant of all organisms in the soil.

Bacteria are divided into two categories on the basis of biological function. The first is autotrophic bacteria, which derive carbon from carbon dioxide or carbonates. The carbon derived from CO_2 or carbonates is utilized for biological mass production. Energy for growth and respiration is derived from two sources, which differentiates the two major bacterial types of auto-trophs—photoautotrophs and chemoautotrophs. The former require light for energy while the latter require inorganics such as nitrite, ammonium, sulfur and ferrous ions. Because of the nitrification and sulfate-producing reactions, the chemoautotrophs are much more important in plant nutrition.

The second bacteria category is the heterotrophs, which, instead of carbon dioxide, utilize organic compounds for both energy and material for biomass. In addition to carbon, biological material must contain nitrogen. Hetero-trophs may be further divided into those organisms that fix atmospheric N_2 and those that utilize inorganic (NO_3) forms. The heterotrophs and chemo-autotrophs may operate under aerobic or anaerobic conditions.

In aerobic soil systems, bacteria are the dominant numbers of micro-organisms. Bacteria have the greatest capacity to grow and multiply; hence, the evolutionary capacity to decompose and the rate of degradation of many organic waste constituents is controlled by soil bacteria populations. Temperature, moisture, aerobic conditions, acidity and inorganic nutrient supply are important in maintaining bacterial growth. Management and design factors in industrial land treatment systems, which improve these environmental factors in the plant-soil system, will enhance waste assimilation.

Actinomycetes

The group of organisms intermediate between bacteria and true fungi are named actinomycetes. Actinomycetes are unicellular organisms that produce a slender, branched mycelium, which may undergo fragmentation or may subdivide to form asexual spores. The individual filaments are 0.5-1.2 μ in diameter, a dimension analogous to that of the bacterial cell. Numerically, abundance of actinomycetes in soil is second only to bacteria, and under high pH conditions may be the dominant organism. The ability to degrade carbonaceous material under dry, hot conditions is dependent on actinomycete activity.

Fungi

Fungi are characterized by a filamentous mycelium forming a network of hyphal strands. Numerically, fungi represent the third most abundant and frequently occurring group of soil organisms, but because of size ($\sim 5\mu$), fungi are the largest mass of the biological matter in the soil system (approximately 550 kg/ha). As a rule, fungi are heterotrophic in nutrition and fungal distribution is determined, therefore, by the availability of oxidizable carbonaceous substrates. The fungal growth is stimulated soon after the soil amendment with ready energy sources such as industrial waste at favorable temperatures under aerobic and moist conditions. Many fungal species can develop over a wide pH range, from highly acid to the alkaline extremes. Fungi are capable of growing at pH as low as 2 to 3, and the numerous strains are active even at pH 9.0 or above. Most filamentous fungi are aerobic and, hence, fungi concentrates predominantly in the surface few centimeters of soil. Most fungi are mesophilic; a very few are thermophilic. Because of the ability to function over wide soil conditions and the heterotrophic nature of fungi, the decomposition of recalcitrant waste constituents is often a fungal activity.

Algae

Like fungi, algae belong to the Phylum Thallophyta, but unlike fungi, many algae are photoautotrophic. Algae may be unicellular or may occur in short filaments with ability to photosynthesize and derive their carbon from CO_2 utilizing solar radiation as a source of energy. The abundance of algae in soil ecosystem varies from 100-10,000 and is limited to the surface of the land, frequently visible to the naked eye. Soil algae are divided into five groups: green algae, bluegreen algae, diatoms, yellow-greens and chlorophyllous flagellates. For autotrophic development, algae obtain water, N, P, S, Mg, K, Fe, Mo and other trace nutrient elements from soils, carbon dioxide from the air, and light energy from the sun. Certain algal genera of cyanophyceae possess ability to fix molecular nitrogen. As a result, those genera have been presumed to colonize the nitrogen-deficient barren surfaces of the rocks first.

Algae are also found in subsurface horizons below the zone of light penetration, primarily depending on either heterotrophic nutrition or mechanisms of dormancy. The role of algae as a soil heterotroph is insignificant. Algae are first to colonize the barren, eroded lands and thereby contribute to the organic carbon content of soil. Algae have played a significant role in soil formation, genesis and development as one of the pioneering inhabitants that brought about biochemical disintegration of rocks. Algae can bind the primary particles of soil, prevent erosion and promote formation of desirable structural units. The nitrogen gains of soil by algal growth are substantial when the supply of water, sunlight, carbon dioxide and other nutrients is plentiful for two to three months.

Protozoa

There are unicellular animals ranging in size from several microns up to one or more centimeters. The soil protozoa are, however, all microscopic. Protozoan population is ubiquitous, ranging typically 10^4-10^5/g of soil. During the active phase, protozoa feed and multiply by asexual fission or reproduce sexually. Under adverse conditions, many protozoa secrete a thick coating about themselves and undergo a resting or cyst state. Encystment allows the protozoa to persist for many years. The growth and energy of protozoa is dependent on utilizing organic compounds or other microorganisms. Energy and CO_2 can also be used by algae-like protozoa.

SOIL MEASUREMENTS

The fundamental soil concepts introduced in this chapter and others introduced throughout this text are often based on certain measurements or tests performed on soils. The designer of a land application system should understand the nature of the soil tests and subsequent data used in developing the assimilative capacity of a plant-soil system. Completely standardized tests for soil and plant properties do not exist. There are variations among states and scientific disciplines. These unresolved differences require that the designer completely specify the testing methods used when preparing a permit application or report on a particular land treatment system. This discussion here delineates the basic concepts, variables and methods in the more important soils and plant tests related to land-based waste treatment.

Texture or Particle Size Distribution

The most fundamental physical characteristic of a soil is the texture or particle size distribution. Porosity, air and water permeability, water-holding capacity, infiltration rate and tilth are among the soil properties influenced by soil texture. The most widely used standard method to determine particle size analysis is by hydrometer (Day 1965).

The apparatus consists of a standard hydrometer (ASTM 1524, with Bouyoucos scale, g/l, an electrically driven mixer with replaceable stirring paddle, ASTM stirring apparatus A) and glass sedimentation cylinder and a brass plunger. The procedure involves mixing 40 g of soil with 500 ml water and 50 ml of 10% sodium metaphosphate solution. After dispersion, the final solution is transferred to a cylinder and the total volume of the solution is adjusted to 1,000 ml. Hydrometer readings are obtained for different settling periods without remixing the suspension between measurements. The principle involved in calculation of corresponding particle sizes is Stokes' law (Day 1965). The percentages sand, silt and clay are determined and the soil phase is then determined from Figure 3.5.

Hydraulic Conductivity

Saturated Soil

Hydraulic conductivity of a saturated soil depends on the cross-sectional area of the pores and, thus, on their size. In saturated flow, the conductivity increases as the fourth power of the pore radius. Soil samples with either disturbed or undisturbed structure are usually held in metal or plastic cylinders, to obtain one-dimensional flow. The soil samples are covered with cloth at one end and soaked in water until saturation point is obtained. Then a constant head of water on the sample is maintained. The water percolated through the soil sample is measured during a known time period. Hydraulic conductivity is estimated from the volume of water passed through soil sample, difference in hydraulic head and temperature of the water (Klute 1965[a]).

Unsaturated Soil

The hydraulic conductivity of an unsaturated soil can be measured by steady-state techniques, which involve the establishment of a flow system in which water content, tension and flux do not change with time. The conductivity is measured by applying a constant hydraulic head difference across the sample and measuring the resulting steady-state flux of water. The soil sample is brought to an unsaturated condition by placing it in a pressure chamber between porous plates or membranes that are permeable to water. A constant-head water supply system is connected to one porous plate, and a constant-head removal system is connected to the other. By adjustment and control of the gas-phase pressure in the chamber, the sample may be brought to various levels of water content. At each level of water content and mean pressure head, a flow is set up from the hydraulic gradient across the sample. The conductivity value obtained is considered to be valid at the mean water content of the soil sample (Klute 1965[b]). Saturated and unsaturated conductivities are essential for evaluating the plant-soil assimilative capacity for wastewater application. The hydraulic properties are thus quite site specific.

Soil Moisture

Saturation (Zero bar tension)

A soil whose pores are completely filled with water is said to be saturated and generally referred to the water in the soil at zero tension. This is the maximum amount of water a soil can hold. Any excess water beyond the saturation point will flood the soil. Saturation moisture content of the soil usually equals the total porsity of the soil. Total porosity equals to [1- (bulk density of soil/particle density of soil)]. Bulk density of a normal soil is about 1.5 g/cm^3 and particle density is 2.65 g/cm^3.

Field Capacity (0.33 bar tension)

This term refers to the water held by the soil at 0.33 bar tension, which is usually called field capacity and determined by saturating the soil and allowing it to drain until the rate of downward movement of water has essentially ceased. The amount of water held by the soil against the gravitational force is the moisture content at field capacity.

Wilting Point (15 bar tension)

The wilting point is generally defined as the soil moisture condition at which the ease of release of water to plant roots is just barely too small to counterbalance the transpiration losses. This term refers to the soil moisture content at 15 bar tension.

The total available water for plants is obtained from the difference in moisture content between field capacity and wilting point. These soil moisture measurements are used in evaluating the potential hydraulic loading at a land application site and corresponding effects on plant growth.

Cation Exchange Capacity

Cation exchange capacity determination of a soil involves removal of all exchangeable cations by leaching the soil with an excess of neutral 1 N ammonium acetate solution and saturating the exchange material with ammonium. This is followed by leaching with 1 N NH_4Cl four times and once with 0.25 N NH_4Cl. The soil is then washed with 99% isopropyl alcohol until all chloride is removed. The ammonium adsorbed on exchange complex is displaced by treating the soil with 10% acidified NaCl. The displaced solution is transferred to a Kjeldahl flask, to which 25 ml 1 N NaOH is added and distilled into a flash containing 50 ml of 2% boric acid indicator solution. The excess acid is titrated with standard 0.05 N HCl to the disappearance of blue color. The cation exchange capacity is then calculated. Many of the assimilative capacity determinations for constituents in industrial waste are based on the CEC measurement.

Microbial Populations

Most common methods for estimating microbial population numbers in soil use an agar-plate method. This method involves the dispersion of the soil or other material in an agar medium to such an extent that individual microbial cells, spores or mycelial fragments have reasonable opportunity, when exposed to suitable conditions, to develop into discrete and macroscopically visible colonies. The necessary degree of dispersion is usually achieved by making series of dilutions of a given soil sample. The diluted samples with an agar medium are incubated at constant temperature, and the number of colonies developed are counted at the end of incubation, and adjusted for dilution factor. This represents the total microbial population of the system (Clark 1965). The capacity of a soil to assimilate organic constituents in waste is dependent on the magnitude of microbial population, hence the need for such tests, especially during land treatment operation.

Specific Microflora

Fungi. These constitute an important part of the microflora of a normal soil. They are particularly active in the initial stages of decomposition of organic residues and actively participate in soil aggregation. A similar procedure described for enumerating the total microbial population is used except the pH of the agar medium is adjusted to pH 4.5, to suppress the bacterial growth.

Actinomycetes. In many soils actinomycetes account for about 10-25% of the total population. The process of counting actinomycetes in soil is complicated because no cultural substrates have been developed that are highly selective for actinomycetes. Most workers prefer an agar medium designed to favor the actinomycetes and to restrict the growth of bacteria.

Nitrifying bacteria. The soil bacteria that oxidize ammonium to nitrite and nitrite to nitrate are chemoautotrophic, *i.e.*, they are able to use inorganic material as energy source and carbon dioxide as carbon source. These properties are used to separate these bacteria during enumeration. In the case of *Nitrosomonas,* dilutions of soil are inoculated into an inorganic medium containing ammonium as the source of nitrogen. If *Nitrosomonas* is present in viable form in the inoculum, growth will occur, and nitrite will be produced. A positive test for nitrite indicates the presence of these organisms. Similarly, medium used for *Nitrobacter* should contain nitrite, and positive test for nitrate indicates the presence of *Nitrobacter.*

Denitrifying bacteria. During bacteria denitrification, there is a disappearance of both energy sources and the nitrate or nitrite used as electron acceptor. The qualitative test for the presence of denitrifiers is based on the increase in alkalinity associated with the formation of gaseous end products. The method used in enumerating the populations is the Most Probable Number (MPN) method.

Similarily, methods are available to isolate specific organisms, which are involved in degradation of specific compounds. The procedures used in enumerating the specific organisms are similar, but the medium used for each organism is different, if one is interested in isolation of that particular organism (Buchanan and Gibbons 1974). Decomposition of particular organics, including priority pollutants and toxic materials, is based on general and specific microorganisms. Thus, to assure that the soil is not degraded such that nitrifiers, fungi, etc cannot exist or that there are microorganisms that degrade industrial organics, the above tests are used.

Heavy Metals

Total elemental analysis of the soils is usually performed after digestion with hydroflouric acid, nitric acid and perchloric acid. This digestion aids in decomposition of resistant soil organic matter, and destruction of crystal lattice. The digestion should be carried out in Teflon®* beakers. The digested material is diluted to desired concentration range and analyzed on an atomic absorption spectophotometer.

Several extraction procedures are used to determine the plant-available metals in the soil system. The most commonly used extractant is sodium acetate, which is used to determine the exchangeable fraction of metals in the soil. Several other extractants used are dilute acid mixture, sodium diothianate, diethylene triaminepenta acetic acid (DTPA), ethylenediaminetetra acetic acid (EDTA) and nitric acid. No standard extraction procedure has been established for heavy metals. Monitoring for heavy metals and determining availability to plants are essential parts of the operation of an industrial land application system.

Organics

Total organics in the soil include native soil organic matter, added plant material, added specific organic compounds, such as pesticides or any other industry-related organic substances disposed on land. Total organic carbon of the soils is usually determined by combustion techniques, which involves conversion of organic carbon to CO_2. To measure the total amount of specific organics, soils are usually extracted with specific solvents and analyzed for a particular organic compound on a gas chromotograph-mass spectrometer system. The usual solvents used in extraction of some of the organics include hexane, ethers, benzene and acetone and the common apparatus used is a Soxhlet extractor. If more than one specific compound

---

*Registered trademark of E. I. du Pont de Nemours and Company, Inc., Wilmington, Delaware.

is present in the soil, the organic compounds are identified in gas chromatograph. These methods should be standardized to meet specific needs since no one method is available at present time. For example, $> 95\%$ PCB present in the soil was extracted using acetone as a solvent, as compared to 70% recovery with hexane and 75% with acetone/methanol mixture (Siedl and Ballschmiter 1976). The decomposition of a specific organic applied to soils can be determiend from the amount of material extracted and analyzed vs time after application.

BENCHMARK SOILS AND SOIL PROPERTIES

Continued emphasis must be placed on the site-specific nature of any industrial waste land treatment system. Ninety percent of the failures in land application systems are directly attributable to the neglect of site data in the design. Soil type is one of the most important site-specific variables. Others are topography, climate, vegetative potential and relation to receiving waters.

Because of the importance of onsite variables, this book cannot propose specific design criteria, but rather must stress a design procedure or methodology. Such a methodology approach is essential to include the complexity of many waste types and site-related variables which are encountered in the multitude of industrial waste land application systems in existence or under contemplation.

Clay Loam and Sandy Loam

To acquaint the reader to typical values of soil properties, two benchmark soils were assembled. These soils are not extremes of soil properties but rather represent two substantially different and sizable soil types. Selected were a clay loam and a sandy loam (Table 3.15). Since these are not extreme examples the reader cannot assume that a particular soil must be between these two example materials. Throughout this book an effort is made to relate the assimilative capacity of industrial waste constituents to these two benchmark soils to determine any effect of soil type.

Table 3.15 Benchmark Soils for Studies of Land Application Systems

Characteristics	Sandy Loam	Clay Loam
Physical Characteristics		
Sand, %	70	33
Silt, %	20	33
Clay, %	10	34
Textural class	Sandy loam	Clay loam
Water-holding capacity at saturation, %	35%	55%
Particle density, g/cc	2.65	2.65
Bulk density, g/cc	1.55	1.25
Pore space, %	41.5	52.8
Chemical Characteristics		
pH	7.5	6.5
Electrical conductivity, μmhos/cm	800	600
Calcium carbonate content, %	2.0	0.0
Organic carbon, %	1.0	2.5
Dominant clay mineral	Illite	Montmorillonite
Sesquioxide, *e.g.*, iron & aluminum oxide, %	4	8
Nitrogen, %	0.08	0.18
Phosphorus, %	0.03	0.07
Calcium, %	1.0	1.0
Magnesium, %	0.8	0.9
Potassium, %	0.5	1.5
Sodium, %	0.1	0.7
Barium, %	0.01	0.05
Iron, %	1.5	2.5
Aluminum, %	4.0	10.0
Titanium, %	0.5	0.8
Zinc, ppm	50	1000
Manganese, ppm	50	500
Boron, ppm	20	50
Chloride, ppm	25	50
Cobalt, ppm	5	15
Copper, ppm	20	50
Molybdenum, ppm	0.5	2
Arsenic, ppm	5.0	10
Lead, ppm	8.0	30
Cadmium, ppm	0.1	0.5
Chromium, ppm	15.0	100
Nickel, ppm	20.0	50
Fluoride, ppm	20.0	50
Bromide, ppm	1.0	5
Beryllium, ppm	0.5	5
Antimony, ppm	2.0	5
Tin, ppm	50.0	10.0

Table 3.15 Continued

Characteristics	Sandy Loam	Clay Loam
Sulfur, ppm	50.0	200
Vanadium, ppm	50.0	20.0
Mercury, ppm	0.01	0.1
Extractable Ions		
NH$_4$-N, ppm	10	5
NO$_3$-N, ppm	30	10
Chloride, ppm	25	45
Calcium, ppm	200	1,800
Magnesium, ppm	120	600
Sodium, ppm	100	460
Potassium, ppm	160	1,800
Cation Exchange Capacity, meq/100 g soil	5	25
Biological Characteristics		
Bacteria, #g^{-1}	10^5	10^6
Actinomycetes, #g^{-1}	10^3	10^4
Fungi, #g^{-1}	10^3	10^4
Algae, #g^{-1}	10^2	10^3
Protozoa #g^{-1}	10^2	10^2
Nature of vegetation	Potato	Cotton
	Corn	Forest-Pine
		Range-grass

REFERENCES

Alexander, M. *Introduction to Soil Microbiology*, 2nd ed. (New York: John Wiley and Sons, Inc., 1977), p. 467.

Black, C. A. *Soil-Plant Relationships*, (New York: John Wiley and Sons, Inc., 1968), p. 775.

Brady, N. C. *the Nature and Properties of Soils*, 8th ed. (New York: Macmillan Publishing Co., Inc., 1974).

Buchanan, R. E., and N. E. Gibbons. *Bergey's Manual of Determinative Bacteriology*, 8th ed. (Baltimore, MD.: Williams & Wilkins Co., 1974).

Clark, F. E. "Agar Plate Method for Total Microbial Count," in *Methods of Soil Analysis*, C. A. Black, Ed. (Madison, WI: American Society of Agronomy, 1965), pp. 1460-1466.

Coleman, N. T. and A. Mehlich. *The Chemistry of Soil pH. The Yearbook of Agriculture (Soil)*, (Washington, DC: U.S. Dept. of Agriculture, 1957).

Day, P.R. "Particle Size Fractionation and Particle Size Analysis," in *Methods of Soil Analysis*, C.A. Black, Ed. Part I. (Madison, WI: American Society of Agronomy, 1965), pp. 545-567.

Flaig, W. *The Chemistry of Humic Substances. The Use of Isotopes in Soil Organic Matter Studies* (New York: Pergamon Press, Inc., 1966), pp. 103-127.

Foth, H. D., and L. M. Turk. *Fundamentals of Soil Science,* 5th ed. (New York: John Wiley & Sons, Inc., 1972).

Franz, H. *Feldbondenkunde* (Vienna: Verlag Georg. Fromme and Co., 1960).

Hockensmith, R. D., and J. G. Steele. "Recent Trends in the Use of the Land-Capability Classification," *Soil Science Soc. Am. Proc.* 14:383-388 (1949).

Kelling, C. S. "Contribution of Organic Matter and Clay to Soil Cation Exchange Capacity as Affected by the pH of the Saturated Solution," *Soil Sci. Soc. Amer. Proc.* 28:517-520 (1964).

Kemper, W. D. "Aggergate Stability," in *Methods of Soil Analysis,* Part 1, C. A. Black, Ed., (Madison, WI: American Society of Agronomy, Inc., 1965), pp. 511-519.

Kirkham, D. "Soil Physics," in *Handbook of Applied Hydrology,* V. T. Chow, Ed. (New York: McGraw-Hill Book Co., 1964), pp. 5-1–5-26.

Klute, A. "Laboratory Measurement of Hydraulic Conductivity of Saturated Soil," in *Methods of Soil Analysis,* C. A. Black, Ed. (Madison, WI: American Society of Agronomy, Inc., 1965a).

Klute, A. "Laboratory Measurement of Hydraulic Conductivity of an Unsaturated Soil," in *Methods of Soil Analysis,* C. A. Black, Ed. (Madison, WI: American Society of Agronomy, 1965b), pp. 253-261.

Longwell, T. J., W. L. Parks and M. E. Springer. "Moisture Characteristics of Representative Tennessee Soils," *Tenn. Agric. Exp. Sta. Bull.* 367 (1963).

Marshall, C. E. *The Physical Chemistry and Mineralogy of Soils,* Vol. 1, Soil Materials (New York: John Wiley and Sons, Inc., 1964).

Page, J. B., and C. L. Willard. "Cropping Systems and Soil Properties," *Soil Sci. Soc. Amer. Proc.* 11:81-88 (1946).

Ross, B. B., D. N. contractor, E. A. Li, V. O. Shanholtz and J. C. Carr. "A Model for Predicting Flood Hazards due to Specific Land-Use Practices," *VPI & SU Water Res. Res. Ctr., Blacksburg, VA Bull.* 99 (1977).

Scheffer, F., and P. Schachtschabel. *Bodenkunde* (Stuttgart: Ferdinand Enke Verlagsbuchhandlung, 1952).

Siedl, G. and K. Ballschmiter. *Isolation of PCB's from Soil–Recovery Rates Using Different Solvent System,* Chemosphere No. 5, pp. 373-376, (Elmsford, NY: Pergamon Press, 1976).

Soil Conservation Service. *Soil Survey Manual,* USDA Handbook 18, (Washington, DC: U.S. Government Printing Office, 1950) p. 228.

Sprangler, M. G. "Engineering Soil Classification," in *Highway Engineering Handbook,* K. G. Woods, Ed., (New York: McGraw-Hill Book Company, Inc., 1960).

University of Illinois Agriculture Experiment Station. "Characteristics of Soils Associated with Glacial Tills in North Eastern Illinois," *Bulletin* 665 (1960).

Whitt, D. M., and L. D. Baver. "Particle Size in Relation to Base Exchange Capacity and Hydration Properties of Putman Clay," *J. Am. Soc. Agron.* 29:703-708 (1930).

Wilde, S. A. *Forest Soils* (New York: The Ronald Press Company, 1958).

Woods, K. B., and C. W. Levell, Jr. "Engineering Description of Soils of North America," in *Highway Engineering Handbook,* K. G. Woods, Ed. (New York: McGraw-Hill Book Company, Inc., 1960).

CHAPTER 4

SOIL HYDRAULIC ASSIMILATIVE CAPACITY

Wade L. Nutter

    Associate Professor of Forest Hydrology
    School of Forest Resources
    University of Georgia
    Athens, Georgia 30602

INTRODUCTION

One of the most important concepts necessary to the successful design and operation of an industrial land treatment system is the hydraulic capacity of the soil to receive and transmit water. The pathways and mechanisms of flow through the soil of precipitation and applied wastewater, in both the saturated and unsaturated phases, determine initial movement and subsequent location of the waste constituents. The objective of slow rate infiltration, whether surface or spray irrigation, is to infiltrate all applied water, whether it be precipitation or wastewater. By contrast, an objective of the overland flow systems is to control the rate of water flow over the soil surface to promote runoff. To meet the objectives inherent to each system certain soil hydraulic variables must be characterized. The principal soil hydraulic variables of interest are the hydraulic conductivity and infiltration capacity. Associated variables are the weathered mantle and geological heterogeneity, soil water storage capacity, depth to groundwater, direction of groundwater flow and topography. Hydraulic overloading of soil is a frequent cause of failure of land treatment systems because it may lead to a rapid leaching of waste constituents, reduction in biological activity (microorganisms, plants, etc.) associated with reduced gaseous exchange (sustained anaerobic conditions), soil erosion and contamination of surface waters.

This chapter will address some of the basic concepts of water movement in soils and the relationship to the hydraulic capacity of industrial land

111

treatment systems. The approach will be to determine at what point the hydraulic loading becomes the land limiting constituent (LLC), to assess impacts of hydraulic loading on the physical and biological components of the land treatment system, and to suggest means of managing the hydraulic loading. The relationship of Chapter 4 is to the total pretreatment-land application design for industrial wastes is depicted in Figure 4.1.

CHARACTERISTICS OF SOIL WATER

Soil is a disperse, three-phase system containing solid, liquid and gaseous states. As a porous medium, soil contains a wide range of pore sizes, which, in turn, affect the soil hydraulic properties. In unsaturated soil the soil pores contain both air and water. The water is held by forces resulting from the surface tension of water, cohesion and adhesion, and molecular electrical attraction. The resulting soil water energy to retain liquid in the soil is termed the soil water pressure potential, or matric potential. This energy potential is less than atmospheric pressure.

The force with which water is held in the soil pore is approximately inversely proportional to the pore radius. As pore radius decreases, the energy required to remove water from the pore increases. Thus, as soil water content decreases, water is evacuated first from the large radii pores and then from progressively smaller pores, such that the matric potential increases, i.e., becomes progressively less than atmospheric pressure. The relationship between soil water content and matric potential is best represented by soil water characteristic curves. Examples of such curves for soils with varying ranges in pore sizes are presented in Figure 4.2. Because soil pores are irregular in shape, and shrinking and swelling may occur when soil is wetted or dried, the characteristic curve for desorption differs from that for adsorption of water. This hysteresis phenomenon presents serious problems in field soil water flow analysis because of the uncertainty associated with measured matric potentials and corresponding soil water contents. In most cases, the hysteresis problem is minimized for wastewater projects because soil drainage or drying is of primary interest. The methodology for developing soil water characteristic curves from undisturbed field samples is presented by Richards (1954).

It is apparent from the soil water characteristic curves in Figure 4.2 that soil texture and structure strongly influence the shape of the curve. A sandy soil has uniformly large pores, and little energy is needed to pull water from the soil. By contrast, a clay soil contains predominantly small pores, and considerable energy is needed to pull water out. Plants growing on both sand and clay soils may exhibit drought symptoms, but for different reasons; the sand, because there is too little water in the soil, and the clay, because

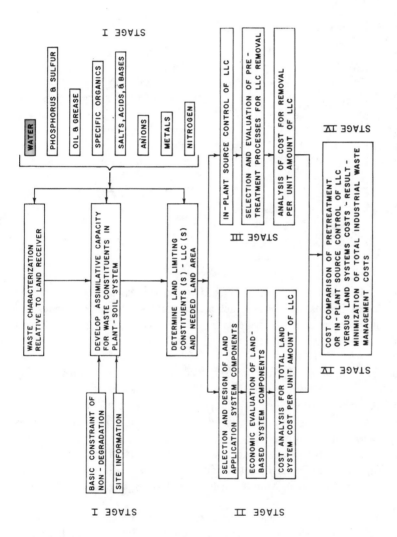

Figure 4.1   Relationships of Chapter 4 to the overall design methodology for industrial pretreatment-land application systems.

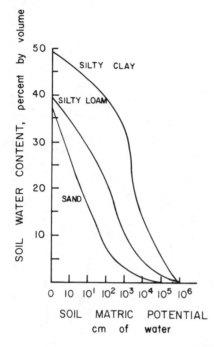

Figure 4.2 Representative desorption soil water characteristic curves.

the water is held too tightly. A well-structured soil has a good distribution of pore sizes because the large pores are contained between aggregates and the small pores within the aggregates. Any soil additions or management practices that improve soil structure will improve the hydraulic properties as well.

Another important feature of soil hydraulic properties illustrated by characteristic curves is the size of pores filled with water at a given matric potential. At saturation, *i.e.,* matric potential equals zero or atmospheric pressure, all the pores are filled with water. As the matric potential increases (becomes less than atmospheric pressure), pores are dewatered. Thus, the largest-sized pore filled with water may be inferred from the characteristic curve. This has significance in determining rates of water movement because the resistance to water moving through the soil is proportional to the spore sizes filled with water.

In addition to the soil water pressure, or matric potential, soil water is subject to several other force fields that cause its energy to differ from that of a pure, free body of water. Since the velocity of soil water is quite slow and the kinetic energy is proportional to velocity squared, the kinetic energy of soil water is considered negligible. Thus, we need only consider the potential at any point within the soil is thus:

$$\Phi_t = \Phi_g + \Phi_p + \Phi_o \qquad\qquad (4.1)$$

where Φ_t =   total soil water potential
       Φ_g =   gravitational potential
       Φ_o =   osmotic potential

The pressure potential is less than atmospheric pressure when the soil is unsaturated and equal to, or greater than, atmospheric pressure when the soil is saturated. The osmotic potential results from the presence of solutes in the soil water and does not significantly affect the mass flow of water. However, solute concentrations resulting from application of wastewater may affect mass movement to and through plant roots. The gravitational potential results from the relative elevation of water with respect to a datum point.

Soil water potential energy may be expressed in several different ways. When expressed as potential energy per unit volume of water, the potential energy has dimensions of pressure. Since hydrostatic pressure may also be expressed in terms of hydraulic head, or the height of a water column corresponding to a pressure, the potential energy per unit weight of water has dimensions of length. The latter expression is convenient and simple for field use. Thus, the total soil water potential energy expressed as total hydraulic head with dimensions of length is:

$$H = H_g + H_p \qquad\qquad (4.2)$$

where   H   =   total hydraulic head
       H_g =   gravitational head
       H_p =   pressure head

The pressure head is negative if the soil is unsaturated, and zero or positive if the soil is saturated. In saturated soil the pressure head is often referred to as the submergence head, a somewhat more descriptive term. In conventional usage, the osmotic potential is not expressed in terms of a hydraulic head. The osmotic potential head is generally not included in total hydraulic head because it is assumed to be negligible when compared to the flow resulting from the gravitational and pressure heads.

The ease with which water passes through the soil is the hydraulic conductivity (K) and is usually expressed as a velocity (LT^{-1}, volume flowrate per unit cross-sectional area perpendicular to flow). Permeability is a term that may be synonymous with hydraulic conductivity, but usually refers to flow in saturated porous media only. The saturated hydraulic conductivity, i.e., when the soil is saturated, is assumed to be a constant if deformation and/or change in the pore size distribution does not occur and the solute concentration does not change. However, as the soil drains and water is contained in progressively smaller pores, the hydraulic conductivity decreases and is a function of the matric potential. The magnitude of hydraulic conductivity values for unsaturated soil may vary as much as 1×10^6. Therefore, hydraulic conductivity (or permeability) must be measured on field samples in the field. Field measurements are restricted to saturated hydraulic conductivity, whereas laboratory measurements can determine both saturated and unsaturated values. Techniques for laboratory and field measurement of hydraulic conductivity are presented by Black (1965).

Since actual measurement of unsaturated hydraulic conductivity is difficult, values may be estimated indirectly by determining the pore sizes filled with water from the characteristic curve. This methodology is described and evaluated by Kunze *et al* (1968).

WATER MOVEMENT IN SOIL

In general, one-dimensional flow through a porous media may be described by Darcy's law as follows:

$$q = K \frac{dH}{dL} \qquad (4.3)$$

where   q   = water flux per unit cross-sectional area
        K   = hydraulic conductivity (or permeability)
        dH/dL = total hydraulic head gradient

Soil water flux changes with time (unsteady flow), and the hydraulic head gradient and hydraulic conductivity vary in space. Therefore, introducing the continuity equation, the general flow equation is

$$\frac{\partial \theta}{\partial t} = \nabla \cdot [K(H_p) \, \nabla H] \qquad (4.4)$$

where $\partial\theta/\partial t$ = change in soil water content with time
        $K(H_p)$ = hydraulic conductivity as a function of the pressure head
        H   = total hydraulic head

In unsaturated soils, where ∇H_g is negligible compared to the strong H_p gradient, the general flow equation may be expressed as follows:

$$\frac{\partial \theta}{\partial t} = \nabla \cdot [K(H_p) \, \nabla H_p] \qquad (4.5)$$

For a one-dimensional horizontal system the general flow equation is, therefore:

$$\frac{\partial \theta}{\partial t} = \frac{\partial}{\partial x} [K(H_p) \frac{\partial H_p}{\partial x}] \qquad (4.6)$$

where   x   = flow path

As is apparent, solving the flow equation for field situations is a complex procedure. Numerical procedures are frequently used, but more often lack of field data to adequately describe the hydraulic conditions results in the application of simplified solutions to the flow equation.

Darcy's law for saturated and unsaturated flow is illustrated in simplified form in Figures 4.3 and 4.4, respectively. Also shown is the relationship of the gravitational, pressure and total hydraulic heads.

For a more detailed discussion of soil water condition, movement and solution techniques of flow equations, the reader is referred to Hillel (1971) and Kirkham and Powers (1972).

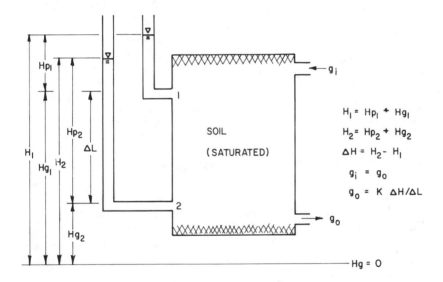

Figure 4.3 Characterization of energy conditions and flow in a saturated soil column.

FIELD WATER MOVEMENT

Under field conditions, soil heterogeneity, topography and evapotranspiration withdrawal cause hydraulic gradients to develop that lead to flow in many different directions within the a vertical profile section. Figure 4.5 schematically outlines the hydrological processes that may occur on a site receiving both wastewater and precipitation. Following infiltration, water is distributed throughout the profile in response to, and at a rate proportional to, the hydraulic gradient and hydraulic conductivity. Water is essentially retained within the soil pores as dynamic storage. Thus, a dry soil initially

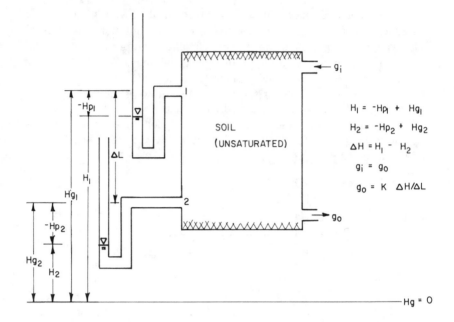

$$H_1 = {}^-Hp_1 + Hg_1$$
$$H_2 = {}^-Hp_2 + Hg_2$$
$$\Delta H = H_1 - H_2$$
$$g_i = g_o$$
$$g_o = K\ \Delta H/\Delta L$$

Figure 4.4 Characterization of energy conditions and flow in an unsaturated soil column.

will retain more water than a wet soil. Withdrawals from storage are continuous, although more rapid in moist soil. In the design of a land treatment system it is important to estimate the rate at which the soil water content decreases and what proportion of the flow goes to evapotranspiration and to lateral vs vertical movement.

In a uniform soil with no slope and with water input exceeding evapotranspiration, the net movement of water is vertically downward. As slope increases, other conditions remaining the same, the net movement of water is lateral down the slope. The degree of lateral movement is a function of soil wetness, soil heterogeneity and slope angle, all affecting energy conditions. How soil wetness affects lateral movement is best illustrated in Figure 4.6. In a relatively homogeneous soil, as the soil drains a lateral flow path develops and approaches a path parallel to the surface. As slope angle increases, the flow becomes parallel to the surface earlier in the drainage sequence, *i.e.*, at soil water conditions (Nutter 1973, 1975).

Water moving laterally will eventually become recharge to a saturated zone and become part of the general groundwater flow. It is important to the design of a land treatment system to estimate how much flow may be lateral flow, because the distance of the flow through the weathered, chemically

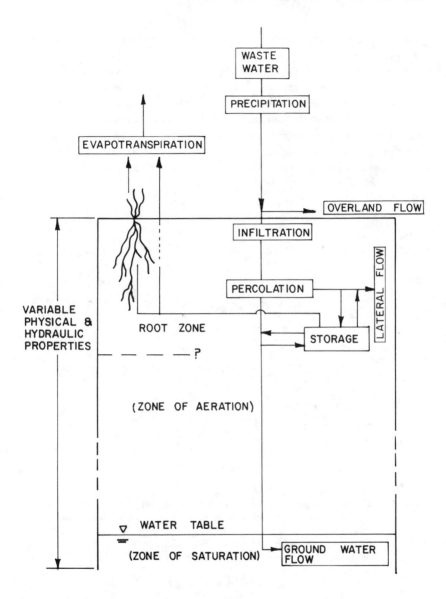

Figure 4.5 Schematic of soil water classification and the distribution of wastewater and precipitation input to soil.

reactive upper soil horizons results in a different water quality than if the water flows vertically downward directly to groundwater. The occurrence of lateral flow may be used to advantage in the design of a land treatment system to provide higher levels of soil treatment control and better monitoring capabilities.

To calculate the quantity, rate and direction of unsaturated soil-water flow requires extensive data and is generally beyond the scope required for the design of most land treatment systems. On the other hand, knowledge of saturated hydraulic conductivities and the hydraulic head gradients associated with saturated conditions can be used to estimate quantity and rate of flow through the simple application of Darcy's law (Equation 4.1). Direction of flow can be estimated from hydraulic gradients. Results of the saturation

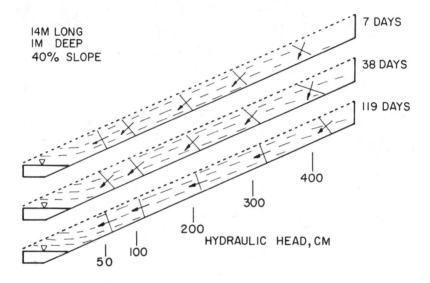

Figure 4.6 A drainage sequence of the covered 14-m long hill slope segment illustrating the reorientation and direction of flow lines during drainage. The dashed lines represent isolines of matric suction. The first dashed line above the water table represents -25 cm and each additional line represents an increment of -25 cm. Flow directions are shown by the arrows.

flow analysis can be extrapolated to unsaturated conditions in subsurface horizons below the root zone or to moist soil within the root zone to estimate direction and relative magnitude of water movement. It is important that soil scientists and hydrologists experienced in interpreting soil water characteristics and movement be involved in the interpretation of subsurface flow patterns. This is especially important in reducing design costs by utilizing estimation techniques derived from field experience.

The Hydrologic Budget

The amount of water that may percolate to groundwater or a stream may be estimated from a hydrologic budget analysis of a land treatment site. For any specified period of time

$$P + W = E_t + D + R \qquad (4.7)$$

where   P   =   precipitation
        W   =   wastewater
        E_t =   evapotranspiration
        D   =   vertical and lateral percolation through the soil profile
        R   =   overland flow

In spray irrigation systems the overland flow is zero by design; and in an overland flow system the percolation is near zero. The hydrologic budget analysis is best applied to periods of time in excess of one week or more. It is nothing more than a bookkeeping method to account for inputs and outputs. For infiltration systems, the percolation may be estimated from a soil water flow analysis previously described, as modified by plant root and soil organism aeration needs. With precipitation and evapotranspiration also known, the wastewater loading may be determined.

HYDRAULIC ASSIMILATIVE CAPACITY

The previous discussion shows that the hydraulic assimilative capacity is site-specific and controlled by the hydraulic properties of the soil and geological structure, topography, vegetation type and management, climate, and nature of the wastewater. Other factors entering into the assimilative capacity determination are use of year-round application, or seasonal application and storage; the necessity and/or feasibility of subsurface drainage; and the quality of groundwater effluent from the land treatment site necessary to meet regulations. Determination of the hydraulic assimilative capacity is focused on two factors: (1) the periodic hydraulic loading, and (2) the rate at which the loading is applied. For instance, a hydraulic loading of 6 cm/wk may be applied at the rate of 0.5 cm for 12 hr/wk.

Data requirements for determining the hydraulic assimilative capacity are varied and should be determined with the aid of a soil scientist and hydrologist familiar with local conditions and land treatment. From the standpoint of soils, the minimum data requirements are saturated hydraulic conductivity, topography, site features and depth of horizon development. If the number of conductivity determinations is limited, the saturated hydraulic conductivities of the most restrictive horizons should be determined. For those horizons in which hydraulic conductivity is not measured, a range in which the actual value may be expected is often available from the Soil Conservation Service (SCS). A general knowledge of soil structure and texture, degree of horizon development, depth of weathering, and past land use are also necessary soil data inputs.

Climatic data are required to determine both the amount and frequency of precipitation events and the occurrence and extent of freezing temperatures. With respect to precipitation, the amount of precipitation that may occur before the combined amount of wastewater and precipitation exceeds the soil assimilative capacity is referred to as the design precipitation. In other words, when the design precipitation is exceeded the wastewater application must cease. Determined on a biweekly or monthly basis often over a 20-25 year period of record, the design precipitation is usually selected to have a return period of 5 or 10 years. The return period is the reciprocal of the probability of occurrence and is determined by ranking the events for a long period of record. A return period of 5 years, for instance, implies that about once every 5 years the design precipitation for the period selected will be equalled or exceeded, and wastewater application must cease. Techniques for determining return periods are presented by Linsley, *et al* (1975).

Precipitation occurrence on a shorter-term basis must also be considered. Knowledge of precipitation amounts in the few days preceding wastewater application is necessary to determine antecedent soil water conditions. Precipitation intensities during the actual period of wastewater application are important in determining hydraulic overload. Design to accommodate these problems is best approached by considering the frequency of occurrence (*i.e.,* number of days per week, month or year) of precipitation events of a certain size and their intensity. Following an evaluation, probabilities may be associated with the number of days wastewater application must cease and the wastewater stored. The periodic precipitation frequency information is available for most design purposes (U.S. Weather Bureau 1961; U.S. Environmental Protection Agency 1977).

Temperature conditions must be evaluated in terms of likelihood of soil freezing and the type of soil frost that may occur. Prolonged freezing temperatures over bare soil result in a soil frost that often has an infiltration and percolation rate much less than that of unfrozen soil. On the other hand, soil frost formation in forest soils with an insulating cover of organic debris is often less extensive than in the open. It forms a granular-type frost that frequently results in little or no reduction in infiltration and percolation rates. Soil freezing data, usually for bare, agricultural soils, are available from State Agricultural Experiment Station field offices.

Evapotranspiration rates must be determined as one of the water outputs. Actual rates for large vegetated areas are unavailable because evapotranspiration is controlled in part by the energy at which water is held within the soil.

If soil water is readily available to a fully vegetated field with a vigorously growing crop, the evapotranspiration may be estimated empirically. Termed the potential evapotranspiration, a number of methods area available for estimating it on a weekly or longer basis. As most methods require wind velocity, vapor pressure and solar radiation data, thus are impractical for most land treatment design purposes because those type of data are not widely available. As an alternative, mean air temperature data, which are widely available, may be used as an index of solar radiation. Techniques employing this latter method are those of Thornthwaite (1948) and Holdridge (1962). Both methods produce similar results, but Holdridge's method is simpler to use. Holdridge's method is as follows:

$$PE_t = 4.9\,\overline{T}_m \qquad\qquad (4.8)$$

where PE_t = potential evapotranspiration
$\overline{T}_m$ = mean monthly air temperature, $°C$.

At sites receiving sufficient wastewater and precipitation to maintain soil water readily available to the plant, the actual evapotranspiration will equal the potential. If soil water is not readily available to the plant, the potential evapotranspiration rate must be adjusted downward to reflect the actual evapotranspiration rate.

Determination of Hydraulic Assimilative Capacity Using the Hydrologic Budget Method

When the hydrologic budget method (Equation 4.7) is used to determine the assimilative capacity, the wastewater hydraulic loading is determined first and the application rate second. Most frequently, the hydrologic budget is solved for monthly intervals, although weekly or biweekly interval solutions are feasible.

Considering the variables in Equation 4.7 estimating the amount of percolation is the most difficult because it requires both judgment and an understanding of soil water flow conditions on the land treatment site. Basically, the percolation in Equation 4.7 is the amount of water that can safely drain through the soil system without damaging the soil, vegetation or soil organisms. The percolation rate must be based on the frequency of wastewater application anticipated, precipitation regime, hydraulic conductivity and its variability with depth, soil water storage characteristics, slope length and angle, the vegetation's root aeration needs, and general water tolerance.

Design experience over a wide range of soil and vegetation conditions has shown that the monthly percolation quantity usually does not exceed 10-15% of the mean saturated hydraulic conductivity for the most limiting horizon, if that horizon is within 2 feet of the soil surface. If the most limiting horizon occurs below that level, the monthly percolation amount should not be more than 20-25% of the mean saturated hydraulic conductivity. For level sites with little free lateral drainage, the computed values for percolation should be reduced by at least 50%. Local knowledge of hydraulic head gradients would be useful to better estimate what reduction in the percolation quantity is justified. The depth, extent and water tolerance of the roots for each species grown on the site must be evaluated in light of the percolation quantity derived.

An example of the solution of a monthly hydrologic budget for wastewater loading is presented in Table 4.1. It is assumed that the system has year-round operation capability. The minimum wastewater loading is 19.1 cm/month in March, and the maximum loading is 33.4 cm/month in July. With a design precipitation return period of 5 years, about once every 5 years the wastewater application must be curtailed because of

Table 4.1   Solution of the Monthly Hydrologic Budget for Wastewater Loading

Month	Design Precipitation[a]	Evapotranspiration[b]	Percolation	Maximum Wastewater Loading
January	10.5	0.5	36.1	26.1
Feburary	16.0	1.3	36.1	21.4
March	19.8	2.8	36.1	19.1
April	15.0	5.8	36.1	26.9
May	15.5	9.4	36.1	30.0
June	17.4	12.7	36.1	31.4
July	16.9	14.2	36.1	33.4
August	17.1	13.2	36.1	32.2
September	17.9	9.7	36.1	27.9
October	15.8	5.3	36.1	25.6
November	11.9	2.3	36.1	26.5
December	16.0	1.0	36.1	21.1

[a]Five-year monthly return period.
[b]Potential evapotranspiration.

hydraulic overload and the excess wastewater stored. A number of wastewater loading options are available. One option is to apply the maximum allowable wastewater loading each month. This option requires coordination between storage and wastewater application to equalize wastewater flow. At the other end of the option spectrum, the hydraulic loading could be established throughout the year at the minimum allowable loading. The latter option requires minimal storage of flow but more land for wastewater application. The actual hydraulic loading selected within the range suggested by the solution must be based on a number of other factors not directly associated with the loading, such as plant operation schedule, site management, land available, costs, intensity of system management desired, etc. A technique for establishing an optimum balance between restricted irrigation and storage is given in Chapter 12.

Determination of Hydraulic Assimilative
Capacity Using a Saturated Flow Model

There are a number of saturated flow models that can be adapted to determine the wastewater hydraulic loading for a site. The situation best lending itself to a saturated flow model is one in which a high water table exists or where groundwater mounding may occur due to wastewater application. The saturated flow model is solved for the optimum wastewater loading, given a precipitation occurrence frequency and a depth of soil that must remain aerated (*i.e.*, unsaturated). Groundwater flow from the site may be to natural outlets, such as streams, or through a subsurface drainage network.

A number of steady-state saturated and transient flow models have been developed. The steady-state models are the easiest to use but are somewhat artificial in that a steady recharge rate is assumed. The assumption is best met for a land treatment system if short periods of time are selected for analysis. An advantage of the transient flow models is that periodic effects of precipitation and wastewater application on water table elevation may be simulated; also, the drainage time required to meet specified water table conditions may be determined. A version of a transient flow model, as developed by Dumm (1964), is presented below.

The differential equation relating water table position, time and hydraulic characteristics of the aquifer for the condition where the site is underlain by a impermeable barrier may be written as follows:

$$\frac{\partial y}{\partial t} = \frac{Kd}{S} \cdot \frac{\partial^2 y}{\partial X^2}$$

(4.9)

where y   = water table height
t   = time
K   = permeability
d   = saturated depth of aquifer to an impermeable barrier below an artificial drain or natural outlet
S   = specific yield
X   = space

Specific yield is the volume of water yielded on drainage of a unit volume of soil. Using a fourth-degree parabola to represent the water table shape and combining Equation 4.9 with Darcy's law (Equation 4.1) results in a complex differential equation that is best solved by numerical or graphical means. If the artificial drain or natural outlet is above the impermeable barrier and the water table fluctuation is small compared to the saturated depth, d,

$$L = [\frac{KDt}{aS}]^{1/2} \qquad\qquad (4.10)$$

where L   = distance between drains or outlets

D   = average flow depth

a   = dimensionless value of KDt/SL^2 from a curve of y/y_0 vs KDt/SL^2, as presented by Dumm (1964).

The definitions of y and y_0 are given in Figure 4.7.

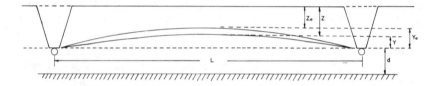

d   =   distance between outlet and barrier

Zo =   depth to water table midway between drains after addition of precipitation and/or rainwater

Z   =   depth to water table midway between drains after drainage

Yo =   height of water table above bottom of ditch midway between ditches, after addition

Y   =   height of water table above bottom of ditch midway between ditches, after drainage

D   =   average flow depth D = d + Yo/Z

Figure 4.7 Schematic representation of conditions for solution of groundwater flow model.

Several alternative approaches are available to determine the wastewater hydraulic loading. If an artificial drainage system is required, the optimum distance between drains can be determined by specifying the range and depth the water table may fluctuate within the root zone, as well as the time necessary to drain from the maximum to minimum water table depths. For example, given the following set of conditions, the required distance between drain outlets can be determined:

$$S = 0.20 \qquad\qquad y_0 = 76 \text{ cm}$$
$$K = 2.9 \text{ m/day} \qquad d = 2.4 \text{ m}$$
$$t = 5 \text{ days} \qquad\qquad D = 3.9 \text{ m}$$
$$y = 60 \text{ cm}$$

The initial conditions place the water table midway between the drains 76 cm above the drain, following the application of 2.5 cm of water. If it takes 5 days to drain to a height of 60 cm above the drains, solution of Equation 4.10 yields a distance between drains of 71.9 m. Assuming a required aerated root zone of 45 cm (z_0 in Figure 4.7), the depth of the drain ($z_0 + y_0$) must be 121 cm. Simulations using past records of observed precipitation, evapotranspiration and wastewater application are possible and are valuable aids in determining the hydraulic assimilative capacity. A number of iterative solutions are generally required to converge on the optimum solution.

Determination of Application Rate

Once the hydraulic loading has been determined, the application rate and sequencing of application must be determined. Important factors to consider are infiltration rates, internal drainage rates within the soil, the type of vegetation and management intensity, type and frequency of soil freezing, and the necessity to control soil wetness to enhance denitrification (See Chapter 11).

The infiltration rate for slow-rate or spray irrigation systems is difficult to determine because infiltration is extremely variable in space and time. Thus, point measurements can serve only as an index. If the site is to be managed for agricultural crops, local agricultural irrigation specialists or irrigation handbooks can provide experience data. Infiltration rates in forests are usually high, exceeding all but the most intense precipitation rates expected. As a guide, the application rate for slow-rate or irrigation systems should not exceed 6-8 mm/hr and, if bare soil is exposed or the site surface soil is compacted through management activities, the application rate should be less.

The design infiltration rate for basin surface irrigation systems is best determined by constructing small basins several meters in diameter and subjecting the soil to periodic flooding. The basins must be large enough and flooding occur for a lengthy period to assure minimal interference from lateral spreading of the water as it infiltrates.

An actively growing crop will maintain high infiltration rates by protecting the soil surface from water droplet impact and by maintaining good soil structure. It is important, therefore, that an intensive vegetation management scheme be part of any land treatment system. Where weather conditions permit, a winter crop should be grown on agricultural sites. If a crop is not planted, a portion of the summer crop residues should be left on the site to protect the soil surface.

Generally, the plant-soil biological system responds better to periodic wetness rather than a continuous high level of wetness. During the dryer periods, root growth occurs and gaseous exchange takes place. It is preferable, therefore, to apply wastewater at a periodic interval that permits sufficient drying between applications and between the random precipitation events. General practice is to apply wastewater on a uniform basis one day each week. Certain situations may exist, such as creating conditions favoring denitrification, that require a more frequent application, e.g., two days per week. Under these circumstances, the design hydraulic loading is uniformly split into two applications.

The application frequency for basin or other surface irrigation systems is a function of hydraulic loading, depth to water table, quality of wastewater and groundwater desired, soil cation exchange capacity, and temperature.

Determination of Storage Requirements

Storage of wastewater must be provided when the site cannot receive wastewater due to inclement conditions; the total hydraulic loading has been exceeded due to high antecedent precipitation; vegetation harvesting and/or regeneration; equipment failure and flow equalization. Discussion in this section will focus on determining storage for inclement conditions and hydraulic overload.

As previously discussed, review of precipitation and temperature records permits a probability of occurrence to be associated with events that may curtail wastewater application. Each record must be examined and the probability of concurrent events established. From this type of analysis and from knowledge of soil drying rates, the number of days of storage can be estimated. The hydrologic budget analysis is one method that can be used as a bookkeeping procedure to estimate the amount of storage necessary to avoid a hydraulic overload.

REFERENCES

Black, C. A., *Methods of Soil Analysis, Part 1* (Madison, WI: American Society of Agronomists, 1965).

Dumm, L. D. "Transient-Flow Concept in Subsurface Drainage: Its Validity and Use," *Trans. Amer. Soc. Agri. Eng.* 142-151 (1964).

Hillel, D. *Soil and Water* (New York: Academic Press, 1971).

Holdridge, L. R. "The Determination of Atmospheric Water Movements," *Ecology* 43:1-9 (1962).

Kirkham, D., and W. L. Powers. *Advanced Soil Physics* (New York: John Wiley and Sons, Inc. 1972).

Kunze, R. J., G. Uehara and K. Graham. "Factors Important in the Calculation of Hydraulic Conductivity," *Proc. Soil Sci. Soc. Am.* 32:760-765 (1968).

Lindsey, R. K., M. A. Kohler and J. C. H. Paulhus. *Hydrology for Engineers,* 2nd ed. (New York: McGraw-Hill Book Co., 1975).

Nutter, W. L. "The Role of Soil Water in the Hydrologic Behavior of Upland Basins," in *Field Soil Water Regime* (Madison, WI: Soil Science Society of America, 1973), pp. 181=193.

Nutter, W. L. "Moisture and Energy Conditions in a Draining Soil Mass," ERC 0875, Environmental Resources Center, Georgia Inst. of Technology (1975).

Richards, L. A., Ed. *Diagnosis and Improvement of Saline and Alkali Soils,* U.S. Department of Agriculture Handbook of 60 (1954).

Thornthwaite, C. W. "An Approach Toward a Rational Classification of Climate," *Geographical Rev.* 38:55-95 (1948).

U.S. Environmental Protection Agency. "Annual and Seasonal Precipitation Probabilities," Environmental Technology Publication Series, EPA-600/2-77-182, Washington, D.C. (1977).

U.S. Weather Bureau. "Normal Monthly Number of Days with Precipitation of 0.5, 1.0, 2.0 and 4.0 Inches or More in the Conterminous United States, Tech. Paper No. 57, U.S. Dept. of Commerce, Environmental Science Services Administration (1966).

U.S. Weather Bureau. "Rainfall Frequency Atlas of the United States for durations from 30 minutes to 24 hours and Regurn Periods from 1 to 100 Years," U.S. Dept. of Commerce, Tech. Paper No. 40 (1961).

CHAPTER 5

PHOSPHORUS AND SULFUR COMPOUNDS

INTRODUCTION

As a part of the stage I design process, the assimilative capacity for phosphorus and sulfur compounds must be determined. Figure 5.1 relates these assimilative capacities to the overall design process, which is the focus of this chapter.

PHOSPHORUS (P)

Compounds containing phosphorus are present in almost every waste considered for land treatment. Total P in municipal raw waste ranges from 6-20 ppm (average 10 ppm) with soluble P nearly one-half to three-fourths of the total P. High phosphorus solid waste may result from cloth mills or textile plants, liquid H_3PO_4 manufacture and other processes employing rock phosphate or iron phosphate ores. Phosphatic basic slags are by-products of steel manufacture from high-P iron ores. Many inorganic phosphate fertilizers have been developed for application to crops, such as superphosphates (monobasic, dibasic and tribasic), calcium ammonium phosphate, ammonium phosphate, diammonium phosphate, nitric phosphates, pyrophosphates, ammonium phosphate nitrate, alkaline orthophosphates and liquid phosphoric acid. Animal manures represent a concentrated form of phosphorus with varying amounts of ortho and total P (Table 5.1). As a builder compound in detergent, phosphate increases the P content of detergent industry wastes.

Agricultural crops commonly contain 0.05-0.5% P as part of phytins, phospolipids, nucleic acids, nucleoproteins, phosphorylated sugars, coenzymes and related structures. Phosphorus may also be present in plant tissue as an internal buffer in the form of inorganic orthophosphate, pyrophosphate, etc., which also participate in ATP (adenosine triphosphate) formation and breakdown to ADP or AMP. In mineral soils, one-half to

131

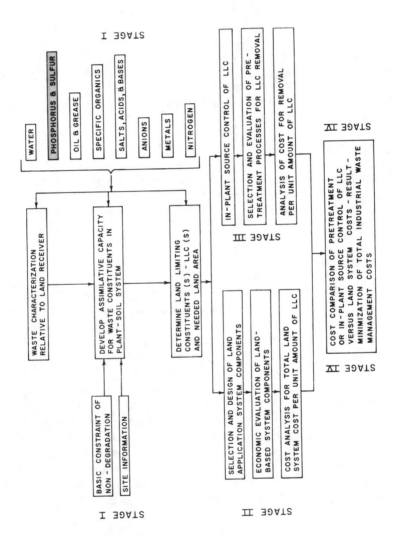

Figure 5.1 Relationship of Chapter 5 to overall design methodology for industrial pretreatment-land application systems.

Table 5.1 Phosphorus Content of Different Animal Wastes (Reddy 1978)

Animal Waste	Total[a] N (%)	P (%)	P:N Ratio	Ortho-P (%)	Ortho-P Total P Ratio
Beef	2.45	0.71	0.29	0.23	0.32
Poultry	6.03	1.79	0.30	0.43	0.24
Swine	3.97	2.29	0.58	1.59	0.66
	Concentration mg/l				
Swine Lagoon Effluent	224	55	0.25	50	0.91

[a]Dry weight basis.

two-thirds of the total P is inorganic, however the proportion varies from 4% for a podzol to 90% for an alpine humus (Cosgrove 1967).

Phosphorus in a plant-soil system undergoes a variety of biological and chemical reactions. Many P pathways transformations are similar to those of nitrogen, but the magnitude of rates are different (Figure 5.2). In contrast to

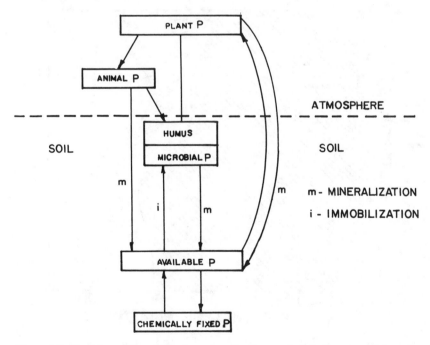

Figure 5.2 Phosphorus immobilization-mineralization cycle in soil system (Alexander 1977).

nitrogen, the phosphorus pool is predominantly inorganic, even in the surface soil, (Figure 5.3).

The term inorganic phosphorus usually refers to phosphate. The exact form of phosphate present in a soil is quite pH-dependent, and as might be expected, the assimilative pathways are also pH-dependent. For phosphate, one has a series of equilibria:

$$H_2PO_4^- \rightleftharpoons HPO_4^{-2} \rightleftharpoons PO^{-3}$$

low pH   neutral pH   high pH

Orthophosphate forms a variety of insoluble precipitates in soil, the exact form depending primarily on soil pH. In acid soils, below pH 6, the compounds are thought to be primarily the iron and aluminum phosphates, strengite $(Fe(H_2PO_4)\,(OH)_2)$ and variscite $(Al(H_2PO_4)\,(OH)_2)$, respectively. In neutral and basic soils the more insoluble compounds that ultimately form are octocalcium phosphate $(Ca_4H(PO_4)_3.3H_2O)$, and fluorapatite $(Ca_{10}(PO_4)_6F_2)$. In general, the amount of phosphate found in soil solution is in agreement with the amount determined from the solubility of the various

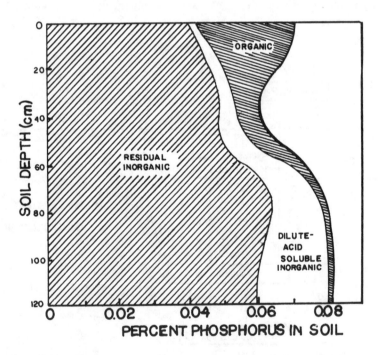

Figure 5.3   Inorganic and organic phosphorus distribution with soil depth (Black 1968).

precipitated compounds provided no phosphate fertilizer or soluble phosphate has been added for at least several years.

PHOSPHORUS ASSIMILATIVE PATHWAYS

Adsorption and Precipitation

Chemical fixation of added inorganic phosphates occurs under all soil pH conditions, with the least fixation in the range from 5.8-6.8 (Figure 5.4). Under acid soil conditions, inorganic phosphates are precipitated as insoluble iron, aluminum, and manganese phosphate, while under alkaline conditions as calcium and magnesium phosphate. Chemical fixation of inorganic and organic phosphates occurs by hydrous oxides, sesquioxides and silicate clays. Therefore, phosphates are relatively immobile in soil systems.

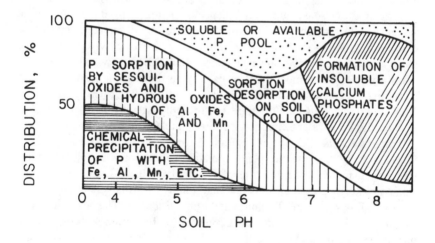

Figure 5.4 Percent distribution of inorganic phosphorus forms at various soil pH values.

Soil pH also affects the P assimilative capacity since the exchange sites or precipitating cations vary with soils having different pH. That is, one must establish the soil pH characteristics and pathways for fixation before determining the plant-soil assimilative capacity.

In addition to precipitation reactions, adsorption on iron and aluminum hydrous oxides is important in fixing P applied to soils. The adsorption reactions are relatively rapid and, hence, under acid soil conditions represent the immediate plant-soil capacity for phosphorus. Phosphate is tightly adsorbed on these sites and not displaced by Cl^- and $SO_4^=$ anions. Several studies have indicated a decrease in P sorption by soils in the presence of organic residues (Reddy *et al* 1977). A number of researchers concluded that during decomposition of organic residues, organic acids form stable complexes with Fe and Al, and, consequently, block P retention in the soil. However, increased retention due to organic residue exchange capacity has also been observed (Harter 1969, Jackman 1955, Rennie 1958).

Phosphorus is held in soils at different energy levels. The ease of P removal to leaching water, plant uptake, microbial activity and surface runoff is determined by the energy of fixation. Various availabilities of P (increased available with increased ease of removal and decreased energy level) have been measured historically by tests on water-soluble P and acid-extractable P.

Water-soluble P is the fraction of P extracted with water and generally represents the soil solution concentration. This is the amount in soil-water and is readily available for uptake processes in the plant-soil system. The amount of P extracted with water depends on the soil-water ratio used in the extraction procedure. To avoid solubilization of precipitates due to increased dilution, a ratio of soil to water close to natural systems should be used (saturation extract); however, most often this narrow soil-water ratio is not possible. As an index of P availability to the growing plants, a simple extraction procedure to determine water-soluble P has been presented by Olsen and Dean (1965). For more routine purposes and large numbers of samples, the 1:10 water extraction is more suitable.

Acid-extractable P is the phosphorus extracted with dilute acids and has been shown to estimate the plant availability index of P in soil. The most widely used method is extraction of P with HCl (0.5 N) and NH_4F (1 N) which are designed to remove easily acid-soluble forms largely calcium phosphates, and a portion of aluminum and iron phosphates. The NH_4F dissolves aluminum and iron phosphates by a complex ion formation with these metal ions in acid solution. This method has been most successful on acid soils. The P concentration obtained by this extraction is interpreted as follows: < 3 ppm very low; 3-7 ppm, low; 7-20 ppm, medium; and > 20 ppm, high. Other extraction procedures found to be successful in acid soils are mixtures of dilute HCl (0.05 N) and H_2SO_4 (0.025 N). In alkaline soils, the most commonly used extractant is sodium bicarbonate (0.5 M) (Olsen and Dean 1965).

For a given soil, whether receiving waste or not, there is an unused capacity to fix P in forms such as are shown in Figure 5.4. An estimate of this unutilized fixation capacity is the phosphorus adsorption maximum.

Phosphate adsorption data on soils are usually described by the Freundlich isotherm or by a Langmuir isotherm. The latter method can be used to calculate adsorption maximum of a particular soil under study. However, this equation is applicable for relatively smaller amounts of adsorbed P, and, consequently, at more dilute equilibrium P concentrations. The Langmuir equation is:

$$C/S = C/S_{max} + (1/K\ S_{max})$$

where C   = equilibrium P concentration, μg/ml
 S   = P adsorbed on soil surface, μg/g of soil
 S_{max} = adsorption maximum of the soil, μg/g of soil
 K   = constant related to the bonding energy.

To obtain data for the plotting the Langmuir isotherm, 5 g air dried soil is shaken in 100 ml 0.01 M CaCl$_2$ containing varying concentrations of P, for a period of 24 hours at a constant temperature. The difference between the amount of P remaining in solution after shaking and the amount initially present was taken as the amount of P adsorbed by the soil. A plot of P adsorbed versus P in solution represents the isotherm. The concentration of P in solution, at which no adsorption takes place (zero adsorption or desorption) is known to be equilibrium P concentration (EPC) of the original soil. Adsorption maximum can be calculated after a linear regression analysis between C/S and C, from which S_{max} can be estimated (Olsen and Watonabe 1957).

This adsorption maximum is related to the unused P retention capacity of a soil. From these data the assimilative capacity can be developed, as discussed later in this chapter.

Mineralization-Immobilization of Phosphorus

The microbial reactions that convert phosphorus organic compounds to inorganic orthophosphate are termed P-mineralization. The utilization of inorganic P as an essential element to form P-compounds in biomass is termed the mineralization rate and is greater under alkaline conditions than under acid conditions. Nucleic acids and nucleoproteins, which constitute 1-10% of total organic P in soils, are readily mineralized after the death of microbial cells. In terms of organic phosphorus compounds found in commonly in soils, nucleic acids are mineralized most readily; phytin is the slowest in degradation, while rate of lecithin breakdown being intermediate. Liming of acid soils to pH 7.5 or more enhances the P mineralization (Brady 1974).

Species of genera *Aspergillus, Penicillium, Rhizopus, Cunninghamella, Arthobacter* and *Bacillus* synthesize phytase, which liberates phosphate from phytic acid or phytin (P-mineralization). Nucleoproteins are dephosphorylated readily by the phosphatase enzymes of many bacteria and fungi. Virtually all microorganisms are involved in P immobilization because of the essential or required nature of phosphorus in microbial growth.

Under an aerobic regime, the flora that immobilize phosphorus utilize between 100 and 200 parts of carbon per unit of P, C:P $\stackrel{\sim}{=}$ 150. Industrial wastes with a C:P > 150 would be expected to generate little orthophosphate and the decomposition of waste organics would be P-limited. In this instance net immobilization results. For wastes with C:P < 150 the decomposition would proceed with adequate P, and excess orthophophate would be generated. This is net P mineralization. From this critical level of 150:1, biological processes occur until ultimately a C:P ratio of 100:1 results. This ratio is characteristic of the organic humus fraction of the soil. The change in C:P is due to carbon loss as CO_2 in microbial growth.

Phosphorus – Plant Relations

Phosphorus ranks next to nitrogen in importance to fertility of soils and crops as a macronutrient. Plants take up phosphorus as $H_2PO_4^-$ and $HPO_4^=$ ions from soils and assimilate these into various cellular constituents. As adenosine triphosphate (ATP), phosphorus is a key element in the accumulation and release of energy for cellular metabolism.

Appraisal of soil phosphorus deficiency has been a major focus of agricultural attention, rather than excess P. Water-soluble concentration larger than 20 ppm P or sodium bicarbonate-extract P concentrations greater than 200 ppm P or sodium bicarbonate-extract P concentrations greater than 200 ppm P in a soil may be in excess of plant needs and may induce Cu, Zn and/or Fe deficiency in crop plants (Bingham 1966).

Crops vary in their requirements of P from 15 kg P/ha for tobacco and 50 kg/ha for cereals, to 60 kg/ha for alfalfa. Phosphorus uptake by Coastal Bermudagrass at a yield of 22.5 metric ton/ha was approximated at 55 kg P removed/ha/yr (Adams et al. 1967). Rye grass (Parks and Fisher 1958) at a yield of 8 metric ton/ha usually contained about 20 kg P/ha/yr. Therefore, a crop combination of Coastal Bermudagrass and rye grass would remove a total of 76 kg P/ha/yr.

A summary of P uptake by plants is shown in Table 5.2. Where phosphorus is potentially a land limiting constituent (LLC), crops with a high P uptake may allow reduction in area requirements.

Leaching and Runoff Losses

The movement of phosphate through soils utilized for land application has been studied in relation to animal and food wastes (Adriano et al. 1975,

Table 5.2   Phosphorus Removal by Some of the Selected Groups

Crop	Average Yield/ha/yr	P Removed (kg/ha/yr)
Alfalfa		35
Corn		
Grain	6,343 l	35
Stover	8,960 kg	15
Cotton		
Lint	1,680 kg	19
Seed	2,520 kg	12
Wheat		
Grain	2,819 l	20
Straw	8,960 kg	5
Rice		
Grain	7,840 kg	20
Straw	7,840 kg	6
Tobacco		
Leaf	4,450 kg	7
Stalks and Tops	4,032 kg	7
Peanuts		
Nuts	4,000 kg	10
Vines	5,000 kg	8
Johnson grass	11 metric tons	94
Coastal Bermudagrass	16.9 metric tons	70
Rye grass	6 metric tons	18
Tall fescue grass	3.1 metric tons	32
Soybeans		
Grain	2,114 l	25
Stalks	7,000 kg	10
Shortleaf Pine	—	15-35

Fiskell and Perkins 1973, Westerman 1977). Leaching has been demonstrated up to nearly 1 m. The mechanism appears to be the steady accumulation of phosphate in the surface zone until the absorption-fixation capacity is exhausted. Then the next soil layer accumulates P. The movement has been more pronounced where the surface horizon is a sandy texture with low absorption capacity. Leaching is usually only observed after prolonged waste applications (5+ years). By comparison to cations such as Ca, phosphate-metal compounds are relatively immobile in soils.

Because of the accumulation at the soil surface, runoff transport of P can be observed from pristine soils, agricultural areas or land application sites. The mode of transport is in the solution phase as well as attached to sediment particles. As P is land applied over time, as with manures, the resulting runoff contains a higher fraction of orthophosphate, because the soil particle capacity to absorb P is reduced. The soluble P, acid-extractable P, and equilibrium phosphorus capacity were all shown to increase at the soil surface after repeated waste application (Reddy et al. 1977). Thus, a greater emphasis is placed on best management practices (BMP) for controlling such runoff. Levels of P in runoff from a variety of phosphorus application rates has been reviewed by Khaleel (1978).

Phosphorus Assimilative Capacity

Despite the considerable knowledge regarding phosphorus in soils, the assimilative capacity for land application usage of plant-soil systems is not well established. Most data exist for deficient or maintenance applications of P. Cases of land application of wastes, such as manures containing substantial phosphorus, have often focused only on nitrogen. The general assumption is made that soils can adsorb such high levels of P that no limitation on loading would exist. Recent studies (Reddy et al. 1978, Westerman et al. 1977, Adriano et al. 1975, Hinesly et al. 1978) indicate that more attention must be directed at the application rate and movement potential of land-applied P. In fact, researchers in Europe, where long-term agricultural effects are documented, emphasize phosphorus as a controlling factor in the land application of wastes.

The assimilative capacity for the P constituents in industrial wastes depends on several primary factors: (1) net immobilization with associated waste organics; (2) the adsorption/precipitation reactions based on soil pH and texture; and (3) crop uptake. These factors are not all inclusive, but represent the existing design data base and the capability to assess important site characteristics of the receiver plant-soil system.

Net immobilization would be determined by assuming the critical C:P ratio of 150:1. For an industrial waste, P, in a ratio of 1:150 of the total organic C would thus be assimilated as biological matter and become a part of the basic soil structure. This immobilized amount of P should be subtracted from the total P waste generation to arrive at a net amount of phosphorus remaining to be assimilated.

Crop uptake values should be determined from Table 5.2 and the crop selected for the land treatment site. Where grasses are used, the total crop harvested is given in Table 5.2, while for nonfood options, the designer must determine how much of the crop is actually removed from the field. As with other nutrients, tree vegetation removes much less phosphorus than grasses.

Adsorption/precipitation of phosphorus often represents the greatest fraction of the plant-soil assimilative capacity but, as previously discussed, there is a finite capacity. The phosphorus adsorption maximum, previously described, is often used to represent the fixation capacity. Phosphorus adsorption maximum is based on the Langmuir isotherm and, hence, is most valid when lower concentrations of P are used. Another measure of the P assimilation capacity is the saturation maximum in which soil is tested in association with high P concentrations (2,500 μg P/g soil). The adsorption/-precipitation is then an estimate of the maximum P removals (Ballard and Fiskell 1974). Results of both these tests are in μg P/g soil and must be analyzed at several depths in the soil. In measurements on 42 Coastal Plain forest soils the saturation maximum was about two to three times the Langmuir adsorption maximum (Ballard and Fiskell 1974).

Sawhney and Hill (1975) observed that a soil saturated with P, as measured by the phosphorus adsorption maximum test, when allowed to rest for six months to one year could then adsorb more phosphorus. Thus, exchange sites are undergoing slow kinetic changes such that new hydrous oxides of Fe, Al and Mg are formed. Also, reactions could be occuring in which Ca, Mg, Fe and Al are being solubilized from within the soil structure or from nonphosphate precipitates and become available to insolubilize phosphates. The exact mechanisms are unclear at this time; however, the effect is to increase the fixation capacity above the measured values of initial soil tests. The increase was estimated to be three- to fourfold as an ultimate P adsorption maximum.

The observations of P movement at land application sites is generally associated with substantial hydraulic application and water movement. Phosphorus movement demonstrates that the kinetics of P adsorption/precipitation are not instantaneous. Thus, when large applications of water are being used (>4 cm/wk), monitoring and cycling of application should be used to enhance and document P assimilative capacity.

For most industrial waste land application systems, the P assimilative capacity is estimated from the P adsorption maximum or the P saturation maximum, μg P/g soil. The soil profile available for phosphorus adsorption is 100-200 cm, instead of the surface 15 cm, thus assessment of P adsorption capacity over this deeper profile. Use of a 100- to 200-cm profile would be restricted where subsurface conditions, such as rock or groundwater, occurred. The μg P/g soil, the bulk density and the usable soil profile are combined to establish the P assimilative capacity from adsorption/precipitation, kg P/ha. With a design site life of 10-50$^+$ years, the annual assimilative capacity, kg/ha/yr, is determined.

The total of crop uptake and adsorption/precipitation capacities constitute the total loading rates for the net phosphorus in an industrial waste. This annual rate is valid over the site lifetime unless soil and soil-water monitoring can establish further adsorption capacity. After the adsorption/precipitation capacity is utilized, the P application rate must be reduced to crop uptake levels. Under such conditions, high phosphorus utilization species can be grown to increase the P assimilative capacity.

SULFUR (S)

Within a biological framework, sulfur (S) is an essential element for microflora and -fauna as well as to macro-species of plants and animals. Sulfur research has identified deficient, as well as excessive, levels with resultant biological responses similar to Figure 10.4. Sulfur is also a ubiquitous element in mineral matter. Sulfur is abundant in nature and thus is often a contaminant in coal, calcium deposits, metal ores, phosphate deposits, etc. The widespread occurrence of S compounds assures that industrial wastes will contain varying levels of sulfur. Wastes from kraft mills, sugar refining, petroleum refining, copper and iron extraction, and several proteinaceous processing facilities, are examples of materials for which a sulfur assimilative capacity would be needed for land application.

The behavior of sulfur and associated compounds (organic and inorganic) in a plant-soil system has a number of similarities to nitrogen and phosphorus cycles previously discussed. The magnitude and importance of soil S pathways varies from those of N and P. A schematic of the sulfur transformations is presented in Figure 5.5. Within the plant-soil system, sulfur is between 75 and 95% organic. Organic sulfur is in two predominant forms: (1) that present in carbon compounds, and (2) that present as esters of sulfate (Figure 5.5). The latter are readily hydrolyzed to sulfate. For inorganic S in soils, the major form is as adsorbed sulfate with a relatively small amount of less oxidized S compounds (Figure 5.5). The addition of organic or inorganic sulfur compounds in an industrial waste represents an increase in respective compartments of the S-pool and can thus participate in established assimilative pathways.

Fate of Sulfur and Sulfur Compounds in a Plant-Soil System

Mineralization-Immobilization

A variety of organic S compounds reach the soil receiver from plant and animal residues or from the purposeful application of industrial waste constituents. These organics undergo microbial transformations in which shorter-chain S-containing species and inorganic S compounds are formed. This process is sulfur mineralization. As might be expected, the rate of mineralization varies considerably with S compound structure (Alexander 1977).

A variety of sulfur compounds are naturally present in soils; therefore, if these are present in an industrial waste, the probability of an established pathway for mineralization is high. A listing of these chemical compounds is assembled by McLaren and Peterson (1967), and includes ring and straight-chain compounds. Other compounds which have been shown to be decomposed by soil microorganisms include thiosulfate, ureasulfate, taurine, cystine,

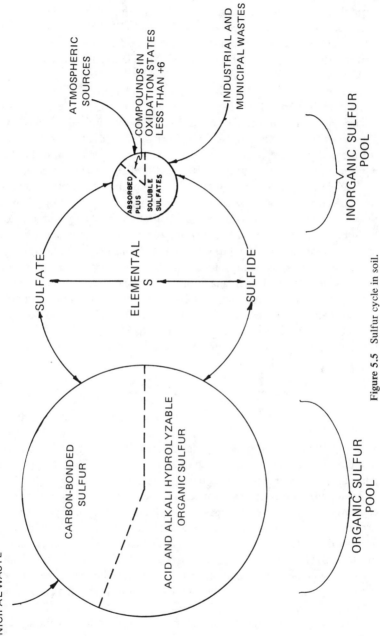

Figure 5.5   Sulfur cycle in soil.

sulfanilac acid, methionine, formaldehyde-sodium-sulfoxylate, sodium-taurocholate, thiamine, thiourea, sulfanilamide, sulfthiozole, ammonium-sulfamate, sulfosalicylic acid and p-thiocresol (Frederick et al. 1957, Gleen 1951, Steinberg 1941). The microorganisms responsible for these mineralization reactions are reviewed by Frederick et al. (1957) and Alexander (1977).

Mineralization of the aggregate organic S pool was found to be 1-3% of the total S in the humid-temperate zone (Alexander 1977). For an A horizon containing 200 ppm S, 5-13 kg S/ha/yr would be mineralized. This rate may be used as a minimum rate characteristic of well-stabilized organic S wastes. Elseewi et al. (1978) compared plant response of alfalfa, clover, and turnip for municipal sludge versus equivalent S additions of gypsum. While this does not establish the mineralization rate, the results indicate no consistent differences in plant yield or S- uptake. Since gypsum is made available to plants by a different microbial pathway than mineralization of an organic material, these comparative results are more qualitative in nature. Interpreting the authors' data for applications of 200-320 kg S/ha, as municipal sludge, the mineralization rate was computed to be in the range of 30-70 kg S/ha/yr.

The end products of organic S mineralization depend on the soil oxygen or moisture status. Under anaerobic conditions, sulfides (often H_2S) are generated with corresponding odorous products. This option should not be considered viable for the plant-soil assimilation of industrial wastes. Thus, a predominantly aerobic environment must be maintained for the mineralization end product to be sulfate. A typical aerobic mineralization sequence is as follows:

$$R\text{-}SH \rightarrow R\text{-}S\text{-}S\text{-}R \rightarrow R\text{-}SO\text{-}OS\text{-}R \rightarrow R\text{-}SO_2H \rightarrow R + HSO_2^- \rightarrow SO_3^- \rightarrow 2H^+ + SO_4^{-2}$$

Formation of two hydronium ions as reaction products along with sulfate lowers the soil solution pH under aerobic conditions. Sulfate is not the exclusive aerobic end product. Frederick et al. (1957) reported loss of volatile S compounds under mineralization of certain organics. Thus disappearance of parent S compound is the measure of mineralization rate.

Immobilization proceeds as microorganisms use inorganic S as an essential element for growth. Microorganisms can incorporate many inorganic and organic S compounds, as well as sulfate, into biomass (Alexander 1977). The mode of sulfate immobilization is not clear, but the intermediates of cystine and methionine are believed to be involved (Freney 1967). Plant root uptake, metabolism and senescence also appear to be major mechanisms for converting inorganic S into the humus fraction (Scharpenseel and Krause 1963).

Under an aerobic regime required for industrial waste land application, the microflora that immobilize S utilize between 200 and 400 parts of C per unit of S. That is, at C:S > 400, net immobilization occurs, and for such a waste little sulfate would be produced. Such a waste is sulfur limited. For wastes with C:S less than 200-400, decomposition proceeds and sulfate is generated. This is net S mineralization. From the critical range of 200-400:1, the biological processes occur until ultimately a C:S ratio is 100-150:1. This ratio is characteristic of the organic humus fraction of the soil. The reduction is characteristic of the organic humus fraction of the soil. The reduction in C:S ratio is due to carbon loss as CO_2 in microbial growth.

Oxidation-Reduction and Solubilization of Inorganic Sulfur Compounds

Many inorganic sulfur compounds (Table 5.3) have been apppplied to plant-soil systems with differing effects on crop yield. The agricultural purpose of such compounds is to dissociate, oxidize to sulfate and to stimulate plant growth or correct alkaline soil pH. Sulfur compounds can range in oxidation state from +6 (sulfate) to -2 (sulfide), and under aerobic conditions, conversions are toward increasing oxidation states. This pathway is predominantly microbial-induced, with chemoautotrophs, primarily *Thiobacillus,* being responsible.

Table 5.3 Sulfur Compounds Used in Agricultural Operations

Ammonium sulfate
Sulfur mixtures with phosphates, urea, lime or ammonium solutions
Ammonium phosphate-sulfate
Potassium sulfate
Potassium-magnesium sulfate
Ammonium thiosulfate
Ammonium bisulfite
Ammonium polysulfide
Sulfur dioxide
Sulfuric acid and iron/aluminum sulfates
Flowers of sulfur
Gypsum, $CaSO_4 \cdot 2H_2O$

Oxidation of elemental S, which appears in a number of S fertilizers, was measured to be about 80 kg S/ha/week (Lint 1914). The aerobic process was found to be more rapid in a sandy loam soil when compared to a clay loam. Thiosulfate, sulfur dioxide and ammonium bisulfite are also oxidized to sulfate when land applied. The forms of sulfur applied to soil can also be reduced under anaerobic conditions. Those mechanisms are reviewed in detail by McLaren and Peterson (1967) and Alexander (1977).

Other salts or sulfate have been evaluated as sources of sulfur (Shedd 1914). For these compounds the solubility product controls the amount of sulfate present at any time and hence the level available to plants. Rantsek (1943) determined that twice as much S was needed to give equivalent crop response when the source was $Al_2(SO_4)_3$ compared with elemental sulfur. Lipman et al. (1921) evaluated gypsum and ferrous sulfate and concluded that on an equal S basis, 1/16 times as much ferrous sulfate produced equivalent crop response. Thus, materials containing sulfate salts should be examined for the solubility and crop effect relative to elemental S. Under certain conditions, applied inorganic sulfur compounds can precipitate in the soil solution and be assimilated. Cations of lead, calcium, and barium would effectively precipitate all stoichiometric quantities of sulfate until solubility product levels are reached.

The microbial oxidation of more reduced inorganic S compounds to sulfate and the dissociation of sulfate salts in soil solution results in an increased level of acid (lowered pH). Various researchers have observed that with inorganic S, greater potassium is liberated for plant use as well as a stimulation of mineralization of soil organic nitrogen. Also, as sulfuric acid is formed, other essential elements such as Ca and Mg are solubilized. As a result, in addition to providing S for microbial and plant use, inorganic S enhances the uptake of other essential elements. However, as increasing levels of sulfate are formed, the soil pH can become prohibitively low.

Adsorption, Leaching and Atmospheric Inputs

Although sulfate is an anion, there exists a greater soil adsorption capacity than for such ions as NO_3^- and Cl^-. The tetrahedral shape of sulfate allows a better geometric fit when hydroxyl groups are displaced and from the iron and aluminum oxides or silicate structure in a soil. However, the sulfate retention is not as strong as phosphate (Barbier and Chabannes 1944, Harward and Reisenauer 1966). Kaolinitic soils, such as in the southeast United States, have high sulfate adsorption capacities. Liming of soil mobilizes the sulfate ions, thus reducing this assimilative capacity. For modeling purposes, the sulfate adsorption has been quantified as a Freundlich or as a Langmuir isotherm (Table 5.4). These results can be used in evaluating sulfate retention and movement in soils from applied wastewaters.

The assimilative capacity for sulfates, whether in industrial wastes or as a result of organic S mineralization, must recognize and anionic character and, hence, leaching of this compound. Lipman and McLean (1931) summarized field data from a number of sites and concluded that seven times as much sulfate is moved from the root zone as is taken up by the plant. The observations that leached soils become S deficient further verifies the mobile characteristics of sulfate.

Table 5.4 Sulfate Retention in Soils

	Freundlich Equation		Langmuir Equation	
	K^a	1/n	k^b	b^c
Aylmore *et al.* (1967)				
API-9 kaolinite	–	–	0.14	1.86
Clackline	–	–	0.062	1.0
Haematite (iron oxide)	–	–	0.14	13.4
Pseudoboehmite (aluminum osyhydroxide)	–	–	0.37	84.2
Chao *et al.* (other soils given in reference)				
Quillayute silt loam	0.13	0.58	–	–
Astoria silt loam	0.14	0.52	–	–
Aiken clay	0.12	0.48	–	–
Knappa silt loam	0.072	0.54	–	–

[a]ml solution/g soil.
[b]l/meq SO_4.
[c]meq SO_4/100 g soil.

A final pathway for sulfur in the plant-soil system is the input from rainfall stripping of the atmospheric sulfur oxides. On an average basis, about one-half the sulfate leached from soils was applied to land through rainfall sulfur compounds (Lipman and McLean 1931). Annual sulfur input varies up to 110 kg S/ha near industrial areas. Pristine areas receive on the order of 1 kg S/ha/yr (Alexander 1977).

Plant Relationships to Sulfur Compounds

The forms of S that are available to the plant are more limited than the potential inputs for microorganisms. In soil systems, plants take up primarily $SO_4^=$ and HSO_4^-, although some S amino acids can be assimilated. Sulfur can also be assimilated by plants through foliar uptake. Within the plant structure, sulfates are quite mobile. The tissue S level has been shown to reflect the deficiency or excess of soil sulfur (Eaton 1966). In water culture data for seven crops, the sulfate concentration in soil solutions in the range of 30 meq/l would produce about a 10% yield reduction. Thus, in plant-soil systems for assimilating industrial wastes containing S compounds, soil solution levels must be kept below approximately 30 meq SO_4/l.

The response of crops to a single does of S has been tested primarily to correct soil sulfur deficiencies. Long-term build-up generally does not occur with low to moderate applications because of leaching. From the crop response data (Table 5.5) it appears that applications of sulfur compounds at 2,000 kg S/ha or less would represent a single-dose assimilative capacity.

Table 5.5 Crop Response to Single Dose of Sulfur Compounds

Application Rate (kg S/ha)	Response	Soil	Crop	Comments	Reference
32	Fertilizing effect on soils	—	Killed weeds but did not effect wheat	H_2SO_4	Rabate (1912)
37	Improved yields	—	Alfalfa	Gypsum	Brown (1917)
63	No effect on yield	—	Corn, oats	Gypsum	Illinois Agricultural Experimental Station (1929)
330	Increased yields	—	Alfalfa	S	Reimer (1914)
1,600–2,200	No effect on germination or growth	Silt loam	Alfalfa	S	Adams (1924)
2,200–4,400	Slight reduction	Clay	Corn	S	Reynolds (1930)
4,700<>9,400	Decreased seedling survival	Alkaline	Pine	S	Young (1938)

Plant growth and uptake of S also represent an assimilative pathway. The magnitude of crop removal of S is estimated in Table 5.6. As crops are harvested at the yields given in Table 5.6, there is net S removal from the system.

Assimilative Capacity for Sulfur Compounds

Long term land application of sulfur constituents of industrial wastes requires that several assimilative pathways be quantified using site data in the design criteria. For inorganic sulfur compounds, several factors are included: (1) neutralization of acid conditions, (2) migration, (3) crop uptake, and (4) immobilization and adsorption. The first factor is based on the continued generation of hydrogen ions during sulfur oxidation which lowers soil pH. One of several techniques must be used to neutralize the acid generated. Chapter 8 details four assimilative design alternatives for acids, which should be reviewed for use with inorganic S additions. Probably the most efficient in terms of the complex behavior of S in soils is to assume complete oxidation to sulfate and production of sulfuric acid, then add the required amounts of base so that a neutralization ultimately occurs. Materials and techniques in Chapter 8 can be used in this regard. Alternatively, soil pH can be monitored and the soil texture determined, then based on Figure 8.9 the corresponding action taken. As a measure of the type of pH correction necessary, Adams (1924) measured the soil pH change with the addition of sulfur (Figure 5.6).

Crop uptake contributes to the plant-soil assimilative capacity and can be estimated from Table 5.6. Immobilization of S can also be substantial where organics are present in the industrial wastes or might be added to the soil. Using the organic carbon to be land applied and the critical C:S ratio for biological growth (300, range 200-400), the amount of S that may be tied up in soil humus can be estimated. This S amount is then subtracted from the total S in the waste to give the net S to be assimilated in land application. A further fixation capacity can be estimated from the S adsorption capacity of soils. Analyses such as used for phosphates should be considered for sulfate adsorption. With a site lifetime, an annual assimilative capacity could then be determined.

In addition to the pathways quantified above, a final factor must be included to determine the overall assimilative capacity for inorganic sulfur compounds. This is the movement laterally to surface water or vertically to groundwater. Again the techniques described for moble salts (Chapter 8) are to be used. Basically the areal application, kg/ha/yr, should not exceed that for which the concentration of sulfate, resulting from inorganic S compound application to land, in receiving waters is elevated beyond a point requiring further water treatment. The leached or drained concentration of allowable sulfates would be 250 mg SO_4/l (McKee and Wolf 1963).

Table 5.6 Approximate Quantity of Nutrients Contained in Various Crops

Crop	Yield kg/ha[c]	Yield units/ac	Nitrogen (N)	Phosphorus (P)[a]	Potassium (K)[b] kg/ha[c]	Sulfur (S)	Magnesium (Mg)
Grains[d]							
Corn	12,544	200 bu	358	52	230	49	74
Grain sorghum	8,064	8,000 lb	291	54	207	43	40
Wheat	5,376	80 bu	168	34	112	25	27
Barley	5,376	100 bu	168	28	140	28	–
Oats	3,584	100 bu	112	22	112	22	22
Rice	7,280	145 bu	151	25	150	20	17
Forage Crops[d]							
Alfalfa	12,544	6 tons	375	35	251	34	34
Clovers	8,064	4 tons	179	20	149	20	26
Grasses (general)	8,064	4 tons	134	20	112	18	17
Coastal bermudagrass	22,328	10 tons	638	72	370	50	56
Orchardgrass	15,805	7 tons	314	48	289	56	–
Oil Crops[d]							
Soybeans (beans only)	3,360	50 bu	207	25	112	11	13
Peanuts	3,360	3,000 lb	246	22	112	28	31
Rapeseed	2,240	2,000 lb	101	18	123	39	–
Fiber Crops[d]							
Cotton	1,400	2.5 bale	140	37	84	26	18

Stimulant Crops[d]						
Tobacco	3,360	2,800 lb	106	12	179	24	27
Sugar Crops						
Sugarcane[d]	67,200	30 tons	135	19	235	90	110
Sugar beets (tops and roots)	67,200	30 tons	179	28	235	55	—
Vegetables						
Potatoes (tops and tubers)	26,800	400 bu	224	27	286	30	27
Cabbage (heads)	44,800	20 tons	146	16	118	41	11
Turnip (tops and roots)	67,200	25 tons	128	25	213	43	25
Onions (tops and bulbs)	44,800	20 tons	134	27	99	28	13

[a] $P \times 2.3 = P_2O_5$.
[b] $K \times 1.2 = K_2O$.
[c] kg/ha $\times$ 0.9 = lb/ac.
[d] All aboveground portions.

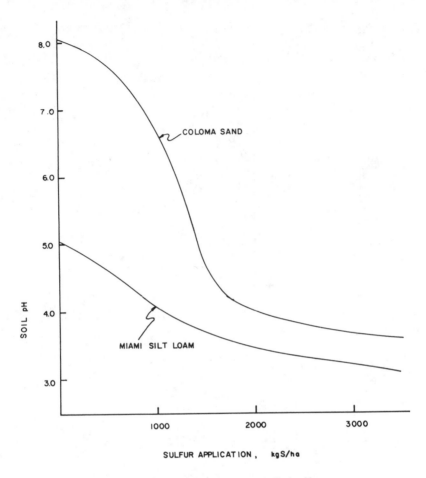

Figure 5.6 Soil pH response to applied sulfur.

Organic sulfur compounds are assimilated along similar pathways; however, an additional step is involved—mineralization. When organic S compounds are present in amounts significant with respect to equivalent sulfate amounts used on land, then design is based on a steady state mineralization. That is, the organic S is assumed to proceed to sulfate based on the amount of S present, and then the assimilative capacity is based on the sulfate, as described previously. This design criterion assumes that the amount of organic S constituents applied is not so large as to prohibit mineralization.

When organic S compounds are present in low amounts, as with trace dyes, priority pollutants or industrial chemicals, the assimilative capacity is based on the rate of mineralization. The resulting sulfate is usually small and within the assimilative capacity for inorganic S compounds. That is, the rate of application should approximate the rate of decomposition of the parent compound. Estimative decomposition rates are given in Table 5.7. These are to be used as first approximations, in the LLC analyses, for a particular industrial waste. Data in Table 5.7 were for application of 20,000 kg organic S compound/ha and thus may have produced extensive lag periods, affecting decomposition results.

Table 5.7 Decomposition of Organic Sulfur Compounds as Measured by Sulfate Formation (10,000 ppm S compound applied to sandy loam) (Frederick et al. 1957).

Compound	Decomposition (kg/wk)
Cystine	810
Na-taurocholate	950
Formaldehyde Sodium Sulfoxylate	620
Thiourea	85
Sulfanilamide	60
Sulfathiozol	50
Sulfapyridine	40
Mercaptoethanol	40
Na-diethylclithiocarbamate	10

Management techniques for land application systems can be used to improve the assimilative capacity for S compounds. Soils with large microbial populations degrade organic S compounds more rapidly (Frederick et al. 1957). Thus between the benchmark soils, the clay loam or soils with substantial organic matter are preferred. Selection of acid-tolerant plant species (Table 8.6) allows greater single doses of S compounds. Use of active monitoring and a pH correction plan should be used to permit only minor deviations below neutral soil pH. In this manner, the pathways for mineralization, oxidation, and plant uptake are enhanced. Of course predominantly aerobic systems are required for the design and operation for wastes containing sulfur constituents.

Land application of S-containing wastes has been utilized for sulfite liquors from kraft mills (Spulnik et al. 1940, Voights 1955, Wisniewski et al. 1955). To determine the S application rates characteristic of acceptable ongoing land application systems, Table 5.8 was prepared. These sulfur land application rates represent a minimum design level, since long-term impact of such systems at the specified application rates have proved successful.

Table 5.8 Sulfur Applications from Representative Land Treatment Systems

Waste Type	Application Basis	Resultant S Application (kg S/ha/yr)	Reference
Sewage Sludge	220 kg N/ha/yr	38	Sommers (1977)
	400 kg N/ha/yr	76	
	600 kg N/ha/yr	114	
Secondary Municipal	1 in./wk	54	Elliot (1977)
Effluent	2 in./wk	108	
Manure	500 kg N/ha/yr		Azevedo and Stout (1974)
Broiler litter		90	
Dairy		55	
Cattle		55	
Swine		125	
Horse		85	
Sheep		40	

REFERENCES

Adams, H. R. "Some Effects of Sulfur on Crops and Soils," *Soil Sci.* 18(2): 111-115 (1924).

Adams, W. E., M. Stelly, H. D. Morris and B. Elkins. "A Comparison of Coastal and Common Bermuda Grasses in the Piedmont Region," *Agron. J.* 59:281-283 (1967).

Adriano, D. C., L. T. Novak, A. E. Erickson, A. R. Wolcott and B. G. Ellis. "Effect of Long Term Disposal by Spray Irrigation of Food Processing Wastes on Some Chemical Properties of the Soil and Subsurface Water," *J. Environ. Qual.* 4:242-248 (1975).

Alexander, M. *Introduction to Soil Microbiology*, 2nd ed. (New York: John Wiley & Sons, Inc., 1977), pp. 82, 353-369.

Aylmore, L. A. G., M. Karim and J. P. Quirk. "Adsorption and Desorption of Sulfate Ions by Soil Constituents," *Soil Sci.* 103(1):10-15 (1967).

Azevedo, J., and P. R. Stout. "Farm Animal Manures: An Overview of Their Role in the Agricultural Environment," *Calif. Agric. Exp. Sta., Ext. Serv. Man.* 44:109 (1974).

Ballard, R., and J. G. A. Fiskell. "Phosphorus Retention in Coastal Plain Forest Soils: I. Relationship to Soil Properties," *Soil Sci. Soc. Am. Proc.* 38(2):250-255 (1974).

Barbier, G., and J. Chabannes. "Sur la retention de l'ion SO_4 dans les soils," *Compt. Rend. Acad. Sci.,* Paris 218:519-521 (1944).

Bingham, F. T. "Phosphorus," in *Diagnostic Criteria for Plants and Soils,* H. D. Chapman, Ed., University of California, Div. of Agric. Sci. (1966), p. 324.

Black, C. A. *Soil-Plant Relationships*, 2nd ed. (New York: John Wiley & Sons, Inc., 1968), pp. 558-653.

Brady, N. C. *The Nature and Properties of Soils,* 8th ed. (New York: Macmillian Publishing Co., Inc., 1974), pp. 465-475.

Brown, G. G. "Alfalfa Fertilizers," *Oregon Sta. Bull.* 141:55-56 (1917).

Chancrin, E., and A. Desriot. "Action of Sulphur as a Fertilizer," *J. Agric. Pract.,* n. ser, 21(14):427-429 (1912).

Chao, T. T., M. E. Harward and S. C. Fang. "Adsorption and Desorption, Phenomena of Sulfate Ion in Soils," *Soil Sci. Soc. Am. Prac.* 26:234-237 (1962).

Cosgrove, D. J. "Metabolism of Organic Phosphates in Soil," in *Soil Biochemistry,* A. D. McLaren and G. H. Peterson, Eds. (New York: Marcel Dekker, Inc., 1967).

Eaton, F. M. "Sulfur," in *Diagnostic Criteria for Plants and Soils,* H. D. Chapman, Ed., Univ. of California, Div. of Agric. Sci. (1966), pp. 444-475.

Elliot, L. F., Ed. *Soils for Management of Organic Wastes and Waste Waters.* (Madison, WI: Soil Science Society of America, ASA and CSSA, Inc., 1977), p 650.

Elseewi, A. A., A. L. Page and F. T. Bingham. "Availability of Sulfur in Sewage Sludge to Plants: A Comprehensive Study," *J. Environ. Qual.* 7(2):213-217 (1978).

Fiskell, J. G. A., and H. F. Perkins. "Selected Coastal Plain Soil Properties," *South. Coop. Series Bull.* 148 (1973).

Frederick, L. R., R. L. Starkey and W. Segel. "Decomposability of Some Organic Sulfur Compounds in Soil, *Soil Sci. Soc. Am. Proc.* 21:287-292 (1957).

Freney, J. R. "Sulfur-Containing Organic," in *Soil Biochemistry,* A. D. McLaren, Ed. (New York: Marcel Dekker, Inc., 1967), pp. 229-253.

Gleen, H. "Microbial Oxidation of Ammonia and Thiocyanate Ions in Soil," *Nature,* 168-117-118 (1951).

Harter, R. D. "Phosphorus Adsorption Sites in Soils," *Soil Sci. Soc. Am. Proc.* 33:630-632 (1969).

Harward, M. E., and H. M. Reisenauer. "Reactions and Movement of Inorganic Soil Sulfur," *Soil Sci.* 101:326-335 (1966).

Hinesly, T. D., R. E. Thomas and R. G. Stevens. "Environmental Changes from Long-Term Land Applications of Municipal Effluents," USEPA Technical Report 430/9-78-003 (1978).

Illinois Agricultural Experimental Station. "Sulfur has Little Value as Fertilizer in Illinois," 43rd Annual Report 1929-30 (1929), pp. 33-34.

Jackman, R. H. "Organic Phosphorus in New Zealand Soils Under Pasture. II. Relation Between Organic Phosphorus Content and Some Soil Characteristics," *Soil Sci.* 79:292-299 (1955).

Khaleel, R., K. R. Reddy and M. R. Overcash. "Transport of Potential Pollutants in Runoff Water from Land Areas Receiving Animal Wastes: A Review," Department of Biological and Agricultural Engineering, North Carolina State University, Raleigh, NC (1978).

Lint, H. C. "The Influence of Sulphur on Soil Acidity," *J. Ind. Eng. Chem.* 6(9):747-748 (1914).

Lipman, J. G., A. L. Prince and A. W. Blair. "The Influence of Varying Amounts of Sulfur in the Soil on Crop Yields, Hydrogen-Ion Concentration, Lime Requirement, and Nitrate Formation," *Soil Sci.* 12(3):197-207 (1921).

Lipman, J. G., and H. C. McLean. "Agricultural Aspects of Sulfur and Sulfur Compounds," *Chem. Met. Eng.* 38:394-396 (1931).

McKee, J. E., and H. W. Wolf, Eds. "Water Quality Criteria," Resources Agency of California State Water Quality Board Publication 3A, (1963), p. 548.

McLaren, A. D., and G. H. Peterson, *Soil Biochemistry* (New York: Marcel Dekker, Inc., 1967), pp. 229-253.

Olson, R. A., Ed. *Fertilizer Technology and Use* (Madison, WI: Soil Science Society of America, 1971) p. 611.

Olsen, S. R., and F. S. Watanabe. "A Method to Determine a Phosphorus Adsorption Maximum of Soils as Measured by the Langmuir Isotherm," *Soil Sci. Soc. Am. Proc.* 21:144-149 (1957).

Olsen, S. R., and L. A. Dean. "Phosphorus," in *Methods of Soil Analysis,* Part 2, C. A. Black, Ed. (Madison, WI: American Society of Agronomy, 1965), pp. 1035-1049.

Parks, W. L., and W. B. Fisher, Jr. "Influence of Soil Temperature and Nitrogen on Rye Grass Growth and Chemical Composition," *Soil Sci. Soc. Am. Proc.* 22:257-260 (1958).

Rabate, E. "The Employment of Sulphuric Acid for the Destruction of Weeds in the Wheat Field," *Prog. Agr. et Vit.* (Ed. l'Est-Centre). V. 33(44):568-572; V. 33(45):591-595; and V. 33(46):629-636 (1912).

Rantsek, J. C. "The Effect of Sulfur on Growth of Roses in Alkaline Soil," *Am. Soc. Hort. Sci. Proc.* 36:973-977 (1943).

Reddy, K. R., *et al.* Biological and Agricultural Engineering Dept., North Carolina State University, Personal communication (1978).

Reddy, K. R., R. Khaleel, M. R. Overcash and P. W. Westerman. "Evaluation of Nitrogen and Phosphorus Transformations in the Soil-Manure System in Relation to Nonpoint Source Pollution," *Agron. Abst.* (1977), p.35.

Reimer, F. C. "Sulphur as a Fertilizer for Alfalfa," *Pacific Rural Press,* 87(26):717 (1914).

Rennie, D. A. "Adsorption of Phosphorus by Four Saskatchewan Soils," *Can. J. Soil Sci.* 39:64-75 (1958).

Reynolds, E. B. "The Effect of Sulfur on Yield of Certain Crops," *Texas Sta. Bull.* 408:24 (1930).

Sawhney, B. L., and D. E. Hill. "Phosphorus Sorption Characteristics of Soils Treated with Domestic Wastewater," *J. Environ. Qual.* 4(3):342-346 (1975).

Scharpenseel, H. W., and R. M. Krausse. *Pfl.-Dueng. Bodenk* 101:11. (CA 59, 9267) (1963).

Shedd, O. M. "The Relation of Sulphur to Soil Fertility," *Kentucky Sta. Bull.* 188:595-630 (1914).

Sommers, L. E. "Chemical Composition of Sewage Sludges and Analysis of Their Potential use as Fertilizers," *J. Environ. Qual.* 6(2):225-231 (1977).

Spulnik, J. B., R. E. Stephenson, W. E. Caldwell and W. B. Bollen. "Effect of Waste Sulfite Liquor on Soil Properties and Plant Growth," *Soil Sci.* 49(1):37-49 (1940).

Steinberg, R. A. "Sulfur and Trace Element Nutrition of *Aspergillus niger. J. Agric. Res.* 63:109-127 (1941).

Voights, D. "Lagooning and Spray Disposal of Netural Sulphite Semichemical Pulp Mill Liquors," *Proc. 10th Ind. Waste Conf.* Purdue Univ., Ext. Series No. 89 (1955), pp. 497-507.

Westerman, P. W., M. R. Overcash, J. C. Burns, L. D. King and F. J. Humenik, "Long-Term Fescue and Coastal Bermudagrass Crop Response to Swine Lagoon Effluent," paper presented at ASAE Meetings, North Carolina State University, June, 1977.

Wisniewski, T. F., A. J. Wiley and B. F. Lueck. "Ponding and Soil Filtration for Disposal of Spent Sulphite Liquor in Wisconsin," *Proc. 10th Ind. Waste Conf.* Ext. Series No. 89 (1955).

Young, H. E. "The Effect of Acidification of Alkaline Nursery Soils for the Production of Exotic Pines," *Queensland Agric. J.* 50:585-600.

CHAPTER 6

OIL AND GREASE CONSTITUENTS OF INDUSTRIAL WASTES

INTRODUCTION

In a wide variety of wastewaters and sludges there are measurable concentrations of oils and greases; hence, the evaluation and design of land treatment systems for complex industrial wastes must include the impact of these constituents. The levels of oil in industrial waste can range from a few parts per million to several percent, so one must establish the soil assimilatory capacity for oils at a particular site to analyze for the land limiting constituent (LLC) in a particular industrial waste. Obviously, oils will control the land area requirements only for certain wastewaters or sludge; while for many other wastes, the oil-based constituents will be adequately assimilated. In addition to controlled application of oil constituents in an industrial waste, there are numerous spills and dump areas that can be reclaimed utilizing the principles established for maintaining an environmentally acceptable land application system.

The concentration and quantity of oils in an industrial waste are rapidly being regulated when the terminal receiver is a stream or surface water discharge. As a waste from both the petroleum and the agricultural processing industries, the oil-based materials are plentiful and, therefore, are well scrutinized in environmental criteria or in surcharge rates prior to treatment. Also, it is estimated that 50% of the 2.5 billion gallons of lubricating oils and greases used annually in the U.S., ultimately become wastes, such as spent crankcase oil and lubrication losses during machine usage. For many of the animal or vegetable-derived oils, a considerable amount is waste after use. One inference of the magnitude of oil-containing wastes is in the present interest by and from inquiries of the U.S. Congress into the economics and legal factors necessary to establish a waste oil industry. With an understanding of the interactions of oils and greases in the plant-soil system, the environmental engineer can evaluate critically the land application alternative for a wide variety of processing and end-use industries or sources of oil-containing wastes.

This chapter describes first the general sources and characteristics of oil-based constituents in industrial wastes. Next the fundamental pathways for assimilation in the plant-soil system are delineated with additional information concerning the management and geoclimatic variables that affect assimilatory capacity. The relation of this information to the total design methodology is given in Figure 6.1. Attention is directed to the response of soil microflora and the effect on crop systems on the land application site. This information provides the basis for the procedures, constraints and criteria in the design of plant-soil systems to effectively treat oil-containing wastes.

PROBABLE SOURCES AND CHARACTERISTICS OF OIL CONSTITUENTS IN INDUSTRIAL WASTES

Oils and grease are generated from three principal sources: (1) crude oil, subsequent petroleum refining and the use of petroleum products; (2) animal-based fats; and (3) vegetable-extracted oils. The basic compositions of these oil and grease sources differ substantially. Petroleum-based constituents are primarily alkanes or paraffins, ranging from 1-40 carbons in size. Substantial cycloalkanes (naphthenes) are also present in certain crudes. By contrast, oils from animals, fish or vegetable sources are esters of glycerol (glycerides). Oils and fats from these various animal and vegetable sources differ in the proportions of carboxylic acids that form the glycerides as well as the degree of hydrogen saturation of these fatty acids (Table 6.1).

The industrial sources of the aninal and vegetable oils and the petroleum-based oils can be included in several general categories. Initial processing of agricultural products generates oils and grease that is often partially recovered and recycled, through rendering facilities, as animal feeds. The unrecovered material appears in the waste stream to be land applied. Waste from a fish processing plant contained 520-13,700 mg/l of fish oil (Chun et al. 1968). Animal and vegetable oils are also a part of industrial wastes from processes utilizing these oils, such as cooking and mass food production, soap manufacturing, margarine production, hydrogenation, drying oils and wax processes.

Petroleum-based oils from the refining and transportation of crude oil and related products are plentiful. Even with considerable treatment of refining wastes and a high degree of product recovery, the oil and grease content of effluent and sludges remains substantial, thus becoming a factor in the design

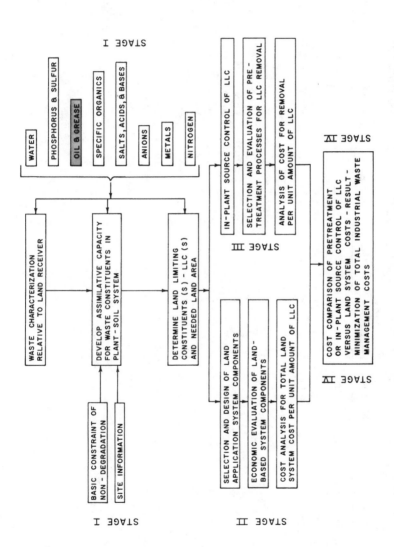

Figure 6.1 Relationship of Chapter 6 to overall design methodology for industrial pretreatment-land application systems.

Table 6.1. Fatty Acid Composition of Fats and Oils (Morrison and Boyd 1962)

Fat or Oil	Saturated Acids							Unsaturated Acids					
								Enoic				Dienoic	Trienoic
	C_8	C_{10}	C_{12}	C_{14}	C_{16}	C_{18}	$>C_{18}$	$<C_{16}$	C_{16}	C_{18}	$>C_{18}$	C_{18}	C_{18}
Beef tallow			0.2	2-3	25-30	21-26	0.4-1	0.5	2-3	39-42	0.3	2	
Butter	1-2[a]	2-3	1-4	8-13	25-32	8-13	0.4-2	1-2	2-5	22-29	0.2-1.5	3	
Coconut	5-9	4-10	44-51	13-18	7-10	1-4				5-8	0-1	1-3	
Corn				0-2	8-10	1-4			1-2	30-50	0-2	34-56	
Cottonseed				0-3	17-23	1-3				23-44	0-1	34-55	
Lard				1	25-30	12-16		0.2	2-5	41-51	2-3	3-8	
Olive			0-1	0-2	7-20	1-3				53-86	0-3	4-22	
Palm				1-6	32-47	1-6	0-1		1-3	40-52		2-11	
Palm Kernel	2-4	3-7	45-52	14-19	6-9	1-3	1-3		0-1	10-18		1-2	
Peanut				0.5	6-11	3-6	5-10		1-2	39-66		17-38	
Soybean				0.3	7-11	2-5	1-3		0-1	22-34		50-60	2-10
Cod Liver				2-6	7-14	0-1	0.5-1	0-2	10-20	25-31	C_{20} 25-32, $>C_{20}$ 10-20		
Linseed				0.2	5-9	4-7				9-29		8-29	45-67
Tung										4-13		8-15	b

[a] 3-4% C_4, 1-2% C_6.

[b] 72-82% eleostearic acid, *cis, trans, trans*-9, 11, 13-octadecatrienoic acid, and 3-6% saturated acids.

wastes. The usual leaking gaskets, motors and lubricating devices are the sources of petroleum-based oils in processing wastes. Spent or replaced oil and grease is also generated from many diverse industrial sources and, consequently, is present in waste effluent or sludges.

In comparing the information on the plant-soil assimilative capacity for the three major types of oils from industrial sources, most research and field experiments have been for petroleum-based oils. On an *a priori* basis, the microbial processes in a soil system would be assumed to break down the glyceride and fatty acid constituents of an animal or vegetable oil at a higher rate than the alkane, cycloalkane and aromatic mixture of a petroleum-based oil. This assumption is based on the probable microbial pathways in which alkanes must first be transformed into fatty acids before complete aerobic decomposition. A more detailed discussion follows later. Thus, in translating concepts, data and design criteria obtained from petroleum-based oils to wastes containing vegetable or animal oils, the latter would be assumed more degradable or capable of land application at the same or higher rates.

The chemical composition of petroleum-based oils is a useful starting point in characterizing oils from the wide variety of industrial waste sources because much of the waste oil is from leakage or has undergone only partial changes during use. Additionally, much information about elemental content is available on crude and refined oils. Characterizations of petroleum oils are presented in Tables 6.2-6.5 for the hydrocarbon classification, elemental content, physical properties and metals composition. Petroleum oils are complex mixtures of hydrocarbons with relatively small amounts of impurities or contaminants such as nitrogen, sulfur and metals. This large ratio of carbon to essential microbial nutrients (N, P, K, S, etc.) has a definite impact on a soil system and must be accounted for in the land application design. The implications and design criteria of the heavy metals in oils are dealt with in a later chapter because the origin of the metals in the total waste is not a factor in determining plant-soil assimilative capacity. General composition and elemental content of petroleum-based oil wastes will be considered similar to those presented for crude and refined oils.

Refinery effluent and sludges are a substantial subset of oil-containing industrial wastes and exemplify the characterization and response of oils in a plant-soil system. Typical characteristics of refinery wastes are summarized in Table 6.6, in which the average value of oil content for 12 refineries was 57 ppm (w/v) with a range of 23-130 ppm (w/v). The composition of this oil fraction is presented in Table 6.7 showing an extractable oil content of 15-39%. The petroleum refineries may also vary considerably in their daily waste composition (Bell 1959, Davis 1967), depending on the efficiency of waste treatment process. The various sources of wastes in a typical refinery are listed in Table 6.8.

Table 6.2 Average Amounts of Major Classes of Hydrocarbons
and Related Compounds Present in Different Petroleums
and Gasolines (ZoBell 1973)

Component	Percentages in	
	Petroleums	Gasolines
Aliphatic or Paraffinic Alkanes		
Hexadecane	15-30	25-68
Cycloparaffinic Cycloalkanes; Naphthenes	30-50	5-24
Cyclohexane		
Cyclohexene		
n-butyl cyclohexane		
Aromatic Benzene and Polynuclear Series	5-20	7-55
Benzene		
Asphaltic Asphaltenes; Heterocyclic Compounds	2-15	0.1-0.5
with Oxygen, Sulfur or Nitrogen		
Naphthalene		
Anthracene		
Olefinic Alkenes or Ethylene Series		
Hexadecane	nil	0-41

Table 6.3 Characteristics of Petroleum Meterials (Plice 1948)

Material	Specific Gravity	Volatile 105°C (%)	Nitrogen (%)	Carbon (%)	Sulphur (%)	C:N
Paraffin Base	0.81	65.6	0.01	85.1	0.07	8,500
Asphalt Base	0.84	22.8	0.05	86.6	0.23	1,730
Basic Sediment	0.86	37.8	0.03	87.2	0.65	2,900

Oil Composition of Refinery Wastes	
	Ponca City (wt %)
Paraffins	21
Aromatics	49
Resin-Asphaltenes	30

Table 6.4 Characterization of Oils Applied to Experimental Plots (Raymond *et al.* 1976)

Parameter	Used Crankcase Oils	Arabian Heavy Crude Oil	Gulf Coast Mix Crude Oil	Home Heating Fuel Oil No. 2	Residual Fuel Oil, No. 6
Gravity, API/60°F	–	27.4	23.6	35.0	16.0
Viscosity, SUS/77°F	–	164.0	324.0	–	–
Viscosity, SUS/100°F	–	103.0	161.0	–	–
Silica Gel Fraction (%)					
Heptane	70.5	45.8	53.6	53.0	37.1
CCl$_4$	9.5	13.3	9.7	3.2	11.2
Benzene	8.6	26.4	22.8	34.7	35.0
Methanol	11.3	14.4	13.8	10.1	16.7
Sulfur	0.3	3.01	0.38	0.15	0.66

Table 6.5 Metals in Oils in Units of μg Metal/g Oil (Raymond *et al.* 1976)

Metal	Used Crankcase Oil Car (μg/g)	Truck (μg/g)	Arabian Heavy Crude (μg/g)	Gulf Coast Crude (μg/g)	No. 6 Oil (μg/g)
Ag	–	–	–	–	–
Al	22	10	–	–	–
B	10	–	–	–	–
Ba	480	36	–	–	–
Cr	21	3	–	–	–
Cu	17	18	–	–	3
Mn	3	–	–	–	2
Mo	10	4	–	–	–
Ni	–	–	23	2	13
Pb	7,500	75	–	–	–
Sn	5	3	–	–	–
V	–	–	65	–	38
Zn	1,500	1,300	21	17	12
Ca	700	780	–[a]	21	17
Mg	1,600	1,100	–	–	21
Na	51	39	–	9	27
P	1,400	1,350	–	–	–
Fe	260	180	3	5	18

Table 6.6 Typical Refinery Waste Characteristics[a] (McKinney 1963)

	Minimum	Maximum	Average
Flow, mgd	1.04	2.62	1.58
Temperature, $^\circ$F	69	100	88
Suspended Solids, mg/l	80	450	350
Sulfides, mg/l	1.3	38	8.8
Phenol, (mg/l)	7.6	61	27
BOD, 5 days, mg/l	97	280	160
COD, mg/l	140	640	320
pH	7.1	9.5	8.4
Alkalinity, mg/l, as $CaCO_3$	77	210	180
Oil, mg/l	23	130	57
Ammonia Nitrogen, mg/l	56	120	87
Phosphate, mg/l	20	97	49
Chromium, mg/l	0.3	0.7	0.5
Chlorides, mg/l	200	960	310

[a]Averages of the minimum, maximum and average values reported by 12 refineries in the API questionnaire on biological waste disposal.

Typical effluent stream requirements for the treated effluent from a petroleum refinery are 15 ppm oil (Bell 1959) with phenols < 0.2 ppm, mercaptans < 0.5 ppm and sulfides < 0.5 ppm. At this concentration, if one takes a land application rate of 10 metric ton/ha, the calculated oil application rate is 0.15 kg/ha of oil. In a sludge with 150 ppm oil, the total oil present in 10 tonnes of waste would be about 1.5 kg. These are extremely low rates for land application indicating that far less pretreatment is needed prior to a plant-soil receiver when compared to that necessary for stream discharge.

In summary, either from petroleum bases or derived from animal or vegetable sources, oils are quite ubiquitous in the industrial wastes from processing and refining as well as from end use in manufacturing processes. The composition of the crude or processed oil is representative of the elemental and compound content of these oils in various effluents and sludges. Utilizing this information, the response of the plant-soil system can be interpreted in establishing the design procedures and criteria for the land treatment of oily wastes.

INTERACTION WITH SOIL SYSTEM

Amendment of soils with a hydrophobic waste causes a change in the physical, chemical and biological processes of a soil system.

Table 6.7 Approximate Composition of Oily Waste From Refinery
(Huddleston and Cresswell 1976)

	Ponce City Oklahoma		Billings Montana
Extractable Oil	39	wt %	15
Water	24		72
Solids	37		13
Total Organic Carbon	53		10
Metal Content		mg/kg	
Silicon	21,000		12,900
Calcium	13,000		12,000
Aluminum	3,200		10,900
Iron	3,000		4,300
Lead	1,400		1,900
Magnesium	880		890
Nitrogen	–		–
Sodium	870		810
Phosphorus	–		–
Potassium	480		–
Zinc	400		690
Chromium	370		–
Barium	240		370
Strontium	180		–
Copper	150		360
Titanium	150		–
Manganese	130		–
Lead	–		50
Nickel	<21		<10
Vanadium	<21		<10
Cadmium	< 3		–

Table 6.8 Sources of Oily Wastes in Refineries (Huddleston and Cresswell 1976)

	Typical Weight and Composition		
	Water	Solids	Oil
Crude Oil Tank Bottoms	15	45	40
Slop Tank Bottoms	64	4	32
API, CPI Separator Bottoms	36	47	17
Biotreatment Solids	96	4	0.4
Coker Blowdown	20	40	40
Water Impoundment Pond Bottoms	40	40	20

Figure 6.2 shows the general pathways of oil disappearance and impacts on the soil system. In overview, oil decomposition constitutes a part of the general soil microbial carbon cycle and not only depends on the nitrogen availability but also influences the transformations of nitrogen, phosphorus, sulfur and other mineral nutrients. In the short term, volatile fractions of added oil can be lost to the atmosphere while the nonvolatile fractions tend to concentrate and solidify initially. These oil fractions are then available for decomposition by microorganisms, despite their initial germicidal effects. In the long term, oils and greases are assimilated in the production of biomass and generation of aerobic end products—carbon dioxide and water. Biodegradation of oils may also lead to improved soil properties.

The prominent effects of oil/soil interaction are physical, chemical and biological properties changes, which are described in the following sections.

Physical Effects

The immediate effect of an oil spill or application on land is to decrease gaseous exchange between soil and air (Stebbings 1970). Oil-saturated soils are not wetted by rainfall easily; hence, at high loading rates, the rainfall runoff characteristics can be changed initially to increase runoff probability (Plice 1948). As noted in the section on long-term effects, soil properties such as decreased bulk density from the assimilation of oils leads to reduced runoff conditions. Fresh applications of oil in soils may impede the water transmission through soils, especially at the soil surface. This is due to (1) hydrophobic nature of oils that form fresh oil-soil mixtures, and (2) the initial accumulation of hydrophobic mucilaginous substances associated with increased microbial activity.

These hydrophobic effects tend to be short term. An application of waste containing oil on a land site results in spreading over an area. For the flow of oil in a porous and permeable soil, the surface tension forces are assumed to be far more predominant than the gravitational forces (Raisbeck 1972). Fluid dynamic principles can be applied to predict the movement of oil in a homogeneous unsaturated soil. Raisbeck (1972) verified the equation:

$$Qv = K t^{-c}$$

where Qv = rate of increase in soil volume, cm^3/min inundated with oil at time t, min

K = a constant

c = $\dfrac{n + 1 - (1/\chi)}{n + 2 - (1/\chi)}$

n, χ = constants dependent on soil properties

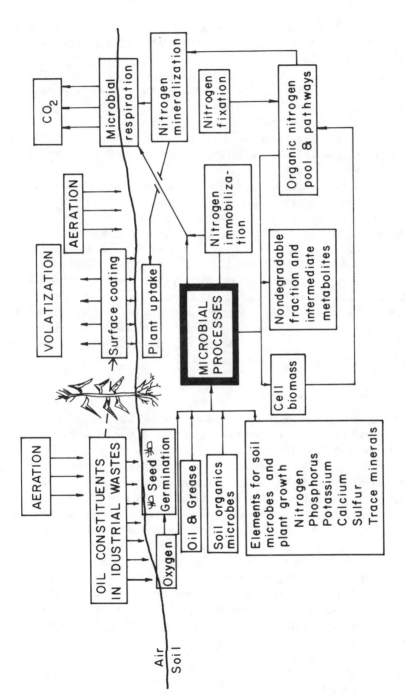

Figure 6.2 Interactions of oily wastes with soil-plant systems.

Brooks (1970) determined the value of n between 3 and 4 for many soils. Taking value of χ between 1 and 2 (Raisbeck 1974), the calculated value of c for soil varies between 0.75 and 0.82.

The equation $Qv = Kt^{-c}$ is valid for movement of oils and kerosene in sandy loam soils (Raisbeck 1974). In warm weather, oil moves vertically in the unsaturated water zone and laterally in the capillary fringe. In cold weather, the principal mode of oil migration is lateral spreading. For reasonable levels of oil application, little leaching occurs, and after being incorporated to 10-15 cm of soil depth, the oil remains in the upper 20 cm of soil (Raymond et al. 1976).

A few months after oil application or spill, the microbial degradation of oils may lead to improved soil physical conditions (Plice 1948). Giddens (1976) reported an increase in water-stable aggregates from 12.9 to 64.4% after the degradation of spent motor oil in a Cecil sandy loam soil when the oil application rate was nearly 6.5% of soil weight. Ellis and Adams, Jr. (1961) mentioned that soil saturated with petroleum gas had lower bulk density and higher water retention capacity then untreated control soils. The significant alterations in soil physical properties after oil additions are described in Table 6.9. For soil physical properties, long-term beneficial effects are the result of biological activity decomposing the applied oil. However, in the short term (2-8 weeks), certain soil physical properties can be adversely changed by oil addition.

Chemical Effects

Crude oil is made up largely of hydrocarbons with a very high C:N ratio. Addition of these carbonaceous oils (Table 6.10) to soils leads to an increase in the soil C:N ratio (Plice 1948). The same holds true with vegetable oils and animal fats when applied to soils as a waste effluent or sludge, unamended with nitrogen. With progressive decomposition of oils, the C:N ratio of the system decreases to between 20:1 and 10:1.

In general, there is a mineral nitrogen deficiency in soils receiving oils, but other essential elements such as P and Ca may also become limiting for plant uptake. The salient changes in soil chemical properties by oil application are given in Table 6.11. By oil pollution of a Nigerian soil, the amount of organic carbon, total nitrogen, exchangeable K, Fe and Mn were increased (Udo and Fayemi, 1975) during decomposition, but levels of extractable nitrate and phosphate and exchangeable calcium decreased. Analogous to these observations, Baldwin (1922) and Murphy (1929) also documented decreased nitrate production in oil polluted soil, at least initially. However, with progressive deacy of oils, there is an increase in total nitrogen and organic matter of soils (Plice 1948) (Table 6.10). The initial apparent decrease in nitrate production may have resulted from nitrogen immobilization by the hydrocarbon metabolizing microbes.

Table 6.9 Influence of Oils and Oily Wastes Decomposition on Soil Physical Properties

Physical Characteristics	Oil Application Rate			Experimental Condition	Remarkable Influence	Reference
	Reporter's Unit	Equivalent in kg/ha	% Soil Wt.			
1. Aggregation	12716 l/ha	10173	0.45	Spent motor oil on Cecil sandy loam—Field Experiment	Increase from 12.9 to 64.4% water stable aggregate	Giddens (1976)
2. Bulk Density	Gas saturation			Gas saturated soils versus normal controls of 0-6 in. and 6-12 in. depths.	Decrease from 1.3 to 0.7 or 1.0 g/cm^3	Ellis, Jr., and Adams, Jr. (1961)
3. Soil Porosity	Gas saturation			Gas saturated soils versus normal controls of 0-6 in. and 6-12 in. depths.	Increase in micropores.	Ellis, Jr. and Adams, Jr. (1961)
4. Water-holding capacity and retention	Gas saturation			Gas saturated soils versus normal controls of 0-6 and 6-12 in. depths.	Marked increase in water-holding capacity and available water.	Ellis, Jr., and Adams, Jr. (1961)
5. Water Infiltration and Transmission	Crude oil				Decrease initially until all the oil transforms to genuine organic matter.	
6. Wind Erodibility	Deeply oiled-oil saturation			Heavy clay and fine sandy loam	Increase in cultivated plots in hot and windy weather	Plice (1948)
7. Overall Soil Physical Condition	Any significant rate			Any soil with scope for oil degradation.	Improves soil tilth and friability after complete oil decay.	Plice (1948)

Table 6.10 Organic Matter and Nitrogen Content of Soils
After Petroleum Oil Application and Incubation (Plice, 1948)

Treatment	Organic Matter (%)	Carbon (%)	Total Nitrogen (%)	C:N After Oil Decay
Check Plots (control)	1.74	1.01	0.05	20
Shallow-Oiled Plots				
Light, 0.1%	1.97	1.14	0.06	19
Medium, 0.5%	2.66	1.54	0.07	22
Heavy, 1.0%	3.71	2.15	0.08	27
Oil-Saturated Plots				
Cultivated	8.23	4.77	0.1	48
Uncultivated	11.45	6.64	0.12	55

Petroleum decomposition in soils led to reducing conditions and lowering of the redox potential into the negative range (Plice 1948, Davis 1967, Jones and Edington 1968). After 18 months of oil decomposition, Kincannon (1972) recorded increased cation exchange capacity of soil from 10 meq to > 30 meq/100 g soil. As a result of increased exchange capacity of the soil system by decomposition of hydrocarbons and fatty acid glycerides, the soil acid/base buffering capacity is increased.

Initial Biological Response

Petroleum Versus Vegetable and Animal Oils

Both hydrocarbon-containing petroleum residues and glyceride-containing animal and vegetable oils are decomposed in a soil system along similar pathways (Figure 6.1). Hydrocarbons degrade to carbon dioxide and water via several intermediates (organic acids, ketones, aldehydes, alcohols and other hydrocarbon derivatives). The pathway of alkaline oxidation occurs as follows:

$$RCH_2CH_3 \rightarrow RCH_2CH_2OOH \rightarrow RCH_2OH \rightarrow RCH_2CHO \rightarrow RCH_2COOH$$

The straight-chain hydrocarbons thus undergo oxidation with formation of alcohols, aldehydes and acids. The n-alkanes are readily metabolized and the primary end products are cell constituents. With few exceptions, the oxidation of alkanes parallels fatty acid metabolism (Figure 6.3) resulting in cell composition changes, particularly the lipid fraction which are directly related to the initial structure of the growth substrate.

The rate at which a molecule undergoes biodegradation depends on its structural relationship to naturally occurring compounds (Gibson 1974). The most common biological alkanes are C_7-C_{36} and methane. Branched chain paraffins with methyl groups occur naturally in waxes of tobacco leaves, wool and sugarcane. It is estimated that 0.02% of higher plants may be considered hydrocarbons. Because of their biological origin and nature, many alkanes are utilized as substrates for growth by many micoorganisms. Both saturated and unsaturated hydrocarbons are attacked. Long-chain hydrocarbons (C_6-C_{28}) are more vulnerable to attack than short chain (C_2-C_5). Thus, increasing molecular weight in the C_6 to C_{28} range favors decomposition. The unsaturated forms are less easily degraded than the saturated molecules and the branched compounds exhibit a lower susceptibility to the action of microbial enzymes than straight-chain hydrocarbons (ZoBell 1973). However, decreased water solubility reduces somewhat this trend of greater susceptibility at higher carbon numbers.

Vegetable oils and animal fats are hydrolyzed by enzyme lipase to produce free fatty acids and glycerol. Glycerol can be metabolized by microbes through the glycolytic pathway, whereas fatty acids are broken down two carbon atoms at a time to provide units of acetyl coenzyme A in fueling the citric acid cycle and the respiratory chain. The fatty acid oxidation by β-oxidation in microbial cells is shown in Figure 6.3.

Relative rates of microbial decomposition of vegetable and mineral oils can be partially compared from available literature data. As the application rate of oil increases in soils at a given fertility level, the percent carbon loss decreases (Table 6.12). There is adequate experimental evidence to illustrate that vegetable oils (triglycerides) are rapidly degraded and require supplemental nitrogen for microbial growth. Because of the chemical and physical similarities with vegetable oils, animal fats (triglycerides) are also expected to be as rapidly metabolized by soil microbes. Petroleum and mineral oils are also attacked by soil microorganisms but their decomposition is variable and somewhat less rapid than vegetable oils (Table 6.13). Not all fractions of crude or waxy raffinate decompose at the same rate. Volatile fractions are lost rapidly (within the first week) by evaporation where, as among nonvolatile fractions, paraffins are degraded at a much faster rate than aromatics, heterocyclic naphthenes and naphthenic acids.

A recent experiment under identical conditions compared the decomposition of vegetable, animal and petroleum oils, as measured by carbon dioxide evolution (Bingham 1977). These soil-oil mixtures were one percent oil of soil weight and were held at room temperature. Results of CO_2 evolution after 21 days were that the animal- and vegetable-based oils were similar in

Table 6.11 Influence of Oils and Oily Wastes Decomposition on Soil Chemical Properties

Soil Chemical Properties	Oil Application Rate			Experimental Condition	Remarkable Influence	Reference
	Reporter's Unit	Equivalent in kg/ha	% Soil wt			
1. Total Organic Matter	1%	22,400	1	Shallow oiled-top 15 cm	Increase % OM from 1.74 to 3.7 in soil with increasing rate of oil added.	Plice (1948)
2. Total Nitrogen	1%	22,400	1	Shallow oiled-top 15 cm	Increase from 0.05 to 0.08% N.	Plice (1948)
3. C:N Ratio				Heavy clay and fine sandy loam	Initially considerable increase which decreases as oil is lost.	Plice (1948) Odu (1972)
4. Redox Potential	Gas-saturated soil				Decrease by high oxygen demand.	Ellis, Jr. and Adams, Jr. (1961) Plice (1948)
5. pH	Gassed soil				Tends to become neutral by increased buffering.	Skinner (1973)
6. Cation Exchange Capacity	23-44% of soil wt	716,800 (average)	32 (average)	Field Experiment at Deer Park, Texas (Houston Ship Channel)	Increase from 10 to > 30 meq/100 g soil.	Kincannon (1972)
7. Available Nutrients	9333 l/ha -31111 l/ha	7466 -24,889	0.33 1.11		Increase Mn, Fe and K solubility but initially decrease in availability of Ca and N. Phosphorus availability may either increase or decrease in short run after oil application, deficiency of N, Mn is during short term, but in long term increase in availability of Mn, Zn. No effect on P, K, Ca, Mg and Pb content of corn plants.	Ellis, Jr., and Adams, Jr. Giddens (1976)

1.1-10.6%	24,640 to 237,440	1.1 to 10.6	Crude oil –Ferruginous tropical soil receiving 90 kg N, 67 kg P, and 45 kg K per ha	Organic C, total N, exch. K, exch. Fe, and exch. Mn increased in soil, while extractable P, NO_3-N and exch. Ca were decreased with increasing oil application rate.	Udo and Fayemi (1975)

soil decomposition capabilities. The petroleum oil was much lower in rate of degradation, thus verifying the *a priori* hypothesis regarding relative decomposition rates.

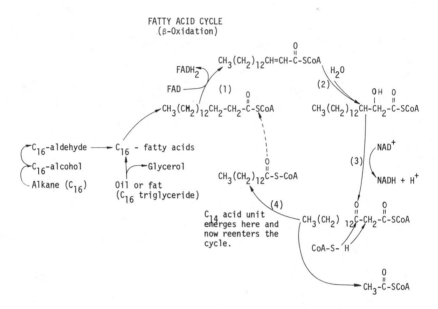

Figure 6.3 Fatty acid metabolism by β-oxidation.

Table 6.12 Microbial Decomposition of Vegetable and Crude Oil in Soils

% Oil of Soil Wt	Rate of Application		
	Equivalent % C(W/W)	Fraction of Added Carbon Lost	Period of Measurement
Vegetable Oil[a]			(weeks)
0.5 (W/W)	0.37	68	7
5.0 (W/W)	3.7	36	7
0.5 (W/W)	0.37	84	12
5.0 (W/W)	3.7 '	47	12
Crude oil[b]			
2.54 (V/W)	1.18	31	7
6.27 (V/W)	4.26	22	7
10.00 (V/W)	6.8	16	7

[a]Smith *et al.* (1975).
[b]Schwendinger (1968).

Table 6.13 Comparative Rates of Microbial Decomposition of
Vegetable and Petroleum-Based Oils

Period (weeks)	% of Added Carbon Lost		Waxy Raffinate
	Vegetable Oil	Crude Oil	
7	36[a]	22[b]	
12	47[a]	12.9[c]	17.7[c]
52		34.9[c]	42.3[c]

[a]Smith et al. (1975)
[b]Schwendinger (1968)
[c]Kincannon (1972)

Baldwin (1922) contended that crude petroleum has a certain amount of germicidal action in a soil system. Despite the antiseptic properties of petroleum, certain bacteria, actinomycetes and fungi in soils are capable of utilizing petroleum hydrocarbons (Jones and Edington 1968) as sources of carbon and energy. Crude petroleum additions to soils lead to a initial decrease in bacterial numbers for the first few days, which is invariably followed by a large microbial increase. The availability of readily degradable biomass from the death of microorganisms appears to be responsible for the reappearance of large numbers of soil microorganisms. With the increased bacterial populations, the oil constituents are also decomposed as microorganism species evolve. The initial germicidal lag and the subsequent elevated levels of soil microflora and fauna occur for varying periods of time, depending on the land-loading rates of oils. Representative values of this lag period prior to active microbial growth are given in Table 6.14.

The effects of hydrocarbons on various biological processes in soil systems (Table 6.15) have been investigated by many workers. Plice (1948) observed that shortly after oil treatment that total number of microorganisms decreased from 8.1-7.3 million/g soil, but two years after contamination the number increased to 94 million/g soil in cultivated plots and 110 million/g soil in uncultivated plots. Odu (1972) reported that within 6 weeks of an oil blowout in Nigeria the microbial numbers increased from 2.6 million/g soil in the control to 37.5 million/g soil in heavily oil-contaminated land. After 6 months of contamination, the population increased to 114 million/g soil. Part of the increase in microbial population is due to assimilation of substrate and hydrocarbons and partly to lethal action of some hydrocarbons on soil fauna which after death are readily metabolized by remaining soil bacteria. Raymond et al. (1976) reported significant increase in hydrocarbon utiilizing microorganisms in all plots treated with either pure hydrocarbon, n-hexadecane, or applied crude oils as growth substrate. These increases were usually sustained throughout the year. They observed that home heating oil No. 2 was the most lethal to nematode fauna at all locations studied, but coincided with the greatest increase in the subsequent hydrocarbon-utilizing microflora. Resistance of the nematocidal effects of oils varied with soil location and oil type. Huddleston and Cresswell (1976) noticed a 10-fold increase in the oil treated soil bacterial population over an 18-month period.

Table 6.14 Representative Lag Periods for Bacteria and
Actinomycetes Growth with Oil Addition to Indicated Media[a]

Lag Period (days)	Experimental Conditions	Reference
3	Motor oil medium with Chesapeake Bay water	Colwell *et al.*(1973)
1	Oil in the inorganic media	Stone *et al.* (1942)
2	Kerosene and crude oils in soils	Dubson and Wilson (1964)
3-4	Budd sandy loam soil from New Jersey in lab at 0.8-2.41 % oil of soil wt (independent of N & P)	Schwendinger (1968)
3-7 and 7-10	Brown sandy loam from Purdue (Indiana) at 1.6 and 4% oil of soil wt, respectively	Baldwin (1922)
7	Crude oil on sandy and clayey soils	Plice (1948)
7	Motor oil on sandy loam soil (0.5% wt)	Evans (1977)
7-11	Motor oil on sandy loam soil (1% wt)	Evans (1977)
11-18	Motor oil on sandy loam soil (4% wt)	Evans (1977)
<1	2% vegetable oil on sandy loam	Evans (1977)
1-2	Animal fat and vegetable oil in a sandy loam soil at a loading rate of 1%	Bingham (1977)
3-4	Motor oil on sandy loam at 1% application rate by soil wt	Bingham (1977)

[a]Fungi grow exponentially in all instances of motor, vegetable and animal oils only 11 days after incubating began (Evans 1977).

In terms of soil microbial response to oil additions, the related research demonstrates a common trend for the variety of soil types, climatological regions and oil application rates. An initial germicidal effect results in a lag period prior to microbial growth and oil decomposition. However, with the evolved populations, the numbers of microorganisms considerably exceeds initial levels and maintains the oil decomposition until the available substrates are exhausted. Thus, both the length of the lag period and the time period of high populations and oil decomposition depend on the oil-loading rates.

Table 6.15 Influence of Oils and Oily Wastes on Soil Biological Properties

Biological Property	Oil Application Rate			Experimental Condition	Remarkable Influence	Reference
	Reporter's Unit	Equivalent in kg/ha	% Soil Wt			
Total Microorganisms	0.93% by wt	20,832	0.93	Billings waste oil land-farming	Tenfold increase in oil treated soil bacterial population over 18 months	Huddleston and Cresswell (1976)
	5 ml/100 g	89,600	4	Laboratory study Brown sandy loam of Purdue under soybean	Decrease in first week by one-half followed by 6-fold increase in second week.	Baldwin (1922)
	0.5%	11,200	0.5	Silt loam corn plot–crude oil amendment	Decrease in first 3 days and then stimulation of microbial number	Baldwin (1922)
	2.0%	44,800	2	Silt loam corn plot–crude oil amendment	Decrease during first 3 days followed by marked increase.	Baldwin (1922)
	Deeply oiled			Sandy and clayey soils–crude oil effects	Decreased shortly after oil treatment only slightly but increased over 10-fold within 6 months to 2 years after oil application.	Plice (1948) Odu (1972) Raymond et al. (1976)
	11.9 m^3 oil/4000 m^2	23,800 kg/ha	1.0625%	At Marcus Hook, PA, Tulsa, OK, Corpus Christi, TX	Significant increases in hydrocarbon utilizing microbes in all treated plots	Raymond et al. (1976)
Respiration	2 ml/100 g	35,840	1.6	Crude petroleum kerosene and mineral oils	Increased respiration 4-fold in first 12 days	Dobson and Wilson (1964)

Table 6.15, continued

Biological Property	Oil Application Rate			Experimental Condition	Remarkable Influence	Reference
	Reporter's Unit	Equivalent in kg/ha	% Soil Wt			
	20 ml/100 g	358,400	16	Crude petroleum and mineral oils	Increased respiration 25-50% over control	Dobson and Wilson (1964)
	500 kg/ha	500	0.022	Waxy cake	Increased respiration 2-fold	Gudin and Syatt (1975)
				Crude oil, mixed fuels and waste sludge (9% oil)	Increased oxygen consumption from 4- to 23-fold	Antoniewski and Schaefer (1972)
		89,600	4% (W/W)	188 ppm N and 82 ppm Padded	N & Paddition increased CO_2 evolution 3- to 4-fold over 3-week period	Schwendinger (1968)
Ammonification	5 ml/100 g	89,600	4		Reduction in ammonification from 30 to 45%	Baldwin (1922)
Nitrification	5 ml/100 g	89,600	4	Brown sand loam	Decrease nearly 80% in first week	Baldwin (1922)
	1.1% or higher	24,640	>1.1	Yahola very fine sandy loam of Red Prairie Div.	Practically checked nitrate production 100%	Murphy (1929)
Immobilization	5%	112,000	5	Soybean & Palm oil in silt loam	Immobilized 688 to 695 ppm nitrogen	Davis (1965) Plice (1948)
Nitrogen fixation	1.0%	22,400	1	Shallow-oiled plots	Certain hydrocarbon-degrading microbes fix atmospheric nitrogen in soils when fertilizer N is not added	Davis (1965) Plice (1948)

Degradative Microbial Species

Most organisms that utilize hydrocarbons as substrate are obligate aerobes, few are obligate anaerobes, and very few belong to a facultative group, which can use sulfae or nitrate as terminal electron acceptor. The literature reveals little potential capabilities of denitrifying organisms in utilization and degradation of hydrocarbons.

The long-term soil system response to oil applications is to support larger than normal microbial populations. These microorganisms are evolved species capable of decomposing constituents of a particular oil material. As a measure of soil capabilities, Jones and Edington (1968) determined that a fraction (1-20%) could degrade specific petroleum-based hydrocarbons (Table 6.16). Ellis, Jr. and Adams, Jr. (1961) reviewed over 100 species of bacteria, yeasts and fungi representing 31 genera that attack one or more types of petroleum hydrocarbons in soils. At present, there are nearly 28 genera of bacteria, 30 genera of filamentous fungi and 12 genera of yeasts that are known to degrade various oil constituents (ZoBell 1973). Prominent species among these are listed in Table 6.17. Specificity of microbes for use of oil substrates depends on the prevalent conditions. For example, *Pseudomonas aeruginosa* utilized n-heptane and n-octane but cannot metabolize aromatic compounds such as benzene, xylene and toluene. By contrast, *Pseudomonas fluorescens* can metabolize all three aromatics but not the n-alkanes (Traxler 1962). *Flavobacterium* can utilize all forms of asphalt but *Bacillus* failed to make use of asphalt-G. *Pseudomonas putida* can utilize benzene, toluene and ethyl benzene as growth substrate (Gibson 1974). *Pseudomonas putida* synthesizes toluene dioxygenase enzyme, which has a broad substrate specificity with a range of relative activity taken as 100% for toluene to 46% for benzene and 5% for cymene. McKenna and Heath (1976) observed that *Pseudomonas putida* and *Flavobacterium* oxidized naphthalene at a very rapid rate whereas only *Flavobacterium* could utilize anthracene at a moderate rate.

Stone *et al.* isolated *Pseudomonas* by enrichment culture at various temperatures (20, 30 and 37°C) with ability to degrade oil. The cultures of *Pseudomonas* so isolated were nonspecific to oil type and were adaptive to the conditions and oil type added to substrate. Hass *et al.* identified *Actinomyces, Mycobacteria* and *Corynebacteria* as principal hydrocarbon utilizers. Jones and Edington (1968) prepared a summary of dominant microbial genera present in the soil and the shale attacking various hydrocarbons. Kincannon (1972) isolated four major microorganisms from soils of Deer Park, TX, that decomposed oil in simulated oil-rich sludges. These were *Corynebacterium, Flavobacterium, Pseudomonas* and *Nocardia*.

As evidence of evolved species for degradation of the oil constituent in industrial waste, bacterial seeding has been demonstrated to accelerate the soil transformation of oil. Schwendinger (1968) found bacterial increase

Table 6.16 Number of Organisms Capable of Growth on Selected Hydrocarbons

Hydrocarbon	Soil Sample Depths, cm			Shale
	5	20	40	
	Percent of total population that metabolized hydrocarbons[a]			
n-Octane	3	4	1	0.5
n-Decane	4	8	2	1-5
n-Dodecane	13	17	20	3
n-Pentadecane	7	10	10	8
n-Hexadecane	13	15	8	8

[a]Sample dilutions were either washed by centrifugation or by membrane filtration. The washed samples or membrane filters were plated on the basal salts agar, 0-5 ml hydrocarbon was introduced into the lid of each sealed plate and thus the carbon source was provided in the vapour phase. Control plates contained no hydrocarbon. Incubation was for 14 days at $25°C$. Percentage of total populations of samples capable of growth on hydrocarbons (Jones and Edington, 1968).

Table 6.17 Dominant Microbial Genera Responsible for Hydrocarbon Oxidation in Soils

Bacteria		Actinomycetes	Fungi
Achromobacter	Methanomonas	Actinomyces	Aspergillus
Aerobacillus	Methanobacterium	Debaryomyces	Aureobasidium
Alcaligenes	Micrococcus	Endomyces	
Arthrobacter	Micromonospora	Nocardia	Cephalosporium
Bacillus	Monila	Proactinomyces	Cunninghamella
Bacterium	Mycobacterium	Saccharomyces	Mycelia
Beyerinckia	Mycoplana		
Botrytis	Pusedomonas		
Candida	Sarcina		
Citrobacter	Serratia		
Cellustomonas	Apicaria		
Colostridium	Spirillium		
Corynebacterium			
Desulfvibrio	Thiobacillus		
Enterobacter	Torula		
Escherichia			
Flavobacterium	Torulopsis		
Gaffkya			
Hansenia	Trichoderma		
	Vibrio		

in the degradation of oil over control. Jobson *et al.* (1974) reported a slight acceleration in the rate of utilization of n-alkane components of chain length C_{20} to C_{25} by application of oil-utilizing bacteria to soils.

In oil-polluted soils, absolute microbial numbers decomposing oils may vary 10^2-10^4/g soil. From Coastal Plain sandy soils of North Carolina, Perry and Cerniglia (1973) isolated several hundred bacterial cultures and a number of yeasts and filamentous fungi that decomposed petroleum hydrocarbons. There are 10^2-10^4 microorganisms/g soil in these North Carolina soils that could utilize the alkanes as carbon and energy sources. Among filamentous fungi, which were most effective in hydrocarbon degradation, are *Cunningham elegans, Penicillium zonatum, Asperguillus versicolor, Cephalosporium acremonium* and *Penicillium ochochlorens*.

Thus, research has isolated specific soil microorganisms that decompose oils. Rate of breakdown depends on the particular hydrocarbon and on the degree of evolution of the total biological population at the site of land application. Acceleration of initial oil decomposition can be accomplished with seeding of oil-consuming microorganisms, whereas the long-term assimilation is more dependent on soil and climatological conditions.

Volatilization and Other Environmental Losses

Oils and greases, when applied to a plant-soil system, must be considered in terms of their nonvolatile and volatile fractions. Nonvolatile components of waste oils tend to stay tightly bound in soil (Huddleston and Cresswell 1976), while volatile fraction may escape to atmosphere depending on temperature and vapor pressure conditions (Schwendinger 1968). Within 7 weeks, the percent of total oil lost from an oil-contaminated soil varied from 46.5-72.2% by microbial degradation and volatilization (Table 6.18). More oil is lost by volatilization than by microbial degradation during the first week at all rates of oil loading and levels of soil fertility. The percent losses of oil by volatilization decreased slightly with increased loading rates, but generally stayed at approximately 37% regardless of nitrogen or phosphorus additions. This is characteristic of a relatively rapid process, compared to the microbial decomposition, and suggests that in determining land application rates the volatile fraction is rapidly lost.

Losses of oil through surface runoff from rainfall or downward leaching were evaluated (Raymond *et al.* 1976). At realistic oil concentrations (1%-2% of soil weight) no significant loss or movement of oil occurred.

Mechanisms and Rates of Degradation of Oils in Soil

Transformations of hydrocarbons in soils are mainly aerobic and oxidative microbial reactions. The methods for evaluating decomposition rates

Table 6.18 Comparison of CO_2 and Volatiles Evolved from Oil-Contaminated Soil over Seven Weeks (Schwendinger 1968)

ml oil/ kg soil	Level[a] Oil:N:P	Treatment	% Oil Evolved as CO_2 by				% Oil Evolved as Volatiles by				Total % Oil Evolved by			
			1 wk	3 wks	5 wks	7 wks	1 wk	3 wks	5 wks	7 wks	1 wk	3 wks	5 wks	7 wks
25	1-1-3	Control	3.9	17.4	24.1	29.0	27.3	27.3	40.9	40.9	31.2	44.7	66.0	69.9
		Bacteria[b]	7.5	19.7	25.9	30.2	27.3	27.3	40.9	40.9	34.8	47.0	66.8	71.1
25	1-3-1	Control	3.6	21.9	28.6	32.1	27.3	27.3	40.9	40.9	30.9	49.2	69.5	73.0
		Bacteria	6.2	22.3	28.7	32.3	27.3	27.3	40.9	40.9	33.5	49.6	69.6	73.2
63	2-2-2	Control	2.0	11.2	16.7	20.6	20.4	24.1	33.3	37.0	22.4	35.3	50.0	57.6
		Bacteria	5.9	16.8	21.0	23.7	20.4	24.1	33.3	37.0	26.3	40.9	54.3	60.7
100	3-1-1	Control	1.6	5.9	8.3	10.5	18.6	25.6	30.2	36.0	20	31.5	38.5	46.5
		Bacteria	3.6	9.9	13.1	15.2	18.6	25.6	30.2	36.0	22.2	35.5	43.3	51.2
100	3-3-3	Control	1.3	9.5	14.8	81.4	18.6	25.6	30.2	36.0	19.9	35.1	45.0	54.4
		Bacteria	4.6	15.6	19.3	21.5	18.6	25.6	30.2	36.0	23.2	41.2	39.5	57.5

[a]For oil 1, 2, 3 means 25.4, 62.7 and 100 ml oil/kg soil.
For N **1, 2**, 3 means 76.3, 188.8 and 300 ppm N added to soil.
For P 1, 2, 3 means 33.3, 82 and 130.8 ppm P added to soil.
[b]*Cellulomonas* sp.

and pathways are classified in four general categories and are delineated below to assure that data and concepts are evaluated realistically.

Products of microbial degradation of oils are carbon dioxide, water, a variety of end product chemicals and biomass. The various test procedures used for measuring microbial degradation of oils in soil-plant systems that utilize these changes are: (1) disappearance of oil (Kincannon 1972, Raymond 1976; (2) microbial growth and biomass (Cooney and Walker, 1973); (3) respiration studies such as oxygen uptake (Liu 1973, and Gudin and Syratt 1975) and carbon dioxide evolution (Jones and Edington 1968, Smith 1974, Schwendinger 1968); and (4) measuring oxidation products such as carboxyls, aldehydes and hydroxy compounds produced by microbial oxidation.

Kincannon (1972) used the disappearance of oil as a criterion for oil loss from soils amended with oil rich sludges (Table 6.19). This study does not permit distinction between oil lost by volatilization and that by microbial degradation. Kincannon (1972) summarized the rate of oil loss from control soil amended with oil-rich sludges as 8.3 kg oil per month per m^3 soil and attributed it all to microbial degradation.

Francke and Clark (1974) demonstrated a biological assimilatory process for the disposal of oil waste products in Oak Ridge, Tennessee. They calculated a degradation rate, as oil loss, of approximately 1.425% of soil wt/ month, which amounted to a loss of 61.4% of added carbon as oils over a 3-month period. In studies at Marcus Hook, Pennsylvania, at an application rate of 11.9 m^3 petroleum oils per 4000 m^2, (i.e., 1.13% of soil weight) the average loss of oils over a year ranged from 48.5%-90.0% of added oil, depending on the type of oil and soil location (Raymond et al. 1976). The maximum rate of degradation was 2.4 m^3 oil/4000 m^2/month, which is equivalent to 0.23% of soil wt/month or 2.8% of soil wt/yr.

In an experiment with ^{14}C-labeled hexadecane, a constituent of petroleum oils, about 93% of carbon was oxidized to carbon dioxide and 7% assimilated into cellular mass (Cooney and Walker 1973). Thus somatic growth or microbial biomass may not be an adequate criterion for hydrocarbon degradation because of the large ratio of total carbon degraded to carbon in new biomass. Errors in evaluating increased biomass are thus critical in this approach.

When the soils were fertilized with nitrogen and phosphorus the rate of oil loss doubled (Table 6.19). Over 150,000 kg of oil were lost per hectare in soils under profitable farming systems and over 7,500 kg oil per year per hectare in soils receiving no fertilizer application. Oil losses during winter months were at a minimum, while most losses occurred during the spring, summer and autumn.

Table 6.19 Oil Loss Rate (Kincannon 1972)[a]

Plot	Rate of Application % of soil wt	25°C Summer-Fall (May to Nov. 1970)	30°C Spring-Summer (Feb. to Oct. 1971)	15°C Winter Nov. 1970 to Feb. 1971	Summer-Fall Winter-Spring Average (for year)	Equivalent Loss % of soil wt/yr
		lb. oil lost/ft^3 of soil/month				
1. Crude Oil		–	–	–	0.63	9.07
A-1	23	0.67	1.12	Minimum	0.67	9.65
A-2	26	0.50	1.37	Minimum	0.70	10.08
A-3	30	0.17	1.25	Minimum	0.53	7.63
2. Bunker C No. 6		–	–	–	1.07	15.41
B-1	44	1.83	1.75	Minimum	1.34	19.30
B-2	40	1.83	1.50	Minimum	1.25	17.71
B-3	36	1.16	0.5	Minimum	0.62	8.93
3. Waxy Raffinate		–	–	–	0.91	13.10
C-1	33	1.83	1.62	Minimum	1.30	18.72
C-2	31	1.50	1.0	Minimum	0.94	13.54
C-3	29	0.83	0.50	Minimum	0.50	7.2

1. Heavily fertilized, 1,100 lb N, 200 lb P_2O_5/ac.
2. Moderately fertilized, 500 lb N, 100 lb P_2O_5/ac.
3. Control check—unfertilized.

[a]Calculated on the basis of 12 months a year and 1 ft^3 soil = 37.8 kg (assuming a soil bulk density of 1.6, 1 ft^3 soil weighs 43.2 kg or 95 lb and the % loss would numerically equal the lb oil loss/ft^3 soil/month.

Schwendinger measured the oil loss from soil by microbial decay as carbon dioxide evolution (Figure 6.4) and found that over 20% of the added oil was lost in 7 weeks when nitrogen and phosphorus were not limiting decomposition. Because of the large fraction of evolved carbon dioxide from degradation of oil the measurements of CO_2 are good estimates of soil stabilization of oil and greases. Coupled with soil measurements for oil constituents, one can obtain both volatilization and microbial degradation losses and thus approximate allowable total oil application rates.

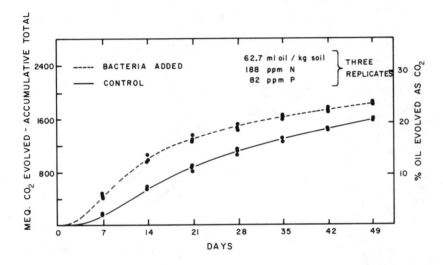

Figure 6.4   Rate of decomposition of medium level of oil in soil at medium levels of N and P fertilizer (Schwendinger 1968).

Several soil factors, environmental parameters and management variables influence the degree and rate of oil decomposition in soil-plant systems, so they must be considered before determining the land-loading capacity and rates for waste oils and oily wastes.

Substrate Level and Oil Concentration. The extent and rate of oil decomposition depends on the amount added to soil (Figure 6.5). If nitrogen is not limiting decomposition, the cumulative rate of oil decomposition was found to increase with amount added (Figure 6.4) through a range of the experiments (0.5-5% of soil weight). Schwendinger (1968) showed that increasing the oil application rate from 40 ml/kg soil to 100 ml/kg soil at medium nitrogen and phosphorus application (188 ppm N and 82 ppm P) increased the magnitude of oil degradation as measured by carbon dioxide evolution. When the concentration of oil in the soil medium is low, all fractions of oil are more likely to be attacked (Raymond *et al* 1976). At high concentration of oil the fractions that are more susceptible to microbial attack are preferentially degraded (ZoBell 1973).

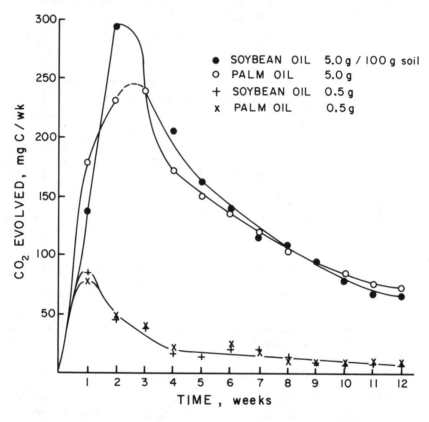

Figure 6.5 Rate of palm and soybean oil decomposition (Smith 1974).

Table 6.20 was developed as a summary of the data from a number of experiments. Decomposition rates depended on the particular experiment. For the lower oil addition rates characteristic of land application rates near the environmentally acceptable assimilative capacity, a rate of 0.2-0.4% of soil weight per month was representative.

Table 6.20 Microbial Decomposition of Various Oil Types in Soils

Oil Type	Decomposition Rate	
	% of Soil Wt/month	kg/ha/month
Crude Oil at 26% of Soil	0.756	16930.7
Bunker C #6 at 40% of Soil	1.284	28765.3
Waxy Raffinate at 31% of Soil	1.092	24453.3
Mineral Oil (a) Crude oil at 2% of soil (b) Crude oil at 8% of soil	0.32 at 2% application rate 0.64 at 8% application rate	7168 14336
Waxy Cake at 0.25% of Soil	0.0025	56
Palm Oil at 2-8% of Soil Weight	0.12-0.76	2688-17024
Soybean Oil at 2-8% of Soil Weight	0.14-0.8	3136-17920

Smith (1974) measured the decomposition rates of waste-cooking oils in a silt loam soil (Figure 6.4) and concluded that oil decomposition proceeded very rapidly during the first 4 weeks. Maximum weekly decomposition was approximately 8 and 2.5 metric ton/ha for 112 and 11.2 metric ton applications, respectively. Unlike petroleum oils, there was no evidence for toxicity to decomposition systems with the high application of vegetable oils and no evidence that a problem would develop with land disposal of waste containing these edible cooking oils.

Type of Oil and Composition. Of the hydrocarbons, straight chain paraffins are more easily degraded than aromatics (Tables 6.21-6.23). The rate of decomposition is usually decreased drastically by branching on the paraffin chain. Aromatic compounds are decomposed at a slower rate than unbranched alkanes (Stone *et al* 1942). Cycloparaffins seem to be poorly utilizable. The effect of the length of side chain on the decomposition of alkylbenzene is shown in Figure 6.6. Earlier, Strawinski and Stone (1940) reported that hydrocarbons in the range of C_{10} to C_{16} were generally attacked more readily by soil microorganisms than those of lower atomic weight. Huddleston and Cresswell (1976) determined that about 82% of paraffins, 60% of aromatics and only 1% of the resin-asphaltenes fraction

Table 6.21 Mass Spectral Analysis of Crude Oils
Effect of Bacterial Attack (Gibson 1974)

Hydrocarbon Type	Volume Percent Total Crude Oil		Volume Percent Paraffin-Free Basis	
	Unaltered	Degraded	Unaltered	Degraded
Paraffins	52.0	11.6	0	0
Noncondensed Cycloparaffins	26.8	29.5	55.8	33.4
Condensed, Cylclopar, 2-Ring	7.8	16.7	16.2	18.9
Noncondensed, Cyclopar, 3-Ring	5.0	12.7	10.4	14.4
Benzenes	5.6	7.6	11.7	8.6
Naphthenebenzenes	0.6	6.0	1.2	6.8
Dinaphthenebenzenes	1.0	4.9	2.1	5.5
Naphthalenes	0.7	4.4	1.5	5.0
Acenaphthenes, Dibzfurans	0.0	1.5	0.0	1.7
Fluorenes	0.1	1.0	0.2	1.1
Phenathrenes	0.2	1.5	0.4	1.7
Naphthenephenanthrenes	0.0	0.6	0.0	0.7
Pyrenes	0.1	0.7	0.2	0.8
Chrysenes	0.0	0.3	0.0	0.3
Perylenes	0.0	0.2	0.0	0.2
Steranes				
4-Ring	0.1	0.7	0.2	0.8
5-Ring	0.1	0.6	0.2	0.7
6-Ring	0.0	0.2	0.0	0.2

were lost over a 22-month period when refinery waste was allowed to decompose in a soil at Ponca City, Oklahoma. The oil initially contained 22% paraffins, 28% aromatics and 50% resin-asphaltenes. The residue after 22 months consisted of over 80% resin-asphaltenes. During 18 months of study, Kincannon (1972) found that much of the oil in soil was transformed to naphthenic acid, which decomposed very slowly. This emphasizes the need for evaluation of long-range effects of naphthenics buildup in soils. The differences in decomposition rate and microbial species due to hydrocarbon type as present in crude, bunker C No. 6, and waxy raffinate oils, are minimal. The average rate of bunker C No. 6 disappearance was twice the crude oil, whereas the waxy raffinate rate was intermediate. McKenna and Heath (1976) delineated the structural limits of polynuclear aromatic hydrocarbons degradability by microorganisms in soil biosphere. In extrapolating the rapid degradation of petroleum oils to other hydrocarbon mixtures the general trends of relative decomposition can be utilized.

Table 6.22 Changes in Chemical Composition of Different Crude Oils
After Microbial Degradation[a] (Gibson, 1974)

Crude Oil Fraction	% Change in Composition			
	Norman Wells	Prudhoe Bay	Lost Horse Hill	Atkinson Point
Asphaltenes	+0.46	+2.69	+5.36	+1.00
Saturates	-0.04	-6.65	-19.41	-0.93
Aromatics	+3.57	+1.30	-0.97	-1.99
NSO's[b]	+5.01	+1.37	+9.74	+1.90

[a]From data of Gibson (1974). Analyses were performed after 10 days growth of a mixed culture of microorganisms on the different crude oils.
[b]Nitrogen, sulfur and oxygen compounds.

Table 6.23 Relative Biodegradability of Various Hydrocarbon Substrates
(Perry and Cerniglia, 1973)

Recalcitrance[a]		Specificity[b]
	Normal Alkanes C_{10}-C_{19}	
	Straight-Chain Alkanes C_{12}-C_{18}	
	Gases C_2-C_4	
	Alkanes C_5-C_9	
	Branched Alkanes to 12 Carbons	
	Alkenes C_3-C_{11}	
	Branched Alkenes	
	Aromatics	
	Cycloalkanes	

[a]The number of microorganisms isolated that would grow on the substrates listed decreased from top to bottom.
[b]Any organism isolated on the compounds farther down the list would generally grow on those above.

Nitrogen Availability and C:N Ratio. In the soil nitrogen cycle, oil can alter major pathways such as ammonification, immobilization, nitrification and nitrogen fixation (Figure 6.2). As early as 1917, Gainey observed ammonia production in soils was decreased by paraffin treatment. Later, Baldwin (1922) found that crude oil application at a 5% rate (v/w) in soils resulted in a 40-45% decrease in ammonification in the first 7 days and a 30-35% decrease in next 10 days or after. These studies were interpreted as inhibition of ammonification by crude oils, although the observations may also result from the intense immobilization by actively growing microorganisms associated with a high C:N ratio.

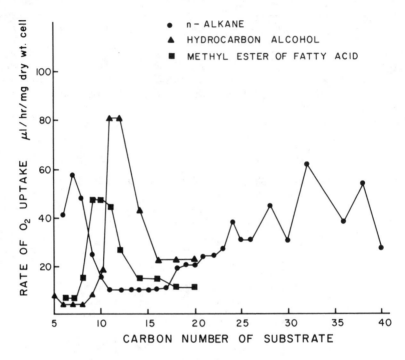

Figure 6.6 Effect of the carbon number in *n*-alkanes, hydrocarbon alcohols and methyl ester of fatty acids on the rate of oxygen consumption by culture CM01 (Liu 1973).

Baldwin (1922) observed that application of crude oil in soil retarded the production of nitrates. Murphy (1929) noted that at a 1% oil level when mixed with the soil, practically all nitrate formation was checked. These observations presumably have been due to intense immobilization of inorganic forms of nitrogen by hydrocarbon utilizers rather than inhibition of mineralization, *per se*. In fact, the activity of hydrocarbon-degrading microorganisms can be increased by application of nitrogen fertilizers (Kincannon 1972, Schwendinger 1968, Jobson *et al.* 1974).

Inorganic nitrogen in the soils is used by heterotrophic microflora for synthesis of biomass. With increasing rate of oil application from 0.1-5%

(W/W) in a silt loam soil, Smith (1974) calculated that N immobilization increased from 2 to 695 ppm for palm oil and 1 to 688 ppm for soybean oil (Table 6.24). Immobilization processes invariably become intense following application of waste oils and oily wastes to soils and dominate the mineralization process during rapid decomposition and microbial growth.

Table 6.24 Nitrate-N Determined in Soil After 12 Weeks
Incubation with Cooking Oils (Smith 1974)

		Palm Oil		Soybean Oil	
Oil	Nitrogen	Soil Nitrate	N Immobilized	Soil Nitrate	N Immobilized
%(W/W)	ppm	– – – ppm, N – – –			
Control	0	76.1	0	76.1	0
0.1	12	86.1	2	86.9	1
0.5	62	39.3	99	34.7	103
1.0	125	3.4	197	4.2	197
5.0	625	6.4	695	12.8	688

As a consequence of the rapid immobilization of nitrates, soils receiving oils may manifest apparent deficiency of nitrogen for plant growth and of other elements, such as phosphorus. These species may also become limiting to microbial proliferation in utilizing hydrocarbons. Schwendinger (1968) illustrated that application of either nitrogen, phosphorus or both to an oil-polluted soil stimulated the rate of oil decomposition (Figure 6.7 and Table 6.18).

Kincannon (1972) discussed the requirements of oil-degrading bacteria and found that the oil degradation rate was doubled by application of nitrogen and phosphorus fertilizers. Huddleston and Cresswell (1976) stated that 50 ppm N and 20 ppm P on a soil basis are sufficient to support the degradation of some oils at their maximum rates in some climates. Bacterial number and utilization of n-alkane were increased by application of nitrogen and phosphorus in the experiments of Jobson *et al* (1974) on soils in Alberta, Canada. Growth of filamentous fungi active in oil degradation of North Carolina Coastal Plain sandy soils was found to be better with ammonium than nitrate as a source of nitrogen (Perry and Cerniglia 1973). This is in agreement with other soil microbes utilizing ammonium preferentially over nitrate. Cooney and Walker (1973) found that activity of

Cladosporium resinae growing on hydrocarbons was enhanced by application of nitrogen. ZoBell (1973) concluded that 50 ppm N level is adequate for maximum growth and biochemical activity of hydrocarbon oxidizers in oil-polluted soils. Gudin and Syratt (1975) observed that the time required for the 50% disappearance of waxy cake was decreased to one-half by nitrogen fertilization when a cover crop was present (Table 6.25).

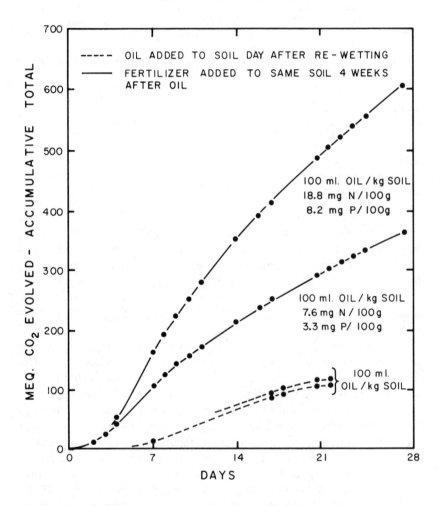

Figure 6.7 Rate of decomposition of oil at selected fertilizer additions (Schwendinger 1968).

Table 6.25   Effect of Cover and Nitrogen Fertilizer on the 50% Disappearance
Time of Waxy Cake Incorporated into Soil (500 kg/ha, Gudin and Syratt, 1975)

| | 50% Disappearance Time, Days | |
	With Nitrogen	Without Nitrogen
With Cover	50	100
Without Cover	125	125

Considerable amount of nitrogen is fixed in soils by nonsymbiotic micro-organisms from molecular nitrogen in the soil atmosphere when low nitrogen, energy-rich substrates are added to soils. These carbonaceous materials, while decomposing, may lower the redox potential of soils for a period of time during which nitrogen is fixed (Plice 1948). Bacteria require roughly 1 part nitrogen for every 10 parts carbon in their growth. By application or spill of hydrocarbons on soils, the C:N ratio is increased (Figure 6.8) and for microbial growth, nitrogen becomes limiting. Under such conditions, those hydrocarbon-degrading microbes that can fix their own nitrogen from the air can compete better for growth (Davis 1967). Gudin and Syratt (1975) reported the relative abundance of leguminous plants on hydrocarbon-contaminated sites because of competitive advantage offered by their symbiotic nitrogen fixing relationship with *Rhizobium* sp. (Table 6.26). The leguminous species thrive better on oil-polluted soils where microorganisms are strongly competitive for available nitrogen. When oil is added to a relatively fertile, fine-textured clay loam, the C:N ratio increases much less than when added to a relatively low-nitrogen, coarse loamy sand (Figure 6.8). Likewise, with a decrease in soil oil level by microbial decay, volatilization, etc., the change in the soil C:N ratio is more pronounced in low organic matter and low-nitrogen soils than in highly fertile and high-nitrogen clay loam. This is evidently demonstrated by calculations of C:N ratio changes as depicted by the slopes of the straight lines for two benchmark soils (loamy sand and clay loam) in Figure 6.8. At 10% oil level, the C:N ratio of a loamy sand is expected to be over 100, while that of a clay loam in the neighborhood of 60. Other soils may fall within this range of 50-120 at 10% level, although at zero oil level, the initial C:N ratio is nearly 10 for all soils. Incorporation of a surface-applied oil (determined as a percent of the surface 15 cm of soil) into the upper 30 cm has the effect of substantially reducing the C:N ratio and, hence, accelerating decomposition.

In summary, the addition of oil to a plant-soil system substantially alters the nitrogen cycle and, hence, the growth of plants and bacteria. The large

C:N ratio that results with oil application shifts reactions in favor of immobilizing all nitrates formed. Supplemental nitrogen (and other needed elements) or long reaction times for recycling of soil nitrogen assure satisfactory breakdown of oils and greases. In this nitrogen-limited situation, species that fix atmospheric N are at an evolutionary advantage.

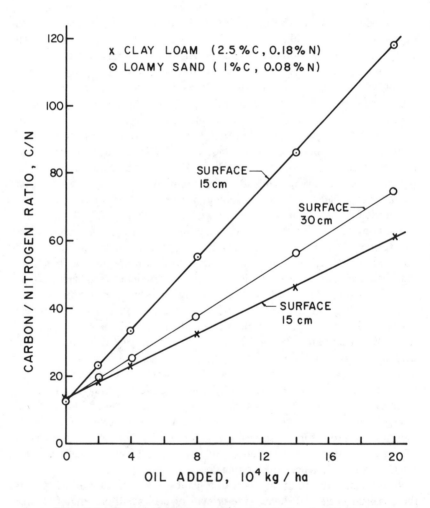

Figure 6.8 Effect of oil addition on C/N ratio of soil surface horizons.

Table 6.26 Percent Cover of Various Groups of Plants
on Polluted and Unpolluted Soils (Gudin and Syratt 1975)[a]

Group	Dunkirk (polluted)	Dunkirk (unpolluted)	Lobsanne (polluted)
Graminae	34.5	63.0	22.0
Leguminosae	41.5	14.0	31.5
Compositae	21.0	4.0	3.0
Others	1.5	28.5	19.0
Bare Ground	7.5	1.5	28.5

[a]At Dunkirk the dominant leguminous plant on the polluted zone was *Medicago lupulina* (26% cover) with *Trifolium pratense* (1.5%), *Trifolium repens* (3.5%) and *Vicia sepium lupulina* (4.5%), *T. repens* (0.5%) and *V. sepium* (0.5%).

Aeration. Microorganisms are very specific in their oxygen requirement while decomposing various hydrocarbons. Jamison *et al.* (1975) described the microbial oxygen requirement as 1.2 g O_2/g cells synthesized. It is estimated that 3-4 g oxygen are required per g carbon oxidized for paraffin oils, *i.e.*, 3 or more oxygen atoms needed for each carbon atom plus associated hydrogen to be completely oxidized to CO_2 and H_2O. For example:

$$C_nH_{2n+2} + \left(\frac{3n+1}{2}\right)O_2 \rightarrow nCO_2 + (n+1)H_2O$$

Less oxygen is required for hydrocarbons that are partially oxidized or unsaturated and metabolized for biosynthesis of cellular materials (ZoBell 1973). Also, less oxygen is required for degradation of animal fat and vegetable oils than hydrocarbons.

Temperature. Microbial degradation of oils has been observed at temperatures ranging from freezing point of seawater (-2°C) to about 70°C (ZoBell 1973). Most species are active in a mesothermic range (20-35°C). According to the temperature categories for growth and function of microorganisms, oil oxidizers can be divided into three general groups:

1. Psychrophilic oil-oxidizing microbes that function at low temperatures for optimum growth and metabolism, usually in the range of subzero to a maximum of 20°C. At -1.1°C, psychrophiles may oxidize mineral oil at a rate of 0.13-0.9 mg/l/day and build a biomass of 1.2 mg/l/day. ZoBell (1973) reported different crude oils to be slowly attacked by bacteria growing at -1.1°C in soil-free culture media. However, in soils, Kincannon (1972) found minimum loss of oil fractions at temperatures below 18°C in a soil that does not possess a psychrophilic population adapted to oil metabolism. Oil-metabolizing organisms that grow well and function at low temperatures are found in oil-polluted soils and waters of northern regions.

The effect of low temperature on oil degradation has been discussed by Walker and Colwell (1974, 1976), Atlas and Bartha (1972), Traxler (1973) and ZoBell (1973). At $0°C$, only representative genera of *Pseudomonas* and *Vibrio* were found to be present in cultures. At $5°C$ and $10°C$ members of genera *Acinetobacter* and *Aeromonas* were also isolated (Walker and Colwell 1974). In general, increased utilization of the model petroleum was observed for cultures incubated at $0°C$. Because of a lower solubility at $0°C$, certain hydrocarbons may be less toxic at the lower temperature; hence, the increased degree of utilization in microbial biosyntheses. Also, a selection of different microbial populations may occur at lower temperatures, which may selectively degrade the petroleum hydrocarbons.

Bacteria accounted for all the petroleum utilization at $0°C$ and $5°C$. However, at $10°C$, yeasts were also responsible. Low temperatures clearly do not block or completely inhibit the autochthonous microbial degradation of oil. A selection for specific mocrobial population capable of carrying out oil degradation at low temperatures does occur.

2.   Mesophilic, oil-degrading microbes work most efficiently in the temperature range of $20\text{-}37°C$. This temperature range has been most frequently studied for oil degradation, but without accounting for temperature variability as a factor when comparing the results of different experimenters. Thus, most of the literature data are for mesophilic microbes, which on an *a priori* basis, are assumed to produce the greatest oil degradation.

3.   Thermophilic, oil-degrading microbes are rare as the optimum temperature for their growth and function is above $37°C$. At high temperatures, such as those exceeding $37°C$, the vapor pressure of volatile fractions of oils may increase to lethal levels for biological activity. Thermophiles exist in unusually hot deserts during summer and may easily be killed by toxic oil levels.

Petroleum-degrading bacteria may vary with seasonal temperature (Walker and Colwell 1976). As shown in Figure 6.9, Kincannon (1972) demonstrated under field conditions that degradation of crude oil, bunker C, and waxy raffinate was minimal during winter months when the temperature remained below $18°C$ for 4 months (November 1970-February 1971). At temperatures above $20°C$, rates of decomposition ranged between 0.55-1.1 lb oil/month/ ft^3, which is equivalent to a degradation rate of 1.23-2.4% oil of the soil weight each month. In this study it must be noted that certain oil fractions may have volatilized at a much faster rate at higher temperatures during summer, autumn and spring than at low temperatures during winter. The volatilization pathway for oils is not well documented at various ambient temperatures.

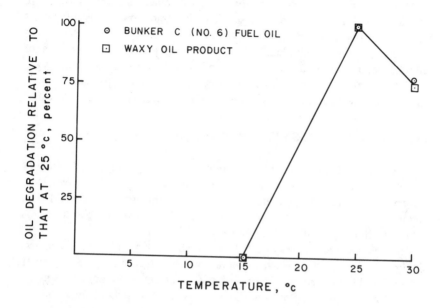

Figure 6.9 Temperature effect on oil degradation (Kincannon 1972).

Water Supply. Microbes active in oil degradation require water for their growth and metabolism. The rate of oil decomposition by microbes depends on the oil-water interface; the greater the oil-water interface, the faster the microbiological decay of oil. Thus, increased soil moisture must be balanced against the restricted oxygen diffusion associated with high-moisture soils. Agitation of oil in an aqueous medium results in emulsification and oil globule formation with increased oil-water interfaces (Davis 1967) and enhanced rate of oil degradation. In land application systems, enhanced degradation can be achieved by mixing oils and oily sludge in soils at a moisture capacity. Huddleston and Cresswell (1976) observed that oil in soil decomposed at a reasonable rate when the water content of soil stayed below 20%, thus approximating the optimum moisture for soil microorganisms.

Other Factors. For most hydrocarbon-utilizing organisms, optimum pH is near neutrality (pH 6-8), which is the normal range for soil systems. Most terrestrial species of microbes that utilize hydrocarbons proliferate at usual soil solution concentrations of salts (EC < 4 mmhos/cm at 25°C) and may not function in extremely saline environments such as found in seawater.

In conclusion, soil conditions that favor sustained microbial growth are the same conditions that promote substantial degradation of oils and greases in land treatment systems. Volatilization occurs along with the evolution of microorganisms, such that oil decomposition proceeds. Elimination of limiting conditions, such as low N, P and other essential nutrients, oxygen deficiencies and extremes of temperature, soil moisture, pH and salinity, increases the degradation of oils and greases in a plant-soil system.

INTERACTION OF OILS WITH CROP SYSTEM

When applied to a plant-soil system, oils and grease constituents of industrial waste can affect seed germination or crop growth and yield. Most information centers on germination and subsequent growth and yield of a variety of crop species in which oil application is followed by planting. Less is known about application to perennial growing crops.

The volatile components of oils have a high wetting capacity and penetrating power (Plice 1948). When in contact with plant seeds, the oil may enter the seed and kill the embryo or prevent the seed from germinating by altering the moisture and aeration conditions necessary for germination. Murphy (1929) noted that surface application of crude petroleum at 4,600 l/ha (or 0.25% of soil weight) did not prevent wheat seed germination but delayed the process, whereas mixing the oil in top 10 cm decreased germination by 36% of the control. At a 23,000 l/ha rate (equivalent to 1.25% of soil weight), surface application of oil reduced the wheat stand by 23% as compared to check plot. A similar amount added 4 inches below the soil surface did not cause stand reduction, but when mixed with soil, germination of seed was prevented. In a land application system the choice exists to have surface application or, by disking, to incorporate an applied oil into the surface soil zone. The effect on germination of mixing oil into the upper soil zone is very pronounced; thus, for consideration of plant germination, the surface application is preferable if higher rates of land loading are needed.

Schwendinger (1968) found a decrease of 25% and 75% in oat seed germination at approximately 3% and 5% application rates of Port Berga crude oil, respectively. Corn seed could germinate without any effect up to 2% crude oil level of the soil weight (Udo and Fayemi, 1975), but at 4.2% crude oil level, the reduction in germination of corn seeds was 50% (Figure 6.10). Up to 1.6% oil of soil weight, crude petroleum did not seem to permanently decrease the growth of corn because the depressive effects on yield could be overcome by addition of nitrogen and dried blood.

However, at oil applications as low as 1.1% soil weight, the yield of corn was reduced to 70% of that of the control. Oil effects on germination are less pronounced than on yield or growth after a fixed time period. This should be noted in comparing available data, which report either yield or percent germination to ensure uniform interpretation. The prohibitive level of oil for seed germination depends on:(1) the type and composition of oil added, (2) the plant species and nature of seed, and (3) the method of oil application in soil.

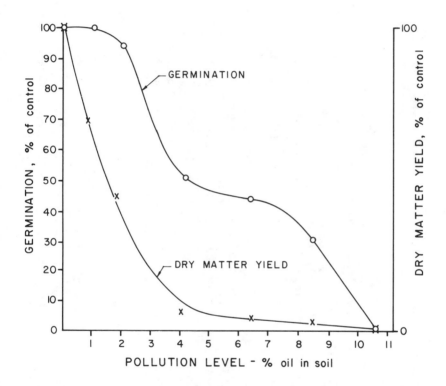

Figure 6.10 Effect of the level of crude oil addition to the soil on the germination of maize grains (Udo and Fayemi 1975).

Table 6.27 shows the effects of various oil applications on the growth and yields of several crops. The table demonstrates the variety of crops that can be used with oil land application. The poor growth of seedlings and plants on oil-polluted soils may be caused by (1) suffocation, such as exclusion of air (Udo and Fayemi, 1975, Stebbings 1970) or exhaustion of oxygen by increased microbial activity (Plice 1948); (2) toxicity from excessive micronutrients, such as increased Mn availability in soils during decay of hydrocarbons (Udo and Fayemi, 1975); (3) reduced water supply (Plice 1948, Carr 1919); (4) induced nitrogen deficiency (Giddens 1976); and (5) direct toxicity (Skinner 1973).

Table 6.27 Influence of Oils and Oily Wastes on Crop Growth and Yield

Test Crop	Oil Application Reporter's Unit	Equivalent % Soil in kg/ha	Wt	Experimental Conditions (soils, climate, oil type, etc.)	Remarkable Influence on Plant Growth and Yield	Reference
Corn	1.1%	24,640	1.1	Crude oil–ferruginous tropical soil (surface)	Growth reduced by 30% of control	Udo and Fayemi (1975)
	2.0%	44,800	2.0	fertilized with N, P & K	Reduction in growth by 50% of control	Udo and Fayemi (1975)
	8.5%	190,400	8.5		Growth occurred only 5% of control	Udo and Fayemi (1975)
	12717 l/ha	10,173	0.45	Spent motor oil–Congaree loamy sand in greenhouse	Yield decrease in first year to 54% of control but 9.8 and 14.4%.	Giddens (1976)
	20 mg/4 kg soil	8,960	0.4	Green house pot experiment N and dried blood added	Crude oil depressive effects were overcome by dried blood supplement & N addition.	Baldwin (1922)
	70 mg/4 kg soil	31,360	1.4		Nearly 70% decrease in yield by crude oil which could be overcome by dried blood and N addition.	
Rye Grass	50 and 100 ml/3 kg	29,860	1.3	Crude oil at 2, 7 and 14 cm depth	20 to 30% yield reduction at 1.3-2.6% oil application rate.	Schwendinger (1968)
		59,590	2.6		Depth of application had no effect on extent of damage.	
Sorghum	9,333 l/ha	7,466	0.34	Mixed with top 10 cm soil layer darso sorghum	Significant decrease in yield after 40 days growth.	Giddens (1976)
	1%	22,400	1.0	Added to a depth of 6 in. on a heavy clay soil	Yield decreased to 42% of a check plot in first season.	Plice (1948)
Cotton	9,333 l/ha	7,466	0.34	Mixed with top 10 cm soil layer.	Yield decrease to one-half of the control in 40 days.	Giddens (1976)

Plant	Rate			Treatment	Effect	Reference
	1%	22,400	1.0	Added to a depth of 6 in. on a heavy clay soil	Yield decreased to 42% of check plot in first season.	Plice (1948)
Soybeans	0.75	16,800	0.75	Greenhouse experiment—sandy peat soil at various rates of oil application.	Growth stimulation, yield increase and greater nodule number	Carr (1919)
	1.75	39,200	1.75	Mixed with top 10 cm soil	Yield decreased to one-half the control.	Carr (1919)
	9,333 l/ha	7,466	0.34	Mixed with top 10 cm soil	Yield increased slightly over control.	Giddens (1976)
	31,111 l/ha	24,888	1.11		Yield decreased to less than one-half of control.	Giddens (1976)
Peanut	9,333 l/ha	7,466	0.34	Mixed with top 10 cm soil layer	Slight decrease in growth and yield.	Giddens (1976)
Tomato, kale	10 ml/2 kg soil	8,960	0.4	Oil put on soil with growing plants (15 cm height), soils fertilized with 56-31-44 (N-P-K) mixed fertilizer	Abundant growth.	Schwendinger (1968)
	20 mg/2 kg soil	17,920	0.8		Excellent growth.	Schwendinger (1968)
	30 ml/2 kg soil	26,880	1.2		Growth was affected, water uptake was slowed.	Schwendinger (1968)
	50 ml/2 kg soil	44,880	2.0	(N-P-K) mixed fertilizer	Yellowing and eventual death of bottom leaves extreme nutrient deficiency.	Schwendinger (1968)
	50 ml/2 kg soil	44,880	2.0	(N-P-K) mixed fertilizer		Schwendinger (1968)
Turnip and beans	11.9 m^3/4000 m^2	23,800	1.06	At 3 locations in Pennsylvania, Oklahoma and Texas with and without fertilization.	Most seeds germinate at all locations but significant inhibition of plant growth.	Raymond et al. (1976)
Wheat	80 ml/4 kg soil	35,840	1.6	Crude oil–dried blood added to Yahola very fine sandy loam soil	Yield reduced to one-half of control	Murphy (1929)

The possible mechanisms of oil effects on seed germination are by penetration of readily mobile fractions into seed and killing the embryo or alternately reducing water and oxygen supply by hydrophobic nature. Both water inhibition and oxidative respiration are essential for germination. During growth of plants, the hydrophobic nature of oils can prevent water uptake by roots, decrease oxygen supply for respiration and may exert some possible toxic effects directly. However, the most frequently noted mechanism of reduced plant growth and yield by excessive oil applications is the decrease in the inorganic nitrogen forms essential for plant nutrition. This decrease results from inorganic N immobilization and denitrification and represents the competition between microflora and plants for available nitrogen. Oil toxicity to the process of ammonification and nitrification does not appear to be the crop growth inhibition mechanism.

To determine the approximate level of oil application that might affect crop yields adversely, the available data were ordered according to increasing percentage of oil in the soil (Table 6.28). The oil level of about 1% of soil weight is the initiation of reduced yields. At oil levels of 1.5-2.0% of soil weight, the reduction in yields is substantial, often exceeding 50%. Germination was essentially the same as the control up to levels of 2.0% oil of soil weight.

These adverse effects on plant growth occur only after fresh application of oil on soils. As the hydrocarbons disappear, the growth of plants is often improved because of increase in soil fertility level (Plice 1948, Udo and Fayemi 1975) and improved soil physical conditions, e.g., moisture retention (Giddens 1976). Carr (1919) noticed enhancement of soybean growth in greenhouse pots receiving oil up to 0.75% of soil weight. Plice (1948) recorded that the yield of darso sorghum beans and peas at 1% oil application rate to soil was 42% of the check plot. In the next cropping season, plots were seeded to wheat, barley and rye. Practically full stands were associated with no significant differences in yield from the control. In the third season, all of the plants on oiled plots gave a 20% higher yield than the check plot. Stimulating effects of petroleum materials continued in successive seasons when hairy vetch, crimson clover and hubam clover were grown. These data suggest that with a rest period after oil application, prior to planting, crop yields may be improved or at least not reduced below those without oil application. The rest period allows soil microbial decomposition; thus, incorporation into the soil would accelerate this decomposition. At higher application rates longer periods prior to planting are necessary not to affect crop yields.

At a certain level, oil in a wastewater can seriously affect a growing crop. For soybeans, oil levels of 4% did not completely kill the crop, although yield reductions were observed (Carr 1919). Data obtained for tomato and

Table 6.28 Crop Effect as Related to Oil Applications

Oil Application (% soil wt/dose)	Observed Response	Crop (method of oil application)
0.34	Yield increased slightly over control.	Soybeans (mixed with top 10 cm)
0.4	Abundant growth.	Tomato and kale (surface applied)
	Need N and P possibly.	Corn
0.45	Yield 54% of control, but increased above control second and third year.	Corn (mixed with soil)
0.75	Growth and yield increased.	Soybeans (incorporated)
0.8	Excellent growth.	Tomato and kale (surface applied)
1.0	Germination 97% of control	Oats (injected into top 6 cm)
1.06	Complete germination, but significant inhibition of growth.	Turnip and beans (incorporated)
1.1	Growth reduced to 70% of control; germination same as control.	Corn (incorporated)
1.11	Yield reduced to less than 50% of control.	Soybeans (mixed with top 10 cm.
1.	Growth was affected.	Tomato and kale (surface applied)
1.25	Surface application reduced wheat stand by 23% of control.	Wheat
1.3-2.6	Yield 70-80% of control, needed N and P to overcome.	Ryegrass (incorporated)
1.6	Yield reduced to 50% of control.	Wheat
1.75	Yield reduced to 50% of control.	Soybeans (mixed with top 6 cm)
2.0	Full germination as in control.	Corn seeds
2.0	Germination 90% of control.	Oats (injection into top 6 cm)
2.0	Yield reduced to 50% of control; germination 95% of control.	Corn (incorporated)
2.0	Yellowing and bottom leaf loss.	Tomato and kale (surface applied)
2.9	Germination 75% of control.	Oats (injected into top 6 cm)
4.2	Yield 8% of control; germination 50%.	Corn (incorporated)
4.9	Germination 28% of control.	Oats (injected into top 6 cm)
8.5	Yield reduced to 5% of control; germination 30% of control.	Corn (incorporated)

kale at 2% by soil weight showed there were symptoms of leaf yellowing and loss. From these and other data it appears that, while some crops are more sensitive than others, oil rates of 2-3% soil weight are levels representative of complete kill of crops.

Crops vary in their sensitivity to fresh application of oil (Schwendinger 1968, Skinner 1973). Plice (1948) observed that cotton and darso sorghum were less sensitive to crude oils than beans and peas. Odu (1972) noticed in Nigeria that yams were the most sensitive to heavy oil pollution, whereas plantain and banana were resistant. Schwendinger (1968) illustrated that growth of beans and rye grass were more sensitive to oil pollution of soil than sorghum. And vegetable crops such as tomato, kale and leaf lettuce could withstand 3% oil by weight in soils. He described the symptoms of plant damage by excessive oil application to soils were similar to those of typical nutrient deficiencies, *i.e.,*, stunted growth and yellowing of bottommost leaves. The tomato was more sensitive than kale or leaf lettuce (Schwendinger 1968).

The harmful effects of excessive oil application could be minimized by heavy fertilization (nitrogen and phosphorus) of soils to overcome the nutrient deficiencies (Schwendinger 1968, Giddens 1976). The nitrogen additions provide nutrients to accelerate soil decomposition of oil thus reducing physical, chemical and biological impact in the soil system. Nitrogen additions also provide nutrients for plant uptake and reduce the competition for available nitrogen, thus reducing the effects on germination and yield.

In discussing the rehabilitation aspects of oil-polluted soils, Gudin and Syratt (1975) stated that plants of the leguminosae family dominated the oil-contaminated sites, whereas in unpolluted areas, plants of the gramineae family were dominant (Table 6.26). This may be because of the legumes can fix their own nitrogen by symbiotic relationship with *Rhizobium*, whereas cereals cannot.

The effects of oil at 10-1,000 ppm application rate specifically on germination pathways and total crop production have not been investigated, but it can be extrapolated from above studies that oil wastes can be utilized for land farming with no expectation of adverse consequences in either short or long term.

LAND APPLICATION ASSIMILATIVE CAPACITY

From our knowledge about the fundamental pathways of oil in a plant-soil system, there are certain major mechanisms involved in the assimilation of oils and greases in an environmentally acceptable manner. Principal losses occur initially by volatilization of constituents in oil wastes. The initial effect on crops and soil microorganisms depends on the amount of oil added.

At moderate-to-high doses, the oil hinders crop germination, growth and yield. Soil microorganisms initially suffer a germicidal effect at moderate-to-high doses. From moderate-to-low doses there is often a beneficial effect on plant yield and a much less pronounced germicidal response by soil microbial populations.

After the initial response to oil application, there is sustained decomposition of the constituents in oils and greases by evolving soil microorganisms. This is the long-term pathway for stabilization and, as delineated, the time necessary to return the plant-soil system to accept more oil addition depends directly on the initial application amount or rate. Certain management variables improve the rate of oil degradation and the initial loading effects, including nitrogen additions, application of other deficient macro- and micronutrients, bacterial seeding, integration with crop planting and growth cycles, method of land application and crop selection.

Observations to date of land application for oil constituents of industrial wastes indicate that for moderate-to-low applications, leaching of oil below the surface 15-30 cm and surface runoff are not significant losses. Therefore, based on the major categories of environmental design constraints (Chapter 2), the oil constituent is governed by the decomposition in soil and plant effects. That is the land application limit is established by the rate of oil and grease decomposition and should be increased or decreased by the use of management variables affecting the rate of stabilization.

This section translates the information on plant-soil fundamentals and the soil response to oil and grease into actual land application rates or the assimilative capacity. As discussed previously, there are differences in the rate of stabilization for animal- and vegetable-based oils and those from petroleum. The petroleum-based oils, hydrocarbons, are expected to be less rapidly degraded by soil microorganisms than the vegetable or animal oils in the glyceride category. With the majority of the available data derived from petroleum-based oils, the assimilative capacity is conservative when adapted for oil and grease constituents from animal and vegetable processing wastes.

Two Principal Controlling Effects

The assimilative capacity for oils and grease is constrained by two principal effects: (1) the maximum single dose that can be applied without adverse environmental impact, and (2) and rate of stabilization of applied material that controls the time between application events. Maximum soil application is based on a percentage of the soil weight (g oil/100 g soil), where the volume of soil used in calculations is the surface 15 cm. When this percentage reaches certain levels, soil microorganisms as well as crops can be adversely affected.

Thus, the concept of applying oils in amounts not to exceed a certain percentage imposes one constraint on the assimilative capacity.

Soil levels of oil that reduce or kill soil microorganisms are in excess of 2-4%. Data in Table 6.15 indicate that after a short time period (one to two weeks), microbial populations and related soil process rates are substantially increased for oil applications in the range of 2-4% of soil weight. By contrast, the single-dose levels of oil that substantially reduce plant growth and yield are 1-2% of soil weight. Thus, conceptually, the single-dose limit would depend on whether there is a crop on the application field. However, because of such factors as rainfall-runoff transport of pollutants, soil erosion and aesthetics, few if any land application sites would be without a vegetative cover for extended time periods. As an exception, where row crops are used there will be periods prior to planting and immediately following harvesting when soil levels may exceed those acceptable for crops. These time periods are probably as short as one month; thus, the soil losses of oil must be sufficient to reduce soil levels to below the critical crop limits prior to planting. With these constraints and time periods, only a slight increase in annual oil application can be realized by taking advantage of periods with no crops. Therefore, in comparison with the maximum acceptable soil levels, the crop tolerance limit determines the single dose assimilative capacity.

With the single-dose level of oil maintained below critical percentages of soil weight, the rate of oil degradation and losses dictates the continuing amounts of oil that can be applied in a land-based waste management system. Decomposition rates are expressed in oil percent of soil weight per month. Knowing these rates, the design engineer can vary the application from frequent loading of small amounts to infrequent application of large oil volumes, the frequency being controlled by the degradation and loss rate in the soil system.

A wide range of decomposition rates have been measured for oil applications on soil systems from 0.025-40% of soil weight (Table 6.20). Assuming that a tolerant crop is grown at the land application site, an approximate degradation rate of 0.2-0.4% oil of soil weight per month represents the expected losses and stabilization. This rate is for time periods when oil degradation is active, i.e., ambient and soil temperatures above 10-15°C. Where no crop is to be grown, higher single-dose levels are allowable and the rate of oil degradation would be greater. However, the impact of prolonged high dose applications is not known and must be approached in controlled studies.

For a particular oil in an industrial wastewater or sludge, short-term experiments can detect the degradation rate at a particular site and the adaptability of microbes to waste utilization. These decomposition studies can be used to evaluate the breakdown of oils and greases over a 6-12 month interval and the effect of various management variables.

Some differences in soil type have been qualitatively noted. Less damage to crop yields was found on clay soils when compared to sandy soils (Plice 1948). In comparison to the benchmark soils delineated in Chapter 2, preferences in land application site selection for oil constituents of wastes would be for soils with properties similar to a clay loam. These soils have better soil moisture-aeration relationships than sandy soils, hence a more favorable environment for microbial growth, which may account for the better oil degradation. These soil differences are qualitative and are to be used only in relative judgments of land sites where oil is the land limiting constituent. The improvement in cost-effective design related to soil properties can be substantial as research develops more refined evaluations of the soil effects on the degradation rate of oil.

As design principles in establishing the plant-soil assimilative capacity, the maximum oil level at a given time in the soil and the rate of degradation and loss are the principal controlling variables. The assimilative capacity of soil can be improved or adversely affected by management and environmental factors associated with a land application site. Several of these design factors evaluated in the following discussion are nutrient additions,, temperature, oxygen demand, bacterial seeding and crop management.

The rate of breakdown in a soil system is primarily determined by microbial utilization and, hence, the degree of contact of oil and soil is an important factor. Incorporation by disking or direct injection increases the mixing of oil in industrial wastes with the plant-soil system. However, as discussed under crop effects, the intimate contact of oils with seeds and root systems can lead to lower yields. Thus, a double effect results from soil incorporation of oil constituents.

Adverse effects of soil mixing can be reduced by not exceeding the critical crop levels in the soil 1.0-1.5% of soil weight and by maintaining lower rates in the soil during the growing season. The latter would require early incorporation and then sufficient time for reducing soil levels of oil constituents prior to planting. In general, the alternative of not exceeding 1.0-1.5% soil weight in a single dose as well as using soil incorporation of waste to enhance the rate of breakdown is the most feasible land application management for oils. The single dose effect of oil applies to use prior to crop planting followed by planting and growth. The single dose effect for oil applied to growing crop foliage would be 3-5 times lower.

Nutrient additions, principally nitrogen, can greatly improve the assimilative capacity of a plant-soil system for the oil constituent of industrial wastes. The degradation rate (oil percent of soil weight per unit time) is controlled by the C:N ratio of a waste. Thus, if primarily oil is land applied, there is a high C:N ratio and degradation is limited by the availability of nitrogen to carbon for microorganism biomass and metabolic energy. With the initial small amount of available nitrogen, the breakdown is rapid; however, when exhausted, the rate of degradation is greatly reduced. The rate is then controlled by the mineralization of nitrogen from dead cell biomass. Thus, addition of nitrogen or other deficient macro- and micro-nutrients permits oil decomposition at a rapid rate characteristic of ideal microbial nutrients, *i.e.*, low C:N ratio. If nitrogen is available in the industrial waste from constituents other than oils and greases, then a more favorable C:N ratio of the total waste may be present and, hence, yield high rates of decomposition.

The quantitiative effect of nitrogen addition to oil applied to a plant-soil system is not conclusive. A definite increase in microbial decomposition occurs with nitrogen addition (Schwendinger 1968, Kincannon 1972, Raymond *et al.* 1976). The percentage increase for large oil application (30-40% by soil weight) was the greatest (50-150% above control), while at lower application (1.5-3.5% by soil weight), the increase rate was 5-80%. Although the data are varied, it would appear that oil application at moderate-to-low rates would be most common, for which the effects of nitrogen supplementation is about a 50% increase over control. Therefore, the degradation rate is in the range of 0.3-0.6% oil of soil weight per month with nitrogen management.

Little is known quantitatively of stimulatory effects of potassium, phosphorus or soil micronutrients on oil decomposition. Under these circumstances, the addition of macro- and micronutrients would be based on soil tests at the application site and the agricultural recommendations for intensive farming operation in that region or locale. With this approach, major deficiencies can be detected and corrected.

While information exists concerning the acceleration of oil degradation due to supplemental nutrients, amelioration of the effects on crop yield and germination from oil addition are not well documented. However, from the basic interaction mechanisms of oil and plant-soil systems, certain conclusions can be reached. A large supply of nitrogen would reduce the competition between soil microorganisms and plant roots for the uptake of mineralized nitrogen. Greater amounts of nutrients can be utilized in plant growth so that yield reductions are not as significant. The presence of supplemental nitrogen enhances oil degradation so that less material is present to reduce oxygen transfer and to be absorbed within the seed. Reduction of these adverse conditions by supplemental nutrients will improve crop growth and yields. In terms of design, nutrients can reduce adverse effects on crops and, thus, increase the allowable oil applications (expressed as oil percent of soil weight). In summary, supplementary nutrient additions enhance oil degradation and probably increase the acceptable single-dose level in a soil.

The first refinement of assimilative capacity for oils and greases for various geoclimatic regions is to approximate the temperature effect on the rate of degradation in the plant-soil system. This factor estimates the monthly and total annual assimilative capacity. Factors not included in the initial loading rate adaptation are temperature effect on maximum allowable soil levels, regional soil types, and regional precipitation effects and soil moisture conditions.

Only preliminary data are available on temperature dependence of oil degradation by soil microorganisms. These data were obtained by observation of the oil disappearance rate under field conditions (Kincannon 1972) and are plotted in Figure 6.8. A relative loss rate was calculated, based on 25°C, so that different average decomposition rates reflecting management and specific soil conditions can be more easily corrected for annual temperature cycles.

The low decomposition rate found below 15°C is equated to zero for months with an average air temperature below 15°C. For months above 15°C, a rate of degradation similar to that in Figure 6.9 is used. The use of zero oil losses below 15°C is based on the results of Kincannon (1972), but the lack of volatilization losses at below 15°C was unexpected. The losses by both microbial decomposition and volatilization should be investigated with research work, since the impacts on design and economics of land application for oil-limited industrial wastes are substantial.

Based on the soil degradation rate in Figure 6.8 and the mean monthly temperature at four U.S. locations (Chicago, Illinois; Philadelphia, Pennsylvania; Galveston, Texas; and Richmond, California), an estimate of regional variation in assimilative capacity was developed (Figure 6.11). These data are necessary site-specific adaptations to estimating land assimilative capacity for oil, but as described, are based on a limited data base for temperature effects.

With the very low biological degradation of oil associated with temperatures below 15°C, many months of the year provide no soil treatment capabilities. The stabilization rates only illustrate general temperature effects, but some broad conclusions are possible. The moderate temperature cycle of the San Francisco-Richmond area limits the annual degradation rate (~0.7% soil wt/yr), and in comparison, is the lowest of the four locations. Again, the major effect is the essentially zero degradation assumed below 15°C. Philadelphia and Chicago have similar climatic patterns of hot summers and prolonged winter periods; hence, on an annual basis they have an intermediate loss rate of land-applied oil (~3% soil wt/yr). The brief high temperature summer periods with the much higher biological activity account for the increase over the San Francisco area. Finally, Houston, which has a short cool season and a sustained hot summer, would provide the greatest soil decomposition (~6% soil wt/yr), nearly an order of magnitude greater than the coolest location. Again, these data extrapolations are preliminary, but indicate some of the alteration in associated regional application of plant- soil systems for oil constituents in industrial wastes. In terms of land areas needed for a given amount of oil constituent assimilative capacity, the regional variations represent a substantial economic impact when expressed in waste treatment cost per barrel of oil processed.

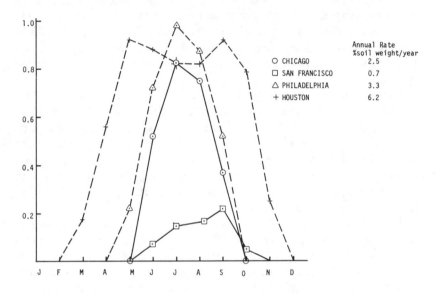

Figure 6.11 Effect of climatic conditions at major refinery locations on the annual pattern of oil decomposition.

When initiating land application of industrial wastes containing oil constituents, an initial lag period may be experienced, since relatively few hydrocarbon-oxidizing microorganisms occur in soils remote from oil fields or related pollution (Jones and Edington 1968). The observations of reducing this lag period by microbial seeding (Schwendinger 1968) can be translated into field practice using several techniques:

1. mixing the previously oil-contaminated soil (that already has high numbers of oil utilizing microbes) with a soil receiving oily wastes for the first time;

2. mixing the enriched culture of an oil-metabolizing bacteria or fungal spores of terrestrial origin with oil-contaminated soil or with oily waste to be disposed on land; and

3. mixing an inoculum of oil-degrading microorganisms from soils to the waste oils and oily wastes and allowing profuse growth of the oil decomposers before application onto the land.

Once the inoculated microorganisms become part of the new habitat and build up in the soils receiving oily wastes, there is no further need for inoculation or introduction of the same species. The economics of such seeding should be weighed against the cost of temporary storage and reduced application rate necessary to achieve acceptable performance at the startup of a land application system.

Agricultural management can be used to enhance oil assimilative capacity, *i.e.*, the selection and production of the more oil-tolerant crop species. With these crops, the single-dose level at which acceptable yields are achieved can be raised above those in Table 6.28. Research must define these improved crop species and tolerable oil levels, and since it reduces the frequency of application, the operating costs can thus be reduced correspondingly.

Long-term operation can be maintained with frequent field cultivation. Cultivation of soil (1) increases the contact of oil waste with the soil by mixing and stirring action, (2) opens the soil for better aeration and improved moisture—soil relationship, and (3) hastens the mineralization of the soil organic matter as well as of the oily wastes. By action of fungi, actinomycetes and bacteria, crude petroleum is coverted to soil organic matter. During this conversion, free living organisms fix large quantities of nitrogen, which may be used by plants later (Plice 1948, Davis 1967). During microbial degradation of oils and hydrocarbons, large quantities of nitrogen are immobilized to meet the needs of rapidly proliferating microorganisms. Addition of an organic amendment such as straw may upset the rate of oil degradation by (1) aggravating the nitrogen deficiency originally created by oil, (2) increasing oxygen demand in the system, and (3) by attracting the attention of many microorganisms as a more preferred substrate over oils. In low-fertility soils and under high C:N ratio conditions, application of nitrogen and phosphorus fertilizers speeds up the mineralization of oils in the wastes.

The oil percent of soil weight that affects crop yield is dependent on the type of crop under consideration. In reviewing the available qualitative and quantitative information, a tentative categorization of oil tolerance for crop species was proposed as follows:

1. The first includes sensitive crops that will be affected by oil application at greater than 0.5% of soil weight, *i.e.*, 10,000 lb oil/ac or 11,200 kg oil/ha at one time. These crops are sugar beet, rape, lawn grasses, yams and other tap root crops such as carrots, turnips and radishes, etc.

2. Slightly tolerant crops that can accept oil application up to 1.5% of soil weight, *i.e.*, 30,000 lb oil/ac or 33,600 kg oil/ha at one time, are included in the second group. This category may include rye grass, oat, wheat, barley, cereals, corn, beans, tomatoes, kale, lettuce, soybeans, mangoes, etc.

3. Moderately tolerant crops that are affected adversely by oil applications greater than 3% of soil weight, *i.e.*, 60,000 lb oil/ac or 67,200 kg oil/ha,

compose the third catetory. Moderately tolerant crops are peas, red clover, cotton, potato, sorghum, coconut, etc.

4. Very tolerant crops are those that can grow and produce well at soil oil levels greater than 3% of oil weight, *i.e.,* > 60,000 lb oil/ac or 67,200 kg oil/ha. This category includes certain perennial leguminous trees, perennial grasses, coastal bermudagrass, and many legumes, banana, plantain, cocoyams, etc.

Those vegetative species in Table 6.28 are predominantly slightly tolerant. With supplementary nutrients, the tolerance of various crops may be improved and, hence, change the appropriate tolerance category. Few detailed studies exist for evaluating moderately and very tolerant species, *i.e.,* there is a need to select economically valuable crops for high oil tolerance.

Table 6.28 should be used to determine the maximum dose rate in a plant-soil system. Application rates for other more tolerant crops must be determined for future research.

Data Needs for Land Application Design

To estimate the assimilative capacity for land application design, certain field data are needed if oils and greases are present in sufficient quantities to be a probable LLC. These data allow comparison of alternative sites to improve the cost-effectiveness by selecting the best site. Soil information, collected in the manner described in Chapter 3 includes (1) soil pH, (2) soil nitrogen levels, (3) macro- and micronutrient deficiencies, and (4) the textural clay, loam, sand, etc., and profile characteristics for the moisture-aeration status of the soil. The waste should be analyzed for C:N ratio, forms of nitrogen and levels of soil macro- and micronutrients. The nature of the oil, *i.e.,* petroleum- or animal/vegetable-based, should be determined. The agricultural efficiency of growing the more oil-tolerant crops should be examined for the land application sites under consideration. Recommendations of local agricultural advisers can be used. Climatic information should be obtained on the mean monthly temperature. These data will allow a better estimate of the plant-soil assimilative capacity for oil constituents in an industrial waste.

Case Situations of Land Application of Oil-Containing Industrial Wastes

Example Case I (Low Oil Sludge)

Assume that corn can be grown safely with damage at oil levels up to 0.5% soil weight and the industrial waste under consideration contains 100 ppm oil. Further, assume the average rate of oil degradation is 0.1% of soil weight per month (*e.g.*, sandy loam); and the initial oil content of the soil is 0.2% of soil weight at the time of application. Calculate how much sludge can be applied per hectare.

At first application, we can apply only 0.5-0.2 = 0.3% oil of the soil weight, which is equivalent to (0.003) (2.24 x 10^6 kg soil/ha 15 cm) = 6720 kg oil/ha, which will be present in (1 kg sludge/100 mg oil) · (6720 x 10^6 mg oil/ha) = 67.2 x 10^6 kg/ha of sludge. Thus, we can apply very high amounts of low oil sludge before the oil concentration would become limiting to corn production. At the assumed degradation rate of oil (0.1% soil weight/month), it is expected that 6720 kg oil/ha will be lost in 3 months from the date of application.

Example Case II (Medium Oil Sludge)

An oil refinery is producing 500,000 gallons of sludge per month with 2% (W/W) oil. Soil has an initial oil content of 0.3% of soil weight. Cotton will be planted 2 months after the sludge application and it can grow for 5 months with full vigor and production at a soil oil level up to 1.0% with no probability of damage, and the oil loss rate from soil is 0.2% of soil weight per month. Calculate the acreage required per year for the sludge disposed by the land application system.

Assuming the specific gravity of sludge to be 1.2, the total sludge produced every year would = 500,000 x 12 x 1.2 x 3.775 = 2.7 x 10^7 kg yr^{-1} = 2.3 x 10^6 kg $month^{-1}$ which is equivalent to 4.5 x 10^4 kg oil/month. Two months before planting we can dispose 0.4% oil over the allowable 1% (soil initially contained 0.3% oil). Thus, we can put 1.4 - 0.3 = 1.1% oil of soil weight in the upper 15 cm of soil, which is equivalent to 1.1 x 2.24 x 10^4 kg soil/ha/15 cm = 2.5 x 10^4 kg oil/ha/15 cm in the first month of application.

If oil is the only limiting constituent for sludge application, the total acreage required for first month would be:

$$\frac{2.3 \times 10^6 \text{ kg}}{2.5 \times 10^4 \text{ kg/ha}} = 92 \text{ ha}$$

In the second month of soil amendment one can put a maximum 1.2 - 0.1 = 1.1% oil, *i.e.*, total acreage required would be the same, 92 ha, as for the first month (one month before planting and the initial oil level being decreased by 0.2%/mo). In the third month we can put a maximum 1.0% oil, that is 2.24 x 10^4 kg oil/ha. Total acreage in the third month would be

$$\frac{2.3 \times 10^6}{2.2 \times 10^4} = 105 \text{ ha}$$

For fourth and fifth months the acreage would be the same, 105 ha, for each month. In the sixth and seventh months, the oil applied in the first and second months is decomposed to the initial soil level and one can then use a repetition of earlier acreage for a second application. So the total acreage required would be:

$$2(92) + 3(101) + 2(101 - 92) = 505 \text{ ha for an annual system}$$

The other way of applying the sludge at 505 ha land would be to spread every month at a rate of:

$$\frac{5 \times 10^5 \text{ gal/mo} \times 3.755 \times 10^3 \text{ cm}^3/\text{gal}}{(10^4 \times 10^4) \text{ cm}^2/\text{ha} \times 505 \text{ ha}} = 0.37 \text{ ha mm}$$

The monthly sludge should not be applied at less than 92 ha land, in which case the application rate would be 2.1 ha mm.

Example Case III (High Oil Sludge and Oils)

Suppose there is an accident and one million liters crude oil has been spilled from an oil pipeline over two hectares of land under an intensive cultivation system. What reclamation procedures would you suggest if a crop is to be planted within three months after the spill?

The areal loading rate is:

(i) $\dfrac{(10^6 \text{ l}) \times (10^3 \text{ cm}^3/\text{l})}{(10^4 \times 10^4 \text{ cm}^2/\text{ha}) (2 \text{ ha})}$ = 5 ha cm crude oil on surface

(ii) $\dfrac{(10^6 \text{ l}) (0.8 \text{ kg/l}) \times 100}{2 \text{ ha} \times 2.24 \times 10^6 \text{ kg/ha} (15 \text{ cm})}$ = 17.9% oil in soil by weight
(assuming special grade of crude crude oil to be 0.8 g/cc)

Crude oils contain a range of boiling point fractions. In permeable soils, dark oils tend to be filtered off in the top foot of oil, whereas the colorless fractions soak in more deeply (Plice 1948). If the situation permits, the excess surface soil should be burned. It is best to leave the dark oils to harden on the surface and scrap them up within a month when they have become friable. Oil penetration into the soil depends on soil permeability, porosity and moisture status of the soil at the time of spillage. Penetration in wet soils is slower than into dry soil. Oils to not penetrate a soil saturated with water.

Within 3 months the objective is to decrease oil percentage from 17.9 to 1.8%. That is, the oily layer of soil must be scrapped off from the 2 ha and spread over 20 ha of land evenly, and the scrapped land must be replaced with at least 15 cm of unpolluted soil to mix with remaining oily soil for dilution.

Heavy fertilization with nitrogen and phosphorus is another aspect of management practice for the 20-ha land evenly distributed with oily soil (1.8% of soil weight) which, in 3 months, will allow decomposition to levels of 1.2% soil permitting plant growth. Optimum moisture and aeration must be restored to promote the microbial degradation of oil. Frequent cultivation would expose oils to atmosphere, facilitate volatilization and open the soils for better aeration and moisture retention in favor of microbial degradation. Under favorable conditions—optimum moisture, aeration and pH—oil can be lost from soils at a rate of 0.2% of higher of soil weight per month. In case the oil loss rate from soils is less, bacterial seeding with oil oxidizing microbial inoculum must be considered in the beginning and pH must be adjusted near neutrality (between 6 and 8 pH). A few hundred pounds of soil from an oil-contaminated site, where microbial population has adapted to oil degradation, can be brought and spread on the lands for rapid degradation of oils. Also, an oil-tolerant crop must be considered for planting such as leguminous red clover, cotton and peas, etc.

REFERENCES

Antoniewski, J., and N. Schaefer. "Recherches sur les reactions des coenoses microbiennes de sols impregnes pars des hydrocarbures, modification de l'actirite respiratoire. *Ann. Inst. Pasteur, Paris* 123:805-819 (1972).

Atlas, R. M., and R. Bartha. "Biodegrdation of Petroleum in Seawater at Low Temperatures," *Can. J. Microbiol.* 18:1851-1855 (1972).

Baldwin, I. L. "Modification of the Soil Flora Induced by Application of Crude Petroleum," *Soil Sci.* 14:465-477 (1928).

Bell, H. S. *American Petroleum Refining,* 4th ed. (Princeton, NJ: D. Van Nostrand Co., Inc., 1959).

Bingham, S. C. "The Relative Rates of Microbial Degradation of Petroleum Oils, Vegetable Oils and Animal Fats," Project Report, Biol. and Agr. Eng. Dept., NCSU (1977).

Brooks J. W. "Environmental Influences of Oil and Gas Development With Reference to the Arctic Slope and Beaufort Sea," Bureau of Sport Fisheries and Wildlife (1970).

Carr, R. H. "Vegetable Growth in Soil Containing Crude Petroleum," *Soil Sci.* 8:67-68 (1919).

Colwell, R. R., J. D. Walker and J. D. Nelson, Jr. "Microbial-Ecology and the Problem of Petroleum Degradation in Chesapeake Bay," in *The Microbial Degradation of Oil Pollutants,* E. G. Ahearn and S. P. Meyers, Eds. LSU–SG-73-01, (1973), pp. 185-195.

Chun, M. J., R. H. F. Young and N. C. Burbank, Jr. "A Characterization of Tuna Packing Waste," *Proc. 23rd Purdue Ind. Waste Conf.,* (1968), pp. 786-805.

Cooney, J. J., and J. C. Walker. "Hydrocarbon Utilization by *Cladosperium resinae.*" in *The Microbial Degradation of Oil Pollutants,* D. G. Ahearn and S. P. Mevers, Eds., Center for Wetland Resources, LSU, Baton Rouge, LA (1973).

Davis, J. B. *Petroleum Microbiology,* (New York: Elsevier Publishing Co., 1967).

Davis, J. B., and J. P. Stanley. "Microbiological Nitrogen Fixation," U.S. Patent, 3, 2101 79 (1965).

Davis, J. B. *Petroleum Microbiology,* (New York: Elsevier Publishing Co., 1976).

Dobson, A. L., and H. A. Wilson. "Respiration Studies on Soil Treated With Some Hydrocarbons," *Soil Sci. Soc. Am. Proc.* 28:536-538 (1964).

Ellis, Jr., and R. S. Adams, Jr. "Contamination of Soils by Petroleum Hydrocarbons," *Adv. Agron.* V(13):197-216 (1961).

Evans, R. "The Effects of Different Loading Rates of Petroleum Oil on Microbial Populations in a Sandy Loam Soil," Project Report, Biol. and Agr. Eng. Dept., NCSU (1977).

Francke, H. C., and F. E. Clark. "Disposal of Oil Wastes by Microbial Assimilation," U.S. Atomic Energy Commission, Report Y-1934, Washington, DC (1974).

Gainey, P. L. "Effect of Paraffin on the Accumulation of Ammonia and Nitrates in the Soil," *Agric. Res.* 10:355 (1917).

Gibson, D. T. "Microbial Degradation of Hydrocarbons," in *The Nature of Sea Water,* E. D. Goldberg, Ed., *Physical and Chemical Sciences Report 1* (1974).

Giddens, J. "Spent Motor Oil Effects on Soil and Crops," *J. Environ. Qual.* 5(2):179-181 (1976).

Gudin, C., and W. J. Syratt. "Biological Aspects of Land Rehabilitation Following Hydrocarbon Contamination," *Environ. Poll.* 8(2):107-112 (1975).

Hass, H. F., M. F.Yantzi and L. D. Bushnell. *Trans. Kansas Acad. Sci.* 4:39-45 (1941).

Huddleston, R. L., and L. W. Cresswell. "The Disposal of Oily Wastes by Land Farming," Presented to the Management of Petroleum Refinery Wastewaters Forum at Tulsa, OK, Sponsored by EPA/API, WPRA & UT, 1976.

Jamison, V. W., R. L. Raymond and J. O. Hudson, Jr. "Biodegradation of High-Octane Gasoline in Groundwater," *Develop. Ind. Microbiol.* 16:305-311 (1975).

Jobson, A., M. McLaughlin, F. D. Cook and D. W. S. Westlake. "Effects of Amendments on the Microbial Utilization of Oil Applied to Soil," *Appl. Microbiol.* 27(1):166-171 (1974).

Jones, J. C., and M. A. Edington. "An Ecological Survey of Hydrocarbon Oxidizing Microorganisms," *J. Gen. Microbiol.* 52:381-390 (1968).

Kincannon, C. B. "Oily Waste Disopsal by Soil Cultivation Process," USEPA Report, EPA-R2-72-110, Washington, DC (1972).

Liu, D. L. S. "Microbial Degradation of Crude Oil and the Various Hydrocarbon Derivatives," in *The Microbial Degradation of Oil Pollutants Center for Wetland Reources,* D. G. Ahearn and S. P. Meyers, Eds. LSU, Baton Rouge, LA (1973).

McKenna, E. J., and R. D. Heath. "Biodegradation of Polynuclear Aromatic Hydrocarbon Pollutants by Soil and Water Microorganisms," Research Rep. No. 113, Univ. of Illinois at Urbana-Champaign, Water Resources Center (1976).

McKinney, R. E. "Biological Treatment of Petroleum Refinery Wastes," American Petroleu

McKinney, R. E. "Biological Treatment of Petroleum Refinery Wastes," American Petroleum Institute, New York (1963), p. 73.

Morrison, R. T., and R. N. Boyd. *Organic Chemistry* (Boston: Allyn and Bacon, 1962).

Murphy, J. F. "Some Effects of Crude Petroleum on Nitrate Production, Seed Germination and Growth," *Soil Sci.* 27:117-120 (1929).

Odu, C.T.I. "Microbiology of Soils Contaminated with Petroleum Hydrocarbons. I. Extent of Contamination and Some Soil and Microbiological Properties after Contamination," *J. Inst. Petrol.* 58:201-208 (1972).

Perry, J. J., and C. E. Cerniglia. "Studies on the Degradation of Petroleum by Filamentons Fungi," in *The Microbial Degradation of Oil Pollutants*, D. G. Ahearn and S. P. Meyers, Eds. Center for Wetland Resources, LSU, Baton Rouge, LA (1973).

Plice, M. J. "Some Effects of Crude Petroleum on Soil Fertility," *Soil Sci. Soc. Am. Proc.* 13:413-416 (1948).

Raisbeck, J. M. Ph.D. Thesis, University of Birmingham, England (1972).

Raymond, R. L., J. P. Hudson and V. W. Jamison. "Oil Degradation in Soil," *Appl. Environ. Microbiol.* 31(4):522-535 (1976).

Schwendinger, R. B. "Reclamation of Soil Contaminated with Oil," *J. Inst. Petrol.* 54:182-192 (1968).

Skinner, R. J. "Contamination of Soil by Oil and Other Chemicals," in *Aspects of Current Use and Misuse of Soil Resources Report, Welsh, Soils Discussion Group No. 14* (1973), pp. 131-141.

Smith, J. H. "Decomposition in Soil of Waste Cooking Oils Used in Potato Processing," *J. Environ. Qual.* 3(3):279-281 (1974).

Smith, J. L., D. B. McWhorter and R. C. Ward. "On Land Disposal of Liquid Organic Wastes through Continuous Subsurface Injection," in *Managing Livestock Wastes*, ASAE Proc. 275, St. Joseph, MI (1975), pp. 606-610.

Stone, R. W., M. R. Fenske and A. G. C. White. "Bacteria Attacking Petroleum and Oil Fractions. *J. Bacteriol.* 44:169-178 (1942).

Strawinski, R. J., and R. W. Stone. "The Utilization of Hydrocarbons by Bacteria," *J. Bacteriol.* 40:461 (1940).

Stebbings, R. V. E. "Recovery of Salt Marsh in Brittany Sixteen Months After Heavy Pollution by Oil," *Environ. Poll.* 1:163-7 (1970).

Traxler, R. W. "Microbial Degradatio of Asphalt," *Biotech. Bioeng.* IV: 369-376 (1962).

Udo, E. J., and A. A. A. Fayemi. "The Effect of Oil Pollution of Soil on Germination, Growth and Nutrient Uptake of Corn," *J. Environ. Qual.* 4(4):537-540 (1975).

Walker, J. D., and R. R. Colwell. "Mercury Resistant Bacteria and Petroleum Degradation," *Appl. Microbiol.* 27(1):285-287 (1974a).

Walker, J. D., and R. R. Colwell. "Enumeration of Petroleum-Degrading Microorganisms," *Appl. Environ. Microbiol.* 31(2):198-207 (1976).

Walker, J. D., and R. R. Colwell. "Microbial Degradation of Model Petroleum at Low Temperatures," *J. Microbial Ecol.* 1:59-91 (1974b).

ZoBell, C. E. "Microbial Degradation of Oil: Present Status Problems and Perspectives," in *The Microbial Degradation of Oil Pollutants*, D. G. Ahearn and S. P. Myers, Eds., Center for Wetland Resources, LSU, Baton Rouge, LA (1973).

CHAPTER 7

SPECIFIC ORGANIC COMPOUNDS

INTRODUCTION

Industrial waste streams often include organic compounds related to the input materials and in-plant processes characteristic of a particular manufacturing facility. Often, within the broad range of organics are specific compounds that are of concern because the compound (1) is present in large amounts, (2) is considered particularly recalcitrant, or (3) at certain concentrations represents a long- or short-term health hazard. As an example, the U.S. Environmental Protection Agency (EPA) has adopted a preliminary list for assessment of waste treatment alternatives (Table 2.1). These are termed priority pollutants. The list is by no means exhaustive since, for different industries, other compounds may be present in more significant quantities.

With the identification of specific organic compounds, based on one or more of the above concerns, the design of an industrial waste land treatment system must try to account for the respective assimilative capacity of these species. Using the basic constraint on land application system design (see Chapter 1) to prevent irreversible plant-soil contamination, the assimilative capacity should be developed for certain specific organic compounds applied to land. Their excessive application will lead to adverse levels in any crops grown presently or in the future and to lowered reliability for long-term, land-based treatment. This basic approach of land application (developed in this book)—to match the assimilative capacity of specific organic compounds found in an industrial waste and to prevent having to dedicate plant-soil areas after use for land application. This is the basis for federal regulations governing such waste management systems (Resource Conservation and Recovery Act).

The principles regarding assimilative capacity are for land treatment systems, not for landfill facilities. The data, pathways, and design criteria for specific organics are aimed at the surface soil treatment of industrial wastes.

221

These rates of application per unit land area are typically several orders of magnitude lower than those for landfills.

In general terms, the design data and decomposition behavior are similar for a number of the specific organics that occur in a diversity of industrial wastes. When an organic species is applied to a plant-soil system in a dose or application event, the soil concentration of that material is immediately raised. The magnitude of the resultant soil concentration depends on the amount applied per unit area and the depth of contact. As the amount applied increases, different plant responses occur. For a growing crop, no effect or a stimulatory effect on yield is often observed at low rates of compound addition (Decock and Vaughn 1975, Buddin 1914). If the rate of application is increased to a medium or high rate, the crop will evidence a decrease in yield response (Figure 2.2). Thus, the single-dose amount of a particular organic can be controlled to minimize adverse plant response by remaining below some critical soil level. In addition to adverse yield, the uptake level or crop concentration of a particular compound can also be used to establish the critical dose level in a plant-soil system of specific characteristics.

Once applied at the critical single dose amount, the organic compound would degrade at some finite, nonzero rate expressed in kg/ha/unit time. No more compound would be applied until soil concentrations had decreased to 10-20% above background levels. More of this constituent then could be applied to the plant-soil system up to the critical soil level. The resultant pattern of applications and soil concentrations is shown in Figure 2.3. This saw-toothed management of a particular organic represents a pattern and depicts the critical design data needed: (1) the critical single-dose levels, and (2) the rate of decomposition.

Additionally, certain classes of organics may be leachable, although the vast majority of organics remain near the soil surface during the decomposition cycle. Thus, for certain organics the soil-water transport potential would be necessary in design.

To establish the assimilative capacity for an organic compound in an industrial waste, a number of possible approaches must be used. However, since the state-of-the-art of plant soil assimilative capacities is just evolving, design criteria are not complete. This situation also exists for the understanding of conventional treatment performance relative to specific organics such as the priority pollutants. With the evolving land-based treatment technology, the emphasis is on understanding the basic pathways for assimilation and the procedures to determine needed data.

The design engineer should first look for specific available data on the organic compounds identified as major components in an industrial waste. Much of the discussion and data in this chapter contain such specific assimilative rates and are thus a source of design data. Literature, particularly in the agronomy field, contains data on other compounds not covered in this book, so is a source of direct data on specific compounds. Without direct references or data concerning the compounds of interest, the designer should examine the numerous pathways presented in this chapter and Chapter 6 as the next possible step. If the desired compound appears as an intermediate of the degradation process, then the probability of significant decomposition is high. One is more confident that microbial potential exists to degrade the chemical to be land applied, and that as a conservative estimate, the time required for stabilization is no more than for the parent compound.

If no direct reference or intermediate status can be established for a compound, the next step is to seek data for closely related chemical compounds. The greater the difference between compounds (*i.e.,* different or more substituents, different stereochemistry, etc.) the less meaningful are the resulting extrapolated assimilative data. The final step is to determine the assimilative capacity experimentally. Site-specific parameters can be included in these studies. If the design engineer must go to this step to obtain needed data, it is comparable to the laboratory kinetic or treatability studies often performed for the use of conventional treatment processes for industrial waste.

These four steps constitute a procedure for determining the assimilative capacity of specific organic compounds in industrial waste. The costs and difficulty increase as successive stages have to be undertaken. One cautionary factor should be remembered when dealing with available literature or test procedures. The data found often reflect a one-time response of the plant-soil system to a compound. Therefore, the evolution of greater assimilative capacity is not tested and results can reflect a limited design.

In this chapter often the only data available on decomposition of a specific compound are those related to laboratory culture experiments. Often the microorganism used is derived from the soil, but still the decomposition rates will not equal those developed in actual soils. The primary purpose of presenting such data is to indicate probable decomposition pathways and order of magnitude, and first estimates of rates. These are a starting point for refining the assimilative capacity data for specific organics as a part of land treatment design. It is important to establish some probability of decomposition of particular compounds in soil.

The rate of decomposition data for a wide span of organic compounds also differs widely. Some materials have half-lives of hours while others require months and years to be decomposed. The objective is to only apply these principal materials of concern to the plant-soil system at the rate of decomposition. With this design criterion, the absolute rate of treatment does not dictate whether a material can be land applied, but does control the rate of application.

Utilizing the assimilative capacity approach, the land site is always be-tween the critical soil level for plant phytotoxic effects and background levels as the organic material decomposes. Thus, within one or two cycles of waste application, the land treatment site is utilizable for some other purposes such as forestation or grass crops. In this manner, a land treat-ment site does not need special closure considerations or financial bonds. All that is needed is strict management and monitoring of operation.

This chapter examines the available information on certain classes of organic compounds of concern in industrial waste land application systems. These compound classes are

1. phenols and phenolics,
2. substituted biphenyls,
3. polyaromatic hydrocarbons (or polynuclear aromatics),
4. nitrosamines,
5. surfactants and detergents, and
6. miscellaneous compounds.

For each category, data, if available, will be presented for three major design factors: (1) the influence on soil properties, (2) the pathways and fate in the soil, and (3) the influence on the vegetative cover. The pathway sections are important to the design process of trying to further improve or enhance the assimilative capacities to make land-based treatment more cost-effective. Finally, a section is included on the consideration of oxygen demand for highly organic wastes and the design for aerobic soil treatment. The relation-ship of Chapter 7 to the overall pretreatment-land application methodology is given in Figure 7.1.

PHENOLS AND PHENOLICS

Phenols, quinones, and associated derivatives are low-to-moderate volatile antiseptics and occur in a variety of industrial wastes, such as those from coke industries; plastic and rubber substitute manufacturing plants; solvents and paints for vehicles; wood preservatives; textiles, medicines, perfumes and pharmaceuticals; pesticides manufacturing; coating, stripping and cleaning agents; petrochemical wastes and petroleum refineries; and processing of aircraft components. Industrial wastes vary in phenol concentration from as low as 3-10 ppm in rubber reclamation process waste to as high as 3,000-10,000 ppm in coke ovens raw waste (Table 7.1). Average phenol concentra-tion in wastewater from 16 petroleum refineries was about 135 ppm (Davis 1967). This could be decreased to 7.8 ppm by wastewater treatment. The im-portance of phenols and phenol derivatives is based on the large number of industries that produce these compounds in wastes and because of the wide

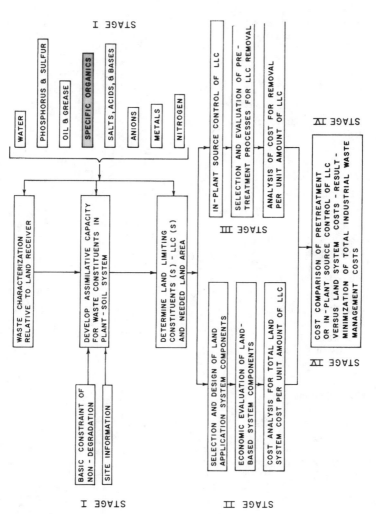

Figure 7.1 Relationship of Chapter 7 to overall design methodology for industrial pretreatment-land application systems.

Table 7.1 Summary of Phenol Concentration Reported in
Industrial Wastewaters (Sittig 1973)

Industrial Source	Phenol Concentration (mg/l)
Coke Ovens	
Weak ammonia liquor, without dephenolization	3,350-3,900
	1,400-2,500
	2,500-3,600
	3,000-10,000
	580-2,100
	600-800
Weak ammonia liquor, after dephenolization	28-332
	10
	10-30
	4.5-100
Wash oil still wastes	30-150
Oil Refineries	
Sour water	80-185 (140 average)
General waste stream	50-80
Poststripping	80
General (catalytic cracker)	40-50
Mineral oil wastewater	100
Petrochemical	
General petrochemical	50-600
Benzene refineries	210
Nitrogen works	250
Tar distilling plants	300
Aircraft maintenance	200-400
Other	
Rubber reclamation	3-10
Orlon manufacturing	100-150
Plastics factory	600-2,000
Fiberboard factory	150
Wood carbonizing	500
Phenolic resin production	1,600
Stocking factory	6,000

range and potentially high concentrations found in industrial waste. Furthermore, the aquatic receiver response to phenols is adverse to the extent that a phenol limit of less than 0.5 mg/l is often imposed. In comparison to the concentrations in Table 2.1, a limit of 0.5 mg/l represents, in many cases, an extremely high level of treatment. Therefore, since the plant-soil system does have a capacity to degrade phenol and phenol-derived compounds, there is an important economic justification to determine the nature and extent of the plant-soil assimilative rates. Phenol or phenol-based compounds that are commonly found in the soil-plant system are: (1) pesticides such as phenoxyalkanoates (*e.g.*, 2,4-D, and 2,4,5-T); (2) cyclic alcohols such as phenols, resorcinol, pyrogallol, cresols, naphthols, etc.; acids like cresylic acid, cinnamic, ferulic, *p*-aminobenzoic, *p*-hydroxybenzoic, syringic and vanillic; (3) quinones and hydroxyquinones; (4) nitrophenols; and (5) pentachlorophenols. These compounds are generated by the following mechanisms or sources:

1. decomposition of naturally occurring vegetation and relatively resistant lignins, and the formation of humus polymers and microbial biomass;
2. deliberate application of certain soil amendments and pesticides that result in formation of phenols and phenolics;
3. the application to soils of wastes containing phenols and their derivatives; and
4. industrial pollution, *e.g.*, near quenching towers of coke industry soils are most polluted with phenols, and somewhat less polluted near cooling towers.

Thus, phenols and phenolics occur in all soils. Typical concentrations in one study were: lignite—0.42 ppm phenolics, peat and red soil—0.25 ppm, and black soil—0.13 ppm phenolics (Purushothaman and Balaraman 1973).

As a general class of compounds, phenols and cresols, as well as quinone and hydroquinones, are relatively nonvolatile compounds, which can inhibit biological activity in the soil when present at sufficiently high concentrations. Therefore, it is important to determine these critical soil levels from available information or, knowing the behavior and needed data, determine the levels from experiments.

Influence on Soil Properties

Phenols and quinones inhibit enzyme activity in soils, especially urease (Douglas and Bremner 1971, Haselhoff 1932) and catalase (Dolgova 1975). As the phenol concentration in polluted soils decreased from 19.6 to 10.3 ppm, the catalase activity increased from 0.5 to 6.6 ml oxygen consumed in a 5-minute exposure (Dolgova 1975). In a control soil sample from a nearby botanical garden, the catalase activity was measured as 18.5 ml O_2 in a 5-minute exposure. The Chernozem sand and clay soils from the least polluted parts of an industrial area were fumigated with phenol in a hermetically sealed chamber using 160 mg/m^3 of phenol for 5 days. Phenol content of the soil was determined after the fumigation, whereas the catalase activity of the Chernozem, clay and sand soils were measured before and after the fumigation. Catalase activity expressed as ml O_2 after 5-minute exposure for Chernozem soil, clay and sand were 17.3, 9.5 and 0.3, respectively, prior to fumigation with phenol and declined to 13.3, 5.5 and 0.0 with fumigation, and soil levels of 104, 96 and 6.2 mg/kg phenol, respectively. Dolgova (1975) also observed a decrease in activities of dehydrogenase, protease and urease enzymes in the same soils receiving 20-50 kg/ha of phenols as a result of coke industry contamination.

The effectiveness of phenols and their derivatives as germicidal agents is well known. The phenol index has been used historically by pharmacists as a means of comparison of germicidal quality (Coffman and Woodbridge 1974). Despite their sterilizing influence, there are certain microorganisms that can withstand relatively high concentrations of phenolics.

Phenols exert an effect on soil microbial numbers that is dependent on the soil concentration or amount added. At low doses (0.01-0.1% of soil wt), the phenol serves as an available substrate and there is a dramatic increase in microbial values. This response is similar to that found with the addition to soil of other hydroxy compounds. As the dose level is increased (0.1-1.0% of soil wt), an increasingly strong inhibitory or sterilizing effect is noted. At these levels, a partial sterilization occurs in which there is a depression in microbial numbers, but not a complete die-off. After a period of time, microbes adapt or phenol is lost through sorptive inactivation or volatilization and a regrowth of population occurs. The milieu of available dead biomass is such that mineralization and excess nitrogen for microbial growth occur. The nonvolatile quinones and hydroquinones exert a prolonged microbial effect at certain soil levels, similar to the response described for phenols.

Doses from M/10 to M kept the soil biological activity in a partially sterilized state during an experiment of 75 days (Buddin 1914). At high doses of phenol, all protozoa were killed and the number of bacteria were reduced to below 5 million/g soil. Compared to volatile antiseptics, phenols can induce a partial sterilization effect at much lower doses in soils (Table 7.2). Also shown in Table 7.2 are the effects on ciliated protozoa (C) amoebae (A) and flagellated protozoa (M). Similar observations were evident with cresol (Figure 72.), quinone (Figure 7.3) and hydroquinone (Table 7.3). Purushothaman and Balaraman (1973) observed that all the phenol derivatives tested (anthranilic, hydroquinone, ferulic, gallic, p-hydroxybenzoic, p-aminobenzoic and cinnamic, *in vitro*) adversely affected the growth of *Rhizobium sp.* at 0.0025 *M* concentration; slightly inhibited at 0.0005 *M*; and significantly reduced in the range of 0.001-0.0025 *M* (Table 7.4). These *in vitro* results indicate greater inhibition than those measured for similar compounds in soils (Buddin 1914). Hydroquinone and p-aminobenzoic were the most inhibitory. With regard to soil physical properties, the presence

Table 7.2 Effect of Phenol on Allotment Soil Population of Bacteria and Protozoa at 16.5% Moisture (Buddin 1914)

Amount Phenol Added to Dry Soil (g/kg)	Bacteria Present (millions/g dry soil)				Ammonium and Nitrate Present (ppm of dry soil)			Effect on Protozoa
	At Start	After 16 Days	After 31 Days	After 74 Days	Initial Ammonium	At Start	After 74 Days	
Untreated = 0	23	28	16	15	4.5	28.5	30.5	C^a A^b M^c
M/200 = 0.47	17	101	59	18.5	6.5	26.5	32	C A M
M/100 = 0.94	12	548	96	28.5	8	29.5	34.5	$(C\ A)^d$ M
M/50 = 1.88	7.5	13.5	114	44	6.5	25.5	26.5	M
M/10 to M	5	4	3	4	4.5	24	23	None Present

[a]C: ciliated protozoa present.
[b]A: amoebae present.
[c]M: flagellated protozoa present.
[d]( ): not always found in cultures.

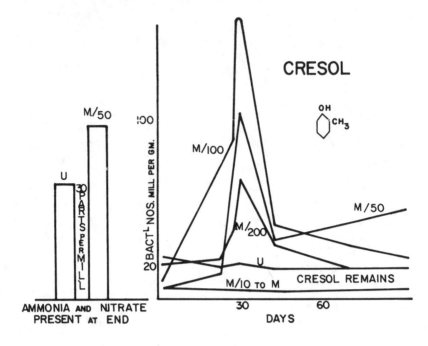

Figure 7.2 Effect of cresol on bacterial numbers, ammonia and nitrate in soils (U-untreated, M-108g cresol/kg dry soil) (Buddin 1914).

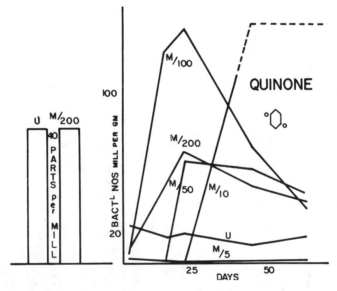

Figure 7.3 Effect of quinone on bacterial numbers, ammonia and nitrate in soils (U-untreated, M-108g quinone/kg dry soil) (Buddin 1914).

Table 7.3 Effect of Hydroquinone on Allotment Soil Population of Bacteria and Protozoa, and the Mineralization of Nitrogen at 17.5% Moisture (Buddin 1914)

Amount Hydroquinone Added to Dry Soil (g/kg)	Bacteria Present (10^6/g soil)			Ammonium and Nitrate Present (ppm)		Effect on Protozoa
	At Start	After 24 Days	After 79 Days	At Start	After 90 Days	
Untreated = 0	20	16	15	29	35	C[a] A[b] M[c]
M/200 = 0.55	4	55	30	26	44	C A M
M/100 = 1.1	1	20	24	24	35	(C A)[d] M
M/50 = 2.2	1	61	45	22	30	M
M/10 to M	0.1	0.1	0.3	22	21	None present

[a]C: ciliated protozoa present.
[b]A: amoebae present.
[c]M: flagellated protozoa present.
[d]( ): not always found in cultures.

Table 7.4 Effect of Certain Phenolic Compounds on the *In Vitro* Growth
of *Rhizobium sp.*[a] (Purushothaman and Balaraman 1973)

Phenolic Compounds	Concentration (M)	O.D. Value
1. Anthranilic Acid	0.0005	0.34
	0.001	0.31
H$_2$N —⬡— COOH	0.0025	0.06
	0.005	0.0
	0.01	0.0
2. *p*-Amino Benzoic Acid	0.0005	0.38
	0.001	0.08
N$_2$H —⬡— COOH	0.0025	0.06
	0.005	0.04
	0.01	0.03
3. Cinnamic Acid	0.0005	0.26
	0.001	0.12
⬡— CH=CH COOH	0.0025	0.12
	0.005	0.03
	0.01	0.0
4. Ferulic Acid	0.0005	0.32
	0.001	0.27
HO—⬡— CH=CH COOH	0.0025	0.19
	0.005	0.15
OCH$_3$	0.01	0.08
5. Gallic Acid	0.0005	0.42
	0.001	0.34
	0.0025	0.20
	0.005	0.08
	0.01	0.02
6. Hydroquinone	0.0005	0.12
	0.001	0.08
HO—⬡— OH	0.0025	0.0
	0.005	0.0
	0.01	0.0
7. *p*-Hydroxy Benzoic Acid	0.0005	0.38
	0.001	0.33
HO—⬡— COOH	0.0025	0.32
	0.005	0.29
	0.01	0.12
8. Control (without phenolic compound)	–	0.51

[a]Data represent average of three estimations.

of phenolic alcohols has been observed to improve the soil by stabilizing soil aggregates (Pokonova 1973).

Fate in Soils

Phenols and phenolics applied to a soil-plant system in wastes or pesticides may be subject to microbial decay, volatilization, photodecay, runoff transport or leaching. In soils, phenols are degraded microbially to catechol by extracellular phenol oxidases. The enzyme occurs abundantly in the organic and phenol-contaminated soils (Dolgova 1973). Most organisms that can metabolize phenols and phenolics are listed in Table 7.5.

Pseudomonas putida utilizes phenols and cresols by the *meta* or α-ketoacid pathway and metabolizes benzoates by the *ortho* or β-ketoadipic acid pathway (Murray and Williams 1973). Both pathways are illustrated in Figure 7.4.

Arthrobacter simplex metabolizes dinitrocresol to methylnitrocatechol, which is further transformed to trihydroxytoluene prior to ring cleavage. *Corynebacterium simplex* converts dinitrocresol to amino-nitrocresol, which is transformed to methylnitrocatechol, methylaminocatechol and trihydroxytoluene, successively. Trihydroxytoluene is then subject to ring cleavage.

Most organisms require one molecule oxygen more for phenol oxidation than catechol degradation. This is compatible with the idea that phenol is oxidized to catechol prior to further transformation of the aromatic ring (Varga and Neujahr 1970).

The abundance in soil of phenol-oxidizing organisms is of special ecological importance in the metabolism of natural and synthetic phenolics. Although phenols are degraded further by the oxidation mechanisms described for *in vitro* conditions, the significance of these processes in relation to adsorption, polymerization or incorporation of phenolic moiety into soil organic matter has not been established.

The phenols and phenoxy compounds can also be degraded by the soil microbes in a cometabolic fashion when a compound is not usable for growth but is still metabolized by a microbe. Differing substrate specificities of enzymes involved in a multistep reaction sequence may lead to accumulation of intermediate metabolites. For example, 2,4-diphenoxyacetic acid can be metabolized to phenols that would not support the growth of the metabolizing organism, but dichlorophenol from 2,4-dichlorophenoxyacetic acid is further degraded and used as a substrate by the same organism. Thus, the first compound would have to be degraded in a cometabolic fashion.

Pentachlophenols and other substituted phenols were studied under upland soil conditions (more characteristic of a land application site) and for flooded soils (Alexander 1972). All the phenolic compounds decomposed in the soil as evidenced by ring cleavage with the rate of degradation being dependent on the chemical structure and substituent location on the ring.

Table 7.5 Microorganisms Degrading Phenols, Phenolics and Like Molecules

Organism	Substrate → Product (where known)	Reference
1. Achromobacter	Dichlorodihydroxybenzene	Kaufman 1974
2. Acinetobacter NC1B9871	Cyclohexanol → adipic	Donoghue and Trudgill 1975
3. Arthrobacter sp.	m-Methoxybenzoate → formaldehyde + m-hydroxybenzoate	
4. Brevibacterium sp	Vanillate → protocatechuate	Raymond and Alexander 1972
	2,3,6 - TBA	Horvath 1971
5. Candida tropicalis	Phenol	Varga and Neujahr 1970
6. Cellulomonas	Coumarin → melilotic + commaric	Func 1971
7. Cephaloascus fragrans	Pentachlorophenol	Madhosingh 1961
8. Corynebacterium simplex	Dinoseb	Gunderson and Jensen 1956
9. Flavobacterium	Butyl-dinitrophenol Phenoxyalkanoics	Kaufman 1974
10. Fusarium oxysporum	4-amino-2-nitrophenol	Madhosingh 1961
11. Hendersonula toruloidea	Dihydroxytoluene, catechol, cresol, orcinol, pyrogallol, etc.	Martin and Haider 1976
12. Moraxella	Phenol	Dolgova 1975
13. Nocardia globerula	Cyclohexanol → acetyl CoA + adipic acid	Norris and Trudgill 1971
14. Norcardia opaca	Phenoxyalkanoics	Kaufman 1974
15. Pseudobacterium lacticum	Methyl - phenol or cresol	Dolgova 1975
16. Pseudomonas liquefaciens	Phenol	Dolgova 1975
17. Pseudomonas putida	Phenol, cresol	Murray and Williams 1974
18. Stachybotrys chartarum	Trihydroxybenzoic, gallic, salicylic, caffeic, dihydroxybenzoic	Martin and Haider 1976
19. Trichoderma virgatum	Pentachlorophenol	Cserjesi 1967
20. Trichosporon cutaneum	Phenol	Varga and Neujahr 1970

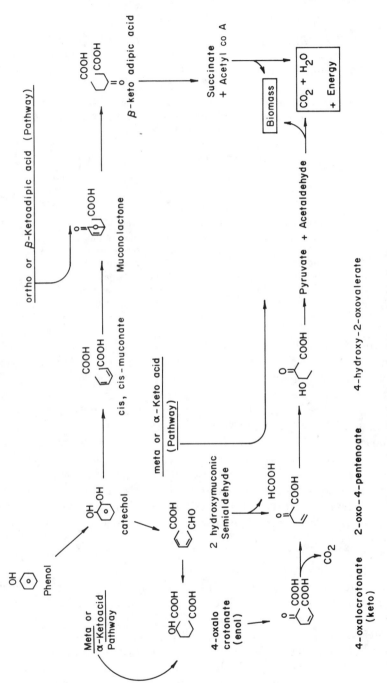

Figure 7.4 Alternative decomposition pathways for phenol.

Phenolic rings with substitution on the 2, 4 and 6 positions degraded within 2 weeks, whereas substitutions at the 3 and 5 positions resulted in delay of ring cleavage to 10 weeks (Table 7.6).

Table 7.6 Chemical Structure of Halogenated Phenols in Relation to the Decomposition by Soil Microorganisms Under Aerobic Condition (Alexander 1972)

Time for Ring Cleavage	
2 Weeks	10 Weeks
Phenol	3-chlorophenol
2-chlorophenol	3-bromophenol
4-bromophenol	2,3-dichlorophenol
4-chlorophenol	3,4-dichlorophenol
4-bromophenol	2,5-dichlorophenol
2,4-dichlorophenol	3,5-dichlorophenol
2,6-dichlorophenol	2,4,5-trichlorophenol
2,4,6-trichlorophenol	2,3,4,6-tetrachlorophenol

On a comparative basis, pentachlorophenol (PCP) as well as other substituted phenols were more rapidly metabolized under conditions of flooded soils than those of upland soils. The half-life for PCP at the 100 ppm level in soils, under upland or drained conditions, was approximately 40 days at 30°C (Figure 7.5). As the soil carbon content increased from 0.1 to 2%, the rate of PCP degradation was increased substantially for both upland and flooded soils (Figure 7.5). Kuwatsuka (1972) also found the degradation to be largely related to soil organic matter content. Soil pH and CEC were only slightly influential on the rate of PCP degradation, while soil texture, clay content, base saturation and free iron oxides were not related to PCP stabilization. These data on half lives for PCP and other substituted phenolics can be used to estimate soil assimilative capacities (kg/ha/yr). Furthermore, in determining any effect of soils, *e.g.*, between the benchmark soils (see Chapter 2), the clay loam (2.5% organic matter) would be more advantageous than the sandy loam (1% organic matter).

The relative degradation rates in soils of phenol and phenolic compounds were studied extensively by Medvedev and Davidov (1972, 1973, 1974), Medvedev *et al.* (1975) and Dolgova (1975). Applied at rates of approximately 1,000 ppm in a Chernozem soil, the rate of decay decreased in the order of hydroquinone > pyrocatechin > *p*-cresol > *o*-cresol > *m*-cresol (Medvedev and Davidov 1973). The authors proposed that the redox potential using chemical agents or the increased level of delocalization of negative charges in the hydrogen atoms of the benzene ring could be used to predict the relative rate of degradation of a series of phenolic or other compounds in soils.

Temperature variations affected the rate of phenol degradation (Medvedev and Davidov 1972, 1974). At 5°C, phenol remained in the soil after 16 days while at 19°C there was complete loss after 6 days. The ability to degrade phenol also improved with successive phenol doses (Medvedev et al. 1975). Initial degradation of phenols in soils has been enhanced by bacterial seeding of Pseudobacterium lacticum and Pseudomonas liquefaciens (Dolgova 1975).

With peachwood or humic acid present, the decay rate of phenolic acids (and benzoic acids) was enhanced (Dolgova 1975). The rate of soil decomposition of benzoquinone was found to be twice that of phenols (Medvedev and Davidov 1974), while naphthol and carbazole were relatively stable in soils.

Medvedev and Davidov (1972) listed the phenolic compounds in the wastes from coke industry and studied each for the duration of the existence in a chernozem soil when applied at a rate of 0.05% of soil weight. The substances in the coke industry waste may be arranged as follows on the basis of ease of biodegradation in the chernozem soil.

	Pyrocatechin = hydroquinone $>$
	Resorcinol = 2,4 − aminoxylene $>$
	Thymol = toluidine = 2,6 − aminoxylene $>$
	2,4 − xylenol $>$
	L − naphthol $>$
EASE OF	phenol $>$
	indol $>$
DEGRADATION	1, 4 − napthoquinone = 3,4-xylenol - n-toluidine $>$
	3, 5 xylenol = m-cresol = o-cresol = n-cresol $>$
	aniline $>$
	2,5 − xylenol $>$
	Fluorine − acenaphthene $>$
	Acridine $>$
	Carbazole = pyrene
	β-naphthol = β-naphthylamine = chrysin

The β-naphthylamine, β-naphthol, n- and o-toluidine, 2,4 − and 2,6 − aminoxylene and indol, in doses of 0.05% of soil weight, degrade to very stable transformation intermediate products, which remain in the soil for greater than 3 months. β-naphthylamine, β-naphthol and chrysin did not distintegrate in the soil, even at small doses (5 ppm in soil). Within a month, therefore,

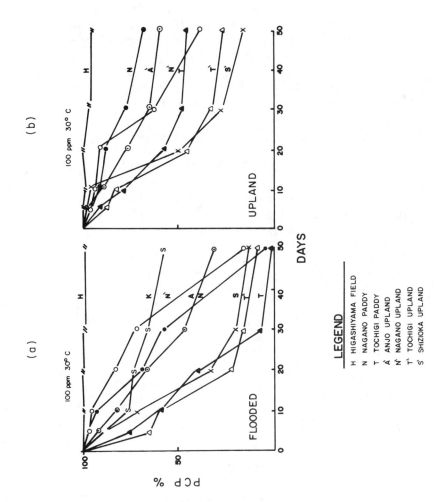

LEGEND

H HIGASHIYAMA FIELD
N NAGANO PADDY
T TOCHIGI PADDY
Á ANJO UPLAND
N´ NAGANO UPLAND
T´ TOCHIGI UPLAND
S´ SHIZIOKA UPLAND

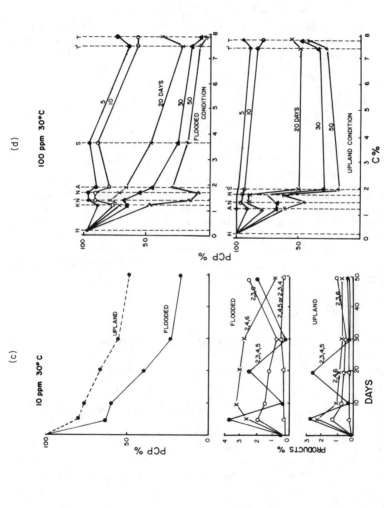

Figure 7.5 Polychlorinated phenol decomposition when applied to upland and to flooded soils at 10 and 100 ppm under temperature conditions of 30°C. Decomposition as a function of time and percent carbon in soil (Kuwatsuka 1972).

these substances should probably be considered recalcitrant under Chernozem soil conditions.

To estimate quantitatively the assimilation rates of phenol and phenolic compounds, research results were expressed in terms of percent degradation occurring in a certain number of days (Table 7.7). The application rates are also given. This table can be used as a starting point for estimating assimilative capacities or conducting experiments to quantify degradation of specific compounds more precisely.

Table 7.7 Rates of Degradation of Phenols, Phenolic Compounds and Derivatives

Phenolic Compound	Application Rate	Degradation Rate	Reference
Phenol		100% in 6 days	Medvedev and Davidov (1972)
		100% in 2 days	Alexander and Aleem (1961)
(benzene ring)—OH	0.05% of soil wt		
o-Cresol	0.05% of soil wt	100% in 8 days	Medvedev and Davidov (1972)
(benzene ring)—CH₃, OH			
m-Cresol	0.05%	100% in 11 days	Medvedev and Davidov (1972)
(benzene ring)—CH₃, HO			
p-Cresol	0.05%	100% in 7 days	Medvedev and Davidov (1972)
HO—(benzene ring)—CH₃			
2,4-xylenol	0.05%	100% in 4 days	Medvedev and Davidov (1972)
CH_3—(benzene ring)—OH, CH₃			

Table 7.7, continued

Phenolic Compound	Application Rate	Degradation Rate	Reference
2,5-xylenol	0.05%	100% in 14 days	Medvedev and Davidov (1972)
3,4-xylenol	0.05%	100% in 9 days	Medvedev and Davidov (1972)
3,5-xylenol	0.05%	100% in 11 days	Medvedev and Davidov (1972)
Thymol	0.05%	100% in 3 days	Medvedev and Davidov (1972)
Pyrocatechin	0.05%	100% in 1 day	Medvedev and Davidov (1972)
Resorcinol	0.05%	100% in 2 days	Medvedev and Davidov (1972)
Hydroquinone	0.05%	100% in 1 day	Medvedev and Davidov (1972)
α-Naphthol	0.05%	100% in 5 days	Medvedev and Davidov (1972)

Table 7.7, continued

Phenolic Compound	Application Rate	Degradation Rate	Reference
β-Naphthol	0.05%	Very little in 90 days	Medvedev and Davidov (1972)
1,4-naphthoquinon	0.05%	100% in 9 days	Medvedev and Davidov (1972)
		Depending on label, about 60% of 3-^{14}C in 1 week & 75% in 12 weeks.	Martin and Haider (1976)
Ferulic acid		Depending on label, about 23% of (2^3-^{14}C in 1 week and 44% in 12 weeks. 30-45% was respired as CO_2.	
Vanillic acid		82% in first week & 96% in 12 days. 65% in first week & 76% in 12 weeks. About (12-71) was respired. 41-78	Martin and Haider (1976)
Veratric		10-54% was respired as CO_2.	Martin and Haider (1976)
Humic		About 13% of 1-^{14}C in 1 week and 18% of 1-^{14}C in 12 weeks.	Martin and Haider (1976)
		About 4% of 2-^{14}-C and ring ^{14}-C in 1 week and about 1% in 12 weeks.	Martin and Haider (1976)

β-Naphthol structure (naphthalene with OH)

1,4-naphthoquinone structure ($O = \text{ring} = O$)

Ferulic acid: $HO-\bigcirc-CH=CHOOH$

Vanillic acid: $HO-\bigcirc-COOH$, OCH_3

Table 7.7, continued

Phenolic Compound	Application Rate	Degradation Rate	Reference
Fulvic		About 11% of 1-^{14}C in 1 week and 15% in 12 weeks	Martin and Haider (1976)
		About 2% of 2-^{14}C in 1 week and 8% in 12 weeks.	
p-hydroxybenzoic acid		100% in 1 week	Haider and Martin (1976)

Among the organic phenolics of the coke industry waste, hydroquinones, phenols, cresols, xylenol, thymol, etc. were decomposed quite readily by the soil microflora. Degradation of phenols occurred without the accumulation of intermediate metabolite (Medvedev and Davidov 1974). Benzoquinones also degrade rapidly and are completely metabolized within 3 weeks when added to soils at 10-100 ppm levels. The capacity of bacteria to utilize phenols at concentrations as high as 0.1% has been recorded (Putilina 1959). When phenol is the only carbon source in the medium the bacteria consumed 92% of it, but in presence of glucose, only 17% of phenol was utilized.

The effect of supplementing N and P to increase the soil degradation of phenols and phenolics has not been investigated. Since these constituents contain little or no N or P (very high C:N or C:P ratio), the response would be expected to be similar to the enhancement of soil assimilative capacity of oils by N and P additions. Further evidence of the need for N and P in decomposing phenols and phenolics is obtained in conventional secondary treatment of industrial wastewaters. A spore-forming bacteria could utilize phenol at the 450 mg/l level with 99% efficiency, when supplied with ample N and P (Webb 1964).

Recently, Haider and Martin (1975) investigated the decomposition rates of various phenolics in a Greenfield sandy loam (Figure 7.6). The rate of decomposition decreased in the order p-hydroxybenzoic > benzoic > caffeic. The caffeic acid in the polymer decomposed much more slowly than free acid without any combination. Haider and Martin (1975) and Martin and Haider (1976) found that the degradation rate depended on the position of label and type of compound.

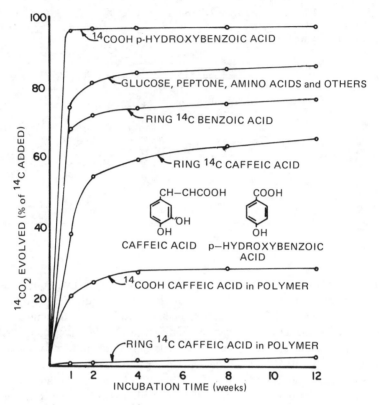

Figure 7.6 Decomposition of ^{14}C-labeled organic compounds (Haider and Martin 1975).

Volatilization and Photodecomposition

Phenols and phenolics have very low volatility, *e.g.*, 0.8 mm for phenol and 0.3 mm for cresols at 25°C. Hydroquinones volatilize very little with vapor pressures < 1 mm at 25°C. During winter, volatilization of phenolics may be negligible, while during summer, the losses due to volatilization may be measurable. Solar radiation may cause photosensitive reactions of phenolics such as the photonucleophilic mechanism of 4-CPA degradation.

Irradiation of Phenols and Phenolics

γ-radiation can destroy the phenol structure in aqueous solution. At phenol concentrations of 10 mg/l, almost complete destruction occurred at a dose rate of 1 million rad (Coffman and Woodbridge 1974). Destruction of phenolics by radiation takes place through oxidative radiolysis in the presence of water. Radiation-initiated reactions are shown by Coffman and Woodbridge (1974) as follows:

1. $HOH \xrightarrow[\text{radiation}]{} H \cdot + \cdot OH$
2. $RH + \cdot OH \longrightarrow R \cdot + H_2O$
3. $R \cdot + O_2 \longrightarrow RO_2^{\cdot}$
4. $RO_2^{\cdot} + RH \longrightarrow ROOH + R \cdot$

Efficiency of a radiation-initiated reaction can be expressed as either G-value, which is the number of oxidation reactions that occur as a result of absorption of 100 eV of energy, or as absorbed radiation dose in "rads" required to achieve the destruction of the phenolics.

Leaching Losses and Runoff Potentials

Phenolic acids vary in their ability to be adsorbed on clay minerals. The rate of adsorption decreases in the order: *p*-hydroxybenzoic > ferulic > syringic > vanillic. The rate and capacity of sesquioxides to retain phenolics far exceed that by kaolinite, illite and vermiculite (Huang *et al.* 1977). Under acid conditions, phenolic acids are adsorbed to a greater extent on clay than under neutral or alkaline conditions. When a sample of sand, clay and chernozem soil was fumigated with phenol in a hermetically sealed chamber using 160 mg/m^3 phenol for 5 days, the phenol accumulation was highest (104 ppm) in chernozem soil, intermediate (96 ppm) in clay and lowest (6.2 ppm) in sand.

There are no experimental data to substantiate the mobility of phenols, phenolics and their derivatives in the soil-plant system. The extent of loss by leaching and runoff at low loading rates (0.1% phenol of soil weight) are expected to be negligible. Surface application under adverse conditions (such as melting snow) in moderate-to-high amounts (1-10% phenol of soil weight) may result in substantial transport of these compounds with water and sediments; however, no quantitative estimate is yet available to this effect.

Influence on Plants

Using excised pea root segments, Decock and Vaughan (1975) demonstrated the maximum growth of segments at 0.05 m*M* concentration with desferal, 8-hydroxyquinoline or caffeic acid; at 0.01 mM of sodium-EDTA, 2, 2'-dipyridyl, chlorogenic, L-dopa, veratric, trans-ferulic, or cinnamic; and at 0.005 mM of 2-hydroxyquinoline or 4,4'-dipyridyl. Stimulation in growth was associated with the cytoplasmic fraction. At a concentration higher than 0.5 mM of desferal or L-dopa; 0.2 mM of sodium-EDTA and 0.1 mM of 8-hydroxyquinoline, 2,2' dipyridyl, chlorogenic or caffeic acid; 0.05 mM of trans-ferulic or cinnamic acid; 0.01 mM of 2-hydroxyquinoline or veratric and 0.005 mM of 4,4'-dipyridyl, the growth of excised pea root segments was negative. Inhibition of growth was associated with lower peroxidase activity. Buddin (1914) noted that soil treatment with 0.2% phenol enhanced the development of fibrous roots as much as the steam-sterilized soil. This was not associated with any marked increase in fruit production when all the plants were fertilized equally. The use of phenol in strength varying from 0.1-0.25% of soil weight has provided consistently good results with tomato plants in pots. The early growth was much better and the crop produced fruits earlier with phenol treatment of soil than in the untreated. The final yield of tomato was slightly better in phenol-treated pots than in the untreated.

Cresol is similar in action to phenol in its enhancing effect on root growth. Dolgova (1975) reported that the application of sodium humate in soil at 10 ppm, or 0.001% rate, increased the activitiy of enzymes in root zone and lowered the plant injury from phenols.

To summarize the constraints on crop growth due to the presence of phenols, it appears that concentrations of 0.25% produce no adverse effects, although applications above this may not necessarily decrease plant yields. More research is needed to obtain more reliable land application systems. For phenolic compounds, not much is known about crop response. Since the upper single dose limit at which crop yields are depressed is one design criteria for land application systems, (see *"Assimilative Capacity for Oils"*), more work is needed. Finally, the effect of various factors such as temperature and moisture on phenol-plant interactions remains an area for further research inquiry.

POLYCYCLIC AROMATIC HYDROCARBONS (PAH) AND SUBSTITUTED PAH

Polycyclic aromatic hydrocarbons (PAH) occur commonly in soils and other geological formations. The relative abundance of PAH has been documented (Blumer and Youngblood 1975 and Giger and Blumer 1974). Polycyclic aromatics, hydrocarbons or polynuclear aromatics (PNA) are formed by natural processes or by the activity of industry and society. Many of the basic PAH formation mechanisms are similar for these two sources, differing only in magnitude of generation. Some common PAH are present in Figure 7.7, ranging from two fused rings up to 5-8 rings of benzene. The accumulation and occurrence of PNA is ubiquitous in the environment and is considered to have influenced evolution.

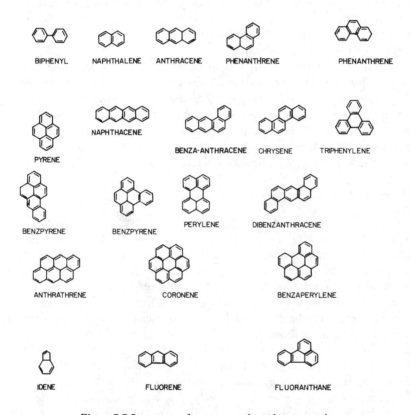

Figure 7.7 Structures of common polynuclear aromatics.

From natural causes, PNA are formed in fires and volcanoes, dispersed by wind and deposited on soil surfaces (Youngblood and Blumer 1975). PNA also occur naturally in the air, soil, water sediments as diterpenes, triterpenes, sterols, plant quinones pigment, etc., which are reduced and form aromatic structures over geologic time. Plant lignins are transformed into humic substances, which proceed to peats. As peat is converted to coal, condensed aromatics (PNA) are formed as a natural process. Aromatic nuclei in bitumen are linked through hydroaromatic ring systems such as 9, 10-dihydro-phenanthrene. Algae and plants can also synthesize PNA (Graft and Diehl 1966 and Borneff et al. 1968).

From the activities of society, the incomplete combustion of coal, wood, petroleum and other organic material is largely responsible for PNA generation. Industrial emissions, automobile exhaust and heating plants are contributors of PNA. Thus, municipal waste containing runoff from streets, etc., solids and liquids from combustion stack gas scrubbers, wastes from coking and wood distillation, coal gasification effluents and residues are examples of wastes that contain PAH, for which assimilative capacity of the plant-soil system needs to be established.

Within the class of PNA are several compounds that are carcinogens, with the usual vector being air particulates and drinking water. Two such species are benzo(a)pyrene and benzo(a)anthracene. The estimated annual benzpyrene emission in the U.S. includes approximately 207 tons from industry, 96.5% of which is from coking operations. Gasoline and diesel combustion add about 44 tons, power generation and heating about 475 tons and, from refuse burning, the largest amount—548 tons yearly.

Microbial Degradation

Microorganisms can be isolated from soil using classical enrichment techniques, which multiply on simple aromatics up to 3 ring-membered polycyclics, *e.g.*, benzene, biphenyl, naphthalene, phenanthrene and anthracene as the sole carbon source. However, higher PAHs such as 5 ring-membered polycyclics are not known to be utilized by microbes as a sole carbon source. The persistence of higher condensed PAHs (*e.g.*, pyrene, anthrathene, coronene, etc.) may be due to slight or no solubility of these compounds in aqueous nutrient solutions. Polynuclear aromatics, which are relatively soluble in water, are also easily degraded. Nevertheless, the higher condensed PAHs are not biologically inert (Groenwegen and Stolp 1976). Higher condensed PAHs can be cometabolized and cooxidized with other organics.

Aromatics are degraded more slowly than aliphatics and alicyclic nonaromatics. Sisler and Zobell (1947) noted wide distribution of PNA-degrading microorganisms in the soil environment. Microbes carry oxygenases that convert PNA to unstable cyclic peroxide, which is reduced to *cis*-glycol (Jeffrey *et al.* 1975). The *cis*-glycol is subject to further oxidation and ring cleavage. Gibson *et al.* (1975) succeeded in showing oxidation products of benzo(a)pyrene and benzo(a)anthracene in a bacteria of *Beijerinckia* type, which was isolated from a polluted stream. Using 100 g of soil continuously perfused with recycling water and air, Groenewegen and Stolp (1976) found that the order of the ease of breakdown was: pyrene, fluoranthene > phenanthrene > fluorene, benz(a)pyrene > benz(a)anthracene > crysene. These tests were of a 4-week duration and the authors recommended that period of 1-2 years might better reflect the potential of the soil to decompose these compounds.

McKenna and Heath (1976) examined the persistence and fate of selected PNAs in the soil environment and determined the structural limits of PNA biodegradation. They isolated *Pseudomonas* by enrichment on naphthalene and *Flavobacterium* for growth on phenanthrene. Both naphthalene and phenanthrene oxidize rapidly with the test organisms. The effect of various factors on utilization of PNAs by microorganisms is discussed below:

1. Effect of the number of fused rings: *Pseudomonas putida* and *Flavobacterium* sp. did not oxidize compounds with more than 3 fused rings at a significant rate. Anthracene, a 3-ring structure, was oxidized at a moderate rate by *Flavobacterium*. The 4-membered ring systems were oxidized at negligible rates (Table 7.8). None of the 4-membered ring compounds underwent oxidation for test conditions with *Pseudomonas putida* and only a small amount of oxidation occurred with *Flavobacterium* sp.

Table 7.8   Oxidation of Polynuclear Aromatic and Saturated Ring Hydrocarbons by Resting Cell Suspensions (McKenna and Heath 1976)

	Rate of Oxidation	
	Pseudomonas putida	*Flavobacterium* sp.
Compound	Naphthalene = 100%	Phenanthrene = 100%
Effect of Number of Fused Rings on Oxidation		
Naphthalene (2 ring)	100	79.8
Anthracene (3 ring)	9.8	36.0
Phenanthrene (2 ring + 1 ring)	67.0	100
1,2-benzanthracene (4 ring)	0	10.1
2,3-benzanthracene (4 ring)	2.7	1.1
Chrysene (4 ring)	0	7.9
Pyrene (4 ring)	1.1	10.1
Triphenylene (4 ring)	0	0
Effect of Alkyl and Phenyl Substituents on Naphthalene Ring Oxidation		
1-methylnaphthalene (2-ring substituted)	41.7	60.0
1-ethylnaphthalene (2-ring substituted)	39.0	36.3
1-phenylnaphthalene (2-ring substituted)	0	0
2-methylnaphthalene (2-ring substituted)	81.0	85.3
2-ethylnaphthalene (2-ring substituted)	70.3	45.0
2-vinylnaphthalene (2-ring substituted)	78.8	40.0
2-phenylnaphthalene (2-ring substituted)	1.4	3.9
1,3-dimethylnaphthalene (2-ring–double substituted)	39.1	0
1,4-dimethylnaphthalene (2-ring–double substituted)	8.8	19.9
1,5-dimethylnaphthalene (2-ring–double substituted)	8.8	0

Table 7.8 Continued

| | Rate of Oxidation | |
| | *Pseudomonas putida* | *Flavobacterium* sp. |
Compound	Naphthalene = 100%	Phenanthrene = 100%
1,6-dimethylnaphthalene (2-ring–double substituted)	1.4	14.4
2,3-dimethylnaphthalene (2-ring–double substituted)	88.0	84.2
2,6-dimethylnaphthalene (2-ring–double substituted)	15.5	62.8
2,6(and 2,7)-Di-tert.-butylnaphthalene (2-ring–double substituted)	0	0
2,3,5-trimethylnaphthalene (2-ring–double substituted)	9.0	17.5
2,3,6-trimethylnaphthalene (2-ring–double substituted)	7.0	9.6
1,2,3,4-tetraphenylnaphthalene (2-ring–double substituted)	0	0
Effect of Position of Methyl Substituent on Oxidation		
Naphthalene (2 ring)	100	79.8
1-methylnaphthalene	41.7	60.0
2-methylnaphthalene	81.0	85.3
Phenanthrene	67.0	100
1-methylphenanthrene	5.0	107
2-methylphenanthrene	23.5	88.0
3-methylphenanthrene	21.8	74.7
Fluorene	32.4	36.7
1-methylfluorene	0	75.6
2-methylfluorene	5.1	41.9
Effect of Saturation on Oxidation of Polynuclear Hydrocarbons		
Naphthalene	100	79.8
1,2-dihydronaphthalene	32.3	59.8
Tetralin	15.1	15.2
cis-decalin	4.4	0
trans-decalin	4.4	0
Phenanthrene	67.0	100
9,10-dihydrophenanthrene	12.0	23.3
1,2,3,4,5,6,7,9-octahydrophenanthrene	9.3	25.5
Perhydrophenanthrene	1.2	0.7
Indene	32.4	14.9
Indane	24.6	13.0
Hexahydroindane	6.8	1.0
Fluorene	32.4	36.7
Perhydrofluorene	0	3.4

2. Effect of ring substituents and position of substitution: Alkyl group, e.g., methyl, ethyl, vinyl substitution on naphthalene ring permitted decomposition at a moderate rate. The substituent at position 2 allowed a faster rate of oxidation than one at position 1. Phenyl group substitution on naphthalene ring decreased the oxidation to a negligible rate. *Pseudomonas putida* was more sensitive to substitution than the *Flavobacterium* sp. (Table 7.8). Substituted naphthalenes containing more than one alkyl or phenyl group also degraded; but, with an increasing number of substituent groups on rings, the oxidation rate was lowered. With increase in size of alkyl substituent, the oxidation rate was also lowered.

3. Effect of saturation: Increased degree of saturation in a given percent hydrocarbon resulted in a decreased rate of oxidation. As long as one aromatic ring was available, the oxidation by microbes occurred at a measurable rate. The perhydro compounds such as decalin and hexahydroindane were relatively resistant to oxidation.

In the biosphere, microbial transformation and degradation may be much slower than in laboratory experiments. Since accumulation of PNA in the environment is significant, the oxidative rate is much less than rate of formation.

Cometabolism or cooxidation studies show that in the presence of appropriate substrate, bacteria can utilize some PNAs, which it fails to utilize without the presence of its growth substrate. For example, *Flavobacterium* can cometabolize benzpyrene while growing on phenanthrene, but *Pseudomonas putida* cannot. Thus, microbes can degrade PNAs without using them as a sole source of C or energy.

Water-soluble PNAs were oxidized at greater initial rates than those with lower water solubility. The water solubilities of selected PNAs is given in Table 7.9. The recalcitrance of PNAs may be a result of (1) a lack of cell membrane permeability, (2) lack of substrate solubility, and/or (3) a greater degree of enzyme specificity.

All the six cultures of bacteria isolated from oil-polluted estuarine water could grow on naphthalene, 2-methylnaphthalene, 1-ethylnaphthalene and 2-ethylnaphthalene, 1,3-, 1,4-, 2,3-, and 3,6-dimethylnaphthalene with the production of metabolites such as methylcatechol, methylnaphthoic acid, dimethylsalicyclic acids, etc. (Figure 7.8, Dean-Raymond and Bartha 1975). Davies and Evans (1964) proposed the degradation scheme of naphthalene by *Pseudomonas* sp. as shown in Figure 7.9. As shown in Figure 7.8 (d), naphthoic acids appear to be the stable end products of metabolism of naphthalenes and dimethylnaphthalenes, when metabolized by marine microorganisms.

Metabolism of phenanthrene and anthracene has been studied by Evans *et al.* (1965) who found hydroxylation followed by ring cleavage, splitting of

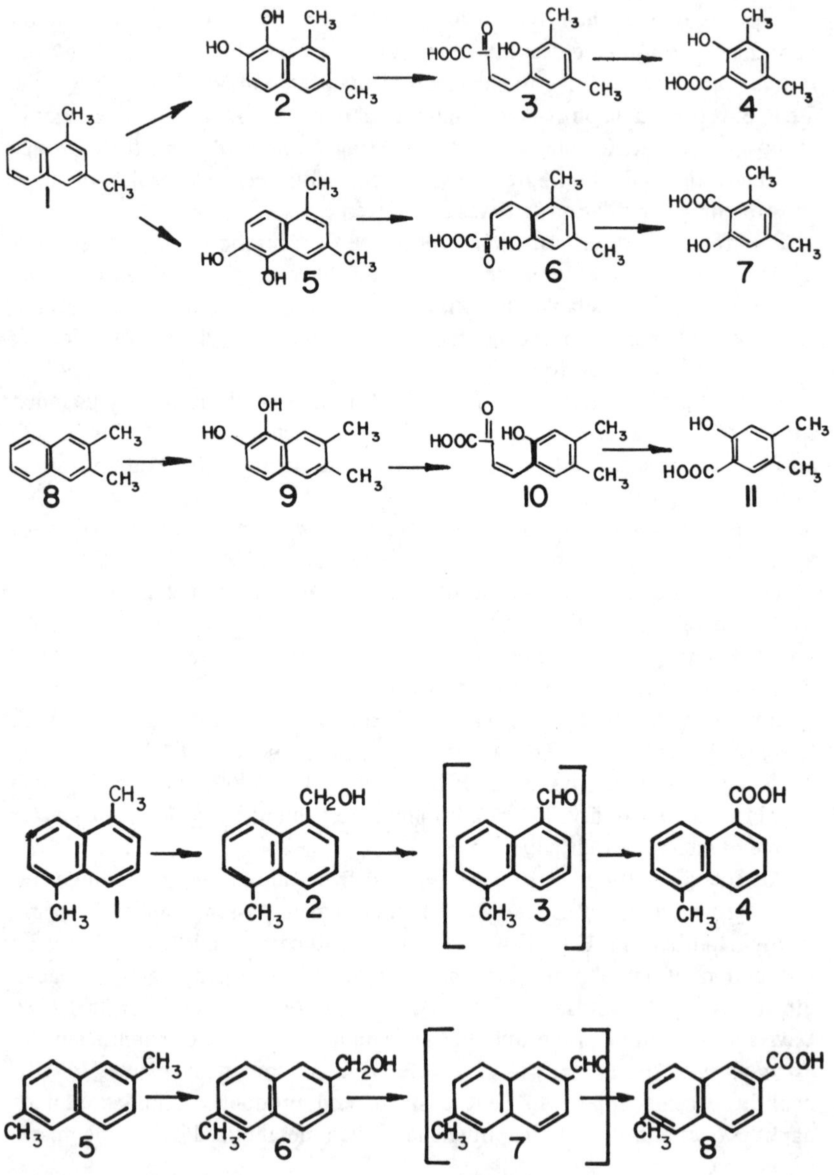

Figure 7.8 Decomposition pathways of 1, 3; 1, 2; 1, 5; and 1, 6 dimethylnaphthalene (Dean-Raymond and Bartha 1975).

Table 7.9   The Solubility of Some Aromatic Hydrocarbons in Water at 25°C (Kelvens 1950)

Substance	Solubility (μg/l)	Substance	Solubility of Substance (μg/l)
Benzol	1,860,000	Triphenylene	43
Toluol	500,000	Pyrene	175
Ethylbenzol	175,000	Chrysene	6
n-propylbenzol	120,000	Benzo(a)anthracene	10
n-butylbenzol	50,000	Naphthacene	1.5 A[a]
Naphthalin	13,500	Dibenz(a,h)anthracene	0.6 A
Phenanthrene	1,600		
Anthracene	75	Fluoranthene	265

[a]A: enrichment value.

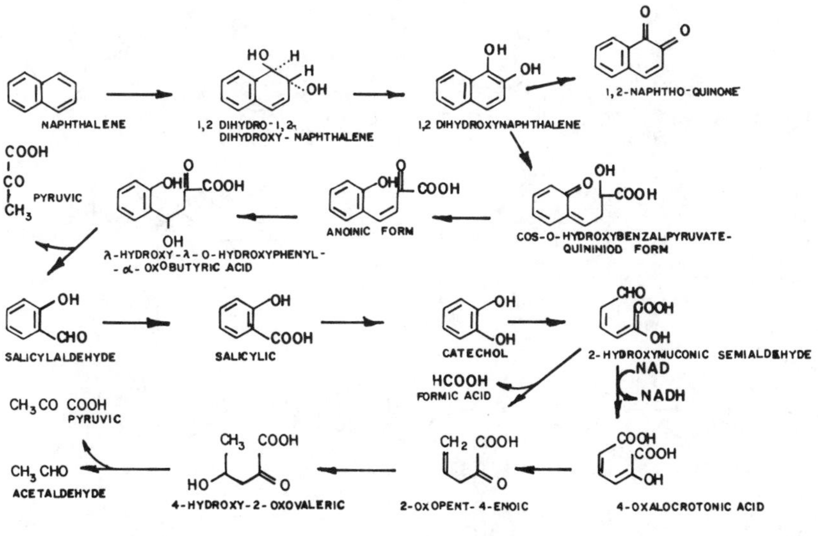

Figure 7.9   Metabolic pathway for naphthalene (Dean-Raymond and Bartha 1975).

a 3-carbon fragment and oxidation to an *o*-hydroxynaphthoic acid (Figure 7.10). This compound is oxidatively decarboxylated to a dihydroxy-naphthalene, which is metabolized by the naphthalene pathway. In marine microorganisms the lack of an enzyme capable of decarboxylating prevented further metabolism of naphthoic acid (Dean-Raymond and Bartha 1975).

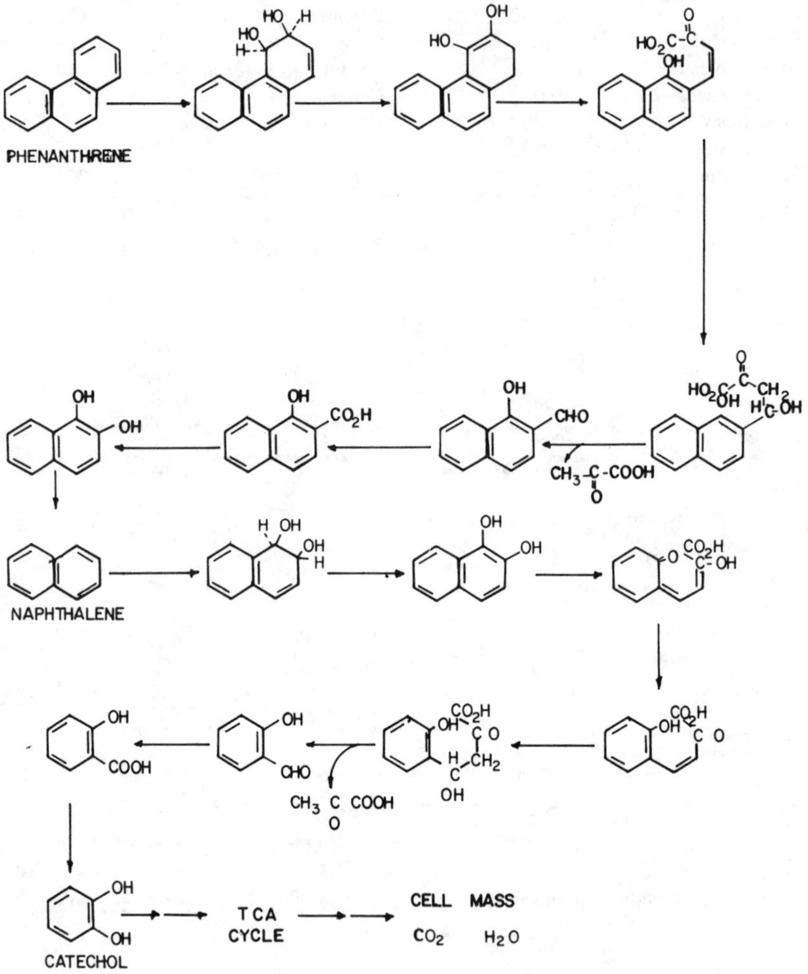

Figure 7.10 Breakdown of phenanthrene through naphthalene (Evans *et al.* 1965).

With respect to polynuclear aromatics, the quantitative assimilative capacity of the soil system has not been established. The existing data indicate that many microorganisms, isolated from soil or from polluted waters, can degrade PAH compounds. The rates of decomposition are dependent on the compound structure. Therefore, the evidence is that in the mixed and evolving microbial system present in a land application system, PNA compounds can be degraded. Furthermore, additions of nutrients, compounds favoring cometabolism, and bacterial seeding may enhance the rate of PNA stabilization. No data on the effect of PNAs on soil physical properties or on the plant effects were found to exist.

NITROSAMINES

Nitrosamines are probably not produced as an industrial chemical except in small research quantities. Therefore, the presence of nitrosamines in industrial wastes is as a by-product or naturally occurring contaminant associated with inputs and unit processes for manufacturing. Nitrosamines occur naturally in certain foods such as bacon, frankfurters and cured products, etc. Among the recommended list of priority pollutants, three nitrosamines are identified: N-nitrosodimethylamine, N-nitrosodiphenylamine, and N-nitroso-di-n-propylamine. In a limited survey of 21 industries by the U.S. Environmental Protection Agency (EDA), the N-nitrosodiphenylamine was found in the timber and leather categories and the N-nitroso-di-n-propylamine was detected in two textile plants at the range of 2-20 mg/m³ (Rawlings 1977).

Nitrosamines have not been found in natural soils or sewage. However, pathways exist for microbes in sewage to convert trimethylamine (TMA) to dimethylamine (DMA) and also to produce dimethylnitrosamine (DMNA). Nitrosamine can also be formed in acid soils from DMA and in acid sewage from the fungicide thiram (Ayanaba *et al.* 1973).

Influence on Soil Properties

The presence of nitrosamines has been studied to determine possible germicidal effects on soil microorganisms (Tate and Alexander 1976). Two soil species, *Bacillus cereus* (a spore former) and *Pseudomonas fragi* (a nonspore former), were tested for bacterial growth with N-nitrosodimethylamine up to levels of 100 ppm. No growth inhibition occurred for *B. cereus* as evidenced by the slope of the growth curve (Figure 7.11). The growth of

P. fragi was decreased at 10 ppm level. A series of 5 other soil species evidenced similar responses to the applied nitrosamine and all bacteria tested could grow in solutions containing even 1,000 ppm N-nitrosodimethylamine. Data are not available on broad classes of nitrosamines in relation to soil physical or chemical properties. Thus, the preliminary evidence, based on the nitrosamine, is that germicidal effects are not substantial at reasonable soil concentrations.

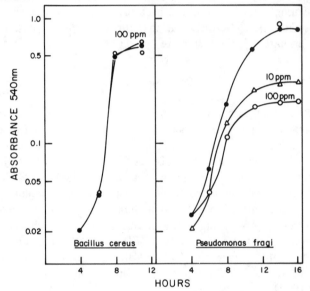

Figure 7.11 Effect of N-nitrosodimethylamine on bacterial growth in culture (Tate and Alexander 1976).

Fate in Soils

For the formation of nitrosamines in soil amended with dimethylamine and nitrite (Mills and Alexander 1976b) and in the decomposition of nitrosamines added to sewage (Tate and Alexander 1975b), when both sterilized and unsterilized conditions are examined similar rates are found. These results indicate that chemical reactions as well as biological phenomena are occurring in the transformations of nitrosamines.

N-nitrosodimethylamine was tested in a Williamson silt loam at several levels of application (Tate and Alexander 1975a). At the high rates of application (250 ppm) there was an initial nitrosamine reduction (3-20 days) and then a period of relatively constant concentration (Figure 7.12). At the lowest initial concentration of 22 ppm, no substantial initial loss occurred, but after a lag period of 35 days, decomposition begins. The higher application rate experiments were not continued long enough to determine the lag period prior to microbial decomposition. Two other nitrosamines were tested in similar tests (Tate and Alexander 1975b) and were found to degrade substantially in the soil (Figure 7.13). The N-nitrosodiethylamine and N-nitrosodipropylamine were applied at 18 and 13 ppm, respectively, and the lag periods were 10-15 days. The shorter lag period could have been due to the lower initial concentration or to the greater degradability of the compounds. Dressel (1976) detected a lag period for dimethylnitrosamine and diethylnitrosamine applied to a sandy soil, but not when applied to a clay soil. Decomposition was substantial after 14-35 days (Figure 7.14). Tests were done with soils both covered and uncovered, and the volatilization loss was determined to be small in comparison to the microbial stabilization (Dressel 1976). Neither experiment by Tate or Dressel examined the decomposition rates under successive nitrosamine application to better determine the assimilative capacity under conditions of land application.

To compare the soil microbial potential for degrading nitrosamines, the previously used Williamson silt loam was investigated under anaerobic flooded conditions (Tate and Alexander 1976). Application of N-nitrosodimethylamine at 17 ppm resulted in an initial drop to 13 ppm in 14 days and then no further loss for the duration of the experiment (63 days). Thus, the initial behavior was similar to the losses in aerobic soil, but even allowing for the extended lag period, no substantial degradation was found. Since most land application systems are designed to remain predominantly aerobic, the behavior of anaerobic systems does not determine the plant-soil assimilative capacity.

Nitrosamines are rarely, if ever, found in field soils. Hence, the assumption can be made that a degradation rate for these compounds exists and that it exceeds the natural nitrosamine formation rate. For industrial wastes containing nitrosamines, the soil system offers a very substantial assimilation capacity when compared to aquatic systems. In a set of experiments, Tate and Alexander (1975b) determined no change in N-nitrosodiethylamine, N-nitrosodimethylamine and N-nitrosodipropylamine when added to lake water over a 110-day period, as compared to the decomposition occurring in soils (Figures 7.12 and 7.13). Therefore, there are certain inherent advantages in the use of a soil receiver system for nitrosamines as compared with the traditional stream receivers.

Leaching

One laboratory study of the soil-water transport of nitrosamine was found (Dean-Raymond and Alexander 1976). Using 100- to 200-g samples of soils with a variety of organic content (1.9-14.8%), the authors reported that

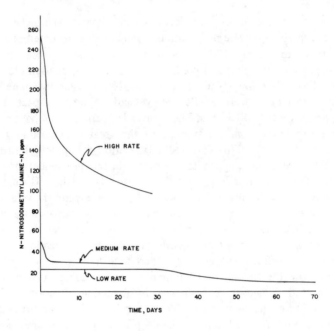

Figure 7.12 Decomposition of N-nitrosodimethylamine as dependent on rate of application to soil (Tate and Alexander 1975b).

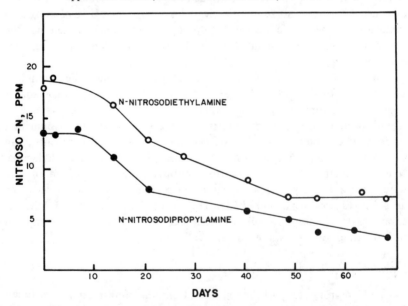

Figure 7.13 Disappearance of nitrosamine compounds (Tate and Alexander 1975b).

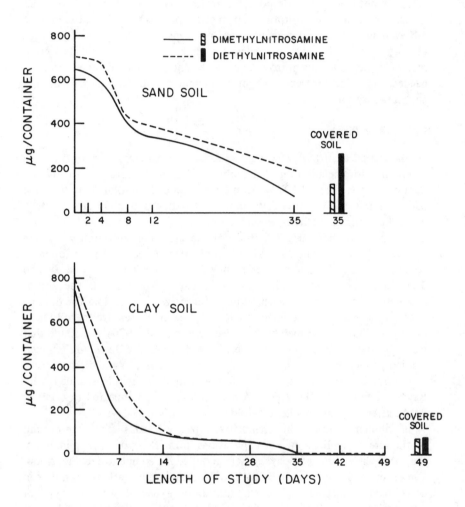

Figure 7.14 Behavior of dimethyl- and diethylnitrosamine in two different soils (800 μg nitrosamine/container (Dressel 1976).

dimethylnitrosamine moved at the same rate as chloride. However, these results are questionable in light of the properties of nitrosamines. Nitrosamines have a definite polarity, so the behavior of other polar organics in soil would indicate adsorption and an effect of organic matter. The rates of nitrosamine solution movement were not reported, but probably a high velocity was used which exceeded the time required for the absorption reaction, thus allowing nitrosamine movement. A more detailed study is needed to define the leaching potential under conditions present in land application systems.

Influence on Plants

The responses of barley, wheat, spinach, lettuce, carrots and tomato were determined for the addition of dimethyl- and diethylnitrosamine to soil at a level of 16 ppb (Dressel 1973). All species contained approximately the same concentration of dimethylnitrosamine (10.5-12.5 ppb plant dry matter), while the levels of diethylnitrosamine were slightly higher (11.5-14 ppb plant dry matter). A more detailed study of the plant concentration with time (Dressel 1976) was conducted. The two nitrosamines studied were found to accumulate to a maximum value at 4-5 days after application and then to decrease to background value within 12 days (Figure 7.15). The author postulated that an evapotranspiration pathway was responsible for the loss from plants, but the experiment could not discern the relative magnitude of evapotranspiration and metabolic losses. Sander et al.(1975) also found that plant-absorbed nitrosamine disappeared with time for Lepidium sativum (garden pepper cress). A final corroboration of the uptake and disappearence of nitrosamines from lettuce was performed with dimethylnitrosamine (Dean-Raymond and Alexander 1976). A peak concentration appeared to occur at 2 days after application with 150-fold reduction after 15 days.

In summary, nitrosamine assimilative capacity of soils appears to depend on the rate of application. For applications in the range of 15-20 ppm in the soil, between 70 and 150 days may be required for complete decomposition. The exact mechanism or the behavior for successive applications has not been determined. Variation in the stabilization is expected with the type of nitrosamine and with the rate of application. Crops harvested after 2-3 weeks following nitrosamine application would be expected to contain few nitrosamines. To reduce the tendency for substantial leaching, the nitroso-compounds should be applied at low rates to allow rapid decomposition. More field monitoring is needed to evaluate the decomposition-leaching phenomena with nitrosamine applications near the assimilative capacity rates developed above.

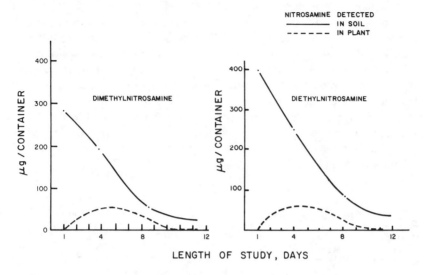

Figure 7.15 Behavior of dimethyl- and diethylnitrosamine in plant-soil system (nitrosamines added at 800 μg/container (Dressel 1976).

BIPHENYLS

Polychlorinated biphenyls (PCB) were first produced commercially in 1930 by U.S. firms and since then have been produced in Japan, France, Germany, Spain, Italy, Soviet Union and Czechoslovakia. PCB possess high heat capacity (fire retardants), high chemical stability and excellent dielectric properties; and, therefore, they have been widely used in a number of industries. Chlorinated biphenyls can carry 1-10 chlorine atoms resulting in 210 possible combinations, whereas chlorinated terphenyls (PCT) can carry 1-14 chlorine atoms with still greater combinations.

Polychlorinated biphenyls (PCB)

Polychlorinated terphenyls (PCT)

PCT have similar uses as PCB but their industrial application is limited. PCT environmental impact is expected similar to that of PCB (Stratton and Sasebee 1976). Industries using PCB and PCT are (1) electrical capacitors, transformers, heat transmitters, vacuum pumps, condensers, gas-transmission turbines and manufacturing facilities requiring compounds with excellent fire-retardant properties, (2) investment casting industries using wax materials, (3) plastics, wrapping paper, carbon paper, noncarbon paper, printing inks, paint, resins, dyes, tires and cooling systems, (4) plasticizers and stabilizers in pesticide sprays, and (5) chemical industries. Commercially used PCB is often referred to by trade names such as Aroclors, which are mixtures of PCB (Table 7.10).

Table 7.10 Approximate Composition of Selected Aroclors (Thurston 1971)

Chlorine Atoms	Wt % Cl	Aroclor[a] 1242	Aroclor 1248	Aroclor 1254	Aroclor 1260
1	18.6	3(0.7)			
2	31.5	13(16)	2		
3	41.0	28(49)	18		
4	48.3	30(25)	40	11	
5	54.0	22(9)	36	49	12
6	58.7	4(1)	4	34	38
7	62.5			6	41
8	65.7				8
9	58.5				1

[a]Numbers in parentheses represent values reported recently by Monsanto.

PCB has been shown to be toxic to birds and mammals. Biphenyls accumulate in livers and eggs of birds at concentrations as high as 900 ppm. The fire-retardant properties of the biphenyl molecule increase with chlorine substitution, but biodegradation decreases with increasing chlorination.

In Japan, "Yusho," i.e., poisoning caused by ingestion of rice oil contaminated with PCB, broke out in October 1968 and resulted in more than 1,200 officially certified cases. The official committee for Food Hygienic Research in Japan set the allowance level for PCB for Japanese adults as 5 μg/kg of body weight. PCB was detected in human milk studies in 1972 and 1973, and 30% of all Japanese babies ingested 75 μg/kg body weight of PCB. Therefore, PCB production in Japan was halted. In the U.S. also, the production and supply of PCB was decreased (NTIS 1972).

Because of the toxicity and accumulation effects of PCB, the assimilative capacity design must be considered conservatively. As will be shown, PCB and PCT decompose in a soil system, probably more rapidly than in any other waste management alternative or receiver system involving biological activity. Therefore, the objective is to establish rates of decomposition and soil levels above which the application of these materials will adversely affect the food chain. These will then allow utilization of the soil assimilative capacity to treat these compounds in an environmentally acceptable manner and with achievable economics. Research must be initiated to study enhancement factors that contribute to rapid degradation of PCB, PCT, PBB and PBT in soils. Such research would not only add to our present knowledge of PCB decomposition pathways by soil microorganisms under various environmental conditions, but may also substantially increase the soil assimilative capacity and develop land application technology as the probable most cost-effective alternative for disposal of biphenyl- and terphenyl-contaminated wastes.

Influence on Soil

The degradation rate of PCB and PCT in a soil system is relatively slow, and the single-dose level that will maintain soil concentrations below critical values is quite low; therefore, only small amounts per unit area would be expected to match the soil assimilative capacity. Certainly the assimilative rate is not zero, since PCT and PCB will decompose at finite rates. At the anticipated low application rates, large stimulation or depression of microbial populations, improvement in soil organic matter and soil physical properties, and alteration of the C:N ratio and soil nitrogen status are not expected. Instead, the PCB or PCT at low concentrations will be degraded by specific soil microorganisms within the overall carbon cycle of the soil system.

Some soil microorganisms will accumulate PCB. For example, *Aspergillus flavus*, a soil microfungus, accumulated PCB in a linear proportion to the product of the concentration of PCB in the medium times the percent chlorination of the PCB (Murado *et al.* 1976). This product was referred to as the "I" index and is shown in Figure 7.16. The presence of the PCB did not affect the capability of the soil microfungus to degrade other compounds such as aldrin. In the normal growth, death and mineralization cycle, these PCB would still become available to microbes that could degrade PCB.

Other soil microbes decompose PCB, thereby establishing a land-based assimilative capacity. *Achromobacter* sp. and *Pseudomonas* sp. have been shown to decompose PCB (Ahmed and Focht, 1973, and Wong and Kaiser 1975); however, the efficiency of such stabilization depends on acclimatizing or evolving the PCB degradation potential. Thus, studies of control soils first receiving PCB have not shown much potential for degrading it. Additionally, the greater the level of chlorination of the biphenyl or terphenyl, the lower the rate of degradation, because of the lack of dehalogenation capacity of most microorganisms. The resultant rates of decomposition vary with the PCB or PCT compound under consideration and must be established to determine the land-based assimilative capacity for these compounds in industrial wastes.

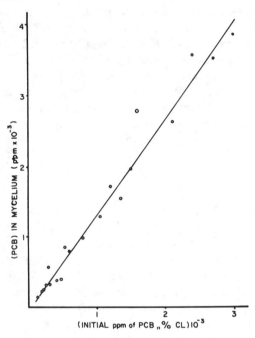

Figure 7.16   Correlation between mycelial accumulation of different Aroclors at initial different concentrations of Cl on biphenyl (Murado *et al.* 1976).

Microbial Decomposition Rates and Fate

Only very few reports are available on the degradation of PCB and PCT by microbes. Kaiser and Wong (1974) described the metabolic products of Aroclor-1242 (mono-, di-, tri and tetrachlorobiphenyls) degradation by bacteria. On average, Aroclor-1242 carried 42% chlorine, which is equivalent to a trichlorobiphenyl molecule. Their study involved lake water bacteria, which were found to degrade PCB at a 0.1% concentration. The bacteria isolated (*Achromobacter* sp. and *Pseudomonas* sp.) are also found in soils. Several of the metabolites were isohexane, isooctane, isoheptane, isononane and benzene derivatives such as dehyl-, isopropyl-, butyl- and n-propyl-benzene. None of the metabolites seem to carry any chlorine. Up to 0.1% Aroclor-1254, bacterial growth was not inhibited, and a slight stimulation was observed with Aroclor-1221 and -1242 at these levels. The stimulation could have been due to the ability of bacteria to utilize the PCB (Wong and Kaiser 1975). Figure 7.17 shows that bacteria could use 0.05% Aroclor-1221 and -1242 as a source of C and energy for growth. No stimulation of bacterial growth was observed in Aroclor-1254 medium, while Aroclor-1260 has also been shown to resist bacterial degradation (Oloffo *et al.* 1972, Wong and Kaiser 1975). The ability of bacteria to degrade PCB decreased with an increase in percent biphenyl chlorination. The results show that not only chlorination of the biphenyl decreases the bacterial degradation but also the position of the chlorine atom in the benzene ring is a determining factor. The degradation rates were found to decrease in the order: biphenyl > 2-chlorobiphenyl > 4-chlorobiphenyl (Figure 7.18).

Lunt and Evans (1970) have shown biphenyl degradation by a soil bacterium to phenylpyruvate. The biphenyl was hydroxylated in the 2,3 position and then a "*meta*-type" cleavage between C3 and C4 occurred. The product was a phenylpyruvate. Bailey and Bunyan (1972) found that PCB with a relatively low degree of chlorination were rapidly metabolized and are not as persistent as the more highly chlorinated compounds. In soils and plants, 2,2'-dichlorobiphenyl converted to phenolic substances.

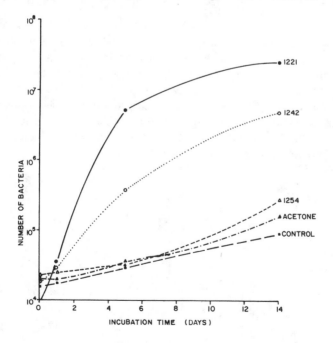

Figure 7.17 Growth curve of bacteria using 0.05% concentration solutions of various Aroclors (Wong and Kaiser 1975).

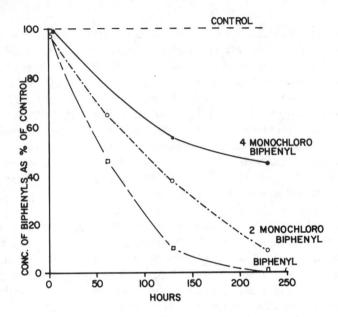

Figure 7.18   Decomposition of biphenyl and substituted biphenyl in growth media (Wong and Kaiser 1975).

The bacterial degradation of PCB isomer depends on the degree of chlorine substitution of the biphenyl and the stereochemical or spatial configuration of the substrate molecule. Ahmed and Focht (1973) selected two species of *Achromobacter* from sewage effluent grown on biphenyl and PCB as sole carbon sources. They studied (using cell suspensions) the decomposition of biphenyl to benzoic and *p*-chlorobiphenyl to *p*-chlorobenzoic acid. Chloride was not produced by either isolate during degradation of the chlorobiphenyls tested. The pathways hypothesized are as follows:

Although p-chlorobenzoic acid was refractory to further degradation by the *Achromobacter* sp., other species in soil may utilize it. *Achromobacter* sp. fail to mineralize the chlorobiphenyls completely to CO_2, H_2O and Cl^- because of the enzyme lacking to dehalogenate effectively. Rates of oxygen uptake during oxidation of di-, tri-, tetra- and pentachlorobiphenyls are shown in Figures 7.19 and 7.20. The adaptation of these microbes to degrade PCB appears to be a major reason for the ability to oxidize a variety of PCB.

Baxter *et al.* (1975) reported that the degradation of commercial mixtures of Arochlors was much more rapid than if present as pure compounds. They used *Nocardia* sp. and *Pseudomonas* sp. to degrade PCB. The decomposition rates for various types of PCB identify three groups of compounds. The first is for species containing two or fewer chlorine atoms which degrade rapidly (fewer than 10 days). Trichlorobiphenyls comprise a second group, which, after a short lag period, degrade rapidly until the percent degraded exceeds 90% and then the stabilization rate slows. The third group includes the more halogenated tetrachlorobiphenyls, which degrade more slowly (Figure 7.21).

Percentage loss of various PCB by two *Nocardia* strains is shown in Table 7.11. In the natural environment it is expected that a number of organisms can degrade PCB, since the two organisms used in this study are common in soil systems. A mixed culture grown on a mixture of PCB types may give more complete biodegradation than do single organisms on single isomers (Baxter *et al.* 1975). Aroclor-1254 in soil at a 10-ppm level during an 11-month period changed in composition—the less chlorinated biphenyls were slowly metabolized, while more highly chlorinated ones were not affected appreciably (Iwata and Gunther 1974). The soils were previously stored under air-dried conditions for four years. With *Pseudomonas putida* and *Flavobacterium* sp. isolated on PNA cultures, little biphenyl oxidation occurred (McKenna and Heath 1976).

Moza *et al.* (1976) identified the metabolites of 2,2'-dichlorobiphenyl-^{14}C in soils and plants:

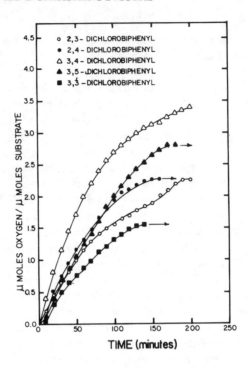

Figure 7.19 Oxidation of dichlorobiphenyls (arrow indicates no further growth) (Ahmed and Focht 1973).

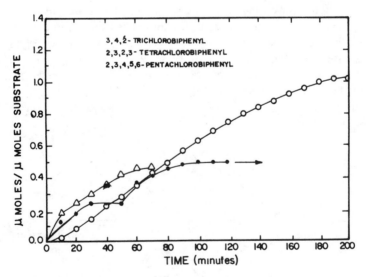

Figure 7.20 Oxidation of tri-, tetra- and pehtachlorobiphenyl (arrow indicates no further growth) (Ahmed and Focht 1973).

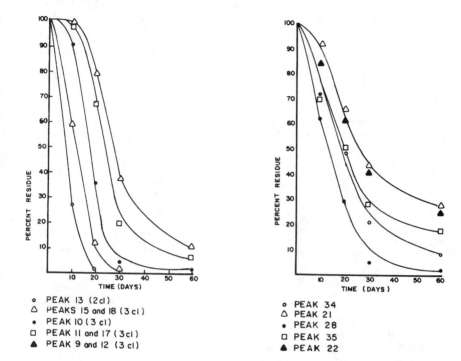

o PEAK 13 (2cl)
△ PEAKS 15 and 18 (3 cl)
• PEAK 10 (3 cl)
□ PEAK 11 and 17 (3cl)
▲ PEAK 9 and 12 (3 cl)

o PEAK 34
△ PEAK 21
• PEAK 28
□ PEAK 35
▲ PEAK 22

Figure 7.21 Degradation of (a) di- and trichlorobiphenyl, and (b) tetrachlorobiphenyl components of Aroclor 1242 exposed to microorganism NCIB 10603 (Baxter et al. 1975).

The metabolites isolated from carrot leaves were free phenols and conjugates of phenol (in 0.012 ppm). This hydroxymetabolite mechanism is in agreement with the report by Wallnöfer et al. (1973), in which gram-negative bacteria were reported to convert biphenyls to hydroxylated derivatives. Lower chlorine-content PCB components disappear from the incubation mixture faster than highly chlorinated ones.

The action of soil fungi, Rhizopus japonicus on 4-chlorobiphenyl and 4,4'-dichlorobiphenyl (Wallnöfer et al. 1973) resulted in formation of a "hydroxy" metabolite, the structure of which remained undetermined.

Volatilization and Photodecomposition

Moza et al. (1976) showed that 53.5% of added 2,2'-dichlorobiphenyl-^{14}C was lost in the first year and about 78.7% in the first two years from the soil-plant system by volatilization. The investigators were not able to distinguish between the loss of PCB or a resultant degradation metabolite;

Table 7.11 Loss of PCBs Over Time from Decomposition by Two Microorganisms,
NCIB 10603 and NCIB 10643 (Baxter 1975)

| | % Degradation Number of Days | |
Compound	NCIB 10603	NCIB 10643
2,4'-dichlorobiphenyl	70/7	60/73
2,4'-dichlorobiphenyl + biphenyl	100/7	100/77
4,4'-dichlorobiphenyl	nil/121	50/15
4,4'-dichlorobiphenyl + biphenyl	nil/121	50/10
2,3-dichlorobiphenyl	67/8	–
2,3-dichlorobiphenyl + biphenyl	64/8	–
3,4-dichlorobiphenyl	80/8	–
3,4-dichlorobiphenyl + biphenyl	100/8	–
2,3,2'-trichlorobiphenyl	50/7	–
2,3,2'-trichlorobiphenyl + biphenyl	95/7	–
2,3,4'-trichlorobiphenyl	94/7	–
2,3,4'-trichlorobiphenyl + biphenyl	100/7	–
2,5,4'-trichlorobiphenyl	nil/73	15/73
2,5,4'-trichlorobiphenyl + biphenyl	60/73	60/73
3,4,3'-trichlorobiphenyl	76/12	–
3,4,3'-trichlorobiphenyl + biphenyl	70/12	–
2,4,6-trichlorobiphenyl	nil/12	nil/84
2,4,6-trichlorobiphenyl + biphenyl	nil/12	nil/84
2,4,2',4-tetrachlorobiphenyl	nil/9	–
2,4,2',4-tetrachlorobiphenyl + biphenyl	nil/9	–
2,4,6,2'-tetrachlorobiphenyl	nil/9	–
2,4,6,2'-tetrachlorobiphenyl + biphenyl	nil/9	–
2,3,4,5,2',3'-hexachlorobiphenyl + +2,3,2'-and 2,3',4'-trichlorobiphenyl	nil	–
2,3,4,5,2',3'-hexachlorobiphenyl + 2,3,2 - and 2,3',4'-trichlorobiphenyl + biphenyl	50/11	–
Aroclor 1242	88/52	70/52
Aroclor 1242	95/100	85/100
Aroclor 1016	96/52	91/52
Aroclor 1016	>98/100	>96/100

hence, the PCB volatilization magnitude could not be ascertained at this time. Vapor pressures of selected Aroclors (Figure 7.22) vary from 10^{-2}-10^{-5} mm Hg at the ambient $20°$-$40°$C temperature range. This is considerably below that described previously for phenol and phenolics. The adsorption on soils would further reduce volatility losses when PCB were incorporated or moved below the soil surface. However, unless an acclimated microbial population evolved that could decompose the PCB, the long degradation times would result in further percent volatilization. Safe and Hutzinger (1971) studied the photolysis of polychlorinated biphenyls in nature. Chlorobiphenyls decomposed relatively easily on exposure to ultraviolet light in the laboratory.

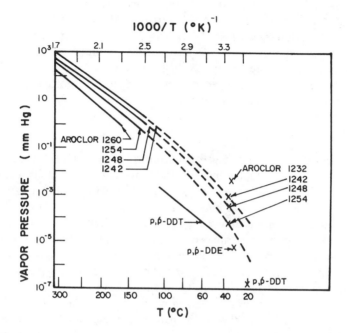

Figure 7.22 Vapor pressure of selected Aroclors, p,p'-DDT, and p,p'-DDE (Rall 1972).

Leaching and Movement

Tucker *et al.* (1975) studied the migration of polychlorinated biphenyls in soil columns coated with Aroclor-1016 and subject to water percolation. The application rate was 2.5% by weight of soil—a very high rate for a land application system but more representative of landfill operations. The soils used were Norfolk sandy loam, Ray silty loam and Drummer silty clay loam; water flowrates for each soil were 0.26, 0.53 and 0.32 l/day H_2O, creating essentially saturated flow. The outcoming PCB was quantitatively adsorbed on a polyurethane foam column, which was analyzed after extraction with acetone.

Soils containing a higher clay content retained PCB. The order in which breakthrough occurred as a function of effluent volume was: (1) Norfolk sandy loam; (2) Ray silty loam; and (3) Drummer silty clay loam. Haque *et al.* (1974) also demonstrated that clay has a high affinity for PCB. Water solubility of PCB is in range of 60-900 ppb. In leached waters the concentration of PCB was less than the solubility of Aroclor-1016 in water. Less than 0.05% of total Aroclor-1016 in soil column was leached from soil during a 4-month period, when 50-100 l H_2O was passed through the soils, which is equivalent to 50-100 ft of rainfall, assuming no runoff or evapotranspiration. The ease of leaching Aroclor-1016 from different soil types was in the order: Norfolk sandy loam > Ray silty loam > Drummer silty clay loam. It was observed that the less chlorinated and more degradable homologs leached from soils more readily than highly chlorinated and resistant organics. Thus, the more resistant types of PCB are retained, permitting greater time periods for microbial decomposition.

Influence on Plants

The occurrence of unchanged polychlorinated biphenyl (PCB) in environmental samples, including human food, is well-documented (Klein and Weisgerber 1976). Moza *et al.* (1976) considered PCB ubiquitous and chose carrots and sugar beets as indicators of xenobiotic accumulators to study their behavior and metabolism in a soil-plant system. When radioactive 2,2'-dichlorobiphenyl was applied to soils (at about 1 ppm), 44.4% of total radioactivity remained in soil up to a 30-cm depth, 2.1% was taken up by carrots and the remaining 53.5% was lost by volatilization. In the second year 21.1% of applied radioactivity was found in the soil, 0.2% in leaching water and < 0.1% in sugar beets and weeds. The remaining 78.7% of radioactivity had volatilized either as the parent compound or as an intermediate. Carrot root uptake exceeded that of sugar beet roots because carrots can take up lipophilic xenobiotics more readily than sugar beet (Moza *et al.* 1976). Carrot roots showed unchanged 2,2'-dichlorobiphenyl in a concentration of 0.24 ppm and phenolic metabolites from this PCB in concentration of 0.012 ppm (Moza *et al.* 1976). Carrots possess outstanding ability to absorb organochlorine pesticides from soils and, therefore, are expected to accumulate PCB as well (Iwata 1974). Carrot peel comprising 14% of carrot root weight carried 97% of PCB residue and very little was translocated in the plant tissue. The less chlorinated biphenyls are translocated to a greater degree from soil into carrot roots (Iwata and Gunther 1974).

To place these vegetable uptake data in perspective, the reader should note that carrots were selected because of the high propensity to uptake lipophilic species such as PCB. Data do not exist for nonhuman food chain crops or other more tolerant vegetative covers that might be used in a land application system. Flora and fauna, responding to PCB in aquatic systems as are used for stream discharge of industrial wastes, have bioconcentration factors of 300-800. However, for the carrot-PCB system of Moza *et al.* (1976), the PCB bioconcentration factor was 2, while the phenolic metabolites were not biomagnified. Although data on the land-based assimilative capacity for PCB (not a landfill situation) has not been firmly established, there appears to be substantial advantage in decreased bioaccumulation by use of this approach, if the design is performed correctly.

In summary, the assimilative capacity for PCB has not been established quantitatively. Studies incorporating PCB with normal agricultural soils have measured some degradation and a substantial volatilization over relatively long time periods. However, other investigators have demonstrated that with microorganisms found in soils, the adaptation or evolution that occurs can lead to a substantial capacity to degrade many PCB. This degradation, particularly of the less chlorinated compounds, is rapid, thereby reducing volatilization losses. While direct experiments are not available, it would appear from the evidence that with bacterial seeding, or *in situ* acclimatization of a land application area, a substantial degradation rate could be achieved, thereby minimizing volatile losses. This would constitute the assimilative capacity.

The resultant interaction of PCB and the vegetative cover used on a land application system can be controlled. Crops with lypophobic uptake capacity, nonfood chain crops, or crops that are returned to the soil promoting further decomposition can be used to reduce the PCB impact. Furthermore, the natural regulation systems in the soil-plant cycle appear to keep biomagnification quite low in comparison to the alternative stream receiver systems.

AROMATICS AND SUBSTITUTED AROMATICS

Compounds in the benzene class are very commonly used in the industrial sector as solvents and reactants. Synthetic textiles, plastics, pesticides and petrochemicals are among the major users of aromatics and substituted aromatics and, hence, would have such compounds in the waste streams. This creates the need to examine the plant-soil assimilative capacity for this organic class. Some of the commonly encountered substituted aromatics are:

1. methyl-, ethyl-, isopropyl-, propyl- and butyl-substituted, *e.g.*, toluene and dimethylbenzene;
2. halogen substituted, such as pentachlorobenzene, hexachlorobenzene, 1,4-dichlorobenzene;
3. nitro substituted, such as nitrobenzene, and 2,4-dinitrotoluene;
4. mixed substitutions, *e.g.*,

toluidine

fungicide (PCNB)

quintozene

tecnazene impurity in fungicide

2,4-methylaminoxylene

methylthiopentachlorobenzene

5. amino substituted, *e.g.*, aniline; and

6. benzidine,

With the recent concern for occupational safety, there is a trend away from benzene as a solvent and toward the use of compounds such as toluene. A variety of the aromatic compounds are also on the current EPA list of priority pollutants.

Influence on Soil

Application of high levels of aromatics can induce changes in the physical properties of soils. Aromatics can induce hydrophobicity in soils, with treated samples not absorbing water during the course of several minutes, and even hours, dependent on the substance introduced (Medvedev and Davidov 1972). Water-repellent qualities were contributed to soil by application of 500 ppm in the soil of aniline, *o-* or *p*-toluidine and 2,4- or 2,5-methylaniline.

Aromatic compounds applied to soils may alter biological activity or microbial species and are often used at high doses for chemical sterilization of the soil. However, depending on the dose level, the initial sterilization may be nil, partial or complete. Therefore, the determination of an assimilative capacity centers on not exceeding critical soil levels, at which large-scale sterilization would occur.

Using a large dose rate of toluene (1.7 g/g dry soil, 3.4×10^6 kg/ha), there was a drastic reduction in bacterial and fungal populations with total elimination of actinomycetes (Vishwanath *et al.* 1975). Toluene at this level was more lethal than chloroform and eliminated most dehydrogenase, saccharase, urease and phosphatase activity (Cerna 1970). *Bacillus* sp., *Aspergillus* sp., gram-negative rods and gram-positive cocci survived this level of toluene addition. In comparing soil types, organic matter appears to provide some protective action for microflora, hence the sterilization effects are more pronounced in light-textured soils than in heavy clay or organically rich soils. Relative to the benchmark soils used in this book, the clay loam ($\sim$ 2.5% organic matter) would provide greater assimilative capacity for aromatics than the sandy loam ($\sim$ 1% organic matter).

As early as 1914, Buddin reported that toluene exerted a partial steriliza-tion effect on soils at a much lower concentration (500 ppm of soil wt) than chloroform (1,200 ppm of soil wt). Ammonia accumulates in the partially sterilized soils in excess over control, especially in soils rich in the organic matter. In low organic matter soils, the rapid development of fungi may lead to a partial utilization of ammonia (Waksman and Starkey 1923). Waksman and Starkey (1923) found no correlation between the increase in bacterial numbers and accumulation of ammonia when soils with different organic matter contents were compared after the application of toluene. Bacteria, fungi, actinomycetes and protozoa were affected by toluene application to soil.

At the microbiological level, toluene and other aromatic compounds affect urease activity (McGarity and Myers 1967, Anderson 1962). The toluene added as a bacteriostat can (1) decrease all extracellular ureolytic activity, (2) increase the cell wall permeability of certain microbes, thereby increasing the intracellular urease activity, or (3) cause lysis of susceptible cells with subsequent toluene inhibition of urease. The effects depend on the level of toluene with the inhibitory effects becoming more pronounced at high applications of toluene (Thente 1970). The toluene influence on enzyme activity is also quite dependent on soil type (Thente 1970). Trichlorodinitro-benzene (TCDNB) killed fungi in soil at 5,000 ppm Olpisan level, while at the 500 ppm level, the fungi were not affected (Naumann 1970). TCDNB does not affect *Penicillium* even at 5,000 ppm Olpisan level. The soil respiration, bacteria and actinomycetes were increased by 5,000 ppm Olpisan, and especially at 500 ppm level Olpisan there was an increased rate of oxygen consumption for a loam soil (Figure 7.23). Toluene is metabolized through the corresponding alcohols and aldehydes to benzoate and catechol, respec-tively, and then by the divergent α-keto acid pathway (or *meta* pathway). The isolates may lose their ability to utilize alcohols, aldehydes and, consequently, the subsequent α-ketoacid pathway may not work. Instead, the isolates may utilize benzoate and benazaldehyde by the alternate *ortho* pathway (β-keto-adipic acid route). Benzene is decomposed first to catechol, after which the decomposition proceeds via the normal soil metabolic pathways (Figure 6.3). While the pathways for decomposition of benzene and toluene are well established, the rate of breakdown has not been measured. One study currently underway will determine the assimilative capacity for toluene (Overcash 1978).

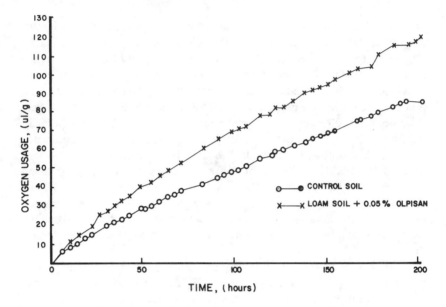

Figure 7.23 Oxygen consumption by microorganisms in response to application of Olpisan to soil (Naumann 1970).

Fate in Soils

Most of the information on the decomposition of aromatics is related to the mechanisms of degradation and are often studied with a single microbial species commonly found in soils. The pathways for aromatic decomposition are given in Figure 7.24. Toluene and other alkyl benzenes appear to be decomposed readily in single species experiments and it is anticipated that, when present at low to intermediate doses in soils, the large mixed microbial population will degrade these compounds readily. The pathways for degradation of alkyl benzenes are common in soil organic cycles. As the dose rate increases, the sterilizing effect described earlier will become substantial.

Aniline, when added to soil at the 500 ppm level, was lost from soil, presumably by degradation, in 13 days, p-toluidine in 9 days, o-toluidine in 3 days, 2,4-methylaniline in 2 days and 2,6-methylaniline in 3 days. Transformation products of aniline did not exist in soil more than 13 days, while that of toluidine and methylaniline persisted well over 90 days (Medvedev and Davidov 1972). The nature of these persistent metabolites was not documented.

Dimethylbenzene (xylene) is degraded by several strains of *Mycobacteria* and *Nocardia* in soils (Skriabin *et al.* 1971). Under cooxidative conditions in presence of hexadecane, these organisms transformed o-, m- and p-xylenes into aromatic acids ranging from monocarboxyl acids to dicarboxyl and

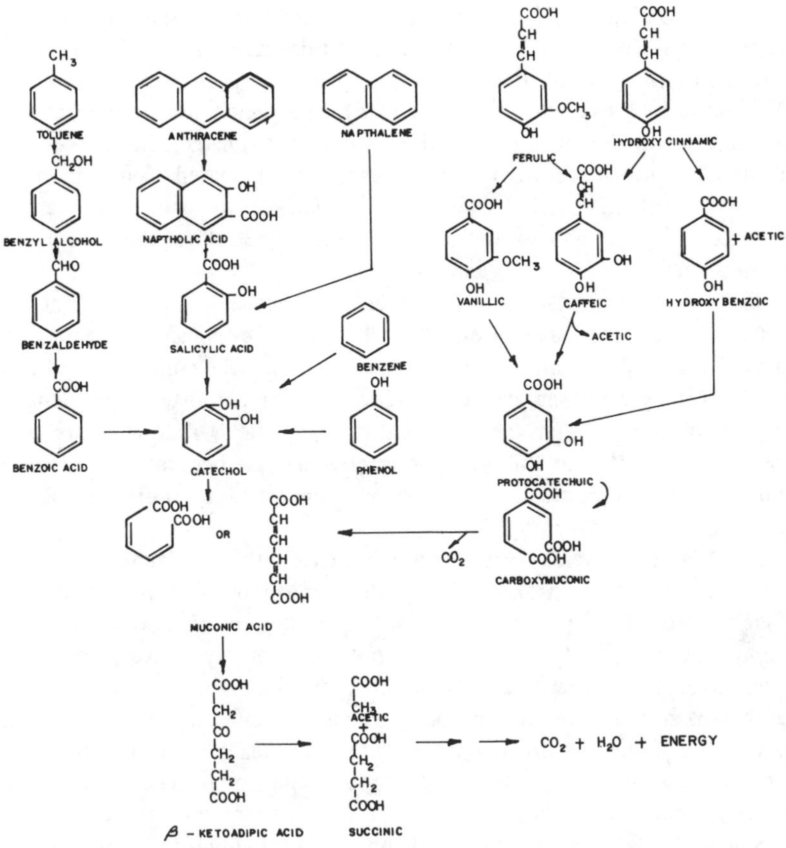

Figure 7.24 Decomposition pathway of various aromatics.

hydroxycarboxyl acids. These species could not metabolize the dimethyl-benzenes in the presence of glucose. Only in the presence of hexadecane was it possible for the microorganisms under study to carry out the transformation. Based on the end product of xylene metabolism, Skriabin *et al.* (1971) put the different species into three groups: (a) those producing monocarboxylic acids, *e.g., Nocardia corallina, N. rubra, N. rubropertincta, N. convalata, N.* species, *Mycobacterium rubrum,* (b) those producing dicarboxylic aromatic acids, *e.g., Nocardia* species and two unidentified strains of *Mycobacteria,* and (c) those yielding hydroxycarboxylic acids, *e.g., Nocardia globerula, N. opaca, N. erythropoli Mycobacterium hyalinum*

BKM-353, *M. mucosum* BKM-353, *Miperrugosum* BKM-850, *M. mucosum* C/X 310 and *Proactinomyces gl berulus.*

Beck and Hansen (1974) conducted experiments on the degradation of quintozene (PCNB) in soils and found a half-life ranging from 699-213 days, with an average of 468 days. Two quintozene application rates were used (10 and 60 kg/ha) with a similar half-life obtained for both rates. The moisture was maintained at 60% of field capacity and temperature 18-20°C. In the laboratory tests, volatilization was nearly eliminated. Hence, these loss rates represent microbial degradation. The half-life for pentachlorobenzene averaged 270 days and for hexachlorobenzene 1,530 days where both compounds were applied at 10 kg/ha.

In field experiments where soils had been receiving quintozene for 2-10 years, samples were taken from 22 soil types to establish the degradation rates. Calculated half-life of quintozene from 22 field samples varied from 117-1,059 days, with an average of 434 days. The half-lives were calculated on the assumption that the degradation in the fields follows a first-order reaction, and although this was not strictly the case because of variation in climate, etc., the values found were comparable to the results from laboratory experiments.

As a result of several years of soil treatments with quintozene compounds, the ratio of BHC:quintozene decreased from 1:100 - 1:17 to 1:20 - 1:5 (Smelt 1976). Pentachloroaniline is also an important metabolic product of quintozene (PCNB). Molds and actinomycetes are very active in the transformations of quintozene. The metabolite pentachloroaniline was found more persistent in soil than quintozene (Beck and Hansen 1974; Dejonckheere *et al.* 1975, Wang and Broadbent 1973). Another reaction product of quintozene in soil is methyl-thiopentachlorobenzene or pentachlorothioanisol (PCTA), which disappears from soil at a faster rate than the quintozene.

Smelt (1976) reviewed the work on a fate of quintozene and benzene hexachloride in soil-plant systems and concluded that laboratory experiments show shorter half-life values than field conditions, partly because the soil temperatures in the field for several months are much lower than those maintained in the laboratory or greenhouse experiments.

Experimental conditions exhibit profound influence on the degradation and loss rate of PCNB and BHC from soils. The degradation rates are higher at higher temperatures. Moist soil is favorable to the degradation of quintozene and HCB residues. In flooded soils, quintozene transformed 2-20 times faster into pentachloroaniline than under normal upland moisture conditions. However, extended flooded conditions are not present in land application systems. If 1% fresh organic matter were added, 79-93% of the PCNB disappeared within 1 month after application (Wang and Broadbent 1973). Thus, the high organic matter clay loam soil would have a higher assimilative capacity than sandy loam of low organic matter level.

Yoshida *et al.* (1973) confirmed that benzidine is stable and not decomposed by an isolated soil bacteria after nine weeks of incubation. Benzidine dyes, however, were degraded by soil bacteria. More data should be collected by expanded research efforts into the persistence and transformation routes of the benzidine and related compounds when applied to field soils.

To estimate the assimilative capacity of substituted aromatics, the data from several sources is summarized in Table 7.12. The application rate in terms of soil concentration is given along with the degradation rates. Values in Table 7.12 can be used as an initial estimate of plant-soil assimilative capacity.

Leaching

Benzene hexachloride (BHC) is a pesticide that moves very little in soils. Residue levels of BHC isomers in the surface soil were much higher than in the subsoil (Kawahara *et al.* 1972). The solubility of quintozene and BHC in water is very low (< 0.44 ppm quintozene and < 0.005 ppm BHC). In addition, there is strong adsorption of these compounds to the organic material in the soil. With the normal precipitation runoff, there is little likelihood of their movement in soil or washing from the soil-plant system. Since quintozene and BHC are essentially nonvolatile, transport by diffusion in vapor phase is insignificant, although it does occur at a very low rate. In an experiment by Bristow *et al.* (1973), quintozene-treated soil was mixed with peat and left standing for one week. After careful separation it was found that organic peat contained 50 times as much quintozene as the treated soil. Thus, soil organic fraction has special affinity to adsorb these hydrophobic chemicals. Mobility tests with soil thin layer chromatography demonstrated that quintozene was not mobile in a soil containing 2.5% organic matter (Heiling *et al.* 1974). Analyses of soil samples collected at the groundwater level from cultivated land treated frequently with quintozene confirmed that quintozene is not transported below a depth greater than plow layer (Smelt 1976). Evaporation and movement of quintozene and BHC is strongly inhibited by higher organic matter content of soil. An increase in soil temperature, however, encourages evaporation.

Adsorption of benzidine at 25°C and pH 3.00 on Na⁻, Li⁻, and Ca-montmorillonite is shown in Figure 7.25 (Furukawa and Brindley 1973). A maximum of 70 mmole benzidine was adsorbed on a 100 g montmorillonite. Loading rates exceeding 70 mmole benzidine/100 g clay did not promote adsorption beyond the stated maximum of 70 mmole/100 g soil. As the pH of the soil system changes, the relative amounts of monovalent and divalent benzidine are shifted. However, the total amount of benzidine (neutral, mono- and divalent) adsorption increases with increasing pH (Figure 7.26).

Table 7.12 Loss Rates (Half-Life and Complete Loss) of Various Substituted Aromatics from Soils Under Different Experimental Conditions

Compound	Application Rate (ppm in soil)	Half-Life (days)	Time for Complete Loss (days)	Experimental Conditions	Reference
Quintozene (PCNB)	5 and 30	213 - 699 (mean 468)	—	Laboratory, loamy sand	Beck and Hansen (1974)
Quintozene	19 - 100	117 - 1059 (mean 434)	—	22 field soils, mostly loamy sands applied over 1 to 10-year periods	Beck and Hansen (1974)
Quintozene	100	140	—	Laboratory, fine sandy loam	Wang and Broadbent (1972)
Quintozene	100	230	—	Laboratory, clay	Wang and Broadbent (1972)
Quintozene	100	290	—	Laboratory, peat muck	Wang and Broadbent (1972)
Quintozene	1,000	29	—	Laboratory, loamy clay	Buser and Bosshardz (1975)
Quintozene	1,000	32	—	Laboratory, clay	Buser and Bosshardz (1975)
Quintozene	1,000	145	—	Laboratory, loamy sand	Buser and Bosshardz (1975)
Quintozene	1 and 6	55 - 71	—	Greenhouse	Dejonckheere et al. (1975)
Quintozene	10	70 - 120	—	Greenhouse	Dejonckheere et al. (1975)
p-toluidine	500	—	9	Laboratory, chernozem	Medvedev and Davidov (1973)
o-toluidine	500	—	3	Laboratory, chernozem	Medvedev and Davidov (1973)
2,4-methylaniline	500	—	3	Laboratory, chernozem	Medvedev and Davidov (1973)
2,6-methylaniline	500	—	3	Laboratory, chernozem	Medvedev and Davidov (1973)
Pentachloroaniline	1 - 10	—	70 - 85	Greenhouse	Dejonkheere et al. (1975)
2,6 dichloro-4-nitroaniline	100	900	—	Laboratory fine sandy clay	Wang and Broadbent (1972)

2,6 dichloro-4-nitroaniline	100	490	—	Laboratory Clay	Wang and Broadbent (1972)
2,6 dichloro-4-nitroaniline	100	410	—	Laboratory peat muck	Wang and Broadbent (1972)
Pentachorothioanisole	1 - 10	—	70 - 85	Greenhouse	Dejonkheere et al. (1975)
Dichloran	4.8	8	8	Greenhouse	Dejonkheere et al. (1975)
Pentachlorobenzene	20	194 - 345	—	Laboratory, loamy sand	Beck and Hansen (1974)
Hexachlorobenzene	20	969 - 2084	—	Laboratory, loamy sand	Beck and Hansen (1974)

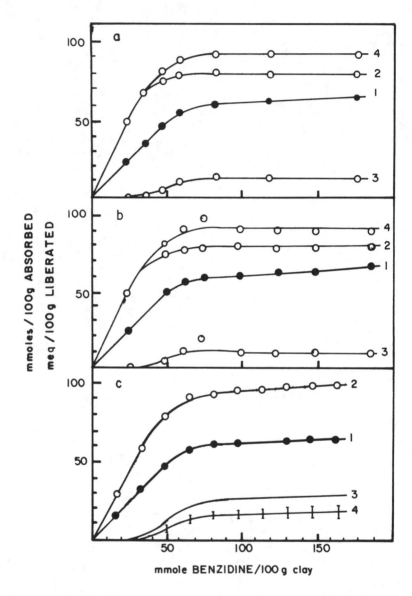

Figure 7.25   (a) Adsorption of benzidine by Na-montmorillonite (curve 1) and liberation of Na$^-$ ions (curve 2), Mg^{2+} ions (curve 3) and Na$^-$ + Mg^{2-} ions (curve 4); (b) adsorption of benzidine by Li-montmorillonite (curve 1) and liberation of Li$^-$ ions (curve 2), Mg^{2+} ions (curve 3) and Li + Mg^{2+} ions (curve 4); (c) adsorption of benzidine by Ca-montmorillonite (curve 1) and liberation of Ca^{2+} ions (curve 2). Monovalent benzidine ions adsorbed derived from curves 1 and 2 (curve 3), and calculated from pH and pK$_{a2}$ (curve 4) (Furukawa and Brindley 1973).

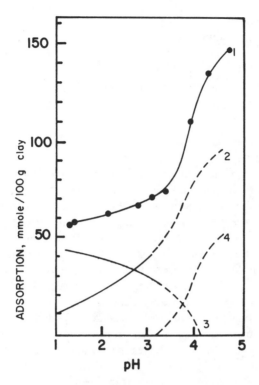

Figure 7.26 Adsorption of benzidine on Ca-montmorillonite at various pH values. Total adsorption (curve 1), monovalent benzidine (curve 2), divalent benzidine (curve 3), estimated neutral benzidine adsorption (curve 4) (Fukukawa and Brindley 1973).

Changes in pH from 6 to 9.3 resulted in a decreased adsorption (Dodd and Ray 1960). Therefore, it appears that at approximately pH - 6 there is maximum benzidine adsorption. At the molecular level, benzidine demonstrates a substantial ability to displace cations such as Ca and Mg and thus would be expected to be adsorbed in proportion to the total cation exchange capacity. In the soil, pH 5-7 range of benzidine would be tightly held in the soil by adsorption and insolubilization, thus allowing increased time for microbial decomposition.

Aniline adsorption is related to the replacing power of anilinium ion over Ca. Not all exchangeable Ca^{2+} ions are replaced by anilinium ions, even when the amount of aniline added is five times the cation exchange capacity. Up to 190 mmole aniline hydrochloride/100 g clay, the amount of replaced Ca^{2+} ions, appear larger than the adsorbed aniline. This result may be due to hydrogen ions participating in the exchange reaction with accompanying increase of the pH of the supernatant solution. Therefore, leaching losses, when benzidine is applied at the application rates matching the biological assimilation, would be expected to be quite low. Aniline is 50-100 times as soluble in water as benzidine and with lower soil adsorption capacity would have to be monitored when applied at the assimilative rate.

Volatilization

Many aromatics such as benzene, toluene, cyclohexane, benzoic acid and other substituted benzene derivatives are volatile. A considerable amount of these volatile aromatics can be lost from the soil-plant system by evaporation at ambient temperatures, depending on the wind speed and method of application in the soil. Chlorinated benzene derivatives are much less volatile and the vapor pressure of chemicals like hexachlorobenzene (1.1×10^{-5} mm Hg at $25°C$) and pentachloronitrobenzene (1.2×10^{-4} mm Hg at $25°C$) is so low that these could be categorized as nonvolatile (Smelt 1976). When applied in high doses, even the nonvolatiles are lost by volatilization, which is encouraged by high temperature, high wind velocities and surface application. In a laboratory experiment, about 80% of added quintozene was lost from a treated soil over a 10-month period when air flow was continuously directed over a 1-cm soil layer. Of this loss, 62% was accountable by volatilization. Increasing organic matter level of soil tends to decrease the volatilization of the hydrophobic nonpolar aromatics. Incorporation of wastes into soil is a managment option available to further reduce volatilization losses.

Influence on Plants

Toluene possesses herbicidal properties (Kamilova *et al.* 1976). With a dosage of 3 kg/ha, toluene caused jute destruction of 38% and raw weight decrease in surface vegetation of 58%. At application of 5 kg/ha, the reductions were 72% and 66%, respectively. Sixty days after treatment, herbicidal activity of toluene decreased significantly; however, at high rates of application (10 kg/ha) toxic effects of toluene continued even after 82 days (Table 7.13). Benzoic acid reduced the growth of tomato plants by 50% at 150 ppm level (Wang *et al.* 1967). For sugar cane, 25 ppm benzoic acid is the critical toxic limit. These plants are, however, very sensitive to applied chemicals and hence may be a conservative reflection of the organic compound application rates that affect yield.

Pentachloroaniline is normally present in plants where quintozene residues are found in soils (Gorbach and Wagner 1967; Wang and Broadbent 1973; DeVos *et al.* 1974). Quintozene is metabolized to pentachloroaniline, which is also found in the lettuce. Quintozene uptake by plants occurs through the root system. Hexachlorobenzene is also taken up by lettuce roots from soil.

Table 7.13 Duration of Toluene Activity in the Soil (Kamilova *et al.* 1976)
(percent based on the control)

Layer Depth (cm)	Toluene, kg/ha					
	3		5		10	
	Item	Raw Weight	Item	Raw Weight	Item	Raw Weight
	After 60 days					
0 - 5	62	42	28	34		
5 - 10	96	88	83	79	66	71
10 - 15	108	93	91	85	73	81
15 - 20	112	108	96	95	96	97
	After 82 days					
0 - 5	104	106	95	91	90	92
5 - 10	95	103	90	97	104	98
10 - 15	104	99	95	95	95	92
15 - 20	100	100	109	101	100	101

The levels of quintozene and HCB were reduced by almost a factor of 10 (to below the legal, allowable limit) by lowering the rate of application from 6 to 0.1% and from 0.17 to 0.007%, respectively, thus indicating that loading rate can be used to control the plant uptake within critical levels. The quintozene/HCB ratio in plants is identical to that of soil. The uptake of dichloran by plant roots follows a pattern similar to that of quintozene (Dejonckheere *et al.* 1975). Parsley plants took up hexachlorobenzene from soils and contained 0.5 ppm BHC, which is 20-fold the legally allowed limit. Lettuce plants also extracted BHC from soil.

Plant type, growth stage, duration of crop growth and metabolism of the aromatic compounds within the plant are some of the factors that influence the level of aromatic accumulation and biomagnification in vegetation and subsequent entry into food chain. Table 7.14 gives the soil levels of quintozene and benzene hexachloride and their relative bioaccumulation by several crops in Germany, France and Holland. Both quintozene and BHC bioaccumulate readily into roots of carrots, grass and head lettuce but not in sugar beet and potato.

Crop:soil concentration ratios for quintozene in potatoes and sugar beets were on average 6-10 times less than for BHC. These differences could result from degradation of quintozene in the plant. Gorbach and Wagner (1967)

Table 7.14 Soil and Crop Levels of Benzene Hexachloride and Quintozene

Crop	Soil Level (ppm)		Crop:Soil Ratio (Average Range)	Reference
Benzene Hexachloride				
Radish	0.05-5		(0.08-2.9)	Wallnöfer et al. (1975)
Parsley	0.02-2		2.6 (0.5-4.6)	Hafner (1975a)
Watercress	0.5-50		0.12 (0.04-0.18)	Hafner (1975a)
Potato	0.03-0.4		0.72 (0.5-1.2)	Smelt (1976)
Carrot	0.05-5.0		1.9 (1.0-2.8)	Wallnöfer et al. (1975)
Sugar beet	0.02-0.44	Roots	0.39 (0.23-0.48)	Smelt (1976)
	0.02-0.41	(leaves)	0.24 (0.05-0.37)	Smelt (1976)
	0.02-0.41	(heads)	0.087 (0.03-0.15)	Smelt (1976)
Head lettuce	0.5-1.9		0.23 (0.16-0.3)	Casanova and Dubroca (1972)
	0.02-0.2	(at 67 days)	9.9 (5.1-16.0)	Dejonckheere et al. (1975)
	0.02-0.2	(at 81 days)	3.6 (1.8-5.8)	Dejonckheere et al. (1975)
Grass	0.07-0.14	(roots)	5.2 (2.7-9.0)	Smelt (1976)
	Same	(0-5 cm tops)	1.7 (1.4-2.4)	
	Same	($>$ 5 cm tops)	0.28 (0.23-0.31)	
	0.001-0.03	(roots of 1-yr-old)	26 (23-29)	Smelt
	Same	($>$ 5 cm tops)	0.09 (0.09-0.09)	
Quintozene				
Potato	1.5-3.2		0.11 (0.03-0.25)	Smelt (1976)
Carrot	0.02-0.16		19 (10-27)	Smelt (1976)
Sugar beet	0.12-0.34	(whole plant)	$<$0.03 (0.01-0.07)	Smelt (1976)
Head lettuce	0.44-6.3	(67 days)	11.0 (5.3-21)	Dejonckheere (1975)
	0.44-6.3	(81 days)	4.4 (1.4-6.0)	Dejonckheere (1975)
	20-69		0.44 (0.34-0.54)	Casanova and Dubroca (1972)
Grass	0.43-2.0	(roots)	3.8 (1.5-5.5)	Smelt (1976)
	Same	(0-5 cm tops)	2.2 (0.2-1.2)	
	Same	($>$ 5 cm tops)	0.03 (0.01-0.03)	

found the transformation of quintozene into pentachloroaniline and penta-chlorothioanisol in potatoes. Kuchar *et al.* (1969) found a similar transformation pattern in cotton plants. To date, nothing is known about BHC degradation in plants (Smelt 1976).

With regard to soil factors, soil organic matter is especially important in the retention and adsorption of aromatics. Crops seem to take up less BHC and quintozene from soils having a higher organic matter content. The addition of 10% peat to a soil reduced the uptake of quintozene by plant parts aboveground by a factor of 2-5 (Bristow *et al.* 1973). Other crops that have been studied for BHC uptake from soils are radish, red radish and spinach, watercress, potato, carrot, sugar beets, head lettuce and grass (Hafner 1975b). Soil type and temperature influence the uptake of BHC by plants, whereas water content of soil is not as important a factor (Hafner 1975a).

Wallnöfer *et al.* (1975) found by field and pot experiments that carrots specifically absorbed BHC in direct proportion to the level of BHC in soil (Table 7.15). Radishes accumulated very small amounts of BHC as compared with carrots.

Table 7.15 Accumulation of BHC from Soil by Growing Carrots and Radishes (Wallnöfer *et al.* 1975)

BHC Concentration (ppm) in Soil		BHC Concentration (ppm) in Plants	
Beginning of Test	End of Test	Carrots	Radishes
0.044	0.028	0.011	Not available
0.4	0.3	0.050	Not available
3.95	2.2	0.50	0.025

Content of absorbed BHC and quintozene is highest in roots, with decreasing trend toward the tops. Potatoes do not biomagnify quintozene and BHC in the tuber. Most of the accumulated quintozene in potatoes stays in the peel, while BHC spreads all over the tuber. Quintozene content of the peel is 5 times higher in potatoes, 7 times in sugar beets and 27 times in carrots, as compared to the peeled portions. Plant parts and seedling located near the soil surface possess BHC as a result of direct adsorption by contact. At soil levels of 0.01-0.1 mg/kg, BHC in plant parts may exceed the standard allowable for cattle feed, 0.03 mg/kg. Quintozene residues disappear rather rapidly both from soils and plants as compared to BHC. Thus, the soil-plant system would have higher assimilative capacity for quintozene than BHC.

In radishes, BHC was present in both the peel and the body of radishes; whereas in carrots, BHC was stored exclusively in the peels or outer skins (Table 7.16). It was assumed by the authors that in carrots, cells of outer epidermis contained lipophilic substances and, hence, stored lipophilic BHC. This is not the case with radish epidermal cells.

Table 7.16 Distribution of BHC in Carrot and Raddish Roots (Wallnöfer *et al.* 1975)

BHC Concentration in Soil (ppm)	BHC Concentration in Plant (ppm)			
	Carrot Roots		Radish Roots	
	Peel	Peeled Root	Peel	Peeled Root
0.05	0.058	–	0.011	0.007
0.05	0.41	0.015	0.048	0.018

A study on the distribution of BHC in soil and sugar beets (Wallnofer *et al.* 1975) showed that at 0.3 ppm BHC in soil, after 73 days of sugar beet growth, whole plant averaged 0.012 ppm BHC, leaves 0.003 ppm, peel of beet 0.053 ppm and peeled beet 0.008 ppm BHC. Young beet plants after 60 days of growth showed higher BHC concentration in all tissues—root and leaves—with an overall average of whole plant at 0.15 ppm. In comparison to lettuce, the application of BHC to sugar beets would be preferred since the bioconcentration is not as high. Thus, crop selection is an important parameter in assuring tolerable concentrations in plants. Data are not available for the usual types of vegetative cover used on land application sites. Because the available data are for highly lipophilic species, it is anticipated that the critical level will be higher in a land application system employing nonfood crops.

SURFACTANTS AND DETERGENTS

Soaps and detergents are increasingly and widely used in the United States as well as throughout the world. Cleaning in the home, in commercial laundries, at car washes and in a wide variety of industrial and food-related washdown procedures represent major sources of surfactants as well as associated compounds used in commercial cleaners. Often the high sodium content of milk or poultry processing plants is from the sodium-based cleaning solutions used widely in industry. The two broad classes of cleaning agents are soaps and detergents. Soaps in solid form are usually sodium salts of natural fatty acids derived from animals or plants. Potassium salts are used with the resulting product called a soft soap. About 10% of the manufacture of cleaning agents for all uses is soaps (EPA 1973).

Synthetic detergents are broadly classed into three categories (1) cationic surfactants (about 5% of total detergents manufacture), (2) nonionics (about 10% of the detergent market), and (3) anionics. Depending on the compounds used to construct detergents, the term "hard" refers to relatively nondegradable detergents and the term "soft" refers to those detergents that are readily degradable (Sebastiani et al. 1971). Since most of the environmental effect of detergents is with the product use and disposal and not with the manufacture, the analysis of common detergents indicates some of the compounds of concern (Table 7.17). The organic fraction is also of concern in the assimilation of surfactant compounds. The wide diversity of organic formulations requires that discussion of detergents be on a basis of the broad categories with some specific example compounds.

Table 7.17 Detergent Analysis (Adams and Gross 1975)

Detergent	%		ppm		
	P[a]	Na Content[a]	Zn	Cu	B
A. Anionic Surfactant with Phosphate Builder	12.3	24.0	8.8	ND	2.3
B. Nonionic Surfactant with Perborate Brightener	8.3	24.5	23.3	ND	19.9
C. Nonionic Surfactant with Sodium Silicate Builder	0.01	30.9	11.0	ND	2.3
D. Homemade Lye Soap	0.11	11.5	1.3	1.2	3.4

[a]Phosphorus and Na contents of detergents A, B, and C were provided by the manufacturers and confirmed by the authors' analyses. Detergent D and trace element contents of all detergents were authors' analyses.

[b]ND: none detectable. Trace elements not listed were below detectable levels. These results were confirmed by an independent analysis by Robert Munter, research associate in soil science, University of Minnesota.

Cationic surfactants are usually quaternary ammonium salts

$$\left[\begin{array}{c} R_1 \\ | \\ R_4 - N - R_2 \\ | \\ R_3 \end{array} \right]^{+} \quad X^{-}$$

where the R groups may be aliphatic or aromatics and X is an anion such as a halide or hydroxide. Common R groups are dimethyl and diallyl moieties. The cationic detergents possess antimicrobial properties.

Nonionic detergents are characterized by the nondissociation in water. A common nonionic consists of a polymer of oxyethylene, $(C_2 H_4 O)$, with both polymer ends attached to an alcohol, e.g., primary alkyl polyoxyethylene alcohol, $(R\text{-}O\text{-}(C_2 H_4 O)_n\text{-}CH_2\text{-}CH_2\text{-}OH)$ or secondary alkyl polyoxyethylene alcohol, $(R\text{-}CH_2\text{-}O\text{-}(C_2 H_4 O)_n\text{-}CH_2\text{-}CH_2 OH)$. Other nonionics are silicone, acylate latex and ampholyte (Batyuk and Samochvalenko 1972).

The largest class of detergents (80-85% of the market) is the anionics, which consists usually of a sodium cation and the active organic anionic surfactant. The anion portion is usually a sulfate or a sulfonate. Sulfate-type detergents result from the general reaction of sulfuric acid and the long-chain alcohols. The product contains a C-O-S bond:

$$\begin{array}{ccc} & H & O \\ & | & \| \\ R - & C - O - S - O - Na \\ & | & \| \\ & H & O \end{array}$$

The R group is most often lauryl alcohol (C_{12}), a primary alcohol, although some secondary alcohols are used. The biological degradation of these two forms of sulfate detergents is very similar (Ryckman and Sawyer 1957). In terms of rates of decomposition, the sulfate anionics are "soft" detergents.

Sulfonate anionics contain a C-S linkage and are formed from the general reaction of sulfuric acid with fatty amines, aliphatics or aromatics. Three main sulfonate formulations are used as detergents.

Sulfonated amide

$$\left[R - \overset{\displaystyle O}{\overset{\displaystyle \|}{C}} - N \left\langle \begin{array}{c} C_2 H_5 \\ \\ \begin{array}{ccc} H & H & O \\ | & | & \| \\ C & C - S - O \\ | & | & \| \\ H & H & O \end{array} \end{array} \right. \right]$$

Sulfonated ester

$$\left[\begin{array}{c} O \quad\;\; H \;\; H \;\; O \\ \| \qquad | \;\; | \;\; \| \\ R-C-O-C-C-S-O \\ \;\; | \;\; | \;\; \| \\ \;\; H \;\; H \;\; O \end{array}\right]^{-}$$

Sulfonated alkyl benzene

$$\left[\begin{array}{c} O \qquad\qquad\quad R \\ \| \qquad\qquad\quad | \\ O-S \underset{\|}{\overset{}{\Longleftrightarrow}} C-R' \\ \| \qquad\qquad\quad | \\ O \qquad\qquad\quad R'' \end{array}\right]^{-}$$

R groups may be H

The sulfonated primary alkyl benzenes have a low solubility and are not used commercially in detergents. The principal anionic detergent is the sulfonated linear alkyl benzene referred to as LAS. As a general category, the LAS are degradable—"soft" detergents.

Besides the active organic fraction of the surface active agent, synthetic detergents contain other compounds that must be assimilated in the plant-soil system. Sometimes these compounds can require greater land areas than the surfactant itself. The first such additional compound is often the sodium portion of the salt. Sodium without the presence of calcium and magnesium cations leads to a sodium imbalance and reduction in soil hydraulic capacity (for more discussion, see Chapter 3). Substitution of degradable cations or a mixture of Ca, Mg and sodium would improve the assimilative capacity of the surfactant complex.

Inorganic species are also added to anionic detergents to improve the commercial properties. Builder compounds such as sodium tripolyphosphates or sodium silicates are added to enhance the cleaning capabilities. To increase the solubility and bulk handling properties, extenders are often added, e.g., sodium sulfate. Brighteners added to detergents are responsible for the presence of boron (Table 7.17). Because of the high level of detergent use, perborate and borax have been banned in Europe. Such compounds have aquatic effects and also must be applied to land only at rates that will preclude adverse anion-salt effects in soils (for more discussion, see Chapter 8).

Since detergent solutions are widely used in industries for cleaning purposes, many industrial wastes would be expected to contain the surfactant compounds. In some industries where sanitation is high, e.g., poultry processing and dairy processing, the detergent compounds can be present in substantial amounts relative to the satisfactory assimilation in soils. Therefore, rates of assimilation for surfactants must be developed. The even wider use of detergents in homes means that municipal effluent and sludges that may be applied to land will also contain detergent species. In the remainder of this section the assimilative pathways and capacity will refer to the organic fraction of the surfactants.

Influence on Soil Properties

Surfactant compounds alter the soil physical properties with the percentage depending on the concentration or amount of applied surfactant. Naturally hydrophobic soils are most affected. One soil characteristic that is improved by the addition of surfactant is the wettability and moisture retention of the water-repellent soils (Osborn et al. 1969). At the concentrations used for field-scale applications ($\sim$ 50 ppm), nonionics improved the water retention efficiency 90-189% over controls with water alone (Osborn et al. 1969). The anionic sulfonated compound ($\sim$ 5,000 mg/l) improved the water retention efficiency 400-500%. For both species of surfactant, increases of 600-1400% over water were found when surfactant solutions of 5,000-20,000 ppm were used.

The rate of water infiltration and the hydraulic conductivity can also be influenced by the presence of surfactants. The water infiltration movement (mm/72 hr) of the surfactant solution (1,000 ppm) was determined for 11 compounds on a Baugh silt loam Ap horizon (Cairns 1972). All compounds had equal or greater rates of solution movement in comparison to water but less than that measured for an NH_4NO_3 solution. The author concluded that the presence of surfactants in the water was not effective in increasing water movement in the slowly permeable soil. However, when the water movement was measured after the surfactant had been applied, there was an increase in the flux rate (Miller et al. 1975b). Two nonionic surfactants were tested and the saturated hydraulic conductivity was increased (Miller and Letey 1975a). Similar results under unsaturated flow conditions were also measured after the application of nonionic surfactants at a concentration of 500 ppm. The effect on wettability and conductivity is restricted to the layers in the soil to which the surfactant can penetrate or be carried by application of water.

Field tests of the improved moisture infiltration and soil physical properties due to surfactant application were conducted in California. A wetting agent was applied at 330 ppm concentration to a soil containing no cover crop (Osborn et al. 1969). During the rainy season the runoff volume was 30% less than a control area, while the resultant soil erosion was reduced by 90%. These measurements reflect the total response of the soil to the influence of a surfactant.

Soil aggregate stability is another important physical property influenced by the land application of surfactants. Introduction of an anionic sulfonated surfactant in varying solution concentrations was examined on a compact, a loose and a medium soil, in which aggregate stability was measured (Cardinali and Stoppini 1974). Up to a solution concentration of 80 ppm the aggregate stability was improved over that for a control solution of distilled water. At 500 ppm there was a 20% decrease relative to control for the compact and loose soils (Table 7.18). The medium texture soil decreased to 65% of control. Their conclusion was that up to a certain concentration ($\sim$ 80 ppm) there was an improvement for all soils tested, and that up to 500 ppm could be applied to certain soils without substantial loss in aggregate stability. Aggregate stability is important to good crop growth (see Chapter 3).

The germicidal effects of cationic, anionic nonionic detergents and soaps has been documented for quite some time (Koch 1881, Eggerth 1929, Baker et al. 1941). However, the critical or needed concentration at which the germicidal action is 50 or 90% effective is not well documented. Therefore, less information is available on the solution or soil levels at which microorganisms are unaffected by the presence of a surfactant. Steinter and Watson (1965) compared surfactant categories and found that cationic compounds were more inhibitory to fungi than anionic species. Nonionic compounds were partially inhibitory above solution concentrations of 100 ppm. For soil microfungi, anionic and nonionic surfactants reduced numbers of microfungi by only 20% at a solution concentration of 1,000 ppm (Lee 1970). For specific fungal species the anionic compound was more inhibitory and resulted in reduction in the number of species (Table 7.19). However, at the 100-500 ppm range, certain species continued to grow well. For soils previously receiving detergents, the microfungi grew better on applied surfactant, indicating strain selection or physiological adaptation (Lee 1970).

The application of an anionic sulfonate and a commercial detergent up to 100 ppm in a sandy sierozem soil resulted in an increase total microbial plate count (Dhawan and Mishra 1974). Carbon dioxide evolution was unaffected up to 100 ppm in soil, and nitrification was the same as control up to a level of 75 ppm. At 100 ppm the oxidation of ammonia was inhibited. Application of a nonionic surfactant at 10 and 20 kg/ha increased the soil biological activity in a loamy chernozem soil, above that of the control (Batyuk and Samochvalenko 1972).

Nematode activity was inhibited with the application of a surfactant at the level of 100-300 ppm in the soil (Miller 1976). Dobozy and Bartha (1975) examined the application of a cationic surfactant and found no reduction in earthworm numbers and weight at a soil level of 20 ppm.

Table 7.18 Aggregate Stability (structural stability index, I) of Several Soils as a Function of Anionic Sulfonated Surfactant Concentration (Cardinai and Stoppini 1974)

1973 Test

Surfactant Concentrations (ppm) (ABS)	Soil SA			Soil A			Soil P			Relative Values Control = 100		
	I	+	++	I	+	++	I	+	++	SA	A	P
Control Solution (distilled water)	13.7	c	B	42.9	b	B	11.2	b	B	100	100	100
16	19.2	a	A	48.7	a	A	10.7	b	B	140	113	96
80	15.7	b	B	45.6	b	AB	17.1	a	A	114	106	153
500	10.8	d	C	35.6	c	C	3.9	c	C	79	83	35
1000	9.2	e	D	1.5	d	D	2.0	d	D	67	3	18
2000	4.4	f	E	1.5	d	D	1.2	d	D	32	3	11

1974 Test

Surfactant Concentrations (ppm) (ABS)	Soil SA			Soil A			Soil P			Relative Values Control = 100		
	I	+	++	I	+	++	I	+	++	SA	A	P
Control Solution (distilled water)	16.2	b	B	77.3	a	A	70.6	a	A	100	100	100
16	12.1	c	C	74.6	b	B	65.2	b	AB	75	96	92
80	20.0	a	A	76.3	a	AB	63.8	b	B	123	99	90
400	9.9	d	D	56.3	c	C	30.3	c	C	61	73	43
500	9.2	d	DE	23.1	d	D	21.4	d	D	57	30	30
1000	7.5	f	E	3.2	d	E	1.7	e	E	46	4	2
1500	8.9	e	DE	1.9	f	EF	1.2	e	E	55	2	2
2000	8.5	ef	DE	0.9	f	F	0.2	e	E	52	1	2

+ Significance to 95% (those with the same letter classification means there is no significant difference).
++Significance to 99% (those with the same letter classification means there is no significant difference).

Table 7.19 Some Characteristics of the Soil and the Fungal Populations in the Untreated Control Soils and Detergent-Treated Soils (Lee 1970)

	Anionic Series		Nonionic Series	
	Control Soil	Anionic Detergent-Treated Soil	Control Soil	Nonionic Detergent-Treated Soil
Water Content (%) Fresh Weight	30	26	24	28
Numbers of Microfungi/g dry soil, 1,000s	131	101	46	38
Reduction in Numbers of Microfungi, % of Control		23		16
Numbers of Species in Total Population	39	25	29	29
Reduction in % of Species		36		0
Number of Dominant Species	6	4	4	6
Percent of Total Isolates Contributed by the Dominant Species	58	75	68	67

In summary, the application of a surfactant to the soil can affect the physical and biological properties in a concentration-dependent manner. Surfactants applied to soil are partitioned between the solid and solution phases. For solution concentrations on the order of 500 ppm, there is a reduction in microbial species, with the degree of inhibition being governed by surfactant type and microbial species. For total soil concentrations on the order of 100 ppm, there does not appear to be substantial adverse response. This information is typically for single application events and, thus, successive inputs of surfactant are needed to determine the capability for microbial adaptation. These critical levels of surfactant are sufficient to produce some improvement in soil moisture properties, but substantial changes result usually at higher application rates.

Fate in Soils

Cationic Surfactants

Evagro compounds are potentially cationic surfactants, which are applied to plant-soil systems to stimulate plant growth (Dobozy and Bartha 1975). Three mixtures of cationic surfactants were studied to determine the degradation and effects on the soil. On a sandy soil at an application of 5 kg/ha, there was a stimulation of CO_2 production relative to the control plots. Quantitative data on cationic compound degradation were not given, but the

CO_2 evaluation had nearly returned to that of the control plot after 42 days. Other evidence in a later section on plant response and the previously presented data on nematodes indicate that degradation also occurs at an application of 10 kg/ha. Data on higher applications were not given, but the authors mentioned other European research in this area. Huddleston and Allred (1967), in reviewing available information, concluded that virtually nothing had been done with respect to cationic surfactants; however, the low percentage of these compounds on the market must also be considered.

Nonionic Surfactants

Compounds in this category consist of a hydrocarbon moiety (n-alkyl or alkyl phenol) and the ethylene oxide chain (4 to > 20 monomers). Biological attack can occur at either portion of the nonionic surfactant, with the rates of degradation being dependent on the structure of each. For ethoxy chains of less than 9 moles, microbial attack is relatively complete (Huddleston and Allred 1967), while ethoxy chains in the 20-mole range are more resistive. The n-alkyl chain is decomposed by terminal group carboxylation and then successive degradation via pathways such as β-oxidation (Figure 6.3). If the alkyl group is highly branched this terminal group decomposition is largely inhibited. The presence of a phenol group is not as inhibitory as the branched-chain counterpart. Phenol near the terminal carbon (1- and 2-phenyl isomers) are more microbially degradable than those located on internal carbons. For alkyl polyoxyethylene nonionics the alkyl group is generally more easily degraded than the ethoxy chain unless there is a high degree of alkyl branching. For alkylphenol polyoxyethylene, the alkyl group is preferentially attacked until branching or a phenol group is encountered; then the rate of polyoxyethylene degradation becomes more favorable. These results were derived from a number of pure culture studies of nonionic surfactant degradation (Huddleston and Allred 1967). However, the relative rates and phenomena occurring would be expected to be similar in the soil microbial system, as a first approach to assimilative capacity.

In a study using 10 soil microfungi, a nonionic surfactant (15-S-9) was used to determine the potential for detergent degradation (Lee 1970). Seven of the ten species utilized the nonionic as a sole carbon source, indicating a high probability of degradation in a field soil. The nonionics were less available as a carbon source than anionic surfactants, but the nonionic did not prevent the growth of any of the 29 microfungal species isolated from soil and tested. Furthermore, a definite adaptation process was observed whereby the rate of degradation of successive applications of nonionic surfactants would be expected to increase.

In greenhouse-scale experiments the degradation of nonionic surfactant (alkyl polyoxyethylene ether) applied to several soils was investigated for various factors important in a land application system, *e.g.*, soil moisture and application amount (Valoras *et al.* 1976a, 1976b). Four soils were tested: Canadian sphagnum peat, and three sandy loams referred to as Glendora, Idylwild and Pachappa. After application there was a lag period, the length of which was dependent on the soil concentration of nonionic (Figure 7.27). For the range of 250-1,000 ppm in the soil, the lag time is on the order of 3-6 days, while at 5,000 ppm, a period of 30 days lapsed before substantial biological activity occurred. This lag could have been due to microbial evolution or some physical effect from the surfactant.

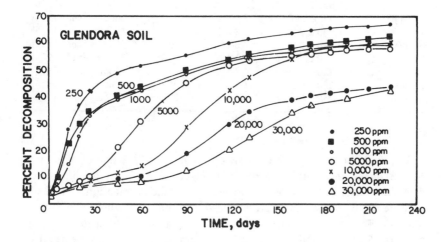

Figure 7.27 Nonionic surfactant decomposition for various initial application levels on soil (Valoras *et al.* 1976a).

Surfactant additions up to 5,000 ppm (∼ 12,000 kg/ha) did not affect the soil organic matter mineralization process. The percent surfactant degradation appeared substantially lower when the applied concentration was above 10,000 ppm. For the range of 250-5,000 ppm surfactant in soils near optimum moisture contents, the rate of degradation in percent decomposition per month is given for the various soils (Table 7.20). Since the decomposition rate is more like an exponential process, the degradation rates for the short-term experiments were higher. For decomposition to the 90% range, a degradation rate of around 6% of applied surfactant per month is a first estimate of assimilative capacity. Note that these results are for an initial application and with the observations of Lee (1970), the rate of degradation would be expected to increase after adaptation of the soil microbial species. With the percent degradation per month, the actual mass of surfactant decomposed can be determined from the application rate (over the range of value 250-5,000 ppm).

Table 7.20 Assimilative Capacity for Nonionic Surfactants

Soil	Application Range (ppm in soil)	Decomposition (% applied surfactant/month)	Duration (days)	Reference
Glendora	1,000-2,000	6.4-7	374	Valoras *et al.* (1976a)
Glendora	1,000-2,000	7-7.2	319	Valoras *et al.* (1976a)
Pachappa	1,000-2,000	8.5-9.4	319	Valoras *et al.* (1976a)
Peat	1,000-2,000	2.8-3.1	319	Valoras *et al.* (1976a)
Pachappa	500-2,000	16-26	28	Valoras *et al.* (1976b)
Glendora	500-2,000	21-48	28	Valoras *et al.* (1976b)
Glendora	250-5,000	4.7-5	473	Valoras *et al.* (1976a)
Glendora	250-5,000	5.4-6.0	307	Valoras *et al.* (1976a)
Idylwild	250-5,000	4.6-7.6	306	Valoras *et al.* (1976a)

As soil moisture is reduced below optimum levels, the microbial activity and surfactant degradation are reduced (Figure 7.28). Up to an application of 5,000 ppm of nonionic on the Glendora soil, the degradation at 10% soil moisture was 63-83% of that determined at 28% moisture. Above this application rate there was a much lower degradation rate and little effect of moisture. This reduction in degradation rate was only about 10% greater than the corresponding reduction in the loss of soil organic carbon (75-87%) found for this same reduction in soil moisture (Valoras *et al.* 1976a). Because of this moisture-dependent phenomenon, the assimilative capacity would be expected to be greater under field conditions for a clay loam soil. Thus, for the two representative soils used in this book, the clay loam would be expected to have a higher nonionic surfactant assimilative capacity than the sandy loam soil.

Anionic Sulfate Surfactants

A wide variety of bacteria have been isolated that will degrade alkyl sulfates, as well as utilize these surfactant compounds as sole sources of carbon (Goodnow and Harrison, Jr. 1972, Williams and Rees, Jr. 1949, Payne and Faisal 1963, Hsu 1965, Dronkers and van der Vet 1964). Many of these bacteria, such as those in the *Pseudomonas* species, are isolated from soils. Lee (1970) isolated eight soil microfungi which, at low surfactant concentration, used alkyl sulfates as a sole carbon source. The anionic surfactant was more available for degradation than a nonionic surfactant, but there was a greater reduction in number of species when the anionic surfactant (solution concentration of 1,000 ppm) was applied to a loam soil (Lee 1970). Because of the surfactant decomposition studies with bacteria and fungi isolated from soils it can be inferred that anionic surfactants will degrade when applied to a plant-soil system.

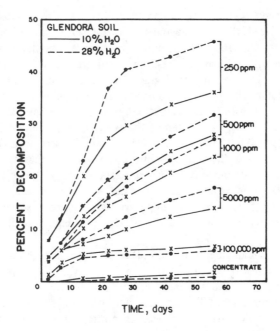

Figure 7.28 Soil moisture effect on decomposition of nonionic surfactant (Valoras *et al.* 1976a).

The pathways of alkyl sulfate surfactant decomposition have been investigated by a number of researchers and thoroughly reviewed by Huddleston and Allred (1967). Decomposition proceeds via two principal routes: (1) the attack at the terminal methyl group with oxidation and subsequent reactions common to the normal fatty acid route, and (2) the attack by alkyl sulfatases with cleavage of the sulfate group and then the normal fatty acid stabilization route. For the normal alkyl groups utilized in sulfated surfactants, the sulfate cleavage is generally more rapid than terminal methyl group oxidation. However, when a sulfonate group is present instead of sulfate, the sulfur compound cleavage is considerably slower.

Very little quantitative research has been done on the soil decomposition of alkyl sulfates in soil systems, possibly because the alkyl sulfates are generally more easily degradable. Hence, attention has been directed at this sulfonated anionic surfactant. Therefore, results on soil degradation of sulfonates would be a conservative estimate with regard to the assimilative capacity from comparable alkyl sulfate surfactants.

Anionic Sulfonate Surfactants

As with the alkyl sulfates, there have been a number of soil microorganisms isolated that can degrade sulfonate surfactants (Goodnow and Harrison, Jr. 1972, Hsu 1965); however, these compounds are generally classed as "hard" or relatively resistant to degradation. Researchers have demonstrated that the rate of degradation of anionic surfactants can be increased substantially when microorganisms are adapted to the compound being degraded (Lee 1970, Swisher 1963b, Setzkorn and Carel 1965, Huddleston and Allred 1964). Demonstrating the evolutionary potential, Huddleston and Allred (1963a) improved the degradation time (90% removal) from 15-16 days to 4-5 days for C_{10} and C_{12} linear alkylbenzene sulfonate when the microorganisms had previously been introduced to the anionic surfactant (Figure 7.29). This phenomenon is important because land application systems are multiple successive applications of anionic surfactants, thus promoting greater assimilative capacity due to adaptation of soil microorganisms.

The alkyl and alkylbenzene sulfonates are degraded along a microbial pathway, the rates of which depend on the moiety structure present in the surfactant. Basically, the sulfonate entity is not cleaved and degradation occurs with terminal methyl group oxidation. Branching at the alkyl terminus severely inhibits degradation as well as critical location of the benzene moiety. This is similar to the situation described for nonionic surfactants. The greatest steric inhibition occurs when the benzene ring is located four carbons from the terminal methyl group (Huddleston and Allred 1967).

A few research studies utilizing alkyl benzene sulfonate (ABS) have been conducted on the degradation in soil systems. Using carbon dioxide evolution as a measure of decomposition, ABS applied to soils (10-100 ppm in soil) produced greater evolution than the control for about 10 days (Dhawan and Mishra 1974). Sebastiani *et al.* (1971) studied the degradation of sodium dodecylbenzene sulfonate (DBS) in a sand soil. After sterilization, a single microorganism—*Micrococcus pyogenes albus*—was used to measure degradation of DBS. The *Micrococcus progenes albus* was adapted to DBS. Degradation reached 90% of the initial DBS in about 14 days (Figure 7.30). This rate of decomposition provides some estimate of the soil assimilative capacity, but the quantitative estimate under field conditions remains to be measured for many anionic sulfonates. Further investigation of the DBS decomposition demonstrated that no toxic intermediates were produced in

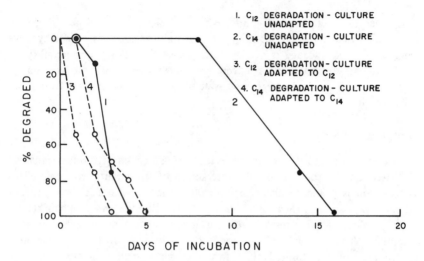

Figure 7.29 Sequential adaptation of Mississippi River water microorganisms to linear alkylbenzene sulfonate, 25°C (Huddleston and Allred 1967).

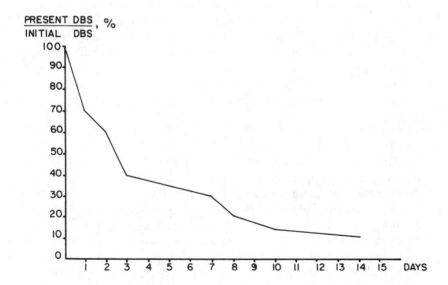

Figure 7.30 Microbial degradation of sodium dodecylbenzene sulfonate by adapted *Micrococcus pyrogenes albus* (Sebastiani *et al.* 1971).

the microbial stabilization (Simonetti *et al*. 1974). Therefore, it can be concluded that pathways for the microbial stabilization exist in soil systems and general knowledge is available to judge the relative degradability of different alkyl-substituted benzene sulfonates. Behavior and assimilative rates in actual soil systems are not available and must be inferred from more basic studies.

Leaching and Runoff

Determination of the impact of surfactants associated with water transport is based on information on the interaction of surfactants with soils. This interaction is primarily an adsorption phenomenon with a fairly significant desorptive capacity (Sebastiani *et al*. 1971, Batuyk and Samochvalenko 1972). Much of the research on surfactant movement in soils has been conducted under saturated flow conditions as opposed to the relatively high percentage of unsaturated flow found in land application systems. The adsorption process for surfactants in most soils cannot be approximated as infinitely rapid, but instead occurs at a quite finite rate. With these kinetics, saturated flow conditions must be slow enough to allow equilibration to establish the movement potential. There are a large number of reports documenting the field movement of surfactants to subsurface waters (review listings by Sebastiani *et al*. 1971). These situations often involve concentrated landfill operations or other heavy applications of such materials. Such data must be reviewed under conditions that reflect land application design criteria.

The surface runoff potential of detergents has not been widely examined. Runoff transport is anticipated to be determined by the relative adsorption to sediment particles. An approach similar to the pesticide transport model (Donigian *et al*. 1977) is useful in projecting movement of an adsorbed species. Therefore, basic adsorption isotherms and water movement data are needed.

Adsorption isotherms for nonionic surfactants in three sandy loam soils were determined by Valoras *et al*. (1976b) and Miller and Letey (1975). The data are shown in Figure 7.31 with comparison of results for one surfactant (soil penetrant) given in Figure 7.32. For the three soils, the relative surfactant adsorption capacity was peat (not shown) > Glendora > Idylwild > Pachappa sandy loam (Valoras *et al*. 1976b). A variety of different adsorption equilibrium curves were determined, depending on the soil and surfactant. Many of the ratios of solution to soil were very high as evident by the soil concentrations, 3,000-5,000 ppm. These are above the rates at which substantial microbial degradation occurs (500-1000 ppm). At the lower levels many of the surfactants would be more completely adsorbed.

Equilibrium adsorption was measured with an alkyl sulfonate surfactant in the range of 400 ppm addition to the soil (Sebastiani *et al*. 1971). For a

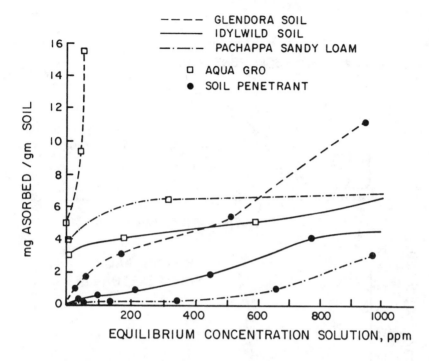

Figure 7.31 Adsorption isotherms for two nonionic surfactants applied to soil (Valoras *et al.* 1976b).

sand, sand-clay and clay-loam, the adsorption was 13%, 26% and 65%, respectively. The time to reach equilibrium was between 24-72 hours, indicating the likelihood of movement where rapid soil-water flowrates are used. Valoras *et al.* (1967b) measured the irreversibly bound soil concentrations of two nonionic surfactants on a Pachappa sandy loam, an Idylwild and a Glendra soil. For Aqua Gro the irreversibly bound material was 8,000, 8,500, and 15,500 ppm, respectively, while for Soil Penetrant, the values were 500, 4,000 and 4,400 ppm, respectively.

Although leaching under conditions of unsaturated flow with active, aerobic microbial populations has not been studied, there are some conditions that may minimize surfactant migration. Relatively low levels of surfactant concentration in soils (500-1,000 ppm) should be allowed as a part of the waste application and then schedule time for decomposition of the parent surfactant. This approach provides a high soil:surfactant ratio, thus, increasing the soil retention of such compounds. The longer retention time is expected to permit greater surfactant stabilization and, hence, less

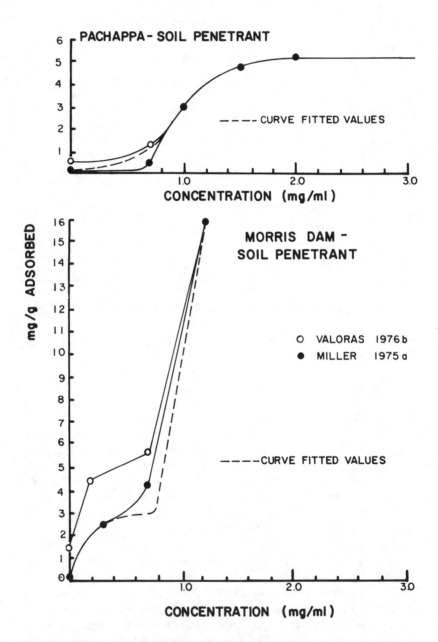

Figure 7.32 Comparison of soil penetrant surfactant isotherms on two soils.

potential for significant leaching. The adsorbed surfactant should be protected from runoff transport by control of erosion. That is, the approach of best management agricultural practices should be used as the most cost-effective means for regulatory control of nonpoint source contributions from land application areas.

Influence on Plants

Three cationic surfactants were studied over a 4-year period (on 3 soils) and as applied to 15 plant species (Dobozy and Bartha 1975). Crop species included sugar beet, tobacco, grapes, tomato and poppy. For every crop the three surfactants stimulated generative growth (*i.e.*, yield) at the field concentrations studied. Detailed results were reported for corn (Table 7.21). At 10 kg/ha, all surfactants promoted increased yield over the control (by 8-17%). Increased yield was obtained over the 5 kg/ha rate for 2 of the compounds (3-5%). In the second year, the effect of the surfactants has been reduced as the compounds are degraded. Therefore, cationic surfactants may exert an adverse effect on crop yield but only at levels in excess of 10 kg/ha.

Three nonionic surfactants were studied in a greenhouse pot experiment for the effect of increasing concentration on the growth of barley (Valoras *et al.* 1976b). A critical application level could be defined at an application that had a 20% yield reduction relative to the control. Using this percent yield reduction as a critical level, the three surfactants produced a range of barley response (Figure 7.33). The concentrations in Figure 7.33 are the applied surfactant solution concentrations converted to soil concentrations of the active surfactant ingredient using the solution volume per unit soil weight (Valoras *et al.* 1976b). Water-In surfactant reached critical soil levels for Pachappa between 1,000 and 2,000 kg/ha while on the Glendora soil, the critical soil level was > 20,000 kg/ha. The decreased toxic effect with the Glendora soil was true for all the nonionics tested and was related to increased adsorption of the surfactant (Valoras *et al.* 1976b) by Glendora soil.

The compound Aqua-Gro produced critical yield reductions at soil concentration levels of 2,000-10,000 kg/ha, 500-1,000 kg/ha and > 40,000 kg/ha for the Pachappa, Idylwild and Glendora soils, respectively. Soil Penetrant was the third nonionic studied and the critical soil levels were 1,000-2,000 kg/ha, < 500 kg/ha and < 1,000 kg/ha for Pachappa, Idylwild and Glendora soils, respectively. Soil Penetrant was found to be more toxic to plant yield (Valoras *et al.* 1976b, Endo *et al.* 1969) than the other surfactants. At these levels of application the barley accumulated 0.22-0.56% of the applied surfactant.

Table 7.21   The Effect of Evagros in the First and in the Second Year
in the Case of Maize Grown on Salty Soil (Dobozy and Bartha 1975)

Dosage (kg/ha)		Percentage Yield Referred to the Nil Application Plots	
		First Year	Second Year, No Surfactant Applied
Evagro C	1	120.6	100.0
	5	112.8	106.5
	10	110.2	106.5
Evagro E	1	115.4	99.1
	5	114.2	104.4
	10	117.2	104.3
Evagro K	1	102.6	101.5
	5	102.6	101.2
	10	107.7	104.9
Average	. . .	111.4	103.2

Soil Penetrant up to 20 ppm in solution culture did not decrease barley
root dry weight but decreased root porosity with increasing concentrations
from 0-100 ppm of nonionic surfactant (Luxmoore *et al.* 1974). At a higher
concentration than 40 ppm in solution phase, the root growth decreased
and the surfactants inhibited root tip meristematic processes.

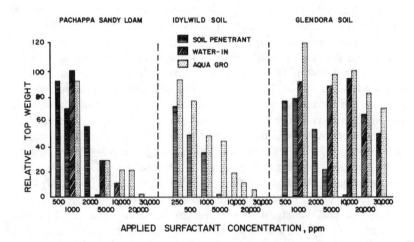

Figure 7.33   Relative top weight of barley grown in soils receiving various applications
of nonionic surfactants (Valoras *et al.* 1976b).

Anionic surfactants were applied to a Daugh silt loam Ap horizon to determine the yield effect on barley (Cairns 1972). The authors did not specify the application in terms of the active surfactant ingredient. Assuming a similar range to the surfactants used by Valoras et al. (1976b), the anionic surfactants produced a critical yield effect at 300 kg/ha, with the exception of one compound (Fenapon). At the time of the second crop (without further surfactant addition), most of the surfactants produced no worse yields than the control. Luzzati (1974) found that on a comparative basis, the nonionic surfactant reduced potato production more than the anionic species.

Judy et al. (1973) investigated the effect of detergent-laden water on corn growth and determined the tolerance level of corn with regard to levels of detergent application on a wide variety of soils. An effort was also made to identify the components of detergents responsible for major effects on plant growth. They used heavy-duty nonenzyme detergent (Bz) and a heavy-duty enzyme detergent (Tx). These were applied in the irrigation water at concentrations of 0, 20, 800, 1,600, 4,800 and 8000 ppm, 1.0, 1.2 and 1.4% solution. Stimulation in corn growth occurred on Davidson soil that received 1,600 ppm Bz (Kroontje et al. 1973) (Figure 7.34) and up to 8,000 ppm Tx (Figure 7.34). On Norfolk f sl, growth was stimulated at 800 ppm Tx. These stimulations were attributed to phosphorus response from detergent. On Davidson soil, yield decreases occurred above 8,000 ppm Bz and above 14,000 ppm Tx. Detergent Bz was more detrimental to plant growth than Tx (Kroontje et al. 1973). Plants grown on Davidson soil could withstand higher detergent level than those grown on Norfolk soil, probably because of larger CEC and higher Mg content of Davidson soil. The harmful effects at higher application rates were correlated to salinity and sodium injury resulted from detergent loading. Judy et al. (1973) concluded that heavy-duty nonenzyme detergent Bz can be applied on land up to 800 ppm on Norfolk and 1,600 ppm on Davidson cl without any damage to corn crop. Heavy-duty enzyme detergent, Tx can be applied at 1,600 ppm level to both soils without any appreciable influence on corn crop, which was in fact benefitted at lower concentrations. The heavy-duty detergents contained 21-21.5% Na, 9.7-10.9% P and traces of Ca, Mg, K and B. The beneficial effects of P nutrition were overcome by harmful effects of Na when detergent loading exceeded 0.16%.

It is difficult to conclude the critical soil concentrations at which surfactants adversely affect crop yield. It does appear that soils with greater adsorptive capacity can permit higher surfactant applications (Table 7.22). The range of values is sufficiently wide that preliminary testing is warranted in determining the crop portion of the assimilative capacity. Surfactant applications in the range of 10-20 kg/ha would not appear to create much yield depression and may even result in improved yields. Surfactant levels

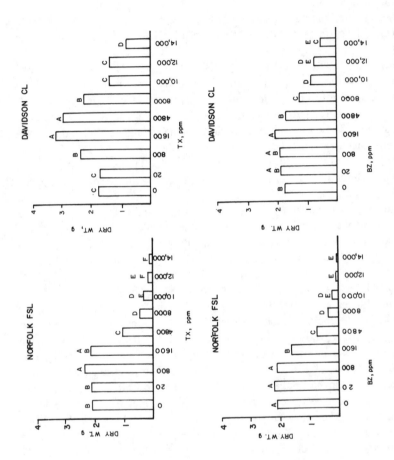

Figure 7.34   Corn yield response to applied detergents (BZ and Tx) in Norfolk and Davidson soils, letters A, B, etc., designate significantly different (5% level) yields (Judy *et al.* 1973).

in the 100-250 kg/ha range are likely to produce a yield reduction. There is a need to determine the tolerance of crop species to achieve optimal assimilative capacity.

Table 7.22 Summary of Surfactant Influence on Crops

Surfactant	Critical Soil Level (kg/ha) Yield < 80% of Control	
Three cationics	>10	
Water-In (nonionic)	$50 <> 250$	Pachappa sl
	$500 <> 1,000$	Glendora
Aqua Gro (nonionic)	$50 <> 250$	Pachappa sl
	$15 <> 25$	Idylwild
	$1,000 <> 3,000$	Glendora
Soil Penetrant (nonionic)	$25 <> 50$	Pachappa sl
	~ 15	Idylwild
	~ 50	Glendora
Three anionics	< 300	

MISCELLANEOUS ORGANIC COMPOUNDS

Cellulose, Hemicellulose and Lignin

Much agricultural research has been undertaken to evaluate the decomposition of such crop residues as rice straw, corn stover and leaves. To unify the knowledge of residue decomposition, refinements were made to examine the pathways for major organic compounds present in plant tissue. Therefore, decomposition rates have been established for cellulose, hemicellulose and lignin (for a complete review see Alexander 1977).

Assuming a first-order decomposition rate as describing the soil processes, the assimilative capacity for three major constituents of waste applied to soils were summarized (Table 7.23). The representative first-order decomposition rate constants were $0.03 d^{-1}$, $0.03 d^{-1}$, and $0.003-0.009 d^{-1}$ for cellulose, hemicellulose and lignin, respectively. Application of these constituents could thus be determined on an annual basis by calculations using projected frequencies of application.

The decomposition constants in Table 7.23 represent a constant temperature of $30°C$. To modify for field conditions, an Arrhenius correction can be used:

$$K_{T_2} = K_s \theta^{(T_2 - T_s)} \tag{7.1}$$

where T_s = Celsius temperature used as the reference value, $30°C$

T_2 = ambient monthly or weekly Celsius temperature at the land application site

θ = temperature correction factor, equal to 1.08 (Reddy *et al.* 1978)

K_s = specific decomposition rate constant at $30°C$

With the temperature correction for the decomposition rate constant (Equation 7.1) the assimilative capacity for these residue components can be determined at site locations in a similar manner to that presented earlier for the oil constituent of an industrial waste.

Because of the carbonaceous nature of these waste constituents and the usually large volumes applied to soils, the nitrogen status and plant yield response can be affected. In addition, at certain levels these waste constituents can exert sufficient oxygen demand to require lower applications to maintain aerobic soil conditions. The nitrogen status and oxygen demand effects on the assimilative capacity for organics applied to a plant-soil system are given in a later section of this chapter.

Table 7.23 Stabilization of Several Organic Fractions in Soil Systems at $30°C$

First-Order Decomposition Rate Constants, Day^{-1}				
Cellulose	Hemicellulose	Lignin	Compound Source	Reference
0.039	0.033	0.0086	Horse manure	Waksman *et al.* (1939)
0.028	0.033	–	Rice straw	Acharya (1935)
0.0017	0.0015	0.0009	Leaf litter	Yamane (1974)
0.14	–	0.0031	Leaves	Hagin and Amburger (1974)
0.022	–	–	Leaves	Hagin and Amburger (1974)
0.030	–	–	Stem	Hagin and Amburger (1974)
0.011	–	–	Stem	Hagin and Amburger (1974)

Solvents (Cyclohexane, Chloroform, Acetone, Formaldehyde, Hexane, Pentane, Heptane, Ether and Pyridine)

The response of soil to addition of solvents depends primarily on the amount of applied material and also on the volatility of the applied compound. At moderate-to-low application rates, these organic solvents are easily assimilated into the plant-soil system. However, as the dose rate increases, there is a greater effect on microbial populations. This effect depends on the compound volatility (Buddin 1914). The volatile species exert an intial soil sterilization (Figure 7.35), in which bacterial numbers are depressed and certain microflora and fauna are eliminated. The solvent concentration at which this sterilization occurs is defined as a critical dose level. Above this

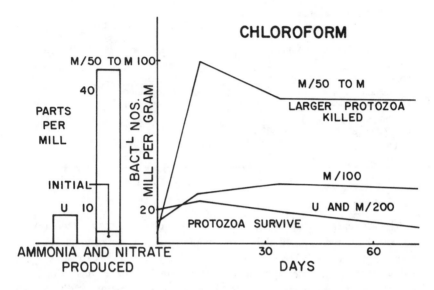

Figure 7.35 Effect of chloroform on bacterial numbers, ammonia and nitrate in soils (U-untreated, M-120 g chloroform/kg dry soil) (Buddin 1914).

amount, the short-term effect is the same, occurrence of partial sterilization. As volatilization losses occur and solvent-metabolizing microbes develop, there is a substained rise in microbial activity. Thus, the initial impact of solvents fades and the soil population then recovers and often exceeds the control. However, as the volatile solvent doses exceed moderate levels, the soil physical properties appear to be adversely affected by the solvent additions, which is an environmentally unacceptable situation.

Organic solvents with low volatility produce little soil sterilization when applied in moderate-to-high levels. Instead, the microbial populations increase substantially over control soils (Figure 7.35). As the dose of this class of organic solvents is increased, there is a long-term reduction in soil microflora, both in terms of numbers and diversity. Thus, critical levels exist, above which there is a long-term adverse soil response.

The differentiation between volatile and nonvolatile compounds is not well defined. Using the response of soil microorganisms and the compounds studied by Buddin (1914), the vapor pressures were listed (Table 7.24). From this simplistic ranking it would appear that solvents with a vapor pressure at 25°C of less than 1 mm water would be in the low volatility category.

Table 7.24   Vapor Pressure of Solvents (Pal *et al.* 1977)

Volatile—Partial Sterilization		Low Volatility	
Species	P,mm (25°C)	Species	P,mm (25°C)
Benzene	94	Phenol	0.8
Toluene	29	*o,m,p*-cresol	0.3
Cyclohexane	99	Hydroquinone	≪ 1
Chloroform	199		
Acetone	283		
Formaldehyde	≫ 760		
Hexane	144		
Pentane	509		

The critical dose at which application of solvent causes a reduction in microbial number immediately after application was determined from the data of Buddin (1914) and Jenkinson and Powlson (1976) (Table 7.25). Below this single-dose application rate there is little or no effect, and microbial populations are stimulated because of the substrate addition. Rate of decomposition or other soil losses would establish the time period between application and, hence, the assimilative capacity. The time necessary for the microorganisms, applied at the critical dose, to reestablish is given in Table 7.25. These times may reflect the degradation or loss rate, but probably represent only a qualitative measure of assimilation rate.

Table 7.25   Response of Soil Microbial Populations to Application of
Various Solvents (Buddin 1914)

Solvent	Critical Soil Level (ppm)	Time Period for Recovery at Critical Dose
Cyclohexane	840	Within 37 days
Hexane	430	Within 19 days
Heptane	10,000	Within 24-63 days
Pentane	7,200	Within 30-53 days
Formaldehyde	150-300	Within 22 days
Chloroform	590	Within 12 days
Chloroform	625[a]	–
Ether	7,400	Within 14 days
Acetone	58,000	Within 12 days
Pyridine	7,900	Within 16-30 days

[a]Jenkinson and Powlson (1976).

Alcohols (Methanol, Ethanol, Isopropanol and n-Propanol)

In an analogous set of experiments (Buddin 1914) the effect of several alcohols on microbial numbers was measured. The alcohols tested have vapor pressures in the more volatile category, but as a class have very little impact on soil microflora. Critical dose levels in soils are given in Table 7.26. These species are easily used by soil microorganisms and, hence, the assimilative capacity is substantial.

Table 7.26   Response of Soil Microbial Populations to Application of
Various Alcohols (Buddin 1914)

Alcohol	Critical Soil Level (ppm)	Time Period for Recovery at Critical Dose
Methanol	32,000	Within 12 days
Ethanol	46,000	Within 11 days
1-propanol	12,000	Within 21 days
2-propanol	>12,000	–

Antibiotics

Antibiotics are present in fermentation spent-beer wastes, pharmaceutical manufacturing wastes and animal wastes. Antibiotics also enter the waste streams from their therapeutic uses and are synthesized by a number of subterranean actinomycetes and fungi. Many antibiotics have been shown to be susceptible to microbiological attack. Their addition to soil has resulted in a complete loss of antimicrobial activity, as shown by experiments with actinomycin, chloramphenicol, chlortetracycline, cycloheximide, griseofulvin, mycophenolic acid, oxytetracycline, patulin, penicillin and streptomycin (Brian 1958, Pramer 1958). This was due to the microbial degradation and chemical inactivation in soils. The mechanisms and types of antibiotic transformations by microorganisms have been reviewed in detail by Sebek and Perlman (1971). These include acylation, phosphorylation, adenylylation, hydrolysis, oxidation, reduction, sulfoxidation, demethylation, deamination and nonspecific degradation. Elmund *et al.* (1971) concluded that the excreted antibiotic residues in animal wastes modify the biological stabilization processes such that the environmental pollution potential is increased. Dietary antibiotics at constant low levels lead to the evolution of drug-resistant pathogens and modify the microflora of the rumen, waste and soil where the animal waste is stabilized by biological decomposition. The concentration of dietary chlortetracycline in the feedlot manure may range from 0.34-14 ppm. These concentrations are so low that the antibiotic assimilation in soil would be rapid and their continuous application would lead to evolution of microorganisms that specialize in the antibiotic degradation or utilization. Soil assimilative capacity for antibiotic stabilization is rather substantially large and, if managed properly by soil incorporation, there appears no environmental problem.

Organic Acids (Formic, Acetic, Pyruvic, Succinic, Benzoic, Cinnamic, Coumaric, Cyanuric, Stearic, Ferulic, Isonicotinic, Salicylic, Vanillic)

Cyanuric acid was reported toxic to pea, barley and radish in water, sand and soil at 600-800, 500 and 1,000 ppm, respectively (Kasugai and Ozaki 1922). The toxic effects were neutralized by $Ca(OH)_2$ and $CaCO_3$. Dihydroxystearic acid occurs in the infertile soil at concentrations as high as 50 ppm and affects growth of wheat seedlings at 50 and 100 ppm concentration in water culture.

Although these acids may accumulate or be present in soils under certain conditions, numerous soil microorganisms can utilize these as carbon and energy sources. Bacteria differ from fungi in their mode of oxidation of benzoic acid (Shepherd and Villanueva 1959). Bacteria oxidize benzoic acid through p-hydroxybenzoic and protocatechuic acid to β-ketoadipic acid through two intermediates, catechol and cis, cismuconic acid. Fungi follow the same mechanism as bacteria up to protocatechuic acid, which is directly oxidized to β-ketoadipic acid without any intermediates.

In soil cultures, trans-cinnamic acid is rapidly decomposed by soil organisms. Soil Pseudomonads degraded cinnamic acid to hydroxyphenylpropionic acid and melilotic acid. Soil fungi can use p-coumaric, ferulic and vanillic acids at high concentrations of 1,000 ppm. Ferulic acid is first oxidized to vanillic acid and then the ring cleavage occurs. The dihydroxystearic acid and others can be adsorbed by soil colloidal fraction, thereby rendering them nontoxic to plant roots. The dihydroxystearic acids are rapidly decomposed by soil microorganisms and are difficult to trace in soil samples that have been stored for a few weeks under conditions of optimum microbiological activity. Numerous soil fungi and bacteria utilize the p-hydroxybenzoic acid as a sole carbon source (Cain 1958).

By means of radioactive labeling of various carbons in acid compounds, the decomposition in a Greenfield loamy sandy was determined (Martin and Haider 1976). A range of breakdown rates were obtained because of the position of the radioactive carbon. In general, the labeled carbon nearest the acid moiety was lost first. Therefore, if one were using disappearance of the parent compound then the first carbon lost would signify decomposition. At applications of 1,000 ppm in the Greenfield loamy sand, the highest and lowest percent loss are shown in Table 7.27 after 7 and 84 days. With regard

Table 7.27 Decomposition of Carboxylic Acids in Soils (Martin and Haider 1976)

Compound	Percentage Decomposition	Time Required (days)
Acetic Acid	52-76	7
Acetic Acid	71-87	84
Acetic Acid	50	5[a]
Acetic Acid	80	22[a]
Pyruvic	47-83	7
Pyruvic	70-93	84
Succinic Acid	52-89	7
Succinic Acid	71-95	84
Glucose	75	7
Glucose	87	84

[a]Sorensen and Paul (1971).

to disappearance of parent compound, acetic, pyruvic and succinic acids were more rapid than glucose. This indicates a high degree of soil assimilative capacity.

Carbon Disulfide

Carbon disulfide (CS_2) has been investigated as a soil fumigant (Kudeyarov and Jenkinson 1976). At high rates of application, the microorganisms are killed, although with a slightly lower effectiveness than $CHCl_3$. However, as the loading rate was decreased to 10 ppm CS_2 in the soil, the germicidal effect was negligible. For a soil with organic C of 2.93%, the oxygen consumption, carbon dioxide respired and nitrate production after CS_2 application were the same as the control. For a soil with organic C of 1.07%, the oxygen consumption was approximately the same as the control, while a slight reduction in inorganic N mineralization was found. These experiments were not repeated to determine the response for successive CS_2 applications. Thus, as a first estimate, a carbon disulfide application of about 20 kg/ha would create little or no adverse effects.

Coumarin

The utilization of the ester coumarin was evaluated for bacteria isolated from soils (Func 1971). Thirty-nine species were capable of utilizing coumarin, of which 25 were *Pseudomonas*, 7 *Cellulomonas* and 7 *Achromobacter*. The decomposition was not related to total bacterial numbers, the nitrogen status or the percent organic pathway for utilization was via o-coumaric acid and melilotic acid and not via the salicyclic and catechol route.

Dimethylamine

In four soils with varying organic matter and pH, the decomposition of dimethylamine, $(CH_3)_2NH$, was evaluated under laboratory conditions (Tate and Alexander 1976). The rates of decomposition were rapid after application of 60-65 ppm of the dimethylamine (Figure 7.36). These data can be used to calculate the assimilative capacity, kg/ha/month. In separate experiments with a single soil, the effect of correcting soil pH from pH 4.6 to 6.8 was to increase the amount of decomposition 15 to 30%. Under anaerobic conditions, the time required for a 50% decomposition was increased from 7 (aerobic conditions) to 29 days, hence emphasizing the need to maintain aerobic conditions at such land application sites.

Phthalic Acid and Phthalate Acid Esters

Tarvin and Buswell (1934) showed that *Rhodopseudomonas palustris* partially degraded phthalic acid. Stahl and Pessen (1953) used *Aspergillus versicolor* Qm 432 and *Pseudomonas aeruginosa* QMB 1408 to study the

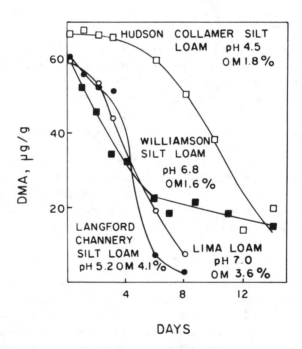

Figure 7.36 Loss of dimethylamine from soils with varying pH and texture (Tate and Alexander 1976).

breakdown of plasticizers. They used *n*-octyl, ethylhexyl and methylheptyl phthalates as the carbon sources for the organisms and found that the three isomers of the octyl ester supported no growth of *A. versicolor* and supported only a small amount of growth of *P. aeruginosa.*

Klausmeir and Jones (1960) looked at microbial degradation of certain plasticizers. The most active cultures were from dibutyl phthalate as carbon source and were classified in genus *Fusarium,* labeled 2P3. The *Fusarium* cultures were capable of growth not only on dibutyl phthalate but with a variety of phthalate ester. Table 7.28 lists the growth rates of *Fusarium* 2P3 on various phthalate esters.

Table 7.28   Utilization of Phthalate Esters by *Fusarium* 2P3
(Klausmeir and Jones 1970)

Ester	pH	Growth (mg)
Control	6.7	—
Dimethyl Phthalate	6.7	1.6
Diethyl Phthalate	4.0	7.1
Dipropyl Phthalate	3.8	4.0
Di-n-butyl Phthalate	4.3	3.5
Di-n-octyl	6.7	0.0
Butyl Isodecyl phthalate	5.9	7.3
Diallyl Phthalate	6.8	0.3
n-octyl-n-decyl Phthalate	6.7	2.0
Isooctyl, Isodecyl Phthalate	6.4	4.4

The only phthalate acid esters that did not support growth of *Fusarium* 2P3 were daillyl and di-n-octyl phthalate. Dimethyl phthalate showed only slight growth. The mechanism for degradation was proposed as a deesterfication of one of the alcohol moeities on the diester, yielding monoester phthalate intact and no deesterfication of the second ester group occurred. Klausmeir and Jones (1960) hypothesized that an extracellular esterase was involved, based on the premise that substrate was available for original growth of the organisms; this *is* conjecture, and the extracellular esterase of the *Fusarium* 2P3 may be more specific than other known esterases. As yet there is no proof whether this hypothesis is correct.

Englehardt *et al.* (1976) studied the microbial metabolism of various dialkyl phthalates. *Penicillium lilachinum* and three bacteria were used in their study. The organisms degraded di-n-butyl phthalate to mono-n-butyl phthalate utilizing the butanol moiety as a carbon source. It is thought that the mono-n-butyl phthalate degrades to phthalic acid but phthalic acid was never detected as an intermediate. They did, however, identify a phenolic metabolite as protocatechuic acid. Eventually, protocatechuic acid disappeared also and they assumed that the degradation of phthalic acid was via 4,5-dihydroxy phthalic acid, protocatechuic acid and *cis, cis*-B-carboxy muconic acid. This pathway is similar to the pathway presented by Ribbons and Evans (1960) for soil *Pseudomonads*. They stated that the mixed populations or organisms found in the soil would be most effective in degrading the phthalate acid esters, but only parts of the degradation were carried out with ease. The hydrolysis of the second alcohol moiety was the rate-limiting step; however, they did not look at the various other phthalate acid esters that have branched alcohol constituents.

Pipecolic Acid

The decomposition of pipecolic acid (2-piperidinecarboxylic acid, $C_6 H_{11}NO_2$) was tested on a chernozem and a brown soil (Guirguis and Vancura 1970). Rates of oxygen evolution were stimulated by previous or simultaneous addition of available carbonaceous compounds. The proposed mechanism for enhanced decomposition was the stimulation of gram-negative bacteria (primarily *Pseudomonas*) and the increased synthesis of enzymes for breaking down pipecolic acid. With pipecolic acid addition of 13 ppm in soil, and the prior addition of cellulose, the pipecolic acid was completely degraded in 5 days. For most of the other supplemental amino acids and less available additions, the decomposition of 60-70% was achieved in 4-5 days under similar conditions. Without additions of carbon substrates, only about 30% of the pipecolic acid was decomposed. Thus, the assimilative capacity could be substantially increased by management of the plant-soil system.

AGGREGATE BIOORGANIC WASTES

A number of wastes can be considered biological in origin, for which the plant-soil assimilative capacity is governed by the overall impact of the waste. That is, there are no specific organic compounds that must be considered as possible LLC. The bioorganic waste material is assumed to be degradable, and since these are biological in origin, land application represents a complete recycle of such materials. This does not imply that all these wastes degrade rapidly. Some naturally occurring compounds are resistant, *e.g.*, lignin, tannins, condensed aromatics and substituted aromatics, etc. However, the lack of deleterious effects due to these biological or naturally occurring substances allows the buildup and degradation of these wastes with the assimilatory capacity dictated by the aggregate effects of the waste, *e.g.*, nitrogen status, oxygen demand, etc.

Wastes from industries such as seafood processing, poultry and meat processing, pharmaceutical fermentation and extraction of agricultural products, dairy products processing, grain mills, canned and preserved fruits and vegetables, sugar processing, and pulp and paper production are examples of materials in which the organic fraction is assumed to be easily recycled to the plant-soil system. For these waste types, other constituents such as salts, acids and bases, water content, etc., must be examined and evaluated because these and not the bioorganics may be the LLC. In addition, industrial wastes from biological treatment sludges do not usually fit in this bioorganic category since other synthetic or less degradable compounds are usually also present. For these conditions, the specific organics must be considered separately.

Oxygen Demand

The assimilative capacity of a plant-soil system for organics is based primarily on the ability to maintain predominantly aerobic conditions for rapid microbial growth and function. The need for aerobic conditions also exists in the stabilization of specific organic compounds. In addition to the microbial demand for oxygen, plant root systems must be sustained under aerobic conditions to ensure adequate growth. The amount of oxygen required varies widely with the plant species. Some plants are more tolerant to wet conditions such as Reed Canary grass and tall fescue. Adverse effects on agricultural crops have been detected for soil atmosphere oxygen partial pressures less than 10% of total pressure, or about 140 ppm O_2.

When the soil system receiving industrial wastes is maintained predominantly aerobic, odor-producing conditions associated with anaerobic decomposition are avoided. Thus, odor nuisance conditions are minimized. Design and operation, which favor aerobic conditions, also promote drying and reduce nuisance associated with mosquito and other pests. Thus, in addition to promoting decomposition, aerobic operation has secondary benefits due to nuisance control.

Anaerobic conditions associated with water-saturated soils and the presence of oxygen-demanding waste organisms, lead to the development of slimes. This growth may seal the soil pores and impede water infiltration. Ponding of water further augments the anaerobic sealing, thus creating nuisance and nonaesthetic conditions as well as impairing the assimilation of wastes. When such conditions occur the design and operation must be reexamined carefully to determine necessary changes. In terms of corrective measures for the soil, waste application should cease, a drying period should be maintained and surface disking after drying should be implemented if possible.

In considering industrial waste land application, the description of the soil as being predominantly, instead of completely aerobic, is appropriate. The waste-soil system has anaerobic-aerobic elements. There are typically anaerobic conditions immediately after waste application which, as liquid drains in the soil and drying occurs, is rapidly converted to an aerobic environment (anaerobic-aerobic cycle). Where nitrogen compounds are present, this cycle can lead to nitrification during the aerobic phase, then denitrification under anaerobic conditions, resulting in total N loss (see Chapter 11). A second set of conditions for anaerobic-aerobic behavior in land application systems is within the soil structure. Certain aggregates or collections of soil particles will lose water at slower rates than the average soil conditions. These are compact structures in which anaerobic conditions can coexist with the predominantly aerobic soil. These are referred to as microsites and have been demonstrated to play a role in sequential aerobic-anaerobic pathways, such as nitrification-denitrification (Greenland 1962).

Therefore, the design of plant-soil receivers for industrial waste must maintain predominantly aerobic conditions to prevent failures and ensure organics assmilative capacity. Oxygen transfer in a plant-soil system occurs in a number of steps (Figure 7-37). originating in the atmosphere. From the atmosphere, oxygen is transferred into the crop canopy by a combination of convective and diffusive fluxes. The relative significance of these fluxes depends on the cover crop present. A well-maintained grass, a row crop or a forested area would have a large degree of wind-induced convection of oxygen to the soil surface. For a tall grass or grain or for a poorly managed system with the cover crop lying on the ground, convection is reduced and diffusion through the crop mat is necessary to get oxygen to the soil surface.

Oxygen at the soil surface must diffuse into the waste layer or waste-soil matrix, depending on the nature of the application and the soil-water migration of waste organics. The transfer mechanism is oxygen diffusion. Because soil pores are smaller than the space between a close-growing or a matted

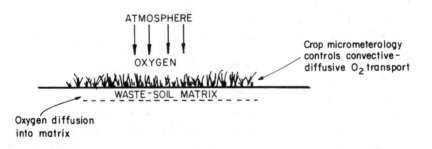

Figure 7.37 Schematic of oxygen transfer to stabilize land-applied organics

crop, the diffusion of oxygen into soils may be the rate-limiting step in satis-fying the waste oxygen demand and maintaining predominantly aerobic soil conditions.

Surface application of industrial wastes containing solids results in a sequence of processes leading to the distribution and assimilation of that material in the plant-soil system. Often, material applied by spray irrigation or with surface spreaders remains on the leaves of any crop present. Depend-ing on the amount of liquid present, material will continue onto the soil surface. After initial distribution, wind and factors of the microclimate oper-ate to dry the waste, which has not entered the soil. After drying, the waste cakes or forms a coating on the soil and vegetative cover. This material is not as available for decomposition, but as rain events occur, the waste is dissipated or washed from the crop. After this mixing of the waste with the soil, greater decomposition and less volatilization occurs. With surface appli-cation, much of the waste assimilation is in the immediate upper soil surface, hence the major differences in soil type are less important. As material is intermixed with the soil, the properties of the soil become more influential. Thus, the assimilation of oxygen demand for surface applications occurs by the convective movement of oxygen to the soil surface and only partially by oxygen diffusion into soils.

Soil incorporation of industrial wastes is the second major mechanism for assimilation of organic wastes. Techniques used are direct surface injec-tion or application followed by disking to incorporate. Under these condi-tions, the oxygen demand is distributed in the soil matrix (Figure 7.37), and so the diffusion of O_2 into the soil limits the rates of assimilation and the ability to maintain aerobic conditions.

An oxygen diffusion model was developed for the flux of O_2 into soils (McMichael and McKee 1966). Fick's law for the diffusion of O_2 into an infinite medium initially at a constant concentration (Cp) was solved and integrated for a fixed time (T) after waste application:

$$N_{O_2} = 2 (C_{O_2} - C_p) \left[\frac{DT}{\pi}\right]^{1/2} \tag{7.2}$$

where: N_{O_2} = the flux of oxygen crossing the soil surface, ML^{-2}

C_{O_2} = vapor phase O_2 concentration above the soil surface, ≈ 300 ppm

C_p = vapor phase O_2 concentration required in soil to prevent adverse yields or root growth, $\simeq 140$ ppm

T = time over which diffusion is occurring

D = effective diffusion coefficient

$$D = 0.6 (S)(D_{O_2}) \tag{7.3}$$

as proposed by Van Bavel (1952)

where: S = air-filled soil pore volume at field capacity

D_{O_2} = oxygen diffusivity in air ($\cong 1.6 \text{ m}^2/\text{d}$)

Solution of the diffusion equations gives the reaeration rates for one day ($T=1$) in various soil types (Table 7.29). The results of Table 7.29 must be modified for the periods of time in which wastewater has been applied to the soil, thus restricting gas diffusion. For a given time period, T_a, between waste applications, $T_a = 1/f_A$ (or the reciprocal of the frequency of application events per unit time) there is a drainage time, T_d, for redistribution of applied liquid to reach field capacity. The ratio of these two times, T_d/T_a, is the fraction of time where gas diffusion is restricted and one minus that ratio is the fraction of time allowed for reaeration. To determine the oxygenation of the soil, the reaeration fraction multiplied by the time period between applications is the reaeration time T used in Equation 7.2. The result is the mass of oxygen per unit area in the time period between application T_a which can be reduced to a per day or per week oxygenation capability by dividing by T_a to match the waste generation.

Table 7.29 Calculated Oxygen Transfer Rates into Various Soils at 0.1 atm Water Tension (Carlile and Phillips 1976)

Soil	O_2 Transfer Rates	
	$g/m^2/\text{day}$	lb/day
Norfolk sl	82	732
Georgeville sil	64	571
Lakeland ls	96	857

The total oxygen demand (TOD) of the wastewater is defined as the total carbonaceous demand (approximated by the chemical oxygen demand, COD), plus the oxygen required to nitrify all ammonia and organic compounds, nitrogenous oxygen demand (NOD):

$$TOC = COD + 4.56 \text{ (Organic N} + NH_4 - N) \tag{7.4}$$

where the organic waste is assumed converted to ammonium and then oxidized as:

$$NH_4^+ + 2O_2 \xrightarrow[\text{activity}]{\text{bacterial}} NO_3^- + H_2O + 2H^+ \tag{7.5}$$

Thus, from a given wastewater analysis for COD and TKN, the land area needed to allow reaeration and adequate plant growth can be determined. The assimilatory capability is a conservative estimate since oxygen transfer to the upper soil layers occurs during the drainage times.

One of the critical elements of Equation 7.2 is the time of drainage and reaeration. For wastes in which little water is involved, the soil surface profile would not be saturated, and thus little allowance for drainage is needed. Under those circumstances, $T = T_A$. For wastes and effluents in which there is substantial weekly application (0.5-2 inches), the drainage time is a function of the soil permeability, size of the application site, topography and the distance to outlets. A hydraulic analysis may be needed to determine the time of drainage and, hence, the remaining time for reaeration prior to the next waste application. Thus, the two benchmark soils at a particular site would have different oxygen demand assimilative capacity due to variations in drainage characteristics.

Example: A land application site is to be reviewed for a pharmaceutical waste (0.01 mgd) of the following composition:

Constituent	Concentration (mg/l)
COD	5,000
NH$_4$-N	700
O-N	500

The soil tests reveal that at field capacity, the vapor-filled pore fraction is 0.22. The plant oxygen requirements are 144 kg/ha/day. Drainage characteristics indicate that after 4 days, field capacity conditions are established and the frequency of application is 24 events per year. Determine the land application area based on oxygen demand.

Nitrogen Status

The application of organic wastes to a plant-soil system can substantially alter nitrogen availability for microbial waste decomposition or for crop growth. That is, if there is a higher ratio of carbon to nitrogen than required for microbial growth ($\approx$23:1), then little nitrogen will be liberated for crop growth. Reduced yields result under such N deficiencies.

To determine the potentially mineralized N in an organic waste the following equation holds:

$$NX = NW - 0.043 \, CW \tag{7.6}$$

where: NX = positive, when there is potentially mineralizable N
NW = total waste nitrogen
CW = carbon content of the waste

For a hypothetical loading of waste, the potentially mineralizable N would be determined (Figure 11.6) based on waste characteristics. For those wastes exceeding C:N = 23, supplemental N would be added to sustain crop growth and to enhance the microbial breakdown of applied organics. This nutrient imbalance resulting from application of organics in industrial waste can also exist for P, K and trace elements needed for crop growth. These parameters must be checked for each waste stream and land receiver.

REFERENCES

Acharya, C. N. "Studies on the Decomposition of Plant Materials. III. Comparison of the Course of Decomposition of Rice Straw Under Anaerobic, Aerobic, and Partially Anaerobic Conditions," *Biochem. J.* 29:1116-1120 (1935).

Adams, R. S., Jr., and M. Gross. "Teaching is not a Research Vacuum. I. Philosophy," *J. Agron. Educ.* 4:123-125 (1975).

Ahmed, M., and D. D. Focht. "Oxidation of Polychlorinated Biphenyls by *"Achromobacter PCB,"* *Bull. Environ. Contam. Toxicol.* 10:70-72 (1973).

Alexander, M. "Microbial Degradation of Pesticides," in *Environmental Toxicology of Pesticides,* F. Matsumura, G. M. Boush and T. Masato, Eds. (New York: Academic Press, 1972), pp. 365-383.

Alexander, M., and M. I. H. Aleem. "Effect of Chemical Structure on Microbial Degradation of Substituted Benzenes," *J. Agric. Food Chem.* 9:44-56 (1961).

Alexander, M. *Introduction to Soil Microbiology.* 2nd ed. (New York: John Wiley & Sons, Inc., 1977).

Anderson, J. R. "Urease Activity, Ammonia Volatilization, and Related Microbiological Aspects in Some South African Soils," *Proc. 36th Congr. S. Afr. Sug. Techn. Assoc.* 97-104 (1962).

Ayanaba, A., W. Verstraete and M. Alexander. "Brief Communication: Possible Microbial Contribution to Nitrosamine Formation in Sewage and Soil," *J. Nat. Cancer Inst.* 50(3):811-813 (1973a).

Ayanaba, A., W. Vestraete and M. Alexander. "Formation of Dimethylnitrosamine, a Carcinogen and Mutagen, in Soils treated with Nitrogen Compounds," *Soil Sci. Soc. Am. Proc.* 37:565-568 (1973b).

Bailey, S., and P. J. Bunyan. "Interpretation of Persistence and Effect of Polychlorinated Biphenyls in Birds," *Nature* 236:34-36 (1972).

Baker, Z., R. W. Harrison and B. F. Miller. "Action of Synthetic Detergent on the Metabolism of Bacteria," *J. Exp. Med.* 73:249-271 (1941).

Batyuk, V. P., and S. K. Samochvalenko. "The Influence of Surface-Active Substances on the Water Physical Characteristics of Soil and the Physiological—Biochemical Characteristics of Plants," *Akadem Hays, Maya, Ochob.* (1972).*

Baxter, R. A., P. E. Gilbert, R. A. Lidgett, J. H. Mainprize and H. A. Kodden. The Degradation of Polychlorinated Biphenyls by Microorganisms," *Sci. Total Environ.* 4:53-61 (1975).

Beck, J., and K. E. Hansen. "The Degradation of Quintozene, Pentachlorobenzene, Hexachlorobenzene and Pentachloroaniline in Soil," *Pestic. Sci.* 5:41-48 (1974).

Blumer, M., and W. W. Youngblood. "Polycyclic Aromatic Hydrocarbons in Soils and Recent Sediments," *Science* 118(4183):53-55 (1975).

Borneff, J., F. Selenka, H. Kunte and A. Maximos. "Experimental Studies on the Formation of Polycyclic Aromatic Hydrocarbons in Plants," *Environ. Res.* 2:22 (1968).

Brian, P. W. "The Ecological Significance of Antibiotic Production," in *Microbial Ecology, 17th Symp. of the Soc. for Gen. Microbiol.* (London: Cambridge University Press, 1958), pp. 168-188.

Bristow, P. R., J. Katan and J. L. Lockwood. "Control of *Rhizoctoria solani* by Pentachloro-Nitrobenzene Accumulated from Soil by Bean Plants," *Phytopathology* 63:808-813 (1973).

Buddin, W. "Partial Sterilization of Soil by Volatile and Nonvolatile Antiseptics. *J. Agric. Sci.* 6:417-451 (1914).

Buser, H. R., and H. P. Bosshardz. "Studies on the Possible Formation of Polychloroazobenzenes in Quintozene Treated Soil," *Pest. Sci.* 6:35-41 (1975).

Cain, R. B. "The Microbial Metabolism of Nitro-Aromatic Compounds," *J. Gen. Microbiol.* 19:1-14 (1958).

Cairns, R. R. "Effects of Surfactants Applied to Samples of Solonetz Soil on Water Penetration and Plant Growth," *Can. J. Soil Sci.* 52:267-269 (1972).

Cardinali, A., and Z. Stoppini. "Detergent Polluted Water and Hydrologic Characteristics of Agrarian Soils. II. Research on Structural Stability," *Perugia Univ. Facolta Di Agraria Annali* 29:351-368 (1974).*

Carlile, B. L., and J. A. Phillips. "Evaluation of Soil Systems for Land Disposal of Industrial and Municipal Effluents," Report No. 118 of the WRRI, UNC (1976).

Casanova, M., and J. Dubroca. "Residues of Pentachloronitrobenzene and its Impurity Hexachlorobenzene in Soils and Lettuce," in *Decomposition of Toxic and Nontoxic Compounds in Soils*, M. R. Overcash and D. Pal, Eds. (Ann Arbor, MI: Ann Arbor Science Publishers, Inc., in press).

Cerna, S. "The Effect of Toluene on Phosphatase Activity in the Soil," *Rostlinna Vyroba* 16(11/12):1285-89 (1970).*

Coffman, L. M., and D. D. Woodbridge. "Effects of Gamma Radiation on Aqueous Solutions of Phenol," *Bull. Environ. Contam. Toxicol.* 11(5):461-466 (1974).

Csarjesi, A. J. "The Adaptation of Fungi to Pentachlorophenol and its Biodegradation," *Can. J. Microbiol.* 13:1243-1249 (1967).

Davies, J. I., and W. C. Evans. "Oxidative Metabolism of Naphthalene by Soil *Pseudomonads.* The Ring Fission Mechanism, *"Biochem. J."* 91:251-261 (1964).

Davies, J. B. *Petroleum Microbiology.* (New York: Elsevier Publishing Co., 1967).

Dean-Raymond, D., and R. Bartha. "Biodegradation of Some Polynuclear Aromatic Petroleum Components by Marine Bacteria," *Develop. Ind. Microbiol.* 16:97-110 (1975).

Dean-Raymond, D., and M. Alexander. "Plant Uptake and Leaching of Dimethylnitrosamine," *Nature* 262C5567:394-396 (1976).

Decock, P. C., and C. Vaughan. "Effects of Some Chelating and Phenolic Substances on the Growth of Excised Pea Root Segments," *Planta.* 126:187-195 (1975).

Dejonckheere, W., W. Steurbaut and R. H. Kips. "Residues of Quintozene, Hexachlorobenzene, Dichloran, and Pentachloroaniline in Soil and Lettuce," *Bull. Environ. Contam. Toxicol.* 13:720-729 (1975).*

Devos, R. H., M. C. ten Noever de Brauw and P. D. A. Olthof. "Residues of Pentachloronitrobenzene and Related Compounds in Greenhouse Soils," *Bull. Environ. Contam. Toxicol.* 11(6):567 (1974).

Dhawan, N., and M. M. Mishra. "Effect of Anionic Detergents on Soil Microbiological Processes," *HAU J. Res., Hissar* 4(1):46-50 (1974).

Dobozy, O. K., and B. Bartha. "Non-pollutant Surfactants Stimulating the Growth of Plant," *La Rivista Italina Delle. Sostanza Grasse* 52:380-382 (1975).*

Dodd, C. G., and S. Ray. "Semiquinone Cation Adsorption on Montmorillonite as a Function of Surface Acidity," *Clays and Clay Min.* 8:237-251 (1960).

Dolgova, L. G. "Biochemical Activity of Polluted Soil," *Pochvovendeni,* 4:113-118 (1975).*

Dolgova, L. G. "Phenoloxidase Activity of Soil Under Conditions of Industrial Pollution," *Pochvovedenic No.* 9:64-69 (1973).*

Donigian, A. S., Jr., D. C. Beyerlein, H. H. Davis, Jr. and H. H Crawford. "Agricultural Runoff Management (ARM) Model Version II." EPA-600/3-77-098. Office of Research and Development, U.S.E.P.A. Athens, GA 293p. (1977).

Donoghue, N. A., and P. W. Trudgill. "The Metabolism of Cyclohexanol by by Acinetobacter NClB9871. *Europ. J. Biochem.* 60;1-3 (1975).

Douglas, L. A., and J. M. Bremner. "A Rapid Method of Evaluating Different Compounds as Inhibitors of Urease Activity in Soils," *Soil Biol. Biochem.* 3(4):309-316 (1971).

Dressel, J. "Uber das Vorkommen von Nitrosaminen in pflanzlichem Material," *Landwirt. Forsch. Sanderh.* 28:273-279 (1973).

Dressel, J. "Relationship Between Nitrate, Nitrite and Nitrosamines in Plants and Soil," *Qual. plant. Pl. Fds. Hum. Nutr.* 25(3/4):381-390 (1976).

Dronkers, H., and A. P. van der Vet. "Investigations into the Mechanism of Microbial Breakdown of Dodecyl Sulfate," proper presented at IV Int'l Congr. Surf. Active Agents, Brussels, 1964).

Edwards, C. A. "Insecticide Residues in Soils," *Residue Rev.* 13:83-132 (1966).

Eggerth, A. H. "The Germicidial Action of Hydroxy Soaps," *J. Exp. Med.* 50:299-313 (1929).

Elmund, G. K., S. M. Morrison, D. W. Grant and M. P. Nevins. "Role of Excreted Chlortetracycline in Modifying the Decomposition Process in Feedlot Waste," *Bull. Contam. Toxicol.* 6(2):129-132 (1971).

Endo, R. M., J. Letey, N. Valoras and J. F. Osborn. "Effect of Nonionic Surfactants on Monocots," *Agron. J.* 61:850-854 (1969).

Engelhardt, G., P. R. Wallnöfer and H. G. Rast. "Metabolism of *o*-phthalic Acid by Different Gram-Negative and Gram-Positive Soil Bacteria," *Arch. Microbiol.* 109:109-114 (1976).

Evans, W. C., H. N. Fernley and E. Griffiths. "Oxidative Metabolism of Phenanthrene and Anthracene by Soil Pseudomonads. *Biochem. J.* 95:819-831 (1965).

Floodgate, G. D. "Biodegradation of Hydrocarbon in the Sea," in *Water*

Pollution Microbiology, R. Mitchel, Ed. (New York: John Wiley and Sons Inc., 1972).

Func, F. "Decomposition of Coumarin in Soil," Abst. of Communications, 9th Annual Meeting of the Czechoslovak Soc. for Microbiology, *Folia Microbiologia* V. 16:152 (1971).

Furukawa, T., and G. W. Brindley. "Adsorption and Oxidation of Benzidine and Aniline by Montmorillonite and Hectorite," *Clays and Clay Min.* 21: 279-288 (1973).

Gibson, D. T., V. Mahadevan, D. M. Jerina, H. Yagi and H. J. C. Yeh. "Oxidation of the Carcinogens Benzo(a)pyrene and Benzo(a)anthracene to Dihydrodiols by a Bacterium," *Science* 189:295-297 (1975).

Giger, W., and M. Blumer "Polycyclic Aromatic Hydrocarbons in the Environment. Isolation and Characterization of Chromatography, Visible, Ultra-Violet and Mass Spectrometry," *Anal. Chem.* 46(12):1663-1671 (1974).

Goodnow, R. A., and A. P. Harrison, Jr. "Bacterial Degradation of Detergent Compounds," *Appl. Microbiol.* 24(4):555-560 (1972).

Gorbach, S., and U. Wagner. "Pentachloronitrobenzene Residues in Potatoes," *J. Agric. Food Chem.* 15(4):654-656 (1967).

Graft, V. W., and H. Diehl. "Uber den naturbedingten Normalpegel kanzerogener polycyclischer Aromate und seine Ursache," *Arch. Hyg. Bakt.* 150:49 (1966).

Greenland, D. J. "Denitrification in Some Tropical Soils," *J. Agric. Sci.* 58:227-233 (1962).

Groenewegen, D., and H. Stolp. "Microbial Breakdown of Polycyclic Aromatic Hydrocarbons," *Tbl. Bakt. Hyg., I. Abt. Orig.* Bl62:225-232 (1976).*

Guirguis, M. A., and V. Vancura. "The Oxidation of Pipecolic Acid in Pre-incubated Soils," *Folia Microbiol.* 15:459-467 (1970).

Gundersen, k., and H. L. Jensen. "A Soil Bacterium Decomposing Organic Nitro-compounds," *Acta. Agric. Scand.* 6:100-114 (1956).

Hafner, V. M. "Hexachlorbevzolruckstände in Gemuse—bedingt durch Aufnahme des Hexachlorokenzols aus dem Boden," *Gesunde Pflanzen* 27(3):37-48 (1975a).*

Hafner, V. M. "Untersuchungen zur Kontamination von gartenerden und landwirtsshaflich Benutzten Boden mit Hexachlorbenzol und Pentachlornitrobenzol," *Gesunde Pflanzen* 5(5):82-95 (1975b).*

Hagin, J., and L. Amburger. "Contribution of Fertilizers and Manures to the N- and P- Load of Waters. A Computer Simulation," Report submitted to D.E.G. (1974).

Haider, K., and J. P. Martin. "Decomposition of Specifically Carbon-14 Labelled Benzoic and Cinnamic Acid Derivatives in Soil," *Soil Sci. Soc. Am. Proc.* 39(4):657-662 (1975).

Haque, R., D. W. Schmedding and V. H. Freed. "Aqueous Solubility, Adsorption and Vapor Behavior of Polychlorinated Biphenyl Aroclor 1254," *Environ. Sci. Technol.* 8:139-141 (1974).

Haselhoff, E. "Grundziig der Rauchschadenkunde. Anlextung für Prüfung und berurteilungder Einwirkung von Rauchabgangen auf Boden und Pflanze," Berlin (1932).

Heiling, C. S., D. G. Dennison and D. O. Kaufman. "Fungicide Movement in Soils," *Phytopathol.* V. 64:1091-1100 (1974).

Horvath, R. S. "Cometabolism of the Herbicide 2,3,6-Trichlosobenzoate," *J. Agric. Food Chem.* 19:291-293 (1971).

328 DESIGN OF LAND TREATMENT SYSTEMS

Hsu, Y.-C. "Detergent Splitting Enzyme from Pseudomonas," *Nature* 207: 385-388 (1965).

Huang, P. M., T. S. C. Wang, M. K. Wang, M. H. Wu and N. W. Hsu. "Rentention of Phenolic Acids by Noncrystalline Hydroxyaluminum and Iron Compounds and Iron Compounds and Clay Minerals of Soils," *Soil Sci.* 123(4):213-219 (1977).

Huddleston, R. L., and R. C. Allred. "Biodegradation of Straight Chain Alkylbenzene Sulfonates Using the River Die-Away Technique," paper presented to the American Chemical Society, Div., of Microbial Chemistry and Technology, New York, September 1963a.

Huddleston, R. L., and R. C. Allred. Microbial Oxidation of Sulfonated Alkyl Benzenes," *Develop. Ind. Microbiol.* 4:24-37 (1963b).

Huddleston, R. L., and R. C. Allred. "The Effect of Structure on Biodegradation of Polythoxylated Nonionic and Ether Sulfate Detergents," paper presented at the Society of Industrial Microbiologists Annual Meeting, August, 1964.

Huddleston, R. L., and R. C. Allred. "Surface-Active Agents: Biodegradability of Detergents," in *Soil Biochemestry,* A. D. McLaren and H. Peterson, Eds. (New York: Marcel Dekker, Inc., 1967), pp. 343-370.

Iwata, Y., and F. A. Gunther. "Uptake of a PCB (Aroclor 1254) from Soil by Carrots Under Field Conditions," *Bull. Environ. Contam. Toxicol.* 11:523-528 (1974).

Jeffrey, A. M., H. J. C. Yeh, D. M. Jerina, T. R. Ratel, J. F. Davey and D. T. Gibson. "Initial Reactions in the Oxidation of Naphthalene by *Pseudomonas putida," Biochemistry* 14:575-584 (1975).

Jenkinson, D. S., and D. S. Powlson. "The Effects of Biocidal Treatments on Metabolism in Soil.—I. Fumigation with Chloroform," *Soil Biol. Biochem.* 8:167-177 (1976).

Judy, J. N., D. C. Martens and W. Kroontje. "Effect of Detergent Application on the Growth of Corn," *J. Environ. Qual.* 2(2):310-314 (1973).

Kaiser, K. L. E., and P. T. S. Wong. "Bacterial Degradation of Polychlorinated Biphenyls. I. Identification of Some Metabolic Products from Aroclor 1242," *Bull. Environ. Contam. Toxicol.* 11:291-296 (1974).

Kamilova. R. M., A. Khikmatov and S. Pazilova. Duration of Toxication of Toluene on Plants in Soil," *Acad. Nauk Uzbekskoi SSR, Dokl.* 1:55-56 (1976).[*]

Kasugai, S. I., and S. Ozaki. "Cyanuric Acid in Soil," *J. Sci. Agric. Soc. (Japan)* V. 232:1-18 (1922).

Kaufman, D. D. "Degradation of Pesticides by Soil Microorganisms," in *Pesticides in Soil and Water,* W. D. Guenzi, Ed. (Madison, WI: SSSA, Inc., 1974).

Kawahara, T., M. Matsui and H. Nakamura. "BHC in Soil of Paddy Field," *Bull. Agric. Chem. Inspec. Stn.* 12:42-45 (1972).[*]

Klausmeir, R. E., and W. A. Jones. "Microbial Degradation of Plasticizers," *Developments in Industrial Microbiology,* S. Rich, Ed. (New York: Plenum Press, 1960), pp. 47-53.

Klein, W., and I. Weisgerber. "PCBs and Environmental Contamination," *Environ. Qual. Safety* 4:237 250 (1976).

Klevens, H. B. "Stabilization of Polycyclic Hydrocarbons," *J. Phys. Chem.* 54:281-298 (1950).

Koch, R. In: *Disinfection and Sterilization,* 2nd Ed., E. C. McCulloch, Ed. (Philadelphia: Lea and Febiger, 1945), p. 1881.

Kroontje, W., J. N. Judy and H. C. H. Hahne. "Effect of Detergent Laden Water on the Growth of Corn," *Bull. 62,* Virginia Water Resources Research Center, Virginia Polytechnic Institute & State University, Blacksburg, VA (1973).

Kuchar, E. J., F. O. Geenty, W. P. Griffith and R. J. Thomas. "Analytical Studies of Metabolism of Terraclor in Beagle Dogs, Rats, and Crops," *J. Agric. Food Chem.* 17:1237-1240 (1969).

Kudeyarov, V. N., and D. S. Jenkinson. "The Effect of Biocidal Treatments on Metabolism in Soil—VI. Fumigation with Carbon Disulphide," *Soil Biol. Biochem.* 8:375-378 (1976).

Kuwatsuka, S. "Degradation of Several Herbicides in Soils Under Different Conditions," in *Environmental Toxicology of Pesticides,* F. Matsumura, G. M. Bouch and T. Masato, Eds. (New York: Academic Press, 1972), pp. 385-400.

Lee, B. K. H. "The Effect of Anionic and Nonionic Detergents on Soil Microfungi," *Can. J. Bot.* 48:583-587 (1970).

Lunt, D., and W. C. Evans. "The Microbial Metabolism of Biphenyl," *Biochem. J.* 118:54-55.

Luxmore, R. J., N. Valoras and J. Letey. "Nonionic Surfactant Effects on Growth and Porosity of Barley Roots," *Agron. J.* 66:673-675 (1974).

Luzzati, A. "The Effect of Detergents on Some Plant Species," *Bolletino dei Laboratori Chimici Provinciali (Bologna)* V. 25(4):60-73; V. 25(6):112-122 (1974).*

Madhosingh, C. "The Metabolic Detoxification of 2,4-Dinitrophenol by *Fusarium oxysporum,*" *Can. J. Microbiol.* 7:553-567 (1961).

Martin, J. P., and K. Haider. "Decomposition of Specifically Carbon-Labelled Ferulic Acid: Free and Linked into Model Humic Acid Type Polymers," *Soil Sci. Soc. Am. J.* 40(3):377-379 (1976).

McGarity, J. W., and G. Meyers. "A Survey of Urease Activity in Soils of Northern New South Wales," *P. Soil* 27:217-238 (1967).

McKenna, E. J., and R. D. Heath. "Biodegradation of Polynuclear Aromatic Hydrocarbon Pollutants by Soil and Water Microorganisms," Water Resources Center, Univ. of Illinois at Urbana-Champaign, Research Report No. 113, UILU-WRC-76-0113 (1976).

McMichael, F. C., and J. E. McKee. "Wastewater Reclamation at Whittier Narrows, State of California, Water Control Board, Sacramento, CA. Publication No. 33 (1966), p.101.

Medvedev, V. A., and V. D. Davidov. "Transformation of Individual Organic Products of the Coke Industry in Chernozemic Soil," *Pochvovedenie* 11:22-28 (1972).*

Medvedev, V. A., and V. C. Daidov. "The Influence of Isomers on the Transformation Rate of Phenols in Chernozem Soil," *Pochvovedenie* 16(5): 122-127 (1973).*

Medvedev, V. A., and V. C. Davidov. "The Rate of Degradation of Phenols and Quinones in Chernozemic Soil According to Data Relating to Oxidation Reduction Potential and Infra-red Spectroscopy," *Pochvovedenie* 1:133-137 (1974).*

Medvedev, V. A., V. D. Davidov and S. G. Mavrody. "Destruction of High

Doses of Phenol and Indole by a Chernozemic Soil," *Pochvevedenie* 6:128-131 (1975).*

Miller, P. M. "Effects of Some Nitrogenous Materials and Wetting Agents on Survival in Soil of Lesion, Stylet and Lance Nematodes," *Phytopathology* 66:798-800 (1976).

Miller, W. W., and J. Letey. "Distribution of Nonionic Surfactant in Soil Columns following Application and Leaching," *Soil Sci. Soc. Am. Proc.* 39:17-22 (1975).

Miller, W. W., N. Valoras and J. Letey. "Movement of Two Nonionic Surfactants on Wettable and Water Repellent Soils," *Soil Sci. Am. Proc.* 39:11-16 (1975).

Mills, A. L., and M. Alexander. "Factors Affecting Dimethylnitrosamine Formation in Samples of Soil and Water," *J. Environ. Qual.* 5(4):437-440 (1976a).

Mills, A. L., and M. Alexander. "n-Nitrosamine Formation by Cultures of Several Microorganisms," *Appl. Environ. Microbiol.* 31(6):892-895 (1976b).

Moza, P., I. Weisgerber and W. Klein. "Fate of 2,2'-Dichlorobiphenyl—[14]C in Carrots, Sugarbeets, and Soil Under Outdoor Conditions," *J. Agric. Food Chem.* 24(4):881-885 (1976).

Murado, M. A., M. C. Tejedor and G. Buluja. "Interactions Between Polychlorinated Biphenyls (PCBs) and Soil Microfungi. Effects of Aroclor-1254 and other PCBs on *Aspergillus flavus* Cultures," *Bull. Environ. Contam. Toxicol.* 15:768-774 (1976).

Murray, K., and P. A. Willliams. "Role of Catechol and the Methyl Catechols as Inducers of Aromatic Metabolism in *Pseudomonas putida*," *J. Bacteriol.* 117(3):1153-1157 (1974).

National Technical Information Service (NTIS). "PCBs and the Environment," Document Com 72-10419, U. S. Govt. Interdepartmental Task Force on PCBs, *Environ. Res.* 5(3):253-362 (1972).

Naumann, K. "On the Dynamics of Soil Microflora after Application of the Fungicide. Olpisan (Trichlorodinitrobenzol), Captan and Thiuram," *Arch. Pflanzenschutz* 6(5):383-398 (1970).*

Norris, D. B., and P. W. Trudgill. "The Metabolism of Cyclohexanol by *Norcardia Globerula* CL-1," *Biochem. J.* 121:363-365 (1971).

Oloffo, P. C., L. J. Albright and S. Y. Szeto. "Fate and Behavior of Five Chlorinated Hydrocarbons in Three Natural Waters," *Can. J. Microbiol.* 18:1393-1398 (1972).

Osborn, J. F., J. Letey and N. Valoras. "Surfactant Longevity and Wetting Characteristics," *Calif. Turfgrass Culture* 19(3):17-18 (1969).

Overcash, M. R. "Assimilation of Toluene by North Carolina Soils," Unpublished resunts (1978).

Pal, D., M. R. Overcash and P. W. Westerman. "Plant-Soil Assimilative Capacity for Organic Solvent Constituents in Industrial Wastes," *Proc. 32nd Purdue Ind. Waste Conf.*, 259-271 (1977).

Payne, W. J., and V. E. Feisal. "Bacterial Utilization of Dodecyl Sulfate and Dodecyl Benzene Sulfonate," *Appl. Microbiol.* 11:339-344 (1963).

Pokonova, Yu., G. A. Kovalchuk and N. P. Matreeva. "Stabilization of Soils by Phenolic Alcohols," *J. Appl. Chem. USSR,* V. 46(9):Pt. 1 (September 1973). Translated February 1974, pp. 2380-2384.

Pramer, D. *Appl. Microbiol.* 6:221-224 (1958).

Purushothaman, D., and K. Balaraman. "Effect of Soil Phenolics on the Growth of Rhizobium," *Current Sci.* 42(14):507-508 (1973).

Putilina, N. T. "Microbes Used in Industrial Purification Installations for Removal of Phenols from Wastewater," *Mikrobiologiya* 28:757-762 (1959).

Rall, D. "Polychorinated Biphenyls-Environmental Impact," *Environ. Res.* 5(3):252-362 (1972).

Rawlings, C. D. Textile Plant Wastewater Toxics Study," paper presented at Textile Wastewater Treatment Conference, Hilton Head, SC, Clemson, SC (1977).

Raymond, D. D., and M. Alexander. "Cleavage of the Ether bond of Phenylmethyl Ethers by Enzumes of *Arthrobacter* sp.," *Pesticide Biochem. Physiol.* 2:270-277 (1972).

Reddy, K. R., R. Khaleel, M. R. Overcash and P. W. Westerman. "A Nonpoint Source Model for Land Areas Receiving Animal Wastes. V. Carbon Transformations, Unpublished results (1978).

Ribbons, S. W., and W. C. Evans. "Oxidative Metabolism of Phthalic Acid by Soil Pseudomonads," *J. Biochem.* 76:310 (1960).

Ryckman, D. W., and C. N. Sawyer. "Chemical Structure and Biological Oxidizability of Surfactants," *Proc. of the 12th Ind. Waste Conf.*, Purdue University, Lafayette, IN, Ext. Series #94 (1957), pp. 270-284.

Safe, S., and O. Hutzinger. "Polychlorinated Biphenyls: Photolysis of 2,4,6, 2′, 4′, 6′-Hexachlorobiphenyl," *Nature* 232:641-642 (1971).

Sander, J., M. Ladenstein, J. LaBar and F. Schweinskerg. In: *n-Nitroso Compounds in the Environment*, P. Bogovski and E. A. Walker, Eds. (Lyon, France: Intl. Agency for Research on Cancer, 1975), pp. 205-210.

Sebastiani, A., A. D. Simonetti and A. Borgioli. "Behavior of Synthetic Detergents in the Soil," *Nuovi Ann. d'Igiene Microbiolozia* V. 22(1):11-27; V. 22(2):81-95; V. 22(4):229-242 (1971).*

Sebek, O. K., and D. Perlman. "Microbial Transformations of Antibiotics," *Adv. Appl. Microbiol.* 14:123-146 (1971).

Stetzkorn, E. A., and A. B. Carel. "The Analysis of Alkyl Aryl Sulfonates by Microdesulfonation and Gas Chromatography," *J. Am. Oil Chem. Soc.* 40(2):57-59 (1963).

Shepherd, C. J., and J. R. Villanueva. "The Oxidation of Certain Aromatic Compounds by the Conidia of *Aspergillus* nidulans," *J. Gen. Microbiol.* 20:VII (1959).

Simonetti, A. D., G. Tarsitani and L. A. Sebastiani. "Behavior of Synthetic Detergents in the Soil. IV. Soft Detergents, Catabolites and Vital Competition," *Nuovi Annali d'igiene Microbiologia* 25(4):262-270 (1974).*

Sisler, F. D., and C. E. Zobell. "Microbial Utilization of Carcinogenic Hydrocarbons," *Science* 106:521-522 (1947).

Sittig, M. *Pollutant Removal Handbook* (Park Ridge, NJ: Noyes Data Corp., 1973).

Skriabin, G. K., E. L. Golovlev, L. A. Goloveva, L. V. Andreev and Z. I. Finkel'shtein. "On the Correlative Link Between the Taxonomic Position of Soil Microbacteria and Their Ability to Transform Aromatic Hydrocarbons," *Akad. Nauk SSSR Dokl.* 200(5):1224-1226 (1971).*

Smelt, J. H. "Behavior of Quintozene and Hexachlorbenzene in the Soil and

Their Absorption in Crops," *Gewasbeschermung* 7(3):49-58 (1976).*

Sorensen, L. H., and E. A. Paul. "Transformation of Acetate Carbon into Carbohydrate and Amino Acid Metabolities During Decomposition in Soil," *Soil Biol. Biochem.* 3:173-180 (1971).

Stahl, W. H., and H. Pessen. "The Microbial Degradation of Plasticizers I. Growth on Esters and Alcohols," *Appl. Microbiol.* 30-35 (1953).

Steiner, G. W., and R. D. Watson. "The Effect of Surfactants on Growth of Fungi," *Phytopathology* 55:1009-1012 (1965).

Stratton, C. L., and J. B. Sosebee, Jr. "PCB and PCT Contamination of the Environment Near Sites of Manufacture and Use," *Environ. Sci. Technol.* 10:1229-1233 (1976).

Swisher, R. D. "Biodegradation Rates of Isomeric Diheptylbenzene Sulfonates," *Develop. Ind. Microbiol.* 4:39-45 (1963a).

Swisher, R. D. "The Chemistry of Surfactant Biodegradation," *J. Am. Oil Chem. Soc.* 40(11):648-656 (1963b).

Tarvin, D., and A. M. Buswell. "The Methane Fermentation of Organic Acids and Carbohydrates," *J. Am. Chem. Soc.* 56:1751 (1934).

Tate, III, R. L., and M. Alexander. "*n*-Nitrosamines: Absence from Sauerkraut and Silage," *Agric. Food Chem.* 23(5)896-897 (1975a).

Tate, III, R. L., and M. Alexander. "Stability of Nitrosamines in Samples of Lake Water, Soil and Sewage," *J. Nat. Cancer Inst.* 54(2):327-330 (1975b).

Tate, III, R. L., and M. Alexander. "Resistance of Nitrosamines to Microbial Attack," *J. Environ. Qual.* 5(2):131-133 (1976).

Thente, B. O. "Effects of Toluene and High Energy Radiation on Urease Activity in Soil," *Lantbrukshogkolans Annater.* 36:401-418 (1970).

Thurston, A. "Quantitative Analysis of PCBs," *PCT Newsletter* No. 3 (July 1971).

Tucker, E. S., W. J. Litschgi and W. M. Mees. "Migration of Polychlorinated Biphenyls in Soil Induced by Percolating Water," *Bull. Environ. Contam. Toxicol.* 13:86-93 (1975).

U.S. Environmental Protection Agency. Development Document for Effluent Guidelines for Soap and Detergent Manufacture Point Source Category, EPA 440/1-74/18 (1973).

U.S. Environmental Protection Agency "Rationale for the Development of BAT Priority Pollutant Parameters, Energy & Mining Branch, Washington (1977).

Valoras, N., J. Letey, J. P. Martin and J. Osborn. "Degradation of a Nonionic Surfactant in Soils and Peat," *SSSA J.* 40(1):60-63 (1976a).

Valoras, N., J. Letey and J. Osborn. "Nonionic Surfactant-Soil Interaction Effects on Barley Growth," *Agron. J.* 68(4):591-595 (1976b).

Van Bavel, C. H. M. "Gaseous Diffusion and Porosity in Porous Media," *Soil Sci.* 73-91-104 (1952).

Varga, J. M., and H. Y. Neujahr. "Isolation from Soil of Phenol-Utilizing Organisms and Metabolic Studies on the Pathways of Phenol Degradation," *Plant Soil.* 33:565-571 (1970).

Vishwanath, N. R., R. B. Patil and G. Rangesevanni. "Dehydrogenase Activity and Microbial Population in Soils Treated with Chloroform and Toluene," *Zbl. Bakt. Abt. II* 130:348-356 (1975).*

Waksman, S. A., and R. L. Starkey. "Partial Sterilization of Soil, Microbiological Activities and Soil Fertility: I, II, and III," *Soil Sci.* 16:137-158; 247-268; 343-358 (1923).

Waksman, S. A., T. C. Cordon and N. Hulpoi. "Influence of Temperature upon the Microbiological Population and Decomposition Processes in Composts of Stable Manures," *Soil Sci.* 47:83-114 (1939).

Wallnöfer, P. R., G. Englehardt, S. Safe and O. Hutzinger. "Microbial Hydroxylation of 4-chlorobiphenyl and 4,4'-dichlorobiphenyl," *Chemosphere* 2:69-72 (1973).

Wallnöfer, P., M. Roniger and G. Englehardt. "Fate of Xenobiotic Chlorinated Hydrocarbons (HCB and PCBs) in Plants and Soils," *Zeit. Pflanzen. Pflanzenschutz* 82(2):11-100 (1975).*

Wang, C. H., and F. E. Broadbent. "Kinetics of Losses of PCNB and OCNA in Three California Soils," *Soil Sci. Soc. Am. Proc.* 36:742-45 (1972).

Wang, C. H., and F. E. Broadbent. "Effect of Soil Treatments on Losses of Two Chloronitrobenzene Fungicides," *J. Environ. Qual.* 2(4):511-515 (1973).

Wang, T. S. C., T. K. Yang and T. T. Chuang. "Soil Phenolic Acids as Plant Growth Inhibitors," *Soil Sci.* 103(4):239-246 (1967).

Webb, F. C. *Biochemical Engineering* (Princeton, NJ: D. Van Nostrand Co., Ltd., 1964).

Williams, O. B., and H. B. Rees, Jr. "Bacterial Utilization of Anionic Surface-Active Agents," *J. Bacteriol.* 58(6):823-824 (1949).

Wong, P. T. S., and K. L. E. Kaiser. "Bacterial Degradation of Polychlorinated Biphenyls II. Rate studies," *Bull. Environ. Contam. Toxicol.* 13:249-255 (1975).

Yamane, I. "Decomposition of Litter of *miscanthus sinensis* During Five Years Under Seminatural Conditions," *Science Rep. Res. Inst. Tohoku Univ. Ser. D.* 25:25-30 (1974).

Yoshida, O., M. Miyakawa, Y. Okada, K. Ohshiro, T. Harada, S. Machida and T. Kato. "The Disintegration of a Benzidine Dye, Direct Dept Balck EX, by *Escherichie coli* and Soil Bacteria," *Med. Biol.* 86(6):361-364 (1973).*

Youngblood, W. W., and M. Blumer. "Polycyclic aromatic Hydrocarbons in the Environment: Homologous Series in Soils and Recent Marine Sediments," *Geochim. Cosmochim. Acta* 39:1303-1314 (1975).

*For translations, see: Overcash, M. R., and D. Pal, Eds. *Decomposition of Toxic and Nontoxic Compounds in Soils* (Ann Arbor, MI: Ann Arbor Science Publishers, Inc., in press).

SOLUTION TO EXAMPLE PROBLEM

Waste: Total oxygen demand (Equation 7.3)

$$
\begin{aligned}
10{,}000 \text{ gal/d} &= 37{,}500 \text{ liter/d} \\
\text{COD} &= 187.5 \text{ kg/d} \\
\underline{(4)\,(\text{TKN})} &= \underline{180 \text{ kg/d}} \\
&\quad\ 367.5 \text{ kg TOD/d}
\end{aligned}
$$

$$= 2{,}570 \text{ kg TOD/wk (applied 24 events/yr = 15 day basis)}$$

Soil Assimilative Capacity (Equations 7.1 and 7.2)

$$
\begin{aligned}
D &= 0.6\,(0.22)\,(1.6 \text{ m}^2/\text{d}) \\
&= 0.21 \text{ m}^2/\text{d} \\
N_{O_2} &= 2\,(300 - 140)\,[\frac{(0.21)\,(15d - 4d)}{\pi}]^{\frac{1}{2}} \\
&= 274 \text{ g O}_2/\text{m}^2/15 \text{ day period} \\
&= 126 \text{ g O}_2/\text{m}^2/\text{week (averaged over year)} \\
&= 1{,}260 \text{ kg O}_2/\text{ha/week}
\end{aligned}
$$

minus the plant needs of 1,010 kg O_2/ha/week

$$= 250 \text{ kg O}_2/\text{ha/week.}$$

Land Area: $\dfrac{2{,}570 \text{ kg TOD/wk}}{250 \text{ kg O}_2/\text{ha/wk}} = 10.3 \text{ ha}$

CHAPTER 8

ACIDS, BASES AND SALTS

INTRODUCTION

The behavior in a land treatment system of this entire class of industrial waste constituents follows a general pathway, with some minor adaptations for the actual compounds present. Based on the broad class of constituents used to determine assimilative capacities which meet the fundamental constraint of nondegradation, acids, bases and salts are primarily mobile parameters (see Chapter 1). A brief description of the behavior of these compounds in a land-based receiver is given prior to the detailed discussion of assimilative capacity. Chapter 3 further explains the basic soil processes described in relation to acids, bases and salts applied to land (Figure 8.1).

In this chapter the term salts refers to the elements traditionally important for agricultural irrigation. This is often referred to as salinity or electrical conductivity of material to be land applied. The sodium, calcium, potassium and magnesium parameters of an industrial waste are included as salts as well as the overall ionic strength. Acids and bases segments include either organic or inorganic compounds, which exert a pH effect in the industrial waste and, hence, on the plant-soil system.

When acids, bases or salts are applied to a plant-soil system, an initial reaction or response occurs. For acids and bases (whether inorganic or organic), a neutralization reaction takes place to an extent dictated by the reactive soil fraction or soil buffering capacity, and by the strength and dissociation of the applied acid or base. Salts also react with the soil in ion exchange and precipitation. The impact of the salt response depends on the soil absorptive capacity (cation exchange capacity, CEC) and the relative proportion of sodium in the wastes.

After the initial plant-soil response, the organic acids and bases would be expected to undergo microbial degradation and thus be converted into soil organic matter or gaseous microbial end products (CO_2). The inorganic

335

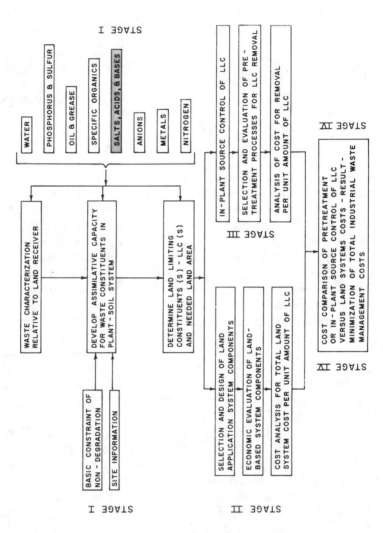

Figure 8.1. Relationship of Chapter 8 to overall design methodology for industrial pretreatment-land application systems.

acids and bases would behave as conservative or nondecomposable species in a similar manner to salts. These inorganic acids, bases and salts will eventually migrate with the movement of water (applied wastewater and rainfall) in a manner analogous to the movement of cations and anions in a natural plant-soil system. Thus, in the long term, the accumulative capacity of soils is exceeded and an ion balance between input and outflow must exist. Outflow of water and associated ions is both lateral to surface waters and vertical to groundwaters.

The assimilative capacity of a plant-soil system for industrial acids, bases and salts centers on the general pathways described above. *The first portion* of the design for waste assimilation is to minimize the dose effect of the initial soil reactions. That is, waste acids or bases are not applied in doses at such a high ratio to the available soil buffering capacity that there is a severe loss of vegetation or soil microbial populations. For salts, an overloading of exchange capacity or sudden shift in sodium balance is to be avoided. Avoidance of these dose effects, which are primarily waste concentration-dependent, means cycling vegetation and waste application or waste pre-treatment.

The second portion of the general assimilative capacity is to ensure in the design that there is a long-term correction for acid, base or certain salt (Na) imbalances; in other words, to amend the waste or soil-waste complex with supplementary chemicals that will maintain functional agricultural capacity of the soil system (again the basic constraint for assimilation of all waste constituents). These supplementary chemicals are a part of the operating cost and program of industrial land application systems in which the acid, bases or salts class of parameters are the land limiting constituent (LLC).

The final portion of the assimilative capacity design is to determine the parameters necessary to keep the impact of the cations and anions on receiving waters within acceptable limits. From Chapter 4 one can approximate the percentage of applied water and rainfall that moves through the soil to surface waters and the remainder to groundwater. Increasing the land area per unit amount of mobile constituent provides lower concentrations, through rainfall dilution, to the receiving water. Therefore, sufficient land must be used so that the concentration of material reaching groundwaters or surface waters does not exceed the respective water quality standards. For groundwater, these would be drinking water standards, while for surface waters, less stringent standards may be in use. Groundwater would require no greater treatment for water use because of the presence of the land application systems. This assimilative capacity is conservative because no allowance is made for aquifer flow or volume and, therefore, some relaxing of this constraint may be appropriate in specific cases.

The recognition of ultimate movement of land-applied inorganic fractions of acids and bases and salt constituents is essential to the assimilative capacity of this category of industrial waste parameters. The large soil concentration of these salts in exchangeable and nonexchangeable form and the nondegradation of these species are important characteristics. This is differentiated from the organics category where migration is extremely low and microbial decomposition is designed to match application rates. In the latter category, movement to surface waters and groundwaters is unlikely, while in the former category the design must be based on such movement.

In considering the above three components involved in the deisgn for assimilation of acids, bases and salts, the capacity of plant-soil systems is well established. Since the industrial waste application is not primarily for water supply or crop irrigation purposes, the previous restrictions on saline concentrations or sodium imbalance can be modified. Within the limit of larger land areas, even very concentrated salt-containing or acid/base wastes can be satisfactorily assimilated in a plant-soil system when adequate design and impact considerations have been made.

The consequences of incorrectly applying acid, base or salt constituents can be inferred from naturally occurring soils in which excessive conditions exist. Salt-affected soils are classified into three groups:

1. *Saline soils:* Sufficient salts are present in the soil-water solution that plant growth is reduced. Plants are stunted, leaf tips are discolored and the soil permeabilities are higher than nonsaline soils. There is good soil structure but the salt content adversely affects germination and growth. Depending on the gradient in salt levels, plants with shallow roots or deep roots may be more tolerant. In quantitative terms, if the saturation soil extract is greater than 4 mmhos/cm in electrical conductivity and less than 15% exchangeable sodium, then it is a saline soil. Definition of these soil measurements is given later in this chapter.

2. *Sodic or Alkaline Soils:* Sufficient sodium is present in exchangeable form that plant yields are reduced. Sodic soils are generally deflocculated or dispersed and thus have altered physical properties that reduce water movement. Necessary moisture for plant growth is thus unavailable. The high sodium levels induce a deficiency in calcium (Ca) and magnesium (Mg), thus affecting plant growth. If sodium contents are sufficiently high that the soil exchangeable sodium percentage exceeds 15, but salinity is less than 4 mmhos/cm, then a soil is alkaline (pH > 8.5).

3. *Alkali-Saline Soils:* If there are excessive amounts of salts (> 4 mmhos/cm) and sodium (> 15% exchangeable Na), then crop reduction is attributed to an alkali-saline soil condition. Typically, these soils have a pH < 8.5. Depending on the leaching phenomenon such soils can remain more saline or become more sodic.

Incorrect design or operation of industrial land application systems, in which acids, bases or salts are the LLC, can lead to soil conditions similar to the above natural soils. Such areas are unproductive and very difficult to recover or reclaim. Therefore, caution must be exercised to ensure a satisfactory assimilation of acids, bases or salts in land application systems.

SALTS

Two principal factors are of concern with salt-containing industrial wastes when applied to a plant-soil system:

1. the total salt content or salinity, and
2. the relative balance between sodium and the rest of the other cations.

The general design approach for the first factor is to ensure adequate salt movement through the soil, while for the second factor, the addition of favorable cations is used to correct sodium imbalance. Prior to the detailed discussion on assimilation of salts, several definitions are necessary.

Equivalents (eq): the weight of an element that releases or picks up 1 mole or Avogadro's number of electrons. It is the molecular weight of an element divided by the valance of the ion when used in calculations related to salinity in soils.

Cation Exchange Capacity (CEC): the sum of all exchangeable cations adsorbed per 100 g of dry soil when the exchange capacity is completely utilized. Units are meq/100 g soil.

Percent Base Saturation (BS): the proportion of the CEC that is occupied by cations other than aluminum and hydrogen. This measurement is made on field soils to determine the unused capacity of the soil for adsorbing cations:

$$BS = \frac{(Ca) + (Mg) + (Na) + (K)}{CEC}$$

where: ( ) = meq/100 g dry soil

The percent base saturation is directly related to soil pH.

Exchangeable Sodium Percentage (ESP); the proportion of the CEC occupied by sodium. The magnitude of ESP denotes the magnitude of adverse soil physical effects due to sodium.

Electrical Conductivity (EC): defined as the reciprocal of electrical resistivity, r, experienced with the flow of current through an aqueous solution when 2 parallel electrodes are immersed in it. Symbolically:

$$EC = \frac{1}{r} = \frac{1}{ohms\ cm} = \frac{mhos}{cm}$$

EC has been directly correlated to salt content and osmotic pressure.

Salt concentration, ppm = 651 x EC, in mmhos/cm
Osmotic pressure, atm = 0.35 x EC, in mmhos/cm
Salt concentration, meq/l = 10 x EC, in mmhos/cm.

Electrical conductivity is measured by use of pipet-type conductivity cell with platinized or platinum electrode attached to a Wheatstone bridge. These are available commercially with operation manual or manufacturer's instructions. Solutions or saturation extracts of soil can be used directly on the conductivity bridge after appropriate calibrations. Based on EC, wastewaters and irrigation waters may be categorized as given below:

EC (wastewater) mmhos/cm at 25°C	Category
<0.25	Low salinity
0.25-0.75	Medium salinity
0.75-2.25	High salinity
>2.25	Very high salinity

Based on the EC of a soil saturation extract, soils are classified as follows:

EC (soil saturation extract) mmhos/cm at 25°C	Soil class
<2	Low salinity, normal
2-4	Medium salinity
>4	Excess salinity

The relationship between EC of a saturated soil extract and percentage of salts in a soil is presented in Figure 8.2. Soil texture is used to demonstrate the variation in this relationship.

Sodium Adsorption Ratio (SAR): is an empirical measure of the sodium imbalance of an industrial waste or soil solution sample. SAR is:

$$SAR = \frac{(Na)}{\sqrt{[(Ca) + (Mg)]/2}} \qquad (8.1)$$

where ( ) = meq/l

The relationship between SAR and ESP is determined empirically (Richards 1954) as:

$$ESP = \frac{(1.55)(SAR) - 5.79}{(0.0155)(SAR) + 0.942} \qquad (8.2)$$

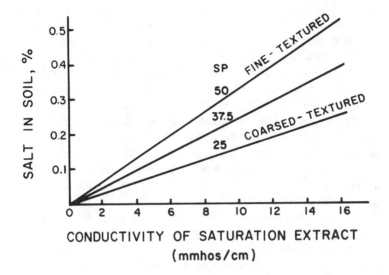

Figure 8.2. Range of conductance values found in saturation soil extracts. The saturation percentage (SP) represents fine texture (50) and coarse texture (25) of soils (Richards 1954).

In the definitions given here and the usage throughout the chapter the reader should carefully note whether what is being referred to is the waste-water, soil water or soil-saturated extract. For more detailed information on the soil mechanisms related to these definitions, see Chapter 3.

Industrial wastes all contain salts just as do all sources of water. Salts are a natural product of water leaching through soils. Thus, in the long-term application of wastes to soils, consideration is necessary to determine whether a particular waste and site are such that salts may be the LLC.

Fate in Soil Systems

Salinity

The clay fraction and orgnaic matter in soils have a net negative charge and, hence, cations are held to the soil particles. Adsorption energies of various cations vary in the sequence:

$$H > Al > Ca = Mg > K = NH_4 > Na$$

Thus, under equal concentration conditions in a soil, the divalent ions (Ca and Mg) would displace the monovalent ones (K, NH$_4$, Na). Resulting soil-water would contain these monovalent ions until such time as nearly complete displacement, after which the soil water would assume the same concentration as the applied waste, corrected for rainfall.

For wastes that contain a disproportionate percentage of a single cation, the Law of Mass Action controls the adsorption process. If there is an equilibrium, such as between various cations and the soil exchange complex, the introduction of a high concentration of one cation in solution will displace other cations from the exchange complex. The displacement continues until a balance is reached among the cations with the solution concentration being increased. In this manner a concentrated potassium or sodium solution could displace Ca or Mg from the soil. This phenomenon has been observed around landfills of sodium hydroxide in which the concentration of Ca is elevated in surrounding wells—a halo effect. Thus, Ca and not Na is made mobile and the effect of the landfill is disguised.

Precipitation reactions also occur in soils and decrease the concentration of cations and anions. For Ca, a principal precipitate is $CaCO_3$, the amount of which is usually limited by the bicarbonate concentration. As Ca and SO_4 approach 50 meq/l a $CaSO_4$ (gypsum) is formed. Other insoluble products are given in Chapter 10. Magnesium can be insolubilized principally as silica compounds. An approximate 5:2 proportion of Ca to Mg was found to precipitate in soil solutions (Eaton 1966).

Anion behavior in soil systems depends on the nature of the chemical species. In general, soil surfaces contain a net negative charge and, hence, anion retention is low. Exceptions to this rule are due to precipitation or replacement or addition to soil structure. Bicarbonate and sulfate are ions for which cation precipitation is common, thus anions are fixed as $CaCO_3$ and $CaSO_4$. At high anionic concentrations, a clay mineral breakdown can occur with the formation of new compounds, e.g., $Al_2 (PO_4)_3$ (Mielenz 1952). Anions with tetrahedral structure (PO_4^{-2}, SO_4^{-2}) are similar in shape to the clay structure (kaolinite) and can displace hydroxyl groups and fit tightly into the soil. For these anions the soil has a very large fixation capacity. Anions such as chloride or nitrate do not have such a structure and so are easily moved through the soil.

Certain site-specific characteristics are important in predicting accelerated effects from saline industrial waste applications to land. Sites with poor natural drainage, either because of soil texture or because of the terrain and distances of water flow to outlets, would not permit as great a movement of salinity. Thus, lower assimilative rates would be used in determining whether salts were the LLC. Rainfall and evaporation at a site are additional important characteristics. These must be balanced with waste concentration and receiving water requirements, to develop the design criteria at a particular site. Differences in net moisture input within the United States have a large impact on determining whether salts are the LLC for a given industrial waste at a potential land application site.

Sodium Imbalance

For an industrial waste land application site, the concentrations of sodium, calcium and magnesium are interrelated controlling factors in the soil system design. The absolute removals of Ca, Mg and Na are small and limited by the unsaturated cation exchange capacity available. The magnitude of available cation storage (% BS) is given in the previous section. After the soil CEC is saturated, then the relative amounts of Na vs Mg and Ca in the wastewater becomes an important long-term land design factor.

The importance of Ca, Mg and Na is due to the effect on soil structure, which is then the major determination for water and wastewater movement. High levels of exchangeable sodium promote deflocculation and swelling of clay particles. The sodium ion, when hydrated, is larger than the lattice space of the clay minerals, and when present in large quantities Na causes deflocculation of such soils. Pore sizes become smaller and water movement is severely restricted. Exchangeable sodium percentage (ESP) is a measure of the Na content in soils relative to the other cations. Water flow (measured as hydraulic conductivity) is substantially reduced by ESP increases (Figure 8.3). A reduction in hydraulic conductivity would lower the water assimilative capacity at a site, thus potentially making hydraulic application the LLC. Such reductions in soil permeability must be avoided by correcting the Na imbalance.

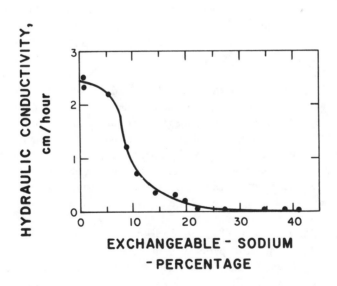

Figure 8.3. Influence of exchangeable sodium percentage on the hydraulic conductivity of a clay loam (Martin *et al.* 1964).

Coarse-textured soils are generally less affected by exchangeable sodium than are fine-textured soils. Soils containing clay of the expanding lattice type, such as montmorillonite, swell and disperse more with sodium adsorption. The degree of swelling is related to the specific surface of soils. High Na in a waste would displace Ca and Mg from the soil exchange sites, thereby preferentially accumulating Na (Marshall 1964). The dynamic condition of the soil throughout the profile must be monitored for the concentrations of Na, Ca and Mg, since localized layers could develop reduced permeabilities. The relation in the sodium adsorption ratio of soil saturation extract with exchangeable sodium percentage (ESP) is shown in Figure 8.4. The ESP of soil at values >5% has great effect on soil hydraulic conductivity (Figure 8.3) and, hence, permeability. At ESP above 20%, the hydraulic conductivity is reduced to almost zero (Martin *et al.* 1964).

Effect on Vegetative Cover

Salinity effects on plants are categorized as (1) osmotic or diffusional relationships, (2) ionic interference, and (3) toxicity of chemical species. High salt wastes and wastewaters when applied to land may raise the osmotic pressure of the soil solution. The result is that a gradient in osmotic potential between soil solution and root cells is lowered, such that there is less water uptake by plants in presence of these high-soluble salts. The visible effects of excess salinity are reductions in both rate of growth and total plant size. Forage and seed yields are also usually reduced. As shown in Figure 8.5, with increasing salinity the available water decreases and so does the plant growth. Salt-affected plants do not respond to applied fertilizers (NH_4NO_3), which further adds to osmotic potential of soil solution and thus aggravates the salinity effects.

The plant response to high levels of Ca, Mg, K and Na is similar to the law of mass action. Elevated Ca reduces the uptake of potassium, presumably by overloading the ion transfer processes into the root. Similar results have been noted for high magnesium or sodium in soils. Thus, aside from crop response to osmotic potential, a plant may be deficient in essential nutrients when a disproportionate quantity of a single cation is land applied. This suggests that supplementary chemicals may be required to maintain soil balance. The final effect of saline waste application, toxicity of chemical species, is discussed in more detail in Chapter 9.

Crop uptake can be used to remove land-applied salts, although the magnitude of this assimilative pathway is much less than that from adsorption and leaching. Experiments with several crops in Texas demonstrate typical crop uptake rates (Longnecker and Lyerly 1974) (Table 8.1). There is a substantial variation in salt species removal among various plants.

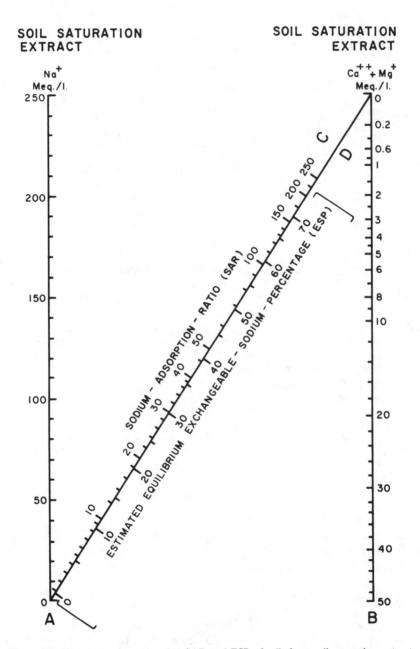

Figure 8.4. Nomogram for estimating SAR and ESP of soils from soil saturation extract analyses (Richards 1954).

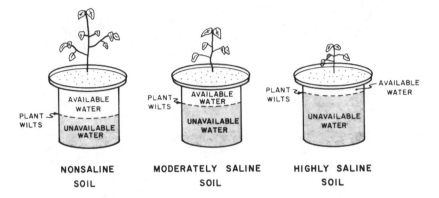

Figure 8.5. Schematic of crop yield effect from soil salinity.

Table 8.1 Approximate kg/ha/yr of Salt Ions Removed from the Land by Various Crops in the El Paso Area (Longnecker and Lyerly 1974).

Crop	Crop Yield (kg/ha)	Sodium	Calcium	Magnesium	Sulfate	Chloride
Sweetclover Hay	9,000	19	175	117	77	37
Sundangrass Hay	11,200	24	38	77	223	75
Alfalfa Hay	9,000	47	67	55	58	62
Barley Straw	2,200	16	9	3	31	17
Corn Silage	33,600	81	65	115	109	115
Barley Grain	11,200	2	1	1	3	8
Sorghum Grain	4,400	7	3	6	9	19
Cotton Seed	1,700	3	2	6	9	18

On sodic and saline-sodic soils, waterlogging may occur and inadequate root aeration may prevail. Toxic symptoms from excess sodium have been described by Lunt (1966). The decrease in yield of various crops in high-sodium soils has been well-documented.

Assimilative Capacity for Salts

Plant-soil assimilation of salts depends on the water movement, the soil exchange properties and the vegetative cover selected. Water movement is highly site specific (see Chapter 4). Moisture balance, the relative lateral and deep movement, and the waste characteristics, would have to be developed before evaluating salinity or Na imbalance assimilative capacity. It is assumed here that the above factors have been determined and can be used in developing design criteria for salts. As another cautionary note, much of the salts assimilative capacity has been developed based on data from alkaline or neutral soils and, therefore, use of data in acid soils should be coupled with monitoring to verify or improve potential rates of salt application.

Sodium Imbalance

There are at least four alternatives available to ensure the satisfactory assimilation of sodium in plant-soil systems. All these alternatives for industrial waste involve the balance between sodium and other cations as expressed by SAR, ESP or the sodium percentage in soil water. This balance prevents preferential sodium buildup in soil, with resultant soil physical and plant growth alterations. The critical SAR value at which soil swelling precludes long-term successful operation depends on the overall clay content as well as the nature of the clay particles. A range of critical SAR values for soils is given in Table 8.2, along with that of certain wastewaters. An average value of SAR = 12 is used in subsequent discussions.

The first and often the most economic alternative is to pretreat the wastewater or sludge stream to obtain SAR $<$ 12 prior to land application. Such a system will ensure that the soil water and also the soil will approach levels in the safe SAR range. Calcium or magnesium additions or substitution for Na in processes can be used to adjust the SAR. Field monitoring should be undertaken to ensure soil equilibration.

A more refined method for estimating the necessary calcium additions has been developed for agricultural crop irrigation (Eaton 1966). To maintain the long-term level in soilwater below 70% of the total cation concentration, three factors must be included. The first is the Ca necessary to keep Na $<$ 70%:

$$\frac{(Na)}{(Na) + (Ca) + (Mg)} = 0.7$$

where ( ) = meq/l, or

$$0.43 \, (Na) - (Ca) - (Mg) = I \tag{8.3}$$

The second factor accounts for the loss of Ca and Mg added to the waste when applied to soil in which carbonate and silica precipitation occurs ($CaCO_3$, Mg-Si compounds). This relationship is empirical:

Table 8.2 Critical SAR Values for Soils and Representative Waste SAR Values

Soil or Wastewater	SAR
Swelling Clay (bentonite)	8 - 10
Nonswelling Clay	20
Pure Sand	750
Loam or Finer Textures	5 - 15
($>10\%$ clay)	
Poultry Manure	0.41
Beef Manure	0.52
Sewage Effluent	4
Processing Plant	
Green beans	4.1
Lima beans	28.7
Tomatoes	3.3
Sweet potatoes (lye-peeled)	91.6
Sweet potatoes (steam-peeled)	2.2
Poultry	1.5

$$[(CO_3) + (HCO_3)] \times 0.7 = II \qquad (8.4)$$

Approximately 70% of this precipitation loss of divalent cations is Ca and 30% is the associated losses of Mg.

The third factor is the relative loss of Ca and Mg in crop uptake to that of Na, which in the long term would preferentially accumulate Na. Again, an empirical relationship from irrigation suggests that an extra constant amount of Ca would compensate for crop uptake:

$$0.3 \text{ meq/l of Ca} - III \qquad (8.5)$$

Another method to account for the relative cation uptake by plants is to determine the mass of Na, Ca and Mg to be taken by the plant (Table 8.1) and add Ca or Mg to match crop needs relative to Na uptake.

The total divalent cation addition to correct sodium imbalance is, thus:

$$I + II + III = \text{desired waste concentration of Ca, meq/l} \qquad (8.6)$$

With the waste volume, the amount of Ca to be added can be determined to pretreat prior to land application.

A second alternative for SAR control is to add calcium or magnesium to the land application site to maintain acceptable soil conditions. This solution is less desirable than industrial waste control, since monitoring must be more intense and field operations are required. The usual soil additive is gypsum rather than agricultural lime ($CaSO_4 \cdot 2H_2O$) because of the former's low cost and greater solubility (Carlile and Phillips 1976).

Other amendments can be applied to soils to displace Na and maintain a favorable SAR (Table 8.3). Limestone alone is nearly insoluble, but as noted, when applied with sulfuric acid or sulfate the Ca becomes soluble. With this alternative the soil physical properties or crop growth are not altered since amendments are made as the soil solution approaches critical values. The soil extract SAR is then reduced and further waste applications can be assimilated.

Gypsum applications are performed with farm spreader equipment. Soil testing after several irrigation or rainfall events will help the land system manager assess SAR correction. A regular schedule of soil additions can then be established to offset the wastewater application. Additionally, as an emergency protection, even if wastewater pretreatment is practiced, gypsum should be available for the accidental sodium spill.

The third alternative is one that requires larger land areas. The basic concept is to consider the soil buildup to an SAR = 12 as the lifetime of the land application site. Often, when the levels of Na, Ca and Mg are low but the SAR exceeds 12, then with continued waste application the soil exchange capacity is filled and SAR approaches critical levels. The time span to reach critical SAR values should be calculated so that preparation to terminate application or to make soil additions can be initiated.

The fourth alternative land management technique for salt applications is to correct a soil that has exceeded the critical SAR value. This is the least desirable alternative since it comes after the soil has been deflocculated and the structure damaged. It is a very slow and tedious process, which results in partial restoration of the soil to the original condition. If the zone of damage is at the soil surface, the corrective problem is much easier since calcium salts can be surface applied and mechanically incorporated in the soil to immediately lower the sodium percentage of the soil adsorption sites. By mechanical manipulation and alternative wetting and drying procedures, the spray field may be restored to partial use in a few months.

The more common problem in soils is that of structure deterioration of the upper B horizon (subsoil) due to high-sodium loadings. The problem then becomes one of applying calcium salts to the soil surface and relying on natural leaching to move the soluble calcium down to the problem zone, some 8-16 inches below the surface. Since movement of water through the deflocculated zone is severely restricted, it may require several years to restore the soil percolation to reasonable rates.

Salinity

Discussion of the assimilative capacity for salts covers two factors: (1) the well-established agricultural need to keep soil-water available to the plant, and (2) the migration of land-applied salt compounds (cations or anions).

Table 8.3 Amounts of Soil Amendments Required to Reduce Exchangeable Sodium to 10% or Less in Exchange Complex per ha m of Soil (Richards 1954)

Exchangeable Sodium (meq/100 g of soil)	metric tons/ha-m					
	Gypsum (CaSO$_4$ + 2H$_2$O)	Sulfur (S)	Iron sulfate (FeSO$_4$ + 7H$_2$O)	Limestone[a] (CaCO$_3$)	Calcium Chloride (CaCl$_2$)	Lime-Sulfur (24% sulfur)
1.00	12.1	2.3	20	7.1	7.8	9.5
2.00	24.2	4.6	40	14.2	15.6	19
3.00	36.3	6.9	60	21.3	23.4	28.5
4.00	48.4	9.2	80	28.4	31.2	38
5.00	60.5	11.5	100	35.5	39	47.5
6.00	72.6	13.8	120	42.6	46.8	57
7.00	84.7	16.1	140	49.7	54.6	66.5
8.00	96.8	18.4	160	54.8	62.4	76
10.00	121	23	200	70	78	95

[a]Required to react with sulfur (S) or iron sulfate (FeSO$_4$ + 7 H$_2$O) if soil is lime-free.

Both factors are based on a mass balance concept with known inputs and regulated outputs.

To eliminate a buildup of salts in the root zone such that the osmotic gradient in the soil liquid reduces the plant ability to obtain water, a leaching or removal of salts must occur. While this problem is more acute in low rainfall regions, the concept is equally valid in more humid areas, and is especially needed when industrial wastes, containing higher salt concentrations than conventional irrigation waters, are to be land applied. The removal or required movement of salts in the plant-soil root zone is referred to as the leaching requirement and is defined as follows (Richards 1954):

$$LR = \frac{D_d}{D_i} = \frac{EC_i}{EC_d} \qquad (8.7)$$

where D_d and D_i = volume of liquid per unit land area (the depth of water) which must flow out of (drained) and which is being applied (irrigated) to a plant-soil system, respectively

EC_d and EC_i = concentration of mobile salts expressed as electrical conductivity or as meq of mobile species/l in the water leaving and being applied to the plant-soil system

Internally consistent units must be used in Equation 8.7 as LR is the fraction of the applied material that must be leached. Also, wastes containing salts do not have to be irrigated but can be applied mechanically in sludge. The effect of rainfall must be included in Equation 1, thus:

$$D_i^T = D_i + D_r \qquad (8.8)$$

$$EC_i^T = (D_r\, EC_r + D_i\, EC_i)/D_i^T \qquad (8.9)$$

where superscript T = total of all liquid inputs, and
subscript r = the properties of rainfall

The amount of applied liquid (waste liquid, low concentration irrigation or leaching water and rainfall) is thus:

$$D_i^T = (EC_d/EC_i^T)\, D_d \qquad (8.10)$$

The amount of water leaving the root zone must be determined using the techniques of Chapter 4, which balance applied liquid, site-specific soil characteristics and water losses to evapotranspiration. Finally, the value of EC_d is established from the tolerance of crops to soil-water concentrations of salts, which is shown in Figure 8.6 (Reeve 1957). Crops that fit in the various

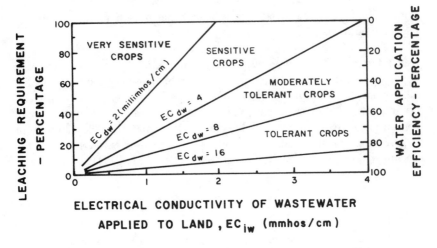

Figure 8.6. Leaching requirement as related to crop salt tolerance (Reeve 1957).

tolerance classes are given in Table 8.4. Thus, the parameters in Equation 8.10 can be determined once the specific land site and crop type are identified.

The crop need for water must also be considered. This is the consumptive need to maintain adequate crop yields and can be obtained from various state agricultural agencies since this need is also quite site-specific. The fraction of applied water available to the plant is $(1-LR)$, then:

$$D_i = D_c \left(1/(1-LR)\right) \tag{8.11}$$

where subscript c = the consumptive plant need.

The irrigation requirement for consumptive need must be compared to the actual water application to determine if more nonwastewater irrigation will have to be applied to maintain the vegetative cover.

The second category of assimilative capacity for salts is the requirement to achieve a given concentration of a particular chemical species prior to entry to groundwater or surface water. The chemical compound is applied in an industrial waste and is considered immediately mobile. The species moves in proportion to water flow with a certain percentage moving in the lateral direction through the soil to surface waters and a certain percentage moving to groundwater. Allowable concentration of the salt species is determined based on the final receiving water. As a first assumption, the allowable level of a salt is taken as the same in groundwater and surface waters. This assumption is removed in later discussion.

From Chapter 4 the designer has the percentages of water movement to groundwater and surface waters based on a site-specific analysis, which is

Table 8.4 Salt Tolerance of Crops (optimum pH)

I. Very Very Salt Tolerant (EC 8-12 mmhos/cm)	II. Moderately Salt Tolerant (EC 4-8 mmhos/cm)		III. Near-Neutral pH Requirements (EC 2-4 mmhos/cm)
Alfalfa (6.2-7.8)	African Violet (6-7)	Spinach (7-7.5)	Apricot (6-7)
Alyssum (6-7.5)	Alfalfa (6.5-8)	Spinach (6-7.4)	Arborvitae (6-7.5)
Asparagus (6-8)	Almond (6-7)	Sorghum (5.5-7.5)	Tobacco (5.5-7.5)
Barberry (6-7.5)	Barley (6.5-7.8)	Sycamore (6-7.5)	Tamarack (5-6.5)
Bermuda Grass (7-8)	Begonia (5.5-7.0)	Sunflow (6.5-8)	Bell Pepper (6-7)
Burnish Bush (5.5-7.5)	Broccoli (6-5)	Tomato (6.5-8)	Black oak (6-7)
Cabbage (7-8)	Calendula (5.5-7.0)	Vetches (7-8.2)	Yam (6-7)
Carnation (6-7.5)	Celery (5.8-7.0)	Wheat (6.5-8)	Cherry (5.5-7)
Carrots (5.5-7.5)	Crab Apple (6-7.5)	Zinnia (5.5-7.5)	Douglas Fir (6-7)
Cauliflower (5.5-7.5)	Cotton (6.5-8)		Hot Pepper (5.5-7)
Chrysanthemum (6.5-8)	Cowpeas (7-8.2)		Lantana (6-7)
Date Palm (7.5-8.2)	Corn (6.5-8)		Poinsettia (6-7)
Garden Beets (6-8)	Cucumber (6.5-8)		Quince (6-7)
Geranium (6-8)	Johnson Grass (6.5-7.5)		Rice (5-6.5)
Ivy (6-8)	Lespedeza (7-8.2)		Reed Canary Grass (5.5-7)
Panic Grass (7-8)	Lily (6-7)		Rose (5-6.5)
Peas (6-7.5)	Lilac (6.0-7.5)		Rye (5-7)
Peach (6-7.5)	Maple (6-7.5)		Soybean (5.5-7.5)
Purple Sage (7-8)	Millet-Sorghum (7-8.2)		Sesbania (5-7)
Rhodes Grass (7-8.2)	Muskmellon (6.0-7.0)		Potato (5-6.5)
Salt Grass (7.5-8.2)	Rhubarb (5.50-7.0)		Sweet Potato (5-7)
Spinach (6-7.5)	Safflower (6.5-7.8)		
Sugar Beets (6.5-8.0)	Snap dragon (6-7.5)		
Sugar Cane (6-8)	Snowball (6.5-7.5)		
Wild Mustard (7-8)	Sweet William (6-7.5)		

critical to determining the salt assimilative capacity. The allowable concentration in the liquid leaving the root zone is specified from regulations such as drinking water standards or other appropriate guidelines. Modifying Equation 8.10 for the conditions in a land application system:

$$D_r + 100 \; Q/A = C_d/C_i^T)D_d \qquad (8.12)$$

where D_r = rainfall input, cm/yr
$\quad Q$ = industrial waste volume, m^3/yr
$\quad A$ = area required for land application, m^2, with 100 being the conversion between m and cm
$\quad C_d$ = drinking water concentration of a particular mobile constituent under consideration
$\quad C_i^T$ = concentration of waste in the aggregate of rainfall and waste application, similar to Equation 8.9
$\quad D_d$ = amount of applied liquid per unit area which moves laterally or vertically out of the root zone.

In Equation 8.12 the only unknown is A, the area required to assure no adverse impact on receiving waters due to mobile salts. The area, A, occurs in C_i^T and D_d and yields:

$$A = \frac{(C_i - C_d)}{D_r[C_d(1-\alpha)-C_r]} \cdot 100 \; Q \qquad (8.13)$$

where α = ratio of evaporative losses to rainfall as determined by geotechnical and vegetation
$\quad C_i$ = concentration of the mobile species in the industrial waste

Equation 8.13 and the preceding analysis can be refined for the conditions in which groundwater and surface water have different allowable levels of the salt species being analyzed. In this instance, the surface water and groundwater requirements are analyzed separately. Into Equation 8.13 C_d is specified for groundwater. The corresponding A_{gw} is the land area necessary to meet groundwater restrictions that the liquid concentrations not exceed levels that would require greater treatment for drinking water. It should be noted that dilution in the groundwater aquifer is not included. The assimilative capacity is increased when flow through an aquifer is included since it further dilutes the concentration of material. With greater site information the analysis can be improved to increase the allowable application rates.

In a similar analysis, A_{sw}, the land area necessary to meet surface water restrictions, can be determined. Then the larger of A_{gw} and A_{sw} is chosen as the salts assimilative capacity. The entire analysis is repeated for each salt species expected to be reasonably mobile in the soil system.

The management and design of a pretreatment-land application system can improve the salts assimilative capacity. One such technique would be to select a more salt tolerant crop species for use as the vegetative cover (Table 8.4). In this manner, leaching requirements and land areas may be reduced. In-plant source control is another important technique for reducing required land area. Often, salt type in an industrial process is indicated by past experience and economics. Substitution for sodium salts can reduce the sodium imbalance problems. Recycle systems can reduce the total salt loads, thus reducing required land areas. These techniques should be employed after a determination that salts or a particular salt compound is the LLC.

ACIDS AND BASES

Neutralization reactions are the first response of plant-soil systems to the application of acidic or basic compounds in industrial wastes. Inorganic acids in wastewaters generally dissociate to a gerater extent than organic acids, and, therefore, organic acid waste constituents are more readily assimilated in the buffering capacity of the soil system. Organic acids are also subject to microbial utilization and degradation. Organic acid decay improves soil physical conditions by increasing the number and stability of soil aggregates and by improving soil structure. Continuous application of acid wastewaters on soils exerts a "washing effect," in which cations are leached (Pal et al. 1978) and the salt concentration is decreased. Consequently, the soil may become deficient in bases and percent base saturation may decrease. This occurrence would certainly be associated with a decrease in soil pH of the total wastewater—soil system.

Exceeding the buffering capacity, the soil pH can be altered in the application of an industrial waste. Changes in soil pH results in a variety of altered relationships, which, in turn, affect the plant-soil assimilative capacity for waste constituents. One relationship that is changed is the nature of the ions present on the soil cation exchange complex. A general description of these ions is found in Figure 8.7. As the pH is lowered, greater amounts of exchangeable and total Al and H^+ dominate the soil complex. If the pH is raised the CEC is dominated by exchangeable bases and greater soil water hydroxyl content occurs (Brady 1974).

The shift in concentration of Al^{+3} demonstrates the soil-water relationship to the soil exchange complex (Figure 8.8) (Magistad 1925). Elevated solution levels occurred under very acid and very basic conditions with a resultant aluminum toxicity to crops. A variety of other shifts in solution concentration or availability to a plant occur when the soil pH is changed (Figure 8.9). For several essential nutrients (Ca, Mg, P), both high and low pH reduce the availability to the plant. The relationship to heavy metals is

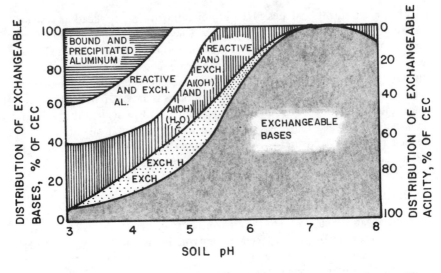

Figure 8.7. Approximate percent distribution of base saturation and exchange acidity at various soil pH values.

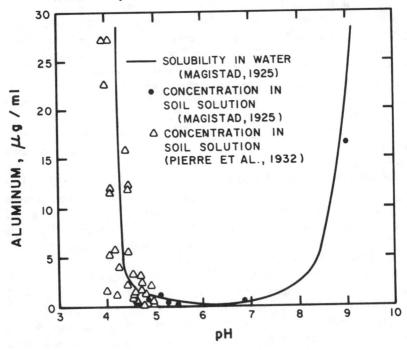

Figure 8.8. Solubility of aluminum in water and Al concentration of soil solution at different pH values.

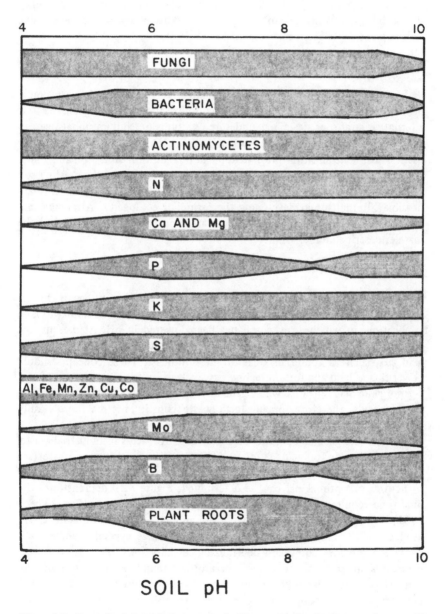

Figure 8.9. General relationship between soil pH, growth of soil microorganisms, nutrient availability and plant roots.

shown, in which generally greater availability to plants occurs as the pH is lowered toward acidity.

To increase the soil assimilative capacity for acids and bases, preference should be given to highly buffered soils and to use of acid wastes on alkaline soils and basic wastes on acid soils (Pal *et al.* 1978).

Assimilative Capacity

In the long term, the addition of acid and base constituents in wastes must be accompanied by other materials to neutralize these constituents. The four alternative strategies described under the sections on sodium imbalance can be used for acids or bases. Instead of SAR as the controlling factor, pH is used. Thus, neutralization prior to application, addition of neutralizing materials after minor alternation in soil pH, utilization of only the existing soil buffering capacity and correction of substantial pH changes in soils are possible alternatives in developing the assimilative capacity. After neutralization the salts produced must be evaluated for assimilative capacity in the manner described under salinity.

Acids

Concepts relating lowered soil pH and the behavior of nutrients, metals and the exchange complex have been developed for conditions similar to land application, in which a given soil has been lowered in pH. At present, the concepts previously discussed are utilized in soils that do not have a naturally acid pH, such as found in areas of the Southeast. These concepts should be verified as land application technology evolves.

Inorganic acids in wastewater generally dissociate to a greater extent than organic acids; therefore, organic acid waste constituents are more readily assimilated in the buffering capacity of the soil system.

For the design and operational alternatives for acids in industrial waste, the neutralization of acids in the soil system is an important phase (see above). The soil pH is measured and the amount of base to return the soil to neutral or optimal pH can be determined. Figure 8.10 demonstrates the limestone requirements to convert a soil from an acid pH to pH = 7 (Peech 1961). The two benchmark soils are given and as noted larger quantities are needed for the clay loam versus the lower exchange capacity sandy loam. Table 8.5 provides equivalent neutralizing capacity for other forms of lime. From this information the system operator can sample a site, determine the amount of required limestone and calculate the required amount of a particular neutralizing compound. If the waste is neutralized prior to land application, the standard acid-base relations can be used to determine the required neutralization amounts.

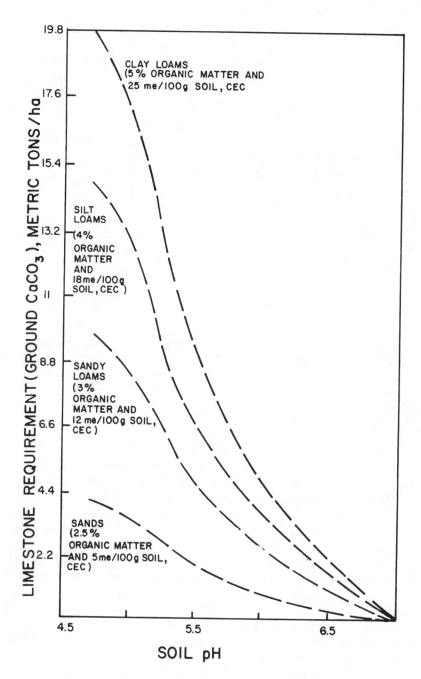

Figure 8.10. Limestone correction of soil pH in New York soils (Peech 1961).

Table 8.5 Relative Neutralizing Power of Different Forms of Lime

Form of Lime	Molecular Weight	Approximate Neutralizing Value (%)	Metric Ton Equivalent to 1 metric ton pure CaCO
Calcium Carbonate	100	100	1
Calcium Hydroxide	74	135	0.74
Calcium Oxide	56	178	0.56
Magnesium Carbonate	84	119	0.84
Magnesium Hydroxide	58	172	0.58
Magnesium Oxide	40	250	0.40

In addition to neutralization of land applied acids, crop selection can enhance the assimilative capacity of soils. Acid-tolerant crops require less neutralization of acid soil pH. Table 8.6 is a partial listing of the pH tolerance of plants (Pal *et al.* 1978).

Bases

In an anlogous manner basic constituents of industrial wastes may raise the soil pH and require acid neutralizations to ensure the long term assimilative capacity. Naturally basic soils occur in the western areas of the U.S. The advantages of organic vs inorganic bases for plant-soil assimilative capacities are similar to the equivalent materials discussed as acids. Aside from sodium or alkaline soils, fewer data are available for the behavior of industrial waste constituents at elevated soil pH.

Correction of alkaline soil pH from the addition of bases in industrial wastes should be directed by soil pH measurements. That is, soil samples are taken and pH measured on a regular basis. Then an acidifying compound is added, such as sulfuric acid or sulfate compounds. After an acclimation period, the soil pH is remeasured and further correction implemented as needed. In this manner curves such as Figure 8.10 can be developed for the acid correction of basic soil. Such curves are site specific, since the soil exchange capacity would determine the needed quantity of amendments for a pH change at each site.

Table 8.6 Acid Tolerance of Crops (optimum pH)

Acid Sensitive (least tolerance)	Slightly Acid Tolerant (moderate tolerance)		Acid-Loving Plants (most tolerance)
1. Alfalfa (6.5-8)	Apple (5-6.5)	Lupine (5.5-6.5)	Aspen (3.8-5.5)
Barley (6.5-8)	Balsam Fir (5-6)	Magnolia (5-6)	Birch (4.5-6.0)
2. Peas (6.5-7.5)	Beech (5-6.7)	Oat (5.5-7)	Blueberries (4-5)
3. Red Clovers (6.5-7.5)	Bent Beans (6-7)	Peanut (5.3-6.6)	Camellias (5-6)
4. Red Beets (6.5-8)	Bluegrass (5.2-6.5)	Phlox (5-6)	Cedar (4.5-5.0)
5. Rhodes Grass (7-8)	Boysenberry (5-6.5)	Pineapple (5-6)	Club Moss (4.5-5)
6. Salt Grass (7.5-9.5)	Corn (5.5-7)	Potatoes (4.8-6.5)	Cranberries (4-5)
7. Sugar Beets (6.5-8)	Field Bean (6-7.5)	Red Clover (5.7-7)	Irish Potatoes (4-5.5)
8. Sweet Clover (6.5-8)	Flax (5-7)	Strawberry	Jack Pine (4.5-5)
9. White Clover (7.5-8)	Hemlock (5-6)	Tobacco (5.3-6.5)	Milkweed (4-5)
10. Wild Mustard (7.5-8)	Heather 4.5-6)	White Oak (5-6.5)	Orchid (4-5)
	Holly, American (5-6)	White Clover (5.7-7.0)	Rhododendron (native) (4.5-6)
	Larch (5-6.5)	Soybean (6-7)	Shpagnum Moss (3.5-5.0)
		Wheat (5.5-7.0)	White & Red Pine (4.5-6.0)

REFERENCES

Black, C. A. *Soil-Plant Relationships,* 2nd ed. (New York: John Wiley & Sons, Inc., 1968).

Brady, N. C. *The Nature and Properties of Soils,* 8th ed. (New York: Macmillan Publishing Co., Inc., 1974), p. 376.

Carlile, B. L., and J. A. Phillips. "Evaluation of Soil Systems for Land Disposal of Industrial and Municipal Effluents," Water Resources Research Institute, University of North Carolina (1976).

Eaton, F. M. "Total Salt and Water Quality Appraisal," in *Diagnostic Criteria for Plants and Soils,* H. D. Chapman, ed., University of California, Division of Agricultural Science (1966), pp. 510-532.

Longnecker, D. E., and P. J. Lyerly. "Control of Soluble Salts in Farming and Gardening," *The Texas Agric. Exp. Sta. Bull.* 876, Revised (1974).

Lunt, O. R. "Sodium" in *Diagnostic Criteria for Plants and Soils,* H. D. Chapman, ed., University of California, Division of Agricultural Science (1966), pp. 409-432.

Magistad, D. C. "The Aluminum Content of the Soil Solution and its Relation to Soil Reaction and Plant Growth," *Soil Sci.* 20:181-225 (1925).

Marshall, C. E. *The Physical Chemistry and Mineralogy of Soils,* Vol. 1, *Soil Materials* (New York: John Wiley and Sons, Inc., 1964).

Martin, J. P., S. J. Richards and P. F. Pratt. "Relationship of Exchangeable Na Percentage at Different Soil pH levels to Hydraulic Conductivity," Soil Sci. Soc. Amer. Proc. 28:620-622.

Mielenz, R. C. "Introduction to Symposium on Exchange Phenomena in Soils," *Am. Soc. for Testing Mat. Publ.* 142:109 (1952).

Pal, D. M., R. Overcash and P. W. Westerman. "Land Disposal of Acidic, Basic, and Salty Wastes from Industries," *Proc. 1977 Nat. Conf. on Treatment and Disposal of Ind. Wastewaters and Residues,* Houston, TX, April, 1977 (1978), pp. 151-159.

Peech, M. "Lime Requirement vs Soil pH curves for Soils of New York State," mimeographed, Cornell University, Ithaca, NY (1961).

Pierre, W. H., G. G. Pohlman and T. C. McIlvaine. "Soluble Aluminum Studies. I. The Concentration of Aluminum in the Displaced Soil Solution of Naturally Acid Soils," *Soil Sci.* 34:145-160 (1932).

Reeve, R. C. "Third Congress of International Commission of Irrigation and Drainage, San Francisco, CA, Question 10R.10, 10. 175-10. 187 (1957).

Richards, L. A., ed. "Diagnosis and Improvement of Saline and Alkali Soils, U.S. Department of Agriculture Handbook 60 (1954).

CHAPTER 9

ANIONS

INTRODUCTION

This class of industrial waste constituents is complex in the assessment of plant-soil assimilative capacity. The anionic species may be elements such as CL^- or I^-, or may typically be an oxide such as BrO_3^-, or AsO_4^{-3}.

To comply with the basic design criteria of nondegradation we must consider two of the three calculational modes associated with waste components (see Chapter 1). That is, toxic anions can (1) accumulate to a critical level at which crop response is adversely affected, or (2) migrate to receiving waters. This first calculational mode is based on the buildup over time of the soil concentration from a single dose of the toxic anionic materials. The second calculational mode is based on water flow, soil-water concentration of toxic anions and allowable receiving water concentrations. Water flow and soil water concentration design techniques are covered in detail in Chapter 4 and 8. In this section, the use of these techniques is described as well as the needed constants. As with the other chapters, emphasis is also placed on the basic processes underway for assimilation within the plant-soil system. This approach is essential to the ability to adapt the design methodology to a given industrial waste and land application site. The relationship of this chapter on the assimilative capacity for specific anions to the total methodology for industrial land treatment is presented in Figure 9.1.

ARSENIC (As)

Much research has been conducted on the application of arsenic compounds to agricultural plant-soil systems, with specific attention to arsenic in the +3 oxidation state (arsenious acid-derived, arsenite, etc.) and in the +5 state (arsenic acid-derived, arsenate, etc.), as well as salts of these

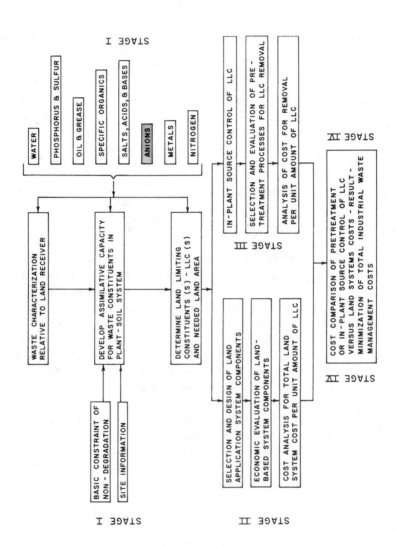

Figure 9.1. Relationship of Chapter 9 to overall design methodology for industrial pretreatment-land application systems.

compounds categorized as soluble arsenic salts or insoluble arsenic salts. Long-term application data and effects on a variety of crops are available, often where sterilization or exceeding the assimilative capacity has occurred. In addition to arsenic compounds for pest control, As is present in super-phosphate in the range of 100-200 ppm, thus providing another soil input (Stutzer 1901).

Arsenic is generally classified as nonessential for microorganisms, plants and animals. However, some experiments have shown As stimulation of plant growth at low application rates. At high levels, As is moderately toxic to plants and highly toxic to mammals.

Pathways for Assimilation

The pathways for As in soil resemble the chemistry of phosphates (Woolson et al. 1971). Most arsenic is fixed in the soil in the form of AsO_4^{-3} by soil sorption to iron, aluminum and calcium, with the predominant complex being Fe-arsenate complexes. Where iron is not present in large amounts, aluminum-arsenate complexes are formed. The adsorption of arsenic compounds and the low solubility of some applied As salts generally reduce the immediate movement of As in soils, thus allowing for further complexing and increased assimilative capacity. Because of the major arsenate sorption pathway, soils with high clay or iron content can assimilate greater amounts of As than coarse, sandy soils.

Soil Solution Levels and Effects

In the soil, As can affect microbial populations. On a low CEC sandy loam, the addition of soluble arsenic salt (Na_3AsO_4) at 15-20 ppm As reduced the ammonification process (Greaves 1913). The reduction was about the same over the studied range of 20-90 ppm. Nitrification is less affected with a reduction in nitrate formation at 90-100 ppm As (as Na_3AsO_4). Using less-soluble compounds on the same soil, the concentration ranges first affecting ammonification were 20-40, >20, 40-80, and >1,120 ppm As (dry soil) for $Pb_3(AsO_4)_2$, Paris green, Zn_3AsO_3 and AsS_3, respectively. Using these same insoluble arsenic compounds, the nitrification process was again less affected with critical concentrations of 640-680, 760, 720-670 and 440-480 ppm, respectively. With a clay soil, no effects on ammonification or nitrification were found up to a soil level of 1,900 ppm As, while a highly organic soil was unaffected up to 7,500 ppm As (McGeorge 1915). The time period for correction of the depressed microbial growth or the significance of a reduction in these microbes has not been addressed.

If an industrial land application site were maintained under anaerobic conditions, microbial pathways have been identified for the production of alkyl arsine, which have a very bad odor and toxicity. Cox and Kamprath (1972) and Epps and Sturgis (1940) have reported on various microbial and field experiments concerning the anaerobic As pathways. Industrial wastes containing substantial As levels must be carefully designed to maintain predominantly aerobic conditions.

Leaching of arsenicals can occur especially on light-textured soils. The adsorption of arsenates was found to exceed that of phosphate, thus indicating that As migration will be less than phosphate movement (Wauchope 1975). From the long-term behavior of arsenic-overloaded orchards, it appears that As remains essentially in the surface zone (Vandecaveye et al. 1936). The loss of water-soluble As with time (Dorman et al. 1939, Woolson et al. 1971) may be due to leaching or soil reactions converting As to less-available forms. Depending on the As compound applied it may take 18 hours to 7 weeks to accomplish the fixation (Vandecaveye et al. 1936).

Crop Tolerance

The response of crops to arsenic applied to soils appears to be correlated with the level of available or acid-extractable As (Rosenfels and Crafts 1939, Reed and Sturgis 1936). Woolson et al. (1971) also found a correlation of crop yield with total soil As. The percentage soluble As of total soil As is presented in Table 9.1. Approximately 3.5% (1.2-8.9%) of the total As is acid extractable, with a lower rate for soils not receiving arsenic applications. Soil type or exchange capacity will affect the ratio of soluble to total As, although the available data are not precise enough to quantify such differences.

Table 9.1 Relationship of Soluble to Total Arsenic in Soils

Total As ppm Dry Soil	Soluble As ppm Dry Soil	Soluble As/Total As (%)	Conditions	Reference
174	3.6	2.1	Moderate to severe toxicity	
96	1.6	1.6	Moderate to severe toxicity	
81.5	0.99	1.2	Orchards	Headden (1908)
1,120	17	1.5	$Pb_3(AsO_4)_2$	Greaves (1913)
1,120	81	7.2	Paris green	Greaves (1913)
1,120	34	3.0	$Zn_3(AsO_3)$	Greaves (1913)
1,120	27	2.4	AsS_3	Greaves (1913)
47	4.2	8.9	Orchards	Greaves (1934)
30	1.7	5.7	Natural soils	Hirai and Kanno (1938)
8	0.1	1.2	Control soil	

The importance of the soil relationships to forms of arsenic is the plant response or toxicity that is related to the available, easily extractable or solution-phase As concentrations. Toxicity of As compounds to crops exerts influence as a severe root and bark irritant, causing root plasmolysis and rotting, rather than as plant uptake of a systemic poison (Coury and Ranzani 1945, Headden 1908). Initial symptoms of As toxicity are wilting of new-cycle leaves followed by retardation of root and top growth. Plants tend to curl, turn pink or light red, and later change to light yellow. Also, As is thought to adversely affect chlorophyll (Stewart and Smith 1922). Distributions of arsenic within the plant are quite dependent on crop species. Jones and Hatch (1945) examined a large number of vegetables on four soils and determined that on the average, the plant leaves were slightly lower in As content than the roots. This observation was also made with a different magnitude of difference by McLean et al. (1944) with vegetables, Clements and Heggeness (1940) with tomatoes, Vandecaveye et al. (1946) with barley and Williamson and Whetstone (1940). Furthermore, in comparing the top portions of crops, the fruit was found to contain very low or undetectable As concentration when compared to the leaves and stems (Clements and Heggeness 1940, Herrmann and Kretzdorn 1939, Jacobs et al. 1970).

Background levels of As in crops was reported in the range of 0.009-3.0 ppm As (dry basis) for potatoes, cabbage, beets, carrots, tomatoes, eggplant, cucumbers and apples (Shtenberg 1941). As the soil As is increased to levels of crop toxicity the leaf or plant content increases. The crop concentration of As has been variable under conditions in which yield is reduced. Some researchers indicate that the plant will not contain greater than the legal food limit of As even when growth is substantially reduced (Vladimirov 1945, Jones and Hatch 1945, Lindner 1943, Grimmett 1937, Pemberton 1934). Other experiments with both unaffected or retarded crops grown on arsenic areas have contained levels far exceeding the legal limit, 3.6 ppm As on a fresh weight basis (Stewart and Smith 1922, McGeorge 1915, Albert and Paden 1931). The discrepancy may be in whether the entire plant, including heavy accumulating roots, is analyzed vs the edible portions. If true, the maintenance of plant yield can be used as a criteria to assure acceptable As application rates for safe food crops.

Response of crops to soil solution levels of As have been measured using water culture techniques. Brenchley (1914) determined the 0.1 ppm As (as arsenious acid) in solution and 10 ppm As (as arsenic acid) were critical levels that first produced yield reductions in barley. This difference in crop response between +5 and +3 As oxidation states has been documented widely in solution and soil culture experiments. Lindner (1943) and Liebig (1966) determined that 1.5-3.6 ppm As and >2 ppm As in soil solution were representative of soils in which fruit trees and other crops had moderate to severe growth reductions. Solution cultures of sudan grass and bush bean with

1 ppm As produced crops with greater than the legal limit of arsenic while at 3 and 18 ppm As, a lethal dose, was present for the bush bean and sudan grass, respectively (Clements and Heggeness 1939). In another study, 5 ppm arsenious acid was strongly poisonous to plant roots (Cobet 1919, Vandecaveye et al. 1936). Nobbe et al. (1884) determined 1 ppm As was injurious to corn, oats, buckwheat and peas produced in solution culture, while Morris and Swingle (1927) found 5 ppm As (as As_2O_3) to reduce oats yield by 30%. From these data on solution concentrations one can conclude that (1) a distinct variation in crop tolerance exists, (2) the +3 oxidation state is more toxic than the +5, and (3) applications of arsenic compounds to soils should not exceed those levels for which the solution-phase concentrations of arsenic (+5) are approximately 1 ppm or approximately 0.1 for the +3 form. Above these levels the probability of crop yield reductions is increased, although the extent of reduction is unknown, especially if more tolerant species are used.

Numerous observations have been made of the crop response to varying amounts of As applied to land. As with other potentially toxic metals or anions, a number of factors influence the plant yield or growth response. Solubility of the arsenic-containing compound and the oxidation state of As are primary factors in predicting vegetative effects. A large number of salts and compounds of As have been investigated. The lower toxicity and, hence, greater plant-soil assimilative capacity of arsenate or +5 As when compared on an equal basis with arsenite or +3 As has been well documented in soil systems (Crafts and Rosenfels 1939, Morris and Swingle 1927, Headden 1908, Kardos et al. 1941). Morris and Swingle (1927) compared 17 As compounds for the relative toxicity to beans and tomatoes (Table 9.2), and the greatest effects were as a rule with more soluble and with +3 As compounds. Headden (1908) found that in salt solutions such as those of soil-water, the solubility of such compounds as lead arsenate and others was increased, thus invalidating solubility product as a useful tool in predicting crop response.

Vegetative species is another major factor in the plant-soil relationships of As. Crop tolerance is discussed later. Total loading rate or soil level of arsenic is also a primary factor in determining the plant-soil assimilative capacity. In Table 9.3 the loading rate of As and the critical levels of As are reported for arsenates and arsenites. From this table one can determine a definite soil exchange capacity effect, which was further documented by Rosenfels and Crafts (1939) (Figure 9.2). The uptake of As was estimated to be 1 kg/ha/yr for barley grown on a soil containing about 34 kg As/ha, which represents about a 4% uptake (Collins 1902).

Management Alternatives for Soil Arsenic

Management alternatives are available to increase the environmentally acceptable levels of As applied to the plant-soil system. Crop selection is

Table 9.2   Relative Effect of Arsenic Compounds on Bean and Tomato, Most Toxic
Listed at Top (Morris and Swingle 1927).

Bean	Tomato
Ammonium Arsenate	Arsenic Metal
Zinc Arsenate	Calcium Arsenite
Arsenic Metal	Ammonium Arsenate
Calcium Arsenite	Mercurous Arsenite
Mercurous Arsenite	Copper Arsenite
Arsenic Sulfide (yellow)	Zinc Arsenate
Lead Arsenite	Arsenic Sulfide (yellow)
Copper Arsenite	Zinc Arsenite
Zinc Arsenite	Arsenic Trioxide
Arsenic Trioxide	Mercuric Arsenate
Calcium Arsenate	Calcium Arsenate
Ferrous Arsenate	Copper Arsenate
Lead Arsenate	Ferrous Arsenate
Mercuric Arsenate	Lead Arsenite
Ferric Arsenite and Ammonium Citrate	Ferric Arsenite and Ammonium Citrate
Arsenic Disulfide (red)	Arsenic Disulfide (red)
Copper Arsenate	Lead Arsenate

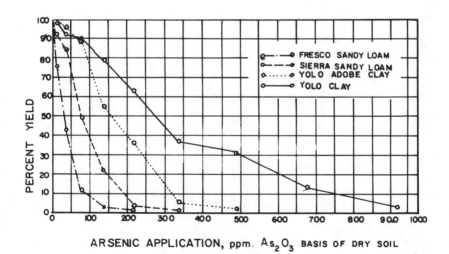

Figure 9.2 Alteration in oat yield with increasing levels of arsenic applied to various soil types (Rosenfels and Crafts 1939).

Table 9.3 Plant-Soil Response to Application of Arsenic Compounds

As Applied (kg As/ha)	Crop	Soil	Response	Reference	Compound
			ARSENATE		
8	Corn	Crowley silt loam	No effect	Picado (1922)	—
10	Rice	Norfolk and Durham sandy loam	20% yield reduction	Meyer (1933)	$Ca_3(AsO_4)_2$
18	Cotton		Decreased yield	Cooper et al. (1931)	$Ca_3(AsO_4)_2$
18	—	Durham coarse sandy loam	50% yield reduction	Paden (1932)	$Ca_3(AsO_4)_2$
18	Cotton	White sand	Reduced vegetative development	Coury and Ranzani (1945)	Na_2HAsO_4
16-20	Oats, Cowpeas	—	No effect	Albert and Paden (1931)	$Ca_3(AsO_4)_2$
40-100	Vegetables	Loamy sand	Retarded growth	McLean et al. (1944)	arsenical sprays
50	Beans	Gravelly loam	No adverse effect on plant weight	Stewart and Smith (1922)	(Na_3AsO_4)
50	Potatoes	Gravelly loam	No adverse effect on plant weight	Stewart and Smith (1922)	(Na_3AsO_4)
0-50	Wheat	Gravelly loam	No adverse effect on plant weight	Stewart and Smith (1922)	(Na_3AsO_4)
0-50	Peas	Gravelly loam	No adverse effect on plant weight	Stewart and Smith (1922)	(Na_3AsO_4)
64	Cotton	White sand	Reduced vegetative development	Coury and Ranzani (1945)	$Pb_3(AsO_4)_2$
75-150	Cotton	Ruston sandy loam (pH 5.0)	Decreased yield	Dorman et al. (1939)	$Ca_3(AsO_4)_2$
94	Cowpeas	Norfolk fine sandy loam	Low germination and and plant growth	Albert and Paden (1931)	$Ca_3(AsO_4)_2$
94	Cowpeas and Soybeans	Greenville sandy loam	25-35% growth depression	Rogers (1932)	$Ca_3(AsO_4)_2$

Rate	Crop	Soil	Effect	Reference	Compound
188	Cowpeas and Soybeans	Cecil sandy loam	Strongly depressed	Rogers (1931)	$Ca_3(AsO_4)_2$
200	—	—	Reduced yield	Rosenbaum (1940)	—
206	Peach	—	Moderate to severe damage	Lindner (1943)	—
280-380	Cowpeas	Davidson clay loam	No yield effect	Paden (1932)	$Ca_3(AsO_4)_2$
320	Average of many crops	Average of many soils	64% of control yield	Woolson et al. (1971)	—
380-560	Cotton	Davidson clay loam	No yield effect	Paden (1932)	$Ca_3(AsO_4)_2$
565	Cowpeas	Davidson clay loam	Depressed	Rogers (1932)	$Ca_3(AsO_4)_2$
600	Cotton	Sarply silty clay loam	No decrease in yield	Dorman et al. (1939)	$Ca_3(AsO_4)_2$
942	Cotton, Rye, Soybean	Davidson clay loam	No effect	Rogers (1932)	$Ca_3(AsO_4)_2$
2,260	Rye	Cecil sandy loam	No effect	Rogers (1932)	$Ca_3(AsO_4)_2$
2,260	Cotton	Cecil sandy loam	Slight growth reduction	Rogers (1932)	$Ca_3(AsO_4)_2$
ARSENITE					
20	Buckwheat	Heavy red clay	Decreased plant weight	McGeorge (1915)	$NaAsO_3$
0-20	Cowpeas	Brown clay	No decreased plant weight	Rogers (1932)	$NaAsO_3$
0-20	Millet	Brown clay	No decreased plant weight	Rogers (1932)	$NaAsO_3$
0-20	Cowpeas	Highly organic silt	No decreased plant weight	Rogers (1932)	$NaAsO_3$
0-50	Millet	Heavy red clay	No decreased plant weight	Rogers (1932)	$NaAsO_3$
0-50	Cowpeas	Heavy red clay	No decreased plant weight	McGeorge (1915)	$NaAsO_3$
45	Peas, Corn, Potato	Plainfield sand	Little or no growth effect	Jacobs et al. (1970)	$NaAsO_3$

Table 9.3 Continued

As Applied (kg As/ha)	Crop	Soil	Response	Reference	Compound
0-300	Buckwheat	Brown clay	No decreased plant weight	McGeorge (1915)	$NaAsO_3$
0-300	Buckwheat	Highly organic silt	No decreased plant weight	McGeorge (1915)	$NaAsO_3$
0-300	Millet	Highly organic silt	No decreased plant weight	McGeorge (1915)	$NaAsO_3$
210	—	Sand and sandy loam	Recommendation for soil sterilization	Crafts and Rosenfels (1939)	AsO_3^{-3}
210	Sorghum and Sugarcane	—	Stunged growth	Mungomery (1938)	AsO_3^{-3}
420-630	—	Loam and silt loam	Soil sterilization	Crafts and Rosenfels (1939)	AsO_3^{-3}
840-1,260	—	Clay loam and clay	Soil sterilization	Crafts and Rosenfels (1939)	AsO_3^{-3}
900	Sugercane	—	Stunted growth and death	Kerr (1930)	AsO_3^{-3}

very feasible, and the large differences in soil As concentrations that produce yield reductions are shown in Table 9.3 and described in a previous discussion. From the work of Morris and Swingle (1927), Morris (1938), Vincent (1944) and Liebig (1966), a relative tolerance categorization was developed (Table 9.4). Selection of very tolerant crops would increase the soil levels,

Table 9.4 Crop Tolerance for Arsenate		
Very Tolerant	Fairly Tolerant	Low Tolerance
Asparagus	Beet	Alfalfa
Carrot	Pea	Barley
Dewberry	Kentucky bluegrass	Cucumber
Grape	Peanut	Lima bean
Pumpkin	Red-top	Onion
Rye	Squash	Pepper
Sudangrass	Orchardgrass	Rice
Tobacco	Strawberry	Snap bean
Tomato	Sugarcane	Cranberry
Turnip	Sweet corn	Clover
Vetch	Oats	Korean lespedeza
Potato	Wheat	
Raddish		

which would represent an acceptable assimilative capacity. Vladimirov (1945) proposed that shallow root species should be avoided in favor of deep-rooted species to increase tolerance to applied arsenic.

Addition of ferrous sulfate has proven to be the best compound for reducing toxic conditions associated with excessive As levels (Vandecaveye et al. 1943; Kardons et al. 1941). Insoluble Fe arsenate formation appears to be the mechanism of detoxification. Other less-effective compounds are aluminum sulfate, various metals chelates, zinc sulfate, organic matter and lime. Application of phosphatic materials may accentuate As toxicity, since many such chemicals contain As and phosphates displace arsenates into solution by mass action. Deeper incorporation of As compounds also reduces As toxicity by increasing the soil adsorptive site ratio to arsenic. Conceptually, pretreatment oxidation from +3 to +5 state can be effective in managing the arsenic assimilative capacity.

Land application of As-containing wastes, intentionally or unintentionally, hans been observed. Poultry manure land application at relatively low rates (0.25-0.5 kg As/ha/yr) has not produced increased As uptake of alfalfa or clover. Much higher rates were land-applied with mud press cake containing As_2O_3 at a rate of 1,440 kg As/hr (Pemberton 1934). There was little or no uptake of As by sugarcane grown at the site, although control of root grub was achieved. Two air pollution situtations have been reported in which As form a smelter adversely affected soil As concentration, crop yield, and produced a surface coating exceeding safe food levels (Swain and Harkins 1908, Rosenbaum 1940).

Assimilative Capacity

The assimilative capacity for plant-soil systems must be determined for a particular soil and crop to be used in a land application system. The high variation in arsenic compound type also demands adaptation of design criteria for specific wastes. As broad estimates, the arsenic assimilative capacity for light-textured soils would be 50 kg As/ha. For loams and silt loams, the assimilative capacity is in the range of 100-200 kg As/ha. Finally, for clays, clay loams and highly organic soils, 500-750 kg As/ha appears to be an appropriate assimilative capacity.

BORON (B)

As an element applied to land, B is an essential material for plants, animals and some microorganisms. Boron is widely distributed in the lithosphere (10 ppm), igneous rock (13 ppm), limestone (18 ppm), sandstone (155 ppm), shale (130 ppm), soil (2-100 ppm), plant leaves (7-75 ppm), sludges (6-1,000 ppm) and irrigation waters (0.3-4.0 ppm). In 200 American soils, Whetstone *et al.* (1942) found an average content of 30 ppm B; the average range of world soils being 20-50 ppm (Aubert and Pinta 1977). Boron is widely used in modern technology and industry from detergent manufacturing to cosmetics, from leather and carpet works to rocket fuels and photography. Borate appears in wastes from industrial and domestic sources. Boric acid and its salts are used routinely in everyday life and, as a result, B occurs in sewage and sludge.

Fate of Boron Compounds in Plant-Soil Systems

Adsorption, Leaching and Insoluble Products

Several forms of B are possible over a wide range of soil pH. The predominant B compound is boric acid, H_3BO_3, under conditions of acid, neutral and slightly alkaline soils. Above pH = 9 the species $H_2BO_3{}^-$ becomes a substantial fraction of the total boron present. Other borate ions such as $B_4O_7{}^{-2}$ are easily hydrolized to H_3BO_3 (Sillen and Martell 1964).

Boron, as the associated acid, is in a form amendable to soil-water solution and migration. However, there are adsorptive mechanisms that reduce the anionic leaching of B. On a comparative basis, B is less mobile than Cl^-, SO_4 or NO_3^-. The principal adsorption pathways for B are in association with aluminum oxides and hydroxides or iron oxides and hydroxides. Other fixation mechanisms include replacement in the clay structure, bonding to organic materials and adsorption to magnesium hydroxides. The pH dependence of boron adsorption is such that for soil pH 5.5-7, the exchangeable aluminum would be the major adsorption site, while in the range of >8, the iron complexes are dominant. In terms of soil tests the apparent adsorption maxima are related to the adsorption on Al and Fe compounds, but the slower kinetic pathway for B entry into the soil crystal lattice would also be expected to increase the assimilative capacity in longer-term experiments.

Nonadsorbed B is subject to leaching or other loss mechanisms. A substantial amount of research has been conducted on B leaching and predictive equations. These have been reviewed by Ellis and Knezek (1972). It was concluded that as a first estimate the use of a Langmuir isotherm and relevant water flow equations were useful for field situations. Constants for the Langmuir equation are given in Table 9.5.

Table 9.5 Langmuir Adsorption Equation Constants for Boron (Tanji 1970)

Soil Texture	Boron Adsorption Maxima, Q (mg B/kg soil)	Boron Adsorption Energy, K (l/kg)
Sandy Loam	10.8	0.046
Silt Loam	20.1	0.038
Clay Loam	6.7	0.088

Precipitation reactions for B compounds do not occur to any substantial degree under conditions similar to an industrial land application system (Sauchelli 1969). However, a number of possible precipitates was listed by Lehr (1972) as naturally occurring compounds. Thus, the plant-soil assimilative capacity for B does not appear to be increased by insolubilization of applied boron.

Plant Response to Boron

The range of soil B, which represents a plant deficient vs a plant toxicity excess is narrow, hence the need to evaluate carefully the plant and soil characteristics at a potential land application site. Plant roots absorb B from the soil solution and it is translocated to the leaves. As the plant matures the constant evapotranspiration of water results in boron concentrating in the leaf. If applied in excess quantities the leaf discoloration and other symptoms become evident (Bradford 1966).

Within a plant the role of boron has not yet been established. The essential nature of B is, however, well demonstrated, as shown in Figure 9.3 for barley (Brenchley 1914). Stimulated yield resulted at less than 0.1 mg boric acid/l. The toxic effects of B are also shown in Figure 9.3 and above 4 mg/l. A literature review on the behavior of B in plants has been assembled by Price *et al.* (1972).

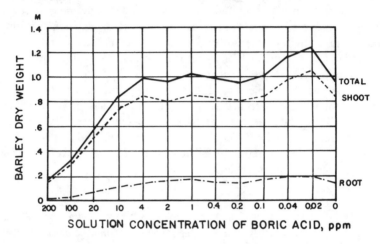

Figure 9.3 Yield (dry weight) of barley grown in solutions containing concentrations of boric acid and nutrient salts (Brenchley 1914).

The general approach used has been to relate response to the B content of soil solution. Measurements are usually performed with a hot water extraction of soil as an estimate of the soil solution available for plant uptake (Cox and Kamprath 1972). The relationship between plant content and the water-soluble B in soil was evaluated by Wear and Patterson (1962) (Figure 9.4). Using a plant level of 200 ppm as a critical level above which B excess symptoms occur, the differences in soil texture are substantial. Thus, greater amounts of B can be applied to silty clay loams since less of the B is available to the plant. Besides the factors in Figure 9.4, a low soil pH produces a shift from soil adsorption to soil solution and may be utilized by the plant or move with the soil-water.

In addition to the soil solution concentration of boron, substantial plant genetic factors can be utilized to improve the assimilative capacity for B. An extensive comparison of plant response (Eaton 1944) has resulted in a classification based on boron tolerance (Table 9.6). Use of the most tolerant species would permit greater assimilative capacity for the boron constituent in industrial waste.

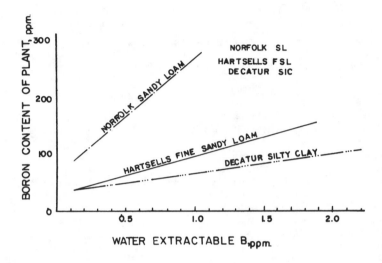

Figure 9.4 Boron content of plant as a function of soil solution concentration (Wear and Patterson 1962).

The uptake of boron by plants and the subsequent harvesting and removal represents an assimilative pathway. In comparison to other anionic species (Table 8.1), the uptake of B is small. Kardos *et al.* (1974) reported corn silage removal of 0.04-0.09 kg B/ha, which was 0.5-10% of the applied boron. Using the normal plant tissue ranges of Bradford (1966), at 10-200 ppm B and a forage crop such as alfalfa, the plant removal would be approximately 1-2 kg B/ha/yr.

Experimental results are available for the soil levels at which boron is toxic to crops. Application over 23 years at 0.1 kg B/ha/yr corrected deficiencies, while 0.8 kg B/ha/yr resulted in some B toxicity for turnip, a sensitive crop (Odelein 1963). Use of 0.5-0.9 kg B/ha/yr for cotton was reported to stimulate cotton maturity (Anderson and Boswell 1968). Mortvedt and Cunningham (1971) recommended 0.25 to 3 kg B/ha/yr as a commonly used application rate with the higher rates associated with boron-tolerant crops. In the range of 5-20 kg/ha, slight initial damage was reported for corn and another unspecified crop (Nakamura 1903, Aguthon 1910).

Assimilative Capacity

In determining the design basis for the assimilative capacity of B in an industrial waste, a number of site-specific characteristics must be evaulated. In addition, the nature of the waste is important because the use of low

Table 9.6   Relative Tolerance of Crops to Boron (Bingham 1973)
(Crops within a category are listed with the most sensitive at the top.)

Boron-Sensitive Plants (0.5-1.0 ppm B)[a]	Boron Semi-Tolerant Plants (1-5 ppm B)[a]	Boron-Tolerant Plants (5-10 ppm)[a]
Apple	Barley	Alfalfa
Apricot	Bell pepper	Artichoke
Avocado	Cabbage	Asparagus
Blackberry	Calendula	Beet
Citrus (orange, lemon, lime, grapefruit)	California poppy	Broadbean
	Celery	Cabbage
Cherry	Corn	Carrot
Cowpea	Cotton	Chard
Elm	Kentucky bluegrass	Cotton
Fig, grass	Lima bean	Date palm
Grape	Milo	Garden beet
Jerusalem Artichoke	Mustard	Lettuce
Kidney Bean	Oats	Mung
Larkspur	Olive	Muskmelon
Lemon	Parsley	Onion
Lupine	Pea	Oxalis
Pansy	Pepper	Palm
Peach	Potato	Sugar beet
Pear, soybean	Radish	Sugar pea
Pecan	Tobacco	Sweet clover
Persimmon	Tomato	Turnip
Persimmon (Japanese)	Vetch	
Plum	Wheat	
Plum, strawberry		
Strawberry		
Violet, flax		
Walnut		
Zinnia		

[a]ppm B in saturation extract of soil.

moisture solids or sludges results in different magnitudes for the assimilative pathways when compared to those active with effluent land application.

For industrial solids and sludges, the assimilative capacity must be computed from several sources. First, the boron adsorption capacity of the soil must be determined. The procedure for this is similar to that described in Chapter 5 for phosphates. A range of reported B adsorption maxima is 10-100 ppm (Bingham and Page 1971, Okazaki and Chew 1968, Singh 1964). The boron adsorption, ppm, is then converted to kg B/ha for the usable soil depth. Then dividing by the site lifetime one arrives at the adsorptive assimilation, kg B/ha/yr. To this is added the crop uptake of approximately 1-2 kg B/ha/yr. The assumption is made that a boron-tolerant crop such as alfalfa or cotton is used at the land application site. The final component of the assimilative capacity is the allowable movement of boron to receiving waters. This analysis requires the water flow, laterally and vertically, from the root zone be evaluated at the land application site. The allowable receiving water concentration is 20 mg B/l (Hoskins 1941). The analysis for acceptable boron migration and the corresponding assimilative capacity, kg/ha/yr, can be performed by direct analogy to salt criteria in Chapter 8. Then with the adsorptive, crop uptake and leaching components, the assimilative capacity is the sum, expressed in kg/ha/yr.

For industrial effleuents the experience with irrigation of agricultural crops can be used. Selection of a crop in the high-tolerance category for boron (Table 9.6) will allow irrigation of industrial effluents with 2-4 mg B/l. In like manner, moderately and low-tolerant categories can receive 1-2 mg B/l and < 1 mg B/l, respectively.

CHLORIDE (Cl)

Chlorine is an essential element for plants as first proposed by Sachs in 1860, but only in 1954 was this shown conclusively (Broyer et al. 1954). Chlorine is present as the chloride anion in soils and industrial wastes. Chloride is highly mobile (i.e., soils have little Cl adsorptive capacity) and in comparison to the sulfate anion is more toxic to plants. Chloride is present in all waters and is especially high in concentration from HCl manufacturing, metal and industrial cleaning, food processing, chilling and insecticide manufacturing industries.

The assimilative capacity for Cl is based on crop uptake and movement to receiving waters within acceptable concentration limits. Crop tolerance varies among species; hence, to maximize the assimilative capacity for industrial land application, the least-affected plant species should be used. Table 9.7 lists the concentration in soil solution at which there is a 25% reduction in crop yield. Since there is little adsorption or precipitation of Cl in soils, there is not likely to be any effect of soil texture on Cl assimilative capacity, that is, no difference between the benchmark soils used in this book. Chloride uptake for several crops is given in Table 9.7.

The leaching requirement for Cl is based on eliminating plant toxic effects (separately from the overall effect of salinity) as well as achieving acceptable Cl concentration in soil-water, leaving the industrial land application site. The area requirements and the leaching assimilative capacity are to be determined in the same manner as described in Chapter 8 for movement of salts. The allowable concentration at groundwater or surface waters is 125 mg Cl/l (McKee and Wolf 1963).

Table 9.7 Concentrations of Chloride in Soil Solutions in the Netherlands Which Caused a Yield Production of 25% Relative to Production on Unaffected Soils (Eaton 1966).

Field Crops	Chloride
Bean (brown and white)	9
Pea	9
Potato (tubers)	26
Broad Bean	34
Onion	34
Flax	51
Red Clover	51
Wheat (spring)	68
Spinach (for seed)	86
Alfalfa	103
Oats	120
Beetroot	120
Barley (spring)	170

Chloride assimilative capacity is thus very site-dependent, since it is dependent on water movement, climate and crop type. The crop uptake and acceptable loading to achieve receiving water limits are added together to give the total plant-soil assimilative capacity, kg Cl/ha/yr.

CYANIDE (CN)

As an anion (CN^- for pH > 8) or the largely undissociated acid (HCN, for pH < 8), cyanides are capable of movement through soils. However, by contrast to most of the other toxic anions, cyanides can be degraded by biological reactions. Thus, the assimilative capacity is a balance between the rates of application, degradating and leaching.

A soil chemoautotrophic organism (an actinomycete) degrades CN by the following reaction (Ware and Painter 1955):

$$CN^- + 4H_2O + O_2 \rightarrow 2OH^- + 2\,NH_3 + 2CO_2$$

This reaction occurs despite the established toxicity to most organisms. In soils the rhodanese enzyme catalyses the loss of cyanide as follows:

$$S_2O_3^{-2} + CN^- \rightarrow SCN + SO_3^{-2}$$

Alfalfa response to cyanides in irrigation water was found to be depressed at > 5 mg CN/l. However, at this level direct consumption of the crop was not permitted because of foliar concentrations in excess of animal lethal doses. However, for nonfood-chain crops this effect would be lessened. For application to land (not specified as spray irrigation), Müller-Neuhaus (1957) allows wastes from coke and gas works containing 10-200 mg HCN/l to be applied to agricultural lands. In a similar manner, discussed for B, Cl and salts, the cyanide migration should be considered. The critical concentration for receiving waters would be 0.01 mg CN/l (WHO 1958).

The land application assimilative capacity is based on the degradation of cyanide through established soil pathways and the selection of sufficient land area to assure acceptable concentrations for the remainder of the CN that is leached. Such calculations are, therefore, quite site-specific. Renn (1955) found that 90-95% of applied cyanides were not in the effluent of soil columns, indicating the magnitude of microbial reactions. The assimilative capacity based on biological stabilization is unaffected between 10 and 35°C, with lower rates below 10°C (Ware 1958).

FLUORINE (F)

The flouride ion is extremely stable, so fluorine is never found in pure form in nature. Fluoride is nonessential to plants and microbes; however, background levels of 100-200 ppm fluoride are not toxic to crops or microbes. Fluorine is essential for animals. High fluoride may accumulate in a soil-plant system artificially, by addition of phosphate fertilizers, from inorganic or organic pesticides and through polluted air contamination near industrial sites (Meyn and Viehl 1941, Kruger 1941).

Fluoride accumulates in the foliage of plants as the growing season progresses. Uptake of fluoride by roots is generally regarded as a passive process controlled not by soil F level but soil type, Ca, Mg and pH. Fluoride can also be taken into the plant through foliar mechanisms. Under acid conditions high fluoride concentration in soils can lead to the accumulation of fluoride in plants. Excessive fluoride can be fixed in acid soil by liming to pH 6.5, as CaF_2 is insoluble. Some varieties of crops are more tolerant to fluoride than others and, therefore, tolerant plant varieties can be screened and selected in fluoride-contaminated areas. Adverse crop effects often are not evidenced before the plant level exceeds that deemed safe for animal or human consumption (Allaway 1977).

Land application of up to 64 kg F/ha (NaF or Na_2SiF_6) did not effect plant development on a loam soil (Bobko and Pryadil'shcikova 1940). Plot experiments with 200 kg F/ha (as CaF_2) showed no adverse crop response for rye, barley, clover or oats (Gautier and Clausmann 1919). In well-limed neutral soils, fluoride compounds were applied at 640, 1,290 and 2,750 kg F/ha (MacIntire et al. 1942). There were no detrimental growth effects nor did plant concentrations increase appreciably. However, on acid soils, 360 kg

F/ha resulted in crop reduction (Prince *et al.* 1949). Hence, the assimilative capacity is greatly improved by correcting the soil pH.

In instances where substantial Ca is not present relative to applied F, the migration of fluoride must be considered. Using the leaching calculations previously described, the assimilative capacity can be determined using a receiving water concentration of 1.0 mg F/l.

Irrigation of water containing fluoride has been classified with respect to plant response (McKee and Wolf 1963). At 10 mg F/l there was no injury to peach, tomato or buckwheat. At 100 mg F/l or higher, severe injury resulted. Thus, industrial effluent concentrations should meet these levels when long-term irrigation is anticipated.

The assimilative capacity for accumulation in soils appears to be a minimum of 360 kg F/ha under acid soils and much higher under well-lined, neutral soils. Fluoride leaching must be monitored and adjustments in land are made to achieve desired leachate concentrations.

IODINE (I)

In relation to plants, I is considered nonessential, while the requirement for iodine in animals and humans is well established. Industrial wastes from manufacture of certain medicines, production of analytical chemicals and for certain types of germicidal or disinfection uses should be examined with regard to the I assimilative capacity.

Iodine is sparingly soluble in water. When applied to soils, I can be adsorbed, converted into iodate (IO_3), leached and/or taken up by plants. The retention of I in soils appears to be related to organic adsorption. Based on the benchmark soils used in this book, the sandy loam would be more easily leached of I than the clay loam (Martin 1966). Using a leaching model such as described in Chapter 8, the concentration of iodate reaching receiving waters can be determined. Drinking water standards for I are not available, although one reference notes no adverse effect on humans at 8 mg I/l (Anon 1953).

Plant uptake of iodine or iodide is low (0.03-2.2% of applied I), but on oxidation to iodate can be increased (Whitehead 1975). Addition of organic matter along with I further reduced the uptake of I by perennial rye. Menzel (1965) defined a relative concentration factor for plants as follows:

$$RCF = \frac{\text{Concentration of element in first harvest of plant}}{\text{Concentration of element added to soil in water soluble form}}$$

The RCF for I^- was between 0.1 and 10 and resembled fluorine (Whitehead 1975). Chloride and Br^- had much higher values of RCF.

The assimilative capacity for iodine is obtained from the dose in the soil at which plant yields are reduced below control and by the rate of leaching loss. Iodine accumulated to 5 kg I/ha represents the upper acceptable soil level. As a conservative estimate this accumulative amount applied over a fixed site life would be the annual assimilative capacity (kg/ha/yr). A detailed water flow design and the percentage of applied I that is in the soluble form could be used to increase this assimilative capacity based on the land application site characteristics.

BROMINE (Br)

Bromine is a relatively strong oxidizing agent yielding bromide ion. In soils, bromide, bromate (BrO_3^-) or bromic acid are the principal forms of bromine. Neither plants nor animals require Br as an essential element, but bromine can substitute in some cases for needed chlorine. Industrial waste containing Br includes photographic emulsions, gasoline additives, dyes, medicines and some disinfectants.

The plant response to Br is quite variable within the limited literature data. Bromine tolerance is dependent on crop type (Martin 1966). A range of soil concentrations at which plant yield was unaffected by the presence of Br is 30-100 kg Br/ha (Newton and Toth 1951, Martin et al. 1956).

SELENIUM (Se)

Se is nonessential for crops, but is necessary in small quantities for animal and human nutrition. In general, plants reach animal toxicity levels of Se (4-ppm dry basis) before an adverse response on crop yield occurs. Thus, monitoring Se content of the vegetative cover in a land application system is necessary to define the actual assimilative capacity.

Detailed reviews of selenium in plant-soil systems have been prepared and should be consulted in relation to the behavior of Se (Trelease and Beath 1949, Ganjae 1966, Louderback 1975). Industries that may contain large quantities of Se in wastes include those producing paints, dyes, glass, electrical components, rubber, alloys and insecticides. Municipal sludges also contain detectable selenium (Furr et al. 1976).

As a first approximation, Se applications up to 9 kg Se/ha would produce safe food-chain crops (Furr et al. 1976, Walsh and Fleming 1952). Standards for effluents used in irrigation have been proposed as follows (Miller 1954):

Selenium Level (mg/l)	Recommendation
0-0.1	No plant toxicity anticipated
0.1-0.2	Usable but possible long-term accumulations—monitoring necessary
0.2-0.5	Doubtful—probable toxic accumulations
>0.5	Unusable

Thus, the accumulation level, kg Se/ha, or the effluent concentrations of selenium, can be used depending on the nature of the waste.

REFERENCES

Aguthon, H. "The Increased Tolerance of Maize to Boron," *Compt. Rend. Acad. Sci. (Paris)* 151(26):1382-1383 (1910).

Albert, W. B., and W. R. Paden. "Calcium Arsenate and Unproductiveness in Certain Soils," *Science* 73:622 (1931).

Allaway, W. H. "Food Chain Aspects of the Use of Organic Residues," in *Soils for Management of Organic Wastes and Waste Waters.* L. F. Elliott, Ed. (Madison, WI: Soil Science Society of America, ASA, CSSA, 1977) pp. 283-300.

Anderson, O. E., and C. F. Boswell. "Boron and Manganese Effects on Cotton Yield, Lint Quality and Earliness of Harvest," *Agron. J.* 60:488 493 (1968).

Anonymous. "Polyiodide Tablets Disinfect Drinking Water in Ten Minutes," *Chem. Eng. News.* 31:2305 (1953).

Aubert, H. and M. Pinta. "Trace Elements in Soils," *Develop. Soil Sci.* 7:5-18; 35-38; 55-62; 69-72 (1977).

Bingham, F. T. "Boron in Cultivated Soils and Irrigation Waters," in *Trace Elements in the Environment,* American Chemical Society, Washington, D.C. (1973), pp. 130-143.

Bingham, F. T., and A. L. Page. "Specific Character of Boron Adsorption by an Amorphous Soil," *Soil Sci. Soc. Am. Proc.* 35:892-893 (1971).

Bobko, E. V., and T. D. Pryadil'shcikova. "The Effect of Fluorine on the Development of Plants," Trudy Vsesoyuz, Akad. Sel'shokhoz Nauk im. Lenina No. 23-24, 33-35, *Khim. Referal. Zh.* 4(7-8):64 (1941).

Bradford, G. R. "Boron" in *Diagnostic Criteria for Plants and Soils.* H. D. Chapman, Ed. University of California, Division of Agricultural Science (1966).

Brenchley, W. E. *Inorganic Plants Poisons and Stimulants.* (New York: Cambridge University Press, 1914), p. XIIO.

Broyer, T. C., A. B. Carlton, C. M. Johnson and P. R. Stout. "Chlorine: a Micronutrient Element for Higher Plants," *Plant Physiol.* 29:526-532 (1954).

Clements, H. F., and H. G. Heggeness. "Arsenic Toxicity to Plants," *Hawaii Agric. Exp. Sta., Ann. Rep.* 77-78 (1939).

Clements, H. F., and H. G. Heggeness. "Arsenic Toxcity to Plants," *Hawaii Agric. Exp. Sta., Ann. Rep.* 79 (1940).

Cobet, R. "The Influence of Arsenious Acid on Growing (Plant) Tissue," *Biochem. Z.* 98:294-313 (1919).

Collins, S. H. "The Adsorption of Arsenic by Barley," *J. Soc. Chem. Inc.* 21:221-222 (1902).

Cooper, H. P., W. R. Paden, E. E. Hall, W. B. Albert, W. B. Rogers and J. A. Riley. "Effect of Calcium Arsenate on the Productivity of Certain Soil Types," *S. C. Sta. Rep.* 28-36 (1931).

Coury, T., and G. Ranzani. "Effects of Arsenic on the Cultivation of Cotton on Sandy Soil," *An Esc. Sup. Agric. Luiz. de Queiroz.* 2:393-422 (1945).

Cox, F. R., and E. J. Kamprath. "Micronutrient Soils Tests" in *Micronutrients in Agriculture,* J. J. Mortvedt, Ed. (Madison, WI: Soil Science Society of America, Inc., 1972).

Crafts, A. S., and R. S. Rosenfels. "Toxicity with Arsenic in Eighty California Soils," *Hilgardia* 12:197-199 (1939).

Dorman, C., F. H. Tucker and R. Coleman. "The Effect of Calcium Arsenate upon the Productivity of Several Important Soils of the Cotton Belt," *J. Am. Soc. Agron.* 31:1020-1028 (1939).

Eaton, F. M. "Chlorine," in *Diagnostic Criteria for Plants and Soils,* H. D. Chapman, Ed., University of California, Division of Agricultural Science (1966), pp. 98-135.

Eaton, F. M. "Deficiency, Toxicity and Accumulation of B in Plants," *J. Agric. Res.* 69:237-77 (1944).

Ellis, B. G., and B. D. Knezek. "Adsorption Reactions of Micronutrients in Soils," *Micronutrients in Agriculture,* J. J. Mortvedt, Ed. (Madison, WI: Soil Science Society of America, Inc., 1972), pp. 59-78.

Epps, E. A., and M. B. Sturgis, "Arsenic Compounds Toxic to Rice," *Soil Sci. Soc. Am. Proc.* 4:215-218 (1940).

Furr, A. K., A. W. Lawrence, S. C. Tong, M. C. Grandolfo, R. A. Hopstader, C. A. Bache, W. H. Gutenmann and D. J. Lisk. "Multielement and Chlorinated Hydrocarbon Analyses of Municipal Sewage Sludges of American Cities," *Environ. Sci. Technol.* 10:683-687 (1976).

Ganjae, T. J. "Selenium" in *Diagnostic Criteria for Plants and Soils,* H. D. Chapman, Ed. University of California, Division of Agricultural Science (1966), pp. 394-404.

Gautier, A., and P. Clausmann. "The Action of Fluorides on Vegetation," *Prog. Agr. et Vitiv.* (Ed. l'est Center) 40(33):153-158; *Compt. Rend. Acad. Sci. (Paris)* 169:115-22 (1919).

Greaves, J. E. "The Influence of Arsenic upon the Biological Transformation of Nitrogen in Soils," *Biochem. Bull.* 3(9):2-16 (1913).

Grimmett, R. E. R. "Arsenic in Soils, Muds, Drainage Waters and Grass," *New Zealand Dept. Agric., Ann. Rep.* 57-62 (1937/1938).

Headden, W. P. "Arsenical Poisoning of Fruit Trees," *Colorado Sta. Bull.* 157:3-56 (1908).

Herrmann, R., and H. Kretzdorn. "The Arsenic Content of Vineyard Soils and the Uptake of Arsenic from Arseniferous Soils by Vines," *Bodenkunde Pflanzenernahr.* 13:169-176 (1939).

Hirai, K., and T. Kanno. "Rarer Elements in Soils. II. Arsenic Contents in Soils," *J. Sci. Soil Manure, Japan* 12:282-286 (1938).

Hoskins, J. K. "Proposed Revisions in Chemical Standards for Drinking Water," *Ind. Eng. Chem. News.* 19:1138 (1941).

Jacobs, L. W., D. R. Keeney and L. M. Walsh. "Arsenic Residue Toxicity of Vegetable Crops Grown on Plainfield Sand," *Agron. J.* 62:588 (1970).

Jones, J. S., and M. B. Hatch. "Spray Residues and Crop Assimilation of Arsenic and Lead," *Soil Sci.* 60:277-288 (1945).

Kardos, L. T., S. C. Vandecaveye and N. Benson. "Causes and Remedies of the Unproductiveness of Certain Soils Following the Removal of Mature Fruit Trees," *Wash. Agric. Exp. Sta., Bull.* 410:25 (1941).

Kardos, L. T., W. E. Sopper, E. A. Myers, R. R. Parizek and J. B. Nesbitt. "Renovation of Secondary Effluent for Reuse as a Water Resource," Environmental Protection Technological Service EPA-660/2-74-016 (1974).

Kerr, H. W. "Damage to Cane Soils by Arsenic," *Cane Growers' Quart. Bull.* 6:189 (1939).

Kruger, E. "Detection of Fluoride in Plants Injured by Fluorine," *Metall Erz.* 38:265-266 (1941).

Lehr, J. R. "Chemical Reactions of Micronutrients in Fertilizers," in *Micronutrients in Agriculture,* J. J. Mortvedt, Ed. (Madison, WI: Soil Science Society of America, Inc., 1972).

Liebig, G. F. "Arsenic," *Diagnostic Criteria for Plants and Soils,* H. D. Chapman, Ed., University of California, Division of Agricultural Science (1966), pp. 13-23.

Lindner, R. C. "Arsenic Injury of Peach Trees," *Am. Soc. Hort. Sci. Proc.* 42:275-279 (1943).

Louderback, T. "Selenium and the Environment," Colorado School of Mines, *Min. Ind. Bull.* V. 18(2):1-14 (1975).

MacIntire, W. H., S. J. Winterburg, J. G. Thompson and B. W. Hatcher. "Fluorine Content of Plants Fertilized with Phosphates and Slags Carrying Fluorides," *Ind. Eng. Chem.* 34:1469-1479 (1942).

Martin, J. P. "Bromine," in *Diagnostic Criteria for Plants and Soils,* H. D. Chapman, Ed., University of California, Division of Agricultural Science (1966) p. 62064.

Martin, J. P., G. K. Helmkamp and J. O. Ervin. "Effect of Bromide from a Soil Fumigant and from $CaBr_2$ on the Growth and Chemical Composition of Citrus Plants," *Soil Sci. Am. Proc.* 20:209-212 (1956).

Martin, J. P. "Iodine," in *Diagnostic Criteria for Plants and Soils,* H. D. Chapman, Ed., Univeristy of California, Division of Agricultural Science (1966), pp. 200-202.

McGeorge, W. T. "The Effect of Arsenite of Soda on Soil," *Hawaii Agric. Exp. Sta. Press Bull.* 50:16 (1915).

McKee, J. E., and H. W. Wolf. *Water Quality Criteria.* 2nd ed. State Water Quality Control Board, State of California, Publication No. 3-1 (1963).

McLean, H. C., A. L. Weber and J. S. Joffe. "Arsenic Content of Vegetables Grown in Soils Treated with Lead Arsenate," *J. Econ. Entomol.* 37:315-316 (1944).

Menzel, R. G. "Soil-Plant Relationships of Radioactive Elements," *Health Phys.* 11:1325-1332 (1965).

Meyer, A. H. "Effects of Calcium Arsenate on Rice," Association of Southern Agricultural Workers, *Proc. 34th, 35th and 36th Ann. Conv.* (1933-1935), p. 241.

Meyn, A., and K. Viehl. "Chronic Fluorine Poisoning in Cattle," *Arch. wiss. prakt. Tierheilk.* 76:329-339 (1941).

Miller, W. M. "Preliminary Report: Summary of Partial Analyses of Wyoming Waters—Salinity and Selenium," University of Wyoming Agric. Expt. Sta. Mimeo. Circ. 64 (1954).

Morris, He. E., and D. B. Swingle. "Injury to Growing Crops Caused by the Application of Arsenical Compounds to the Soil," *J. Agric. Res.* (U.S.) 34(1):59-78 (1927).

Morris, O. M. "The Tolerance of Various Orchard Cover Crops to Arsenical Toxicity in the Soil," *Proc. 34th Ann. Meeting Washington State Hort. Assoc.* (1938), pp. 110-112.

Mortvedt, J. J., and H. G. Cunningham. "Production, Marketing and Use of Other Secondary and Micronutrient Fertilizers," in *Fertilizer Technology and Use*, R. A. Olson, Ed. (Madison, WI: Soil Science Society of America, Inc., 1971).

Müller-Neuhaus, G. "Industrielle Atwasserprobleme unter Berücksichtigung besonderer Verhältnisse in Rheinisch-Westfalischen Kohlenrevier," *Gluckauf* 93:684 (1957).

Mungomery, R. "Problems of Arsenic applications to Soils," *Cane Growers' Quart. Bull. (Queensland)* 5:151-152 (1938).

Nakamura, M. "The Effect of Boric Acid in High Diultion on Plants," *Bull. Col. Agric. Tokyo Imp. Univ.* 5(4):509-512 (1903).

Newton, H. P., and S. J. Toth, "Response of Crop Plants to Iodine and Bromine," *Soil Sci.* 73;127-134 (1951).

Nobbe, F. P., P. Baessler and H. Will. "Toxic Action of Arsenic Lead and Zinc in Plants," *Landw. Vers. Sta.* 30:380-423 (1884).

Odelein, M. "Long-Term Field Experiments with Small Applications of Boron," *Soil Sci.* 95:60-62 (1963).

Okazaki, E., and T. T. Chew. "Boron Adsorption and Desorption by Some Hawaiian Soils," *Soil Sci.* 105:255-259 (1968).

Paden, W. R. "Differential Response of Certain Soil Types to Applications of Calcium Arsenate," *J. Am. Soc. Agron.* 24:363-366 (1932).

Pemberton, C. E. "(Report on) Entomology," *Proc. 54th Ann. Meeting Hawaiian Sugar Planters' Assoc.* (1934), pp. 19-26.

Picado, C. "Arsenic, a Catalytic Fertilizer," *Comp. Rend. Soc. Biol. (Paris)* 87(39):1338-1339 (1922).

Price, C. A., H. E. Clark and E. A. Funkhauser, "Functions of Micronutrients in Plants," in *Micronutrients in Agriculture*, J. J. Mortvedt, Ed. (Madison, WI: Soil Science Society of America, Inc., 1972).

Prince, A. L., F. E. Bear, E. G. Brennan, I. A. Leone and R. H. Daines. "Fluorine: Its Toxicity to Plants and its Control in Soils," *Soil Sci.* 67:269-277 (1949).

Reed, J. R., and M. B. Sturgis. "Toxicity from Arsenic Compounds to Rice on Flooded Soils," *J. Am. Soc. Agron.* 28:432-436 (1936).

Renn, C. E. "Biological Properties and Behavior of Cyanogenic Wastes," *Sew. Ind. Wastes* 27:297 (1955).

Rogers, W. B. "Response of Some Common Field Crops to Various Rates of Application of Calcium Arsenate to Several Soil Types," *Proc. Assoc. S. Agric. Workers* 33:30-31 (1932).

Rosenbaum, H. "Injury to Plant Growth from Arsenic in the Soil," *Bodenkunde Pflanzenernähr* 19:248-52 (1940).

Rosenfels, R. S., and A. S. Crafts. "Arsenic Fixation in Relation to the Sterilization of Soils with Sodium Arsenite," *Hilgardia* 12(3):201-223 (1939).

Sauchelli, V. "Trace Elements in Agriculture" (New York: Van Nostrand Reinhold Co., 1969).

Shtenberg, A. I. "Natural Arsenic Content of Some Vegetables and Fruits," *Vopr. Pitan.* 10(5-6):29-33 (1941).

Sillen, L. G., and A. E. Martell. "Stability Constants of Metal Ion Complexes," Spec. Pub. No. 17, The Chemical Society, London (1964).

Singh, S. S. "Boron Adsorption Equilibrium in Soils," *Soil Sci.* 98:383-387 (1964).

Stewart, J., and E. S. Smith. "Some Relations of Arsenic to Plant Growth, I and II," *Soil Sci.* 14(2):111-126 (1922).

Stutzer, A. "Is the Arsenic in Superphosphates Harmful?" *Dent. Landw. Presse* 28(9):61 (1901).

Swain, R. E., and W. D. Harkins. "Arsenic in Vegetation Exposed to Smelter Smoke," *J. Am. Chem Soc.* 30:915-928 (1908).

Tanji, K. K. "A Computer Analysis on the Leaching of Boron from Stratified Soil Columns," *Soil Sci.* 110:44-51 (1970).

Trelease, S. F., and O. A. Beath. *Selenium* (Burlington, VT: Champlain Printers, 1949).

Vandecaveye, S. C., G. M. Horner and C. M. Keaton. "Unproductiveness of Certain Orchard Soils as Related to Lead Arsenate Spray Accumulations," *Soil Sci.* 42(3):203-215 (1936).

Vincent, C. L. "Vegetable and Small Fruit Growing in Toxic Ex-orchard Soils of Central Washington," *Wash. Agric. Exp. Sta. Bull.* 437:31 (1944).

Vladimirov, A. V. "Influence of Nitrogen Sources in the Formation of Oxidized and Reduced Organic Compounds in Plants," *Soil Sci.* 60(4):265-275 (1945).

Walsh, T., and G. A. Fleming. "Selenium Levels in Rocks, Soils, and Herbage from a High Selenium Locality in Ireland," *Int. Soc. Soil Sci. Trans.* (Comm II and IV) II:178-183 (1952).

Ware, G. C., and H. A. Painter. "Bacterial Utilization of Cyanide," Nature 175:900 (1955).

Ware, G. C., "Effect of Temperature on the Biological Destruction of Cyanide," *Water Waste Treat.* 5(6):537 (1958).

Wauchope, R. D. "Fixation of Arsenical Herbicides, Phosphate and Arsenate in Alluvial Soils," *J. Environ. Qual.* 4(3):355-358 (1975).

Wear, J. I., and R. M. Patterson. "Effect of Soil pH and Texture on the Availability of Water-Soluble Boron in the Soil," *Soil Sci. Soc. Amer. Proc.* 26:543-546 (1962)

Whetstone, R. R., W. O. Robinson and H. C. Byers. "Boron Distribution in Soils and Related Data," *U.S. Dept. Agric. Tech. Bull.* 797 (1942).

Whitehead, D. C. "Uptake by Perennial Rye Grass of Iodide, Elemental Iodine and Iodate Added to Soil as Influenced by Various Amendments," *J. Sci. Fed. Agric.* 26:361-367 (1975).

Williamson, K. T., and R. R. Whetstone. "Arsenic Distribution in Soils and its Presence in Certain Plants," *U.S. Dept. Agric. Tech. Bull.* 732 (1940).

Woolson, E. A., J. H. Axley and P. C. Kearney. "The Chemistry and Phytotoxicity of Arsenic in Soils. I," *Soil Sci. Soc. Am. Proc.* 35:938-943 (1971).

World Health Organization. *International Standards for Drinking Water,* Geneva (1958).

CHAPTER 10

METALS

INTRODUCTION

The metals category contains a broad collection of elements, typically those in groups 4, 5 and 6 of the Periodic Table. The common characteristic of these elements is their tendency to accumulate in plant-soil systems and not to readily migrate from the upper soil zone. Hence, the assimilative capacity is usually determined from a critical soil level, for which uptake by vegetation presents a food chain problem or where plant growth is affected adversely. By establishing a desired site lifetime and with knowledge of the critical soil amount of concentration, the annual metal assimilative capacity can be established for the LLC evaluation of an industrial waste, Figure 10.1.

Heavy metal wastes are generated by a number of mining, ore refining, metal producing and electroplating industries. The steel industry, galvanizing processes for metals and pipes, the auto industry, tannery wastes, pesticides and fungicides manufacture and use result in various proportions of heavy metals entering waste streams. The elements included in the metal category represent some essential and nonessential elements to both plants and animals.

This chapter is structured to accomplish several objectives. First, the designer or engineer must understand the basic mechanisms by which metals exert an impact in a plant-soil system, so the first section presents the pathways by which metals are assimilated in terrestrial systems. Next, a detailed discussion of specific metals is presented. The purpose of that section is to describe the data base and the resulting assimilation rates (kg/ha) for metals. Such application amounts range from some precise values to broader ranges of values based on available data, which vary with the specific metal. Finally, some engineering concepts are given for the design of industrial wastes containing metals along with research needs related to metals applied to land.

389

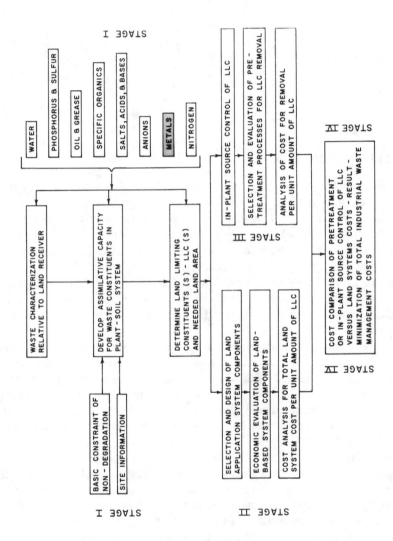

Figure 10.1. Relationship of Chapter 10 to overall design methodology for industrial pretreatment-land application systems.

Heavy metals occur in the soil solution phase, in association with organic molecules as metal-organic complexes, in colloidal and suspension phase, in sparingly soluble solid forms as precipitates, and/or simultaneously in several of these forms. Heavy metals vary in their level or concentration in wastewaters (Table 10.1), sludges (Table 10.2), soils (Table 10.3) and other sources of material causing metal pollution (Table 10.4). These metals are essentially conservative and, therefore, at any waste level an accumulation occurs in the ultimate receiver system. If critical soil concentrations of these metals are exceeded by uncontrolled industrial waste application, then the basic constraint of nondegradation has been violated and the design would be considered incorrect. Municipal sludge generally contains heavy metals from domestic sources as well as contributing industrial sources (Table 10.2) and is frequently land applied. Industrial waste from certain manufacturing categories may contain much higher levels of metals and, hence, may require greater land areas than municipal wastes over the design years of application. The concentration of heavy metals in treated wastewater effluent is generally low because most metals are contained in the sludge (Overcash and Humenik 1975). However, if a zero-discharge approach is to be achieved and both effluent and sludge are to be land applied, then separation into a liquid and high solids stream will not affect the land requirements. Such separation may not be cost-effective.

Table 10.1 Comparison of the Amount of Toxic Metal that Would be Applied with a Typical Secondary Treated Sewage Effluent vs That Added under the 1972 Irrigation Water Standards (National Academy of Science National Academy of Engineering 1974, Bouwer et al. 1974)

Element	1972 Irrigation Water Standard		Typical Effluent	
	Concentration (mg/l)	Amount[a] (kg/ha/yr)	Concentration (mg/l)	Amount[a] (kg/ha/yr)
Cd	0.01	0.02	0.005	0.1
Cu	0.20	4.0	0.10	2.0
Ni	0.20	4.0	0.02	0.4
Pb	5.0	100	0.05	1.0
Zn	2.0	40	0.15	3.0

FATE OF METALS IN SOIL SYSTEMS

Soil Characteristics Pertinent to Metals

Effective removal of heavy metals from wastewaters depends on certain characteristic soil properties and must be accounted for in determining the loading rate of metal-carrying wastes:

Table 10.2 Municipal Sludge Summary of Metal Constituents
(Page 1974, Sommers 1977)

Component	Sample Type	Number	Range (mg/kg)	Median (mg/kg)	Mean (mg/kg)
Mn	Anaerobic	81	58- 7,100	280	400
	Aerobic	38	55- 1,120	340	420
	Other	24	18- 1,840	118	250
	All	143	18- 7,100	260	380
	Page (1974)	300	60- 3,900	500	-
B	Anaerobic	62	12- 760	36	97
	Aerobic	29	17- 74	33	40
	Other	18	4- 700	16	69
	All	109	4- 760	33	77
	Page (1974)	300	6- 1,000	50	-
As	Anaerobic	3	10- 230	116	119
	Aerobic	-	-	-	-
	Other	7	6- 18	9	11
	All	10	6- 230	10	43
	Page (1974)	300	2- 18	5	-
Co	Anaerobic	4	3- 18	7.0	8.8
	Aerobic	-	-	-	-
	Other	9	1- 11	4.0	4.3
	All	13	1- 18	4.0	5.3
	Page (1974)	300	2- 260	10	-
Mo	Anaerobic	9	24- 30	30	29
	Aerobic	3	30- 30	30	30
	Other	17	5- 39	30	27
	All	29	5- 39	30	28
	Page (1974)	300	2- 1,000	5	-
Hg	Anaerobic	35	0.5-10,600	5	1,100
	Aerobic	20	1.0- 22	5	7
	Other	23	2.0- 5,300	3	810
	All	78	0.5-10,600	5	733
	Page (1974)	300	0.1- 50	5	-
Se	Page (1974)	300	-	1	-
Ba	Page (1974)	300	40- 4,000	1,000	-
	Page (1974)	300	20- 400	50	-
Pb	Anaerobic	98	58-19,730	540	1,640
	Aerobic	57	13-15,000	300	720
	Other	34	72-12,400	620	1,630
	All	189	13-19,700	500	1,360
	Page (1974)	300	15-26,000	500	-
Zn	Anaerobic	108	108-27,800	1,890	3,380
	Aerobic	58	108-14,900	1,800	2,170
	Other	42	101-15,100	1,100	2,140
	All	208	101-27,800	1,740	2,790
	Page (1974)	300	72-49,000	2,000	-

Table 10.2, continued

Component	Sample Type	Number	Range mg/kg	Median	Mean
Cu	Anaerobic	108	85-10,000	1,000	1,420
	Aerobic	58	85- 2,900	970	940
	Other	39	84-10,400	390	1,020
	All	205	84-10,400	850	1,210
	Page (1974)	300	52-11,700	500	–
Ni	Anaerobic	85	2- 3,520	85	400
	Aerobic	46	2- 1,700	31	150
	Other	34	15- 2,800	118	360
	All	165	2- 3,520	82	320
	Page	300	10- 5,300	50	–
Cd	Anaerobic	98	3- 3,410	16	106
	Aerobic	57	5- 2,170	16	135
	Other	34	4- 520	14	70
	All	189	3- 3,410	16	110
	Page (1974)	300	1- 1,500	10	–
Cr	Anaerobic	94	24-28,850	1,350	2,070
	Aerobic	53	10-13,600	260	1,270
	Other	33	22-99,000	640	6,390
	All	130	10-99,000	890	2,620
	Page (1974)	300	20-41,000	200	2,620
Ag	Page (1974)	300	5- 150	10	–
Sn	Page (1974)	300	40- 700	100	–

Soil Inorganic Components and Minerals

The inorganic solid fraction of soil is complex in both mineralogical and chemical composition. One common method of characterizing in inorganic fraction is according to particle diameter, including sand (2.00-0.02mm), silt (0.02-0.002 mm) and clay (<0.002 mm) fractions (Brady 1974).

The sand and silt fractions are mostly primary minerals, the identity and amounts of which reflect the parent material from which the soil originated. The most abundant minerals classified as sand and silt are quartz (SiO_2) followed in importance by feldspars, which are sodium (Na), potassium (K) or calcium (Ca) aluminosilicates, *e.g.*, $NaAlSi_3O_8$. Other primary minerals (silicates) include pyroxenes (Ca, Mg, Fe, Al) $(Si, Al)_2O_6$, amphiboles, Ca $(Mg, Fe)_5$ $Si_8O_{22}(OH)_2$, and olivines $(Mg, Fe)_2$ SiO_4. Examples of non-silicate minerals are calcium, magnesium and iron carbonates ($CaCO_3$, $MgCO_3$, $FeCO_3$) and sulfur (S)-containing compounds such as gypsum ($CaSO_4$) and pyrite (FeS). In many of these minerals, heavy metal cations are substituted for small quantities of other elements. Formulae of compounds have been listed only to indicate that even the simple sand and

Table 10.3 Composition of Soils[a] (Bowen 1966)

Element	Mean Dry Soil–Range (ppm)	Element	Mean Dry Soil–Range (ppm)
Ag	0.1 (0.01-5)	Mg	5,000 (600-6,000)
Al	71,000 (10,000-300,000)	Mn	850 (100-4,000)
As	6 (0.1-40)	Mo	2 (0.2-5)
B	10 (2-100)	N	1,000 (200-2,500)
Ba	500 (100-3,000)	Na	6,300 (750-7,500)
Be	6 (0.1-40)	Ni	40 (10-1,000)
Br	5 (1-10)	O	490,000
C	20,000	P	650
Ca	13,700 (7,000-500,000)	Pb	10 (2-200)
Cd	0.06 (0.01-0.7)	Ra	8×10^{-7} (3-20×10^{-7})
Ce	50	Rb	100 (20-600)
Cl	100	S	700 (30-900)
Co	8 (1-40)	Sb	(2-10?)
Cr	100 (5-3,000)	Sc	7 (10-25)
Cs	6 (0.3-25)	Se	0.2 (0.01-2)
Cu	20 (2-100)	Si	330,000 (250,000-350,000)
F	200 (30-300)	Sn	10 (2-200)
Fe	38,000 (7,000-550,000)	Sr	300 (50-1,000)
Ga	30 (0.4-300)	Th	5 (0.1-12)
Ge	1 (1-50)	Ti	5,000 (1,000-10,000)
Hf	6	Tl	0.1
Hg	0.03 (0.01-0.3)	U	1 (0.9-9)
I	5	V	100 (20-500)
K	14,000 (400-30,000)	Y	50 (25-250)
La	30 (1-5,000)	Zn	50 (10-300)
Li	30 (7-200)	Zr	300 (60-2,000)

[a]The figures refer to oven-dried soils. Soils near mineral deposits have been omitted in computing ranges. There are insufficient data for Ag, Be, Cd, Ce, Cs, Ge, Hf, La, Sb, Sn, Tl and U, and the values quoted for these elements may require revision.

silt fractions of soil are complex in terms of chemical structure and composition. Virtually all the chemical elements in wastewater are normal components of the solid fraction of every soil sample. However, elements in wastewater are usually in a soluble form or become soluble during biodegradation. By contrast, the minerals cited above are relatively insoluble and many play important roles in soil chemistry only in long-term soil formation. Jenne (1968) suggested that hydrous oxides of manganese and iron were very important in adsorption of heavy metals. The clay-sized fraction has a greater surface area of higher chemical activity than sand and silt particles. Several minerals in the clay fraction have a permanent negative charge for adsorption of exchangeable cations and a pH-dependent charge, which

Table 10.4 Typical Concentrations of Some Trace Metals in Source Materials
Causing Metal Pollution of Soil and Air (List 1972)

Metal	Limestones (ppm)	Superphosphate (20%) (ppm)	Sewage sludge (%)	Coal (ppm)	Petroleum (ppm)	Urban Air ($\mu g/m^3$)
Antimony	0.2	0-100	ND	0.5-5	30-107	0.05-0.06
Arsenic	1	2.2-1199	0.009	2-25	0.05-1.1	0.005
Barium	120	0-100	0.052	20-3000	750-1000	0-1.5
Beryllium	<1	–	ND	0.1-1000	–	0.0001-0.0006
Bismuth	–	–	ND	0.0001-0.0002	–	0.001-0.008
Cadmium	0.04	7.3-170	ND	0.2-0.5	0.0015-0.018	0-0.017
Cerium	12	20	–	–	–	–
Cesium	0.5	–	–	1.3	–	–
Chromium	11	66-243	0.140	5-60	–	0-0.048
Gallium	4	–	–	5.5	–	–
Germanium	0.2	–	ND	0.0017-0.0048	0.05	–
Lead	9	7-92	0.169	2-20	–	0.1-2.8
Lithium	5	0.04-1.6	ND	0.5-25	–	–
Mercury	0.04	–	ND	0.07-33	0.02-30	–
Niobium	0.3	11.5-44.5	–	1-5	–	–
Rubidium	3	5	–	15	–	–
Scandium	1	30	–	3	–	–
Selenium	0.08	0-1.5	–	4-7	0.03-1.4	–
Silver	0.05	15-20	0.01-0.09	0.5-2	0.0002	–
Strontium	610	25.9-36.6	0.01-0.09	0.07-0.15	–	0.05
Tellurium	–	19.5-22.6	ND	0.5-2	–	–
Thallium	1.7	0.2	0.01-0.09	–	–	0-0.05
Tin	0.5	3.2-4.1	0.047	1-10	–	0-0.13
Titanium	400	43-270	0.001-0.009	500-2000	–	0.0.315
Vanadium	20	2.3-180	0.001-0.009	10-50	0.004-0.3	0.004
Zirconium	19	50	0.001-0.009	7-250	–	–

determines the retention of both cations and anions as a function of pH. The most common crystalline clay minerals are layer silicates such as kaolinite, halloysite, montmorillonite, illite and mica vermiculite. Other non-crystalline clay minerals are allophane, sesquioxides and zeolites, which can fix and adsorb both anions and cations specifically and nonspecifically. Because of their cation exchange properties, all minerals adsorb heavy metal cations from wastewater applied to soils. The sorption of heavy metals by mineral matter decreases with increasing acidity of the system (Anderson 1977a, 1977b).

Soil Organic Fraction

The native soil organic matter results primarily from the decomposition of plant material. On a weight basis, organic matter has a higher cation and anion exchange capacity than minerals and contributes significantly to the total exchange capacity of soils. Heavy metal cations are strongly adsorbed by organic matter, which reduces the mobility of these chemicals in soils. On the other hand, certain organics form soluble metal-organic complexes, which may increase the mobility of adsorbed metal cations and also enhance the solubility of inorganic precipitates. Sillanpaa (1962) studied the solubility of Mn, Pb, Co, Zn, Ni and Cu as a function of soil organic matter content. The solubility of Cu, Ni and Zn organic complexes increased to a maximum of 10% of total amount even at 75% organic matter level. In the case of Mn, percent solubility of total amount increased over 30% at the 50% organic matter level. In addition, organic matter is a more effective sorbent for metals than the mineral matter in acid environments (Anderson, 1977a, 1977c).

Chemical Reactions of Heavy Metals in Soil System

Ion Exchange Reactions

The ability of soil minerals to exhibit ion exchange properties is due primarily to the substitution of small amounts of divalent for trivalent cations, and trivalent for quadravalent cations, within the crystal structure of the soil minerals. The location within the mineral crystals, and the degree and types of substitutions that occur, differ from one mineral to another. This type of exchange capacity is relatively permanent. However, the mineral fraction of soils also possesses an element of exchange capacity, which depends upon the pH of the soil. This type of ion exchange is reflected in the retention of both cations and anions.

The nature of pH-dependent exchange capacity of the mineral phase is not well understood but is thought to result primarily from the dissociation

of hydrogen ions (H^+) from structural elements at the crystal edges of alu-
minosilicate minerals (Aubert and Pinta 1977). At pH values of 8 or more,
the contribution of pH-dependent changes to the total cation exchange capacity can be considerable. At pH values below 4 or 5, the retention of anions
such as sulfates (SO_4^{-2}) can be significant. However, in the pH range from
5 to 7 (characteristic of many soils), the retention of anions by this mechanism is small to insignificant.

In many soils organic matter is as significant as the mineral phase in
determining ion exchange capacity, particularly for the surface soil at any
location. For organic and peat soils, cation exchange due to organic matter
predominates. Although a considerable amount of effort has been devoted to
investigating the exchange properties of soil organic matter, the exact organic
components that participate in exchange reactions are not well characterized,
nor is the process itself understood. Organic matter in soils contains a variety
of both acidic and basic chemical groups potentially available for cation and
anion exchange, respectively. The basic groups appear to have little activity,
as evidenced by the relatively low anion exchange capacity of soils, except
at very low pH. At normal soil pH values, acidic groups of organic matter
are weakly dissociated, providing opportunity for cation exchange. At lower
pH values the cation exchange capacity (CEC) of organic matter decreases,
because fewer acidic groups are dissociated. Also, at lower pH, considerably
more iron, aluminum and heavy metal cations are present in solution. These
types of cations interact strongly with exchange groups of organic matter
and effectively block sites normally available to other types of cations. The
CEC of organic matter increases significantly with pH. At pH values near
neutral, the CEC of organic matter ranges from 50-400 meq/100 g, with
values of more than 200 meq/100 g being common. Thus, on a unit weight
basis, the organic fraction can contribute more than the clay fraction to the
total CEC of a soil.

Based on the preceding discussion, it is clear that the cation exchange
characteristics of soils depend primarily on the types and amounts of
materials in the clay fraction, the nature and amount of organic matter, and
on soil pH. Sandy soils typically have CEC values of 2-4 meq/100 g of soil,
sandy loams 4-10, loams 10-15, silt loams 10-25, and clay and clay loams
20-60 meq/100 g. However, one must use care in such generalizations since,
for example, clay soils low in organic matter with a kaolinitic clay fraction
may exhibit CEC values as low as 4 meq/100 g. Although the CEC of soils
increases with pH markedly over an extreme pH range, the increase in CEC
resulting from increasing soil pH from 5 to 7 probably would not exceed
30% of the original value.

Figure 10.2 represents a schematic view of ion exchange on a negatively
charged electrostatic surface. Assume a mineral soil has a cation exchange

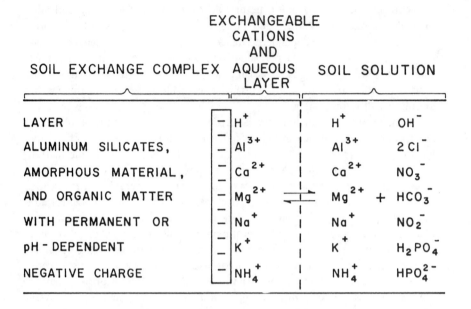

Figure 10.2. Representation of equilibrium between soil exchange complex and soil solution (Murramann and Koutz 1972).

capacity of 25 meq/100 g dry weight at 80% base saturation with Ca^{+2} as the predominant exchangeable cation. If about 25% of wet bulk soil volume is soil solution, then 100 g of wet soil would contain about 18 ml of solution, assuming the bulk density of 1.4 for soil and 1.0 for soil solution. The solution of a normal soil rarely exceeds a concentration of 0.01 mol/l dissolved salts. If the dissolved salt present is assumed to be $CaCl_2$, this concentration represents about 400 ppm Ca^{+2} and 730 ppm Cl^-. If the exchangeable Ca^{+2} associated with the solid soil phase were released to the solution, the resultant concentration would be an additional 22,000 ppm Ca^{+2}. From another point of view, 1 ha·20 cm of soil would contain about 11 metric tons of exchangeable Ca^{+2}. Thus, the exchange capacity of soil is substantial.

Soils with a high cation exchange capacity also have the ability to remove heavy metal cations from solution due to adsorption by the soil inorganic and organic components. However, exchange capacity should not be equated to the capacity of a soil to remove chemicals from wastewater. Since the exchange capacity of a soil is already saturated with common cations such as Ca^{+2} and Na^+, retention of wastewater chemicals or heavy metals will be accompanied by the release or displacement of these cations into solution.

Sorption, "Catch-All" and Fixation Processes

Cations such as NH_4^+, and K^+ and Co^{+2} are sorbed or fixed permanently or irreversibly by selective mechanisms characteristic to the ion and sorption surface. For the general case, sorption processes can be represented by the following equilibrium (Murrmann and Koutz 1972):

$$Insoluble - x \longleftarrow Sorbed - x \rightleftharpoons Soluble - x$$

where x represents the component sorbed from solution. Sorption processes are quite rapid. Although the component adsorbed tends to equilibrate with the solution phase, it cannot be readily displaced by common cations and anions. With time, the sorbed component is slowly converted to even more insoluble forms. Elements with ionic radii between 0.52 and 0.93 Å can assume the octahedral positions as Al, Fe and Mg in a 6-coordination. Manganese, Zn, Cu, Co, Ni and Cr are well within these radii limits and can thus be trapped in the alumniosilicate minerals (Aubert and Pinta 1977). Larger ions such as lead and cadmium are less effective in fixation by clay but can be retained by organic matter. Cobalt, Cr and Ni with smaller radii are preferably fixed in octahedral positions, and their distribution is correlated to the clay fraction and to secondary mineral distribution in the soil profile (Anderson 1977c).

Precipitation

When the concentrations of cations and anions in the soil solution become sufficiently high, mutual association between specific types of cations and anions in solution will occur to form solid chemical compounds with limited solubility. The concentration levels at which precipitation will begin depend on the individual compounds in question. Concentrations of dissolved constituents from wastewater along with the soil solution cannot exceed the concentrations determined by the solubility product of the corresponding compounds, except where formation of the solid phase proceeds very slowly or where formation of soluble chelates (to be discussed later) increases the apparent solubility of the solid phase. As long as the solubility product (Table 10.5) is exceeded, the solid phase will continue to form, reducing the concentration of the common ions in the solution to the critical levels determined by the solubility product (Murrmann and Koutz 1972). If the concentrations of common ions in solution are reduced below the critical levels by dilution, adsorption of ions, etc., the solid phase dissolves to maintain an equilibrium condition. Many of the metals found in industrial wastes would precipitate in soil and, thus, insolubilization represents a major mechanism for metal assimilative capacity.

Table 10.5 Solubility Product Constants for Compounds which Could Form from Chemicals in Wastewater at Room Temperature

Substance	Equilibrium	Solubility Product Constant
Carbonates		
Barium Carbonate	$BaCO_3 = Ba^{++} + CO_3^{--}$	1.6×10^{-9}
Cadmium Carbonate	$CdCO_3 = Cd^{++} + CO_3$	5.2×10^{-12}
Calcium Carbonate	$CaCO_3 = Ca^{++} + CO_3$	6.9×10^{-9}
Cobalt Carbonate	$CoCO_3 = Co^{++} + CO_3$	8×10^{-13}
Cupric Carbonate	$CaCO_3 = Cu^{++} + CO_3$	2.5×10^{-10}
Lead Carbonate	$PbCO_3 = Pb^{++} + CO_3$	1.5×10^{-13}
Magnesium Carbonate	$MgCO_3 = Mg^{++} + CO_3$	4×10^{-5}
Manganous Carbonate	$MnCO_3 = Mn^{++} + CO_3$	9×10^{-11}
Nickelous Carbonate	$NiCO_3 = Ni^{++} + CO_3$	1.4×10^{-7}
Silver Carbonate	$Ag_2CO_3 = 2Ag^+ + CO_3$	8.2×10^{-12}
Strontium Carbonate	$SrCO_3 = Sr^{++} + CO_3$	7×10^{-10}
Zinc Carbonate	$ZnCO_3 = Zn^{++} + CO_3$	2×10^{-10}
Chlorides		
Lead Chloride	$PbCl_2 = Pb^{++} + 2Cl$	1.6×10^{-6}
Mercurous Chloride	$Hg_2Cl_2 = Hg_2^+ + 2Cl$	1.1×10^{-18}
Silver Chloride	$AgCl = Ag^+ + Cl$	2.8×10^{-10}
Fluorides		
Barium Fluoride	$BaF_2 = Ba^{++} + 2F$	2.4×10^{-5}
Calcium Fluoride	$CaF_2 = Ca^{++} + 2F$	1.7×10^{-10}
Lead Fluoride	$PbF_2 = Pb^{++} + 2F$	4×10^{-8}
Magnesium Fluoride	$MgF_2 = Mg^{++} + 2F$	8×10^{-8}
Strontium Fluoride	$SrF_2 = Sr^{++} + 2F$	7.9×10^{-10}
Phosphates		
Variscite	$Al(H_2PO_4)(OH)_2 = Al^{3+} + H_2PO_4^- + 2OH^-$	3×10^{-31}
Strengite	$Fe(H_4PO_4)(OH)_2 = Fe^{3+} + H_2PO_4^- + 2OH^-$	1×10^{-35}
Octocalcium Phosphate	$Ca_4H(PO_4)_3 = 4Ca^{2+} + H^+ + 3PO_4^{3-}$	1×10^{-47}
Fluorapatite	$Ca_{10}(PO_4)_6(FH)_2 = 10Ca^{2+} + 6PO_4^{3-} + 2F^-$	4×10^{-119}
Hydroxyapatite	$Ca_{10}(PO_4)_6(OH)_2 = 10Ca^{2+} + 6PO_4^{3-} + 2OH$	2×10^{-114}
Hydroxides		
Aluminum Hydroxide	$Al(OH)_3 = Al^{+++} + 3(OH)$	5×10^{-33}
Cadmium Hydroxide	$Cd(OH)_2 = Cd^{++} + 2(OH)$	2.0×10^{-14}
Chromic Hydroxide	$Cr(OH)_3 = Cr^{+++} + 3(OH)$	7×10^{-31}
Cobaltous Hydroxide	$Co(OH)_2 = Co^{++} + 2(OH)$	2.5×10^{-16}
Cupric Hydroxide	$Cu(OH)_2 = Cu^{++} + 2(OH)$	1.6×10^{-19}
Ferric Hydroxide	$Fe(OH)_2 = Fe^{+++} + 3(OH)$	6×10^{-38}
Ferrous Hydroxide	$Fe(OH)_2 = Fe^{++} + 2(OH)$	2×10^{-15}

Table 10.5, continued

Substance	Equilibrium	Solubility Product Constant
Lead Hydroxide	$Pb(OH)_2 = Pb^{++} + 2(OH)$	4×10^{-15}
Magnesium Hydroxide	$Mg(OH)_2 = Mg^{++} + 2(OH)$	8.9×10^{-12}
Manganese Hydroxide	$Mn(OH)_2 = Mn^{++} + 2(OH)$	2×10^{-13}
Mercuric Hydroxide	$HgO + H_2O = Hg^{++} + 2(OH)$	3×10^{-26}
Nickel Hydroxide	$Ni(OH)_2 = Ni^{++} + 2(OH)$	0.6×10^{-16}
Zinc Hydroxide	$Zn(OH)_2 = Zn^{++} + 2(OH)$	5×10^{-17}
Sulfates		
Barium Sulfate	$BaSO_4 = Ba^{++} + SO_4$	1.5×10^{-9}
Calcium Sulfate	$CaSO_4 = Ca^{++} + SO_4$	2.4×10^{-5}
Lead Sulfate	$PbSO_4 = Pb^{++} + SO_4$	1.3×10^{-5}
Strontium Sulfate	$Sr SO_4 = Sr^{++} + SO_4$	7.6×10^{-7}
Sulfides		
Cadmium Sulfide	$CdS = Cd^{++} + S$	6×10^{-27}
Cobalt Sulfide	$CoS = Co^{++} + S$	5×10^{-22}
Cupric Sulfide	$CuS = Cu^{++} + S$	4×10^{-36}
Ferrous Sulfide	$FeS = Fe^{++} + S$	4×10^{-17}
Lead Sulfide	$PbS = Pb^{++} + S$	4×10^{-26}
Manganous Sulfide	$MnS = Mn^{++} + S$	8×10^{-14}
Mercuric Sulfide	$HgS = Hg^{++} + S$	1×10^{-50}
Nickelous Sulfide	$NiS = Ni^{++} + S$	1×10^{-22}
Silver Sulfide	$Zg_2S = 2Ag^{+} + S$	1×10^{-50}
Zinc Sulfide	$ZnS = Zn^{++} + S$	1×10^{-20}

Chelation and Metal-Organic Complex Formation

Coordinate bonding between a metal and a chelating ligand may lead to formation of soluble complexes that can move readily through the soil profile, and that may render the metal ion more amenable to plant root uptake. Thus, chelation of toxic heavy metals is not desirable for reasons of increased metal mobility. However, the magnitude of complexed metal movement for materials applied at low rates characteristic of land application sites has not been established. Stability constants of metal organic complexes may govern their availability (Lisk 1972). The order of chelate stability is Hg>Cu>Ni>Pb>Co>Zn>Cd>Fe>Mn>Mg>Ca.

Small amounts of metals are bound to relatively specific sites, which form higher stability chelates (Bouwer 1974). Larger amounts of metals are held much less firmly and are similar to cation exchange in behavior. The affinity of bonding of various cations with humic acid was in the decreasing order Cu>Fe>Zn>Ca (Randhawa and Broadbent 1965).

The equilibrium relationship of the metal ion and chelating ligand is shown in Figure 10.3, which has a characteristic formation constant. The activity of free ligand L^{4-} is common to each reaction. With the constants for all the equilibria, it is possible to calculate the fraction of the chelating agent present that will be associated with each metal (Lindsay 1975).

Metal organic chelates in sludge may decay in soils through biochemical processes and thus free the metal to be precipitated, sorbed or retained by other mechanisms. This is a similar process to the decomposition of more resistant organic materials and associated metals present in industrial wastes.

Oxidation-Reduction Reactions

Most of the oxidation-reduction reactions that occur in soils are biochemical, related to the processes by which microorganisms degrade organics applied to soil. Under aerobic conditions, many types of microorganisms in soil use molecular oxygen to oxidize organic substrates in obtaining energy and in forming the various metabolic by-products. However, when the oxygen

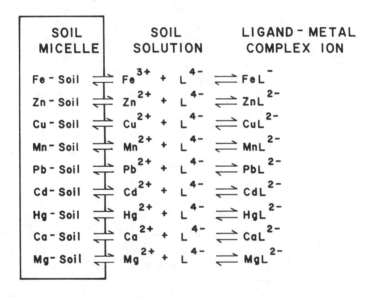

L^{4-} REFERS TO LIGAND OR COMPLEXING AGENT WITH 4 VALENCE OR COMPLEXING SITES, E.G. EDTA (ETHYLDIAMINE TETRA ACETATE)

Figure 10.3. Schematic of ligand-metal chelate equilibria in soils (Lindsay 1975).

demand rate exceeds the rate of oxygen transfer into the soil at the soil/
atmosphere interface, microorganisms must find a substitute for oxygen in
metabolic processes. In soils, these substitutes for oxygen mainly include
forms of nitrogen (NO_3^-) sulfur (SO_4^{2-}), iron (Fe^{3+}) and manganese (Mn^{4+}).
Several metals exist in different oxidation states and can also undergo changes
with redox fluctuations of soil-plant systems.

The chemistry of these elements varies considerably depending on whether
they are present in oxidized or reduced form (Mitchell 1967). For example,
nitrate nitrogen (NO_3^-) in soils is freely mobile, but when reduced to ammonium (NH_4^+) nitrogen it is in a form that can be immobilized by cation
exchange and adsorption processes. Iron in the oxidized form (Fe^{3+}) is
virtually immobile in soils due to limited solubility, whereas the reduced
form (Fe^{2+}) is relatively soluble. Thus, while oxidation/reduction is not a
process by which soils retain wastewater chemicals, it is extremely important
in determining the behavior and properties of certain types of metals, *e.g.*,
Cr, Mn and Fe (Mitchell 1967). Thus, if soils undergo a change in redox
status, certain metals will change valence and influence the mobility and
assimilative capacity.

Interaction of Heavy Metals with Soil Microorganisms

Certain metals such as iron, manganese and copper are essential in trace
amounts for soil microorganisms in metabolism and synthesis of biomass.
Effects of nonessential metals on growth and activities of soil microbes has
not been well studied. Drucker *et al.* (1974) indicated that silver and mercury
at 1 ppm were the most toxic to carbon dioxide evolution and total biomass
production. Porter (1946) cited the order of decreasing toxicity of metals
to bacteria as $Ag > Hg^{+2} > Cd > Ni^{+2} > Zr > Tl > Pb > Be > Cr^{+3} > Ba > Sr > Li$. Thus,
for metals with the highest microbial toxicity, the site-specific soil characteristics (clay and organic content) will substantially alter the assimilative
capacity by metal sorption to reduce toxicity.

Plant Uptake Mechanisms of Heavy Metals: Requirements and Toxicities

Processes of heavy metal ions movement to root surfaces and subsequent
uptake mechanisms and translocation within the plants are speculative and
have not been completely delineated for such critical metals as Cd. Agronomic research must focus on obtaining empirical data and develop models
that adequately predict the long-term impact of metals on soil-plant systems.
In the absence of more complete crop response understanding, the assimilative capacities for metals are conservative and adequate monitoring is essential.

Movement of Metals to Plant Roots

Barber (1974) accounted for three processes that move metals to plant roots: interception, mass flow and diffusion. Root exudates and contact may dissolve the insoluble metals (Bromfield, 1958) and may result in uptake by plants. Contact exchange interception was originally proposed by Jenny and Overstreet (1939) and has since been supported in successive research. As water moves to plant roots for uptake and transpiration, the solute metal ions move with it to the surface of roots, thus a mass flow occurs. The amount of metal ions transported in this fashion will depend on (a) the concentration of metal ions in soil solution, (b) amount of water moving to roots for plants needs of metabolism and evaporative cooling by evapotranspiration, and (c) plant response to metal ion. Various metal ions respond differently. Some increase, some decrease and others remain unchanged in concentration near the root. Diffusion of metal ions occurs in the direction of a concentration gradient. When root adsorption of metal ions exceeds that supplied by interception and convection, the difference is attributed to diffusion. Predominant factors for metal ion concentration in soil solution and transport to roots are pH or soil, chelating agents and transpiration rate.

Adsorption of Metals by Plant Roots

Moore (1972) reviewed the mechanisms of trace metals uptake by plants. He discussed three major models of ion transport: (1) electrochemical potentials, (2) Donnan equilibria, and (3) Carrier theory.

Chaney (1975) described the uptake kinetics of several metal ions by plants and concluded that little is known about the entry site of micronutrients in relation to root development. Maximum iron uptake occurs along a short distance about 1-10 cm behind the root tip. Interactions may occur among metal chelates for carrier sites on roots. To a metal ion, the root appears like a tube surrounded by ion exchange surfaces filled with a solution of chelators and other metals.

Tiffin (1972) found that Cu and Ni have a similar pattern of chelation in root and xylem sap. He pursued identification of Ni and Cu carriers in peanut and found two compounds (both nonprotein amino acids), which carry Ni a distance similar to Cu during electrophoresis. These amino acids bind Cu more strongly than Ni and have a heavy metal transport role in root cytoplasm as well as in the xylem sap.

Chaney (1975) summarized the metal uptake as follows:

1. Metals like Fe form strong, slowly exchanging chelates and may be moved intact from epidermis to leaves.
2. Metals like Cu and Ni form similar chelates with organic complexing agents such as Fe, but the ability of the copper chelate to enter the xylem is much less than that of iron chelates.

3. Metals like Zn and Mn remain soluble in the "chelate pool" and, thus, chelates hold very little control on their movement.
4. Metals like Pb may be precipitated by phosphate or may form a mixed chelate-precipitate with the ion exchange surfaces in the root.
5. Metals like Hg may bind to very stable slowly exchanging macromolecules or ion-exchange surfaces and not be soluble enough to reach the xylem parenchyma cells for release to the xylem.

Total metal accumulation by plants depends on many factors.

Nature of Plant. This includes species, growth rate, root size and depth, transpiration rate and nutritional requirements. Some plants may accumulate high levels of specific metals such as Se and Co and, conversely, other specific metals are largely excluded by plants. Plant species differ markedly in accumulation and requirement for different metals (Allaway 1968, Chaney 1973, Leeper 1972, Lisk 1972, Page 1973, Tiffin *et al.* 1973, Hodgson 1963, Hodgson 1970). Most plants exclude toxic trace metals from their seeds and fruits. Most root crops exclude these elements from the edible parts, with some exceptions. Carrot accumulates large quantities of Cd, Pb, etc. from metal-enriched soils. This variation in crop response has led to research to develop hybrids that will exclude metals and, hence, allow beneficial use of land application sites containing higher levels of metals.

Soil Factors. These refer to pH, organic matter content and nature, nutrient status and amount of metal ions, and certain anions like phosphate, sulfate and sulfides, clay content and type. The absorption of metals may be promoted in the presence of organic-complexing agents and pH in the acid range for which heavy metals are solubilized to a greater extent in soil solution. The heavy metal uptake may be retarded by (1) precipitation reactions, in which these elements become insoluble, such as sulfides, phosphates or carbonates, and (2) partial fixation in the soil system on clays and organic colloids, especially in neutral to alkaline reaction range.

Metal ions in aqueous systems undergo extensive hydrolysis and complex formations, adding to the difficulty of predicting the chemical behavior of metals in soils. Generalizations from one heavy metal to another are hazardous unless detailed chemical properties and behavior in soil-plant systems have been investigated. Instead, agricultural experience and field data should be used to assess the plant-soil assimilation of metals.

Environmental and Management Variables. These are factors such as temperature, moisture, sunlight, amendments and fertilization and may operate by altering any of the above plant and soil factors, thereby influencing the heavy metal accumulation by plants.

Modes of Heavy Metal Toxicity and Plant Tolerance

An element becomes toxic above a threshold concentration limit when it injures the growth or metabolism of an organism or plant. Even essential elements become toxic at levels higher than the optimum for growth and follow the idealized curve depicted in Figure 10.4. There is an optimum range of soil concentration, which is quite narrow for some elements. Nonessential elements show no effect below a critical soil concentration, above which potential toxicity effects become evident (Figure 10.4). Bowen (1966) attempted to classify the elements based on their toxicity as:

1. very toxic elements, which show injurious effects on test organisms at concentrations below 1 ppm in the solution, e.g., Ag^+, Be^{+2}, Hg^{+2}, Sn^{+2} and possibly Co^{+2}, Ni^{+2}, Pb^{+2} and $CrO_4^=$.
2. moderately toxic, which show inhibitive effects at concentrations between 1 and 100 ppm in nutrient solutions. This group includes arsenate, borate, bromate, chlorate, permanganate, molybdate, antimonate, selanate, aluminum ion, barium ion, cadmium ion, chromium, ferrous, manganous, zinc, etc.
3. scarcely toxic, which rarely show effects at levels exceeding 1,800 ppm, e.g., Cl^-, Br^-, I^-, Ca^{+2}, Mg^{+2}, K^+, Na^+, Rb^+, Sr^{+2}, Li^+, NO_3^-, $SO_4^=$, Ti, etc.

Toxicity due to heavy metals may be any one or more of the following mechanisms:

1. Most metals at toxic levels are enzyme inhibitors. Copper and mercury have a strong affinity for the reactive sites on enzymes. These metals are also chelated by organic molecules prior to penetrating the cell membranes.
2. Mercury, Pb, Cu, Be, Cd and Ag have been found to inhibit alkaline phosphatase, catalase, xanthine oxidase, ribonuclease enzymes (Lisk 1972).
3. Heavy metals like Al, Ba and Fe may form precipitates with $PO_4^=$, $SO_4^=$, etc. or may complex into a chelate with an essential metabolite and prevent their further metabolism.
4. Heavy metals may catalyze the degradation of essential metabolites such as decomposition of adenosine triphosphate in the presence of lanthanum.
5. Heavy metals may react with cell membranes, may alter their permeability and may affect other properties, for example, Au, Cd, Cu and Fe^{+2}, and may rupture cell membranes.
6. Some heavy metals may compete and substitute for essential functional roles of other essential metals and may carry out the function completely, e.g., Li replacing Na, Cs replacing K, and Ba replacing Ca.
7. Certain heavy metals may act as antimetabolites, e.g., selenate, arsenate, chlorate, bromate and borate, and may occupy sites normally used for phosphates, sulfates and nitrates.

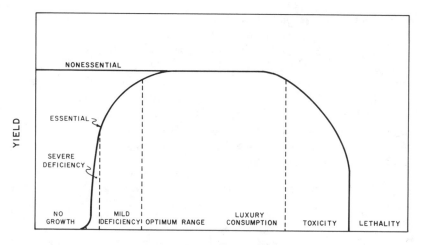

NUTRIENT CONCENTRATION

Figure 10.4. Diagram of growth as a function of concentration of essential and non-essential elements or compounds.

Very little is known about mechanisms by which crops differ in metal tolerance (Antonovics *et al.* 1971, Chaney 1974, Leeper 1972, Patterson 1971, Webber 1972). Although the responsible mechanisms of heavy metal tolerance in plants have not been elucidated fully, they are known to be under genetic control (Turner 1969). Tolerance for one element does not usually confer tolerance to other elements. In future, in may be necessary to screen and select the crop genotypes based on their selective avoidance capabilities of potentially toxic and nonessential heavy metals in uptake and accumulation.

Both metal transport as well as metal tolerance phenomena depend on chelation equilibria involving both soluble and fixed chelators. Phytotoxic effects of metals and plant tolerance depends on:

1. Type and amount of metal in the solution phase.

2. Soil factors such as pH, cation exchange capacity, soil organic matter content and nutrient-fertility status, which determine metal loading rates. Metal loading rates that are safe at pH 7.0 for all crops become lethal to many crops at pH 5.5. Soil pH influences metal movement and availability to roots as well as uptake processes. Soil organic matter binds some metals strongly and others loosely. Good and balanced supply of essential nutrients may provide plants with vigor to tolerate specific heavy metals.

3. Plant factors such as species genotype, stage of growth, rooting depth and characteristics, annual vs perennial plants, etc. Reuther and Labanaukas

(1966) noted that where high levels of copper, zinc or nickel accumulated in orchards or vineyards, the perennial crops simply altered their rooting depth to obtain adequate nutrients from soil zones low in metals. Annual cover or replacement crops suffered severely because of their rooting characteristics, which did not permit avoidance of toxic surface horizon. Antonovics and Breadshaw (1971) examined the comparative tolerance of different ecotypes to different metals and concluded the natural selection of tolerant preexisting genetic combinations. Plant species vary in evolving ecotypes tolerant to one metal and not to the other. For example, celery may tolerate high copper but not high zinc or nickel.

4. Certain management practices such as fertilizing, liming or other amendments may alleviate or accentuate toxic effects of metals as well as influence plant vigor and stand for tolerance of toxic effects.

5. Environmental variables such as sunlight, temperature and rainfall influence the transport and transformations of heavy metals in the soil biosphere, the plant growth and vigor, and the interactions between plant response and heavy metal supply.

Obviously, the behavior of metals in soils and the effects on microorganisms or plants are complex and very much dependent on site characteristics. This is essential to understanding and controlling the environmentally acceptable assimilation of metals in a plant-soil receiver system. The design for land application of industrial wastes containing metals is, thus, based primarily on field data from agricultural lands, which have received metals through irrigation, fertilization, pesticide and herbicide application, or from air pollution or waste sources. Other data from municipal sludge land application sites focusing on five primary metals (Cu, Zn, Pb, Cd and Ni) are also useful in establishing plant-soil assimilative capacities.

Using field data and current knowledge of metal responses, conservative metal assimilative capacities are established. Monitoring the vegetative cover becomes critical if heavy metals are the LLC, since long-term data are not always available. Such monitoring may indicate that the industrial site life can be extended beyond presently expected periods.

Differentiation is necessary between metals factors, with respect to the human food chain vs the presence of nonfood chain crops, since land can be used profitably for either. Research is needed on nonfood chain alternatives where metals are the LLC. However, the concept of private vs public lands, presumably dedicated to waste disposal, does not appear to be acceptable, because long-term guarantees and wildlife access are not addressed adequately. Such dedicated areas violate the basic constraint necessary for land-based treatment systems.

BEHAVIOR OF SELECTED METALS IN SOIL-PLANT SYSTEMS

The assimilative capacity of metals in the land receiver is dictated primarily by the critical soil levels (kg/ha) at which adverse crop or food chain responses occur. Metals are only removed from the soil system by slow, nearly irreversible, reactions, by which metals become a part of the soil structure (see previous discussion in this chapter), and by very low net removal through harvested vegetation. The other reactions and processes occurring in the soil simply convert metals from one state or form to another, with a substantial difference in environmental effect. Thus, to a large extent metals accumulate in the surface zone of the soil with little soil-water migration. The accumulation will reach defined or critical levels as presently known, and then the land should not be used for further industrial waste application. The time frame for reaching such critical levels is a management decision and may range from 10-1,000 years. Further waste application would convert that land out of other potential uses and would require land dedication. Such practices would be incompatible with the basic environmental constraint with land treatment design (Chapters 1 and 2).

Metal assimilative capacity thus depends primarily on establishing these critical soil levels. From research and field use of industrial waste land application, the data indicate that crop response and uptake are the controlling factors in establishing these critical soil concentrations of metals. Discussion of the fate of metals in plant soil systems is centered, therefore, on the metal effects of the vegetative cover, and mobility of metals in the soil-plant system, including the factors affecting the mobility.

Antimony (Sb)

Antimony is nonessential to all life forms and activity and has been reported to be moderately toxic. It occurs generally in very low concentrations in municipal sewage and sludges, but concentrations in the industrial sludges can be as high as 900 ppm. Stack dust from copper smelters contains 2% Sb and has caused contamination of the Seattle-Tacoma area in Washington state. Within 5 km of a smelter, about 200 ppm Sb has been measured in surface soil (Crecelius et al. 1974). In background control soils, Sb ranges from 1-30 ppm as compared to contaminated areas, with elevated levels from 50-200 ppm Sb. Antimony compounds are used in rubber, textile, fireworks, paint, ceramic, glass, leather and dyeing industries. In soils, antimony is strongly sorbed by kaolinite and sesquioxides under acid conditions but under neutral to alkaline conditions it appears to be more soluble (CAST 1976).

Naturally occurring forms of antimony are sulfides (stibinite) and oxides (cervanite Sb_2O_4 and valentinite, Sb_2O_3). Antimony usually occurs in the

+3 or +5 valence state and is very strongly precipitated as Sb_2O_3 or Sb_2O_5.

Antimony potassium tartrate was found to hinder substrate utilization of protozoa when solution concentrations exceeded 15 mg Sb/l (Bringmann and Kuhn 1959). In a study of water culture using beans and corn, Wober (1920) found that at 62 mg Sb/l, bean plant weight was 45% of control while at 83 mg Sb/l, corn plant weight was 77% of control. The antimony form used was $Sb_2O_7{}^{-4}$. Thus, concentrations in soil solution below these values are necessary to achieve adequate plant yields. In a soil (80% humus, 10% peat and 10% sand), greater tolerance to antimony was found. Up to 835 ppm Sb dry soil basis (as Sb_2O_3), bean plant weight was unaffected. For corn, a soil level of 418 ppm Sb caused a plant weight of 75% of control. Thus, an approximate assimilative soil level for Sb is <800 kg/ha.

Arsenic (As)

Arsenic is a nonessential element and occurs predominantly as oxides or sulfides in industrial wastes. For this reason, arsenic compounds are better classified as potentially toxic anions (see Chapter 9). Arsenic metal itself was utilized in a plant-soil system (Morris and Swingle 1927). At levels of 1,250 ppm As (dry soil basis), the arsenic metal was one of the three most injurious (of 17 As compounds) to bean and tomato. Thus, on a relative basis, the assimilative capacity should be lower for arsenic than for arsenic anionic compounds described in Chapter 9 for the respective soil types.

Cadmium (Cd)

Soils contain on the order of 0.06 ppm Cd naturally. Igneous rocks consist of 0.2 ppm Cd with a Zn/Cd ratio of 900:1. The chemistry of Cd is similar to that of Zn, and the two elements occur together in zinc ores at a ratio of about 1:2500 (0.04%). Cadmium is added to municipal wastewaters from a variety of sources, but concentrations in sludges are not sufficient to cause phytotoxicity (Page 1973).

The estimated Cd addition to Wisconsin soil in the form of fertilizer contaminants was about 2,150 kg/yr, while Cd in sludges could add about 1,700 kg, if all Wisconsin municipal sludges were land-applied (Keeney 1976). However, due to differences in the areal rate of application of these Cd sources, sludge disposal on land could result in a much higher Cd concentration in the soil than the fertilizers (Keeney 1976). Cadmium reaches soils from smelting, metal plating operations, lithography, engraving, soldering welding, cadmium-contaminated waters, metal pipes, ore refining for zinc, copper and lead, mining dusts and fumes, fertilizer impurity, organic wastes and sludges, products of gasoline and soil, tire wear and rubber products, and combustion of coal and cigarette smoke (Lisk 1972). Slight contamination of Cd occurs in areas surrounding lead smelters (deKoning 1974). Cadmium levels in various products and minicipal sludges are given in Tables 10.2 and 10.4.

Cadmium has not been shown to be essential for plants and has a crop effect curve similar to that in Figure 10.3. Cadmium is injurious to animals at levels as low as 2-5 ppm in the diet. Environmental levels of Cd have steadily increased because of its industrial uses, and threshold values for long-term toxic effects are not yet known with certainty. Cadmium is exclusively toxic and relatively easily taken up by plants. Small increases in Cd intake may be a hazard to man in the long term; but, because of the ubiquitous nature of Cd, slight increases may be inevitable.

Plants contain cadmium concentrations up to 0.1 ppm on a fresh weight basis (Bowen 1966). Certain fungi may contain 1 ppm cadmium, 10-fold that of plant levels. Many plants absorb cadmium from soil easily (Page 1974). Plants accumulate about 10-fold more cadmium than in soil on a percentage basis (Hodgson 1970). Increased accumulation of cadmium, an impurity, by plants fertilized with superphosphate has also been reported (Lisk 1972).

In soils cadmium behaves like zinc. Where the cadmium level of soil is high due to waste application and the available zinc level is low, crops may accumulate cadmium in plant-toxic amounts. Soil additions of cadmium at 4.5 kg/ha/yr for two continuous years raised Cd content of corn leaves from 0.06 to 0.82 ppm, of corn grain from 0.03 to 0.16 ppm, of soybean leaves from 0.15 to 0.71 ppm and of soybean grain from 0.13 to 0.72 ppm. Accumulation of cadmium in soybeans has been reported by Jones et al. (1973). Andersson (1977c) found that Cd is readily taken up by lettuce followed by carrot, tomato > rape, raddish > mustard, corn > cucumber, sunflower, pea, bean > wheat and oats in a decreasing concentration. Analysis of seeds revealed that Cd concentration in cultivars of wheat differed due perhaps to genetic reasons. The Cd concentration in leafy vegetables is a good indicator of its adverse effects on the food chain. Melsted (1973) suggested a tolerance limit of 3 ppm Cd in such agronomic crops; however, the consequences of Cd in crops produced for fiber, animal-fed crops, ornamental species or tree production has not been established. Such research is critically needed because of the primary impact on human health involved.

Zinc fertilization increases the amount of Zn in the soil and reduces the competitive crop uptake of cadmium (Chaney 1974). Adsorption of cadmium by radishes was diminished by raising soil pH from 5.9 to 7.2 (Giordano and Mays 1976a). Cadmium reaching soil by aerial contamination is not available for plant use immediately, thus indicating little foliar uptake. Organic matter from soil or waste may fix Cd as a chelate, thus reducing Cd uptake by plants. As a result, application of cadmium-free superphosphate and liming are pertinent in reducing available Cd in soils.

Recommendations have been made that total cadmium application should never exceed 6.7 kg/ha if the Zn:Cd ratio is greater than 100 and should never exceed 3.4 kg/ha if the Zn:Cd ratio is less than 100 (EPA 1976). EPA guidelines also suggest that Cd loadings should not exceed 1 kg/ha/yr (or 0.9 lb/ac/yr) from liquid waste and not more than 2 kg/ha/yr (or 1.8 lb/ac/yr) from dewatered waste. Cadmium metal additions apply only to soils adjusted to pH 6.5 or greater and when the waste pH is 6.2 or higher. Maximum cumulative cadmium application should not exceed 5 kg/ha for a soil with cation exchange capacity (CEC) less than 5 meq/100 g soil, 10 kg/ha for a soil with CEC 5-15 meq/100 g soil, and 20 kg/ha when the CEC exceeds 15 meq/100 g soil (EPA 1976). This is the assimilative capacity for Cd. It is further suggested that sludge with Cd contents greater than 25 ppm should not be applied to privately owned land unless their Cd:Zn ratio is <0.015.

Literature on cadmium in soil-plant-animal systems has been reviewed extensively by Allaway (1968), Hodgson (1970), Lagerwerff (1972) and Murphy and Walsh (1972), and studied extensively by Giordano et al. (1975), Giordano and Mays (1976b) and Chaney et al. (1978).

Chromium (Cr)

The role of Cr in plants has not been ascertained and evidence indicates that it is a nonessential element. However, in animals and man, chromium has been found to be essential (Mertz et al. 1974). The chemistry of chromium in plant-soil systems is both interesting and illustrative of the inherent assimilative advantage of soil systems.

The +6 oxidation state of Cr is most frequently present in industrial wastes as chromate (CrO_4^{-2}) or dichromate ($Cr_2O_7^{-2}$). In this form, Cr is toxic with the degree of impact being greatest in aquatic systems or soil solutions, then in sandy soils and finally in clays or organic soils. In addition to toxicity, the Cr anion (in the hexavalent state) is very mobile in a soil system. However, under conditions of an aerobic land application system, the conversion of Cr from +6 to +3 occurs easily. In the trivalent state Cr is very unavailable to plant uptake or leaching, with the probable insoluble forms being hydroxides or oxides. Therefore, the land application of chromate wastes results in an environmental shift from a toxic hexavalent form discharged to streams to a low-impact trivalent material in a plant-soil system. However, as with any system, the land assimilative capacity is quite finite and should not be exceeded in the design criteria or operation.

The soil adsorption capacity, without regard for plant effects, was determined to be 40 and 70 ppm Cr (dry soil) for two soils (Vanselow 1951, Turner and Rust 1971). Above this level Cr would move to the next Cr-unsaturated zone, unless insoluble precipitates were formed. At critical levels in soils, chromium in solution becomes sufficient to affect plant growth and microflora. The impact of Cr is dependent on the chemical form. In the +6 state, dichromate is more toxic than chromate (Voelcker 1921, Koenig 1911) and the +6 state is more toxic than the +3 state (Gemmell 1972). Under acid soil conditions there is a conversion from chromate to dichromate, thus suggesting an inherent advantage to lime correction of low pH soils.

When Cr is applied to land or is present naturally in soils, there is a pathway for plant uptake. The mode of plant uptake appears to be as a chelated compound (Patel and Wallace 1975). A number of studies with various crops have shown that Cr is not substantially transported to the upper plant portions, even when soil levels are high enough to produce toxicity (Myttenaere and Mousny 1974, Wallace et al. 1976, Vanselow 1951, Soane and Saunder 1959). Background levels of Cr in plants are about 0.1 ppm Cr, dry basis (Lisk 1972, Pietz 1978a). As soil chromium levels increase beyond the toxic level, certain plants can contain concentrations from 4->200 ppm Cr dry basis (Pratt 1966).

The total Cr:solution Cr ratios of Cr in soils receiving chromium waste have not been investigated so it is difficult to compare plant response data of water culture and field soils. However, total Cr:solution data will be obtained with future research, hence the need to bracket known plant responses. Arnon (1937) and Warington (1946) tested barley and lettuce, respectively, and found that 0.05 and 0.1 ppm Cr in water culture, respectively, increased yields. In the range of 0.1 < 1.0 Turner (1971) noted that the level of Cr produced a yield reduction for soybeans. At levels of 5 ppm Cr, the toxic effects on bush beans were quite evident (Wallace et al. 1976). One can conclude that chromium applied to soils of various adsorptive capacities must be restricted to levels that keep soil solution or easily extractable Cr less than 0.1-1.0 ppm Cr.

Studies have been conducted with a variety of Cr compound applications to plant-soil systems. These data are assembled in Table 10.6 in ascending order of Cr application. The variability is related to the type of Cr compound, since the conversion from +6 to +3 is confounded in these tests and since several soil types were used. It appears that in general, single-dose applications of chromate or dichromate should be less than 50 kg/ha of Cr, while the trivalent could be applied in a single dose up to 100 kg/ha of Cr. After such application, unavailable complexes will form with time allowing further applications. Then when Cr levels reach the 500-4,000 kg/ha of Cr range, the assimilative capacity would be more nearly exhausted. Applying Cr with organics, such as in sludge, allows greater single doses, since the Cr becomes available at a slower rate. In this instance, the rate of availability and rate of formation of insoluble complexes are more nearly equal (CAST 1976).

The consequences of Cr application to land with respect to the food chain appear to be small (Underwood 1971). Since the normal mode for contact of the more toxic hexavalent Cr is via exposure to chromate dust, and since

Table 10.6 Plant-Soil Response to Land Application of Chromium.

Cr Application Rate	Crop	Soil	Effect	Compound	Reference
0.006	Fescue & alfalfa	Silt	No increase in plant Cr	Sludge	Stucky and Newman (1977)
0.04	Potato	Cr-deficient	Increased yield	—	Bertrand and DeWolfe (1968)
4.8	Mustard	Pure sand	Decreased yield	CrO_4	Gemmell (1972)
7 < <32	Mustard	Sand & peat	Decreased plant weight	$Na_2Cr_2O_7$	Gemmell (1972)
32<	Cucumbers	—	Probable toxicity	$K_2Cr_2O_7$	Reinhold and Hausrath (1940)
50<<100	Wheat	Field	Reduced yield	K_2CrO_4	Voelcker (1921)
50<<100	Wheat	Field	Reduced yield	$K_2Cr_2O_7$	Voelcker (1921)
55	Rye	Silt l & sandy l	No increased plant Cr concentration	Sludge	Kelling et al. (1977)
100	Wheat	Field	No effect	$CrCl_3$	Voelcker (1927)
100	Wheat	Field	No effect	$Cr_2(SO_4)_3$	Voelcker (1921)
100	Wheat	Field	No effect	Chromic acid	Voelcker (1921)
100 < <200	Bush bean	Yolo loam	Decreased yield	$Cr_2(SO_4)_3$	Wallace et al. (1976)
110	Sorghum Sudan	Silt l & sandy l	No increased plant Cr concentration	Sludge	Kelling et al. (1977)
128 < <640	Mustard	Sand & peat	Reduced yield	Na_2CrO_4	Gemmell (1972)
135-530	Corn	—	No effect on yield or plant Cr concentration	Sludge	Clapp et al. (1976)
300	Orange	—	Reduced yield	—	Vanselow (1951)
400 <	—	Sand	Reduced yield	—	Schweneman (1974)
833	Corn	—	No increased plant Cr concentration	Sludge	Henesly (1976)
4,000 <	Mustard	Sand	Reduced yield	$Cr_2(SO_4)_3$	Gemmell (1972)

there is a strong conversion to trivalent Cr in soils, the food chain elevation of Cr from industrial land application appears to offer little risk. However, if the plant-soil assimilative capacity is exceeded, the risk will increase. Beneficial effects of additional nutritional Cr from sludges does not appear large (Elliot 1977); however, with inorganic Cr, which more prevalent in industrial wastes, some beneficial effects may result.

Management options that improve the plant-soil assimilative capacity include: (1) addition of organics to reduce Cr solution concentrations (CAST 1976, Gemmell 1972); (2) elevation of pH to shift dichromate to chromate, (3) maintenance of predominantly aerobic conditions to convert +6 Cr to the +3 state, and (4) site selection for soils with greatest assimilative capacity.

Besides municipal sludge containing Cr, there have been a number of industrial wastes considered for or unintentionally applied to land. The chromite or chrome iron ore extraction industry has been indicated several times for making areas unusable by heaping wastes on land (Breeze 1973, Gemmell 1972). Chrome-tanning wastes have been tested for land application at about 400 kg Cr/ha (Hamence and Taylor 1948, Gericke 1943). Chromium in steel-making slag was found to be amenable to moderate rates of land application (Gericke 1943).

Cobalt (Co)

This heavy metal occurs naturally in soils at levels indicated in Table 10.3. Extreme levels of cobalt occur naturally in serpentine or basalt-derived soils. The Co range approximately parallels the iron content of soils with a Fe:Co ratio of 1,000. Cobalt has been shown to be essential in trace amounts to nitrogen-fixing organisms, to animals as vitamin B_{12} and as a prosthetic group element for several enzymes. In higher amounts it is very toxic to plants and moderately toxic to mammals.

Soils of the following origin may be low in cobalt and are suitable for amendment with cobalt-rich organic wastes:

- soils derived from granite
- highly calcareous soil
- Atlantic Coastal Plain soils
- peat soils
- acid, highly leached, sandy soil

Crop response to cobalt has been determined for several plant types. Incremental levels of Co (as $CoCl_2 \cdot 6H_2O$) applied on soil in which forage was growing resulted in plant uptake response. At 2.2 kg/ha, the herbage had 0.2 ppm Co as compared to 0.07 ppm for the control. Use of 11.2 kg/ha increased the Co herbage level to 0.8 ppm, and at 90 kg/ha the plant herbage concentration was 3 ppm. Cobalt applied to sudan grass and citrus crops at

220 kg/ha did not decrease yields and resulted in plant concentrations of 3-6 ppm. At a soil solution concentration 1 ppm, Co toxicity was detected for corn and beans, thus the relation of soil and soil solution levels of Co is a factor in establishing plant tolerances.

The crop Co level at which fed animals would receive a deficient intake of cobalt is approximately 0.07 ppm. Animal diseases such as Coast disease (Australia), pining (Scotland) and salt sick (Florida) have been attributed to such Co deficiencies. At the upper tolerance level, cattle fed in excess of 50 mg Co/100 lb body weight ($\sim$60 ppm Co in feed) evidences toxic effects. For rats, a feed concentration of 25 ppm showed no toxic effects, so there is a wide range between the deficient and toxic levels of Co in animal diets.

In a soil system, 2-25 ppm of molybdenum can counteract the toxic effect of cobalt. As the soil pH becomes acid, the solubility of Co increases, thus reducing the soil level at which no adverse effects are produced. Liming to keep pH near neutral reduced the Co uptake by oats (Vanselow 1966). Leaching of cobalt also reduces soil levels.

Copper (Cu)

Both plants and animals require copper in trace amounts as part of their nutrition and normal physiological functions, hence it is classified as an essential metal. In higher amounts it offers moderate toxicity to higher animals. Copper is very toxic to algae, fungi and seed plants. Excess copper decreases decomposition rates, phosphate activity and P mineralization rate in soils. Use of copper at high levels and subsequent data on the plant-soil response are much more limited than the behavior under Cu deficiency conditions. Large applications of Cu are associated with sludges, swine manure from Cu-supplemented diets (Overcash et al. 1978), use of Bordeaux mixtures on crops, and from air emission sources of copper smelting on nearby land areas. In soils, Cu is adsorbed to sites containing iron and manganese hydrous oxides and is made labile under acid conditions. At a pH of 7-8, Cu is tightly bound in the soil. Copper is also bound to organic matter forming many insoluble structures.

Crop response to Cu reflects the essential nature of this metal. When soil levels are in the range of 0-5 ppm, Cu would be deficient for most crops. Such deficiencies can result from prolonged rainfall leaching of sandy soils or from the strong binding of Cu to organic material, thus being less available to plant growth.

Normal crop tissue concentrations of Cu are 5-20 ppm (CAST 1974). The response of plants to high applications of Cu depends on the availability of copper to root uptake and to the time frame used for application. Observations of celery production areas with up to 50 years of Cu accumulation emphasized the water-extractable measure of Cu (Fiskel and Westgate 1955).

The levels at which an adverse (chlorotic) crop response was observed were 0.7-1.1 kg/ha of Cu as extracted by water. This corresponded to 390-1,340 kg/ha of applied Cu over a 50-year period, or about 8-27 kg/ha/yr. For citrus crops, soils containing 110-220 kg/ha of Cu were found to have reduced yields and produced chlorosis by induction of iron deficiency.

In an interpretation of crop response to heavy metals, copper was suggested to be twice as toxic as zinc (Cunningham *et al.* 1975). The symptoms of copper toxicity include leaf chlorosis, stunted roots and reduced growth (Reuther and Labanauskas 1966). Copper phytotoxicity normally will occur before plant levels of Cu reach the range of upper food chain susceptibility (National Academy of Sciences 1974). The major mechanism for copper uptake by plants is via the root network, even when foliar application (airborne material from smelter) occurred (Beavington 1975).

In a study of inorganic copper ($CuSO_4$) and chelated copper (CuEDTA), the organic-associated Cu was more available to the plant (Wallace and Mueller 1973). The experiment was performed on calcareous loam soils (pH 7.5), for which copper solution availability is low. Thus the chelated form of Cu was more easily taken up and at a single-dose application of about 110 kg/ha of Cu, a slight yield depression of bush beans was found. Plants were shown to differ in the amounts of Cu uptake (Beavington 1975). White clover contained about 50% greater Cu levels than a perennial grass (*paspalum*) in the vicinity of a smelter. Soil-water-extractible levels of 40 ppm Cu were found to reduce yields of clover (Purves 1968).

Management can be used to reduce the phytotoxicity and migration of Cu in a land treatment site. Correction of soil pH to a range of 7-8 will place more of the Cu in unavailable precipitates or soil exchange sites. As organic matter of an industrial waste is added, the Cu-organic complexes formed will generally reduce the food chain impact of this heavy metal species. Supplementary carbonaceous residues may even be added to enhance the organic matter accumulation.

Crops contain Cu generally at low levels because of the phytotoxic effects described previously. On an injested basis, sheep are most sensitive to Cu levels. As ruminants accumulate Cu in body tissue, such animals probably should be monitored if fed elevated Cu diets. Some control of Cu accumulation and animal toxicity can be exerted by increasing molybdenum diet levels. Monogastric species are not substantially affected by diets less than up to 20 ppm Cu.

The assimilative pathway for Cu appears to be by a combination of inorganic adsorption and organic binding to make the metal less available for plant uptake. When applied in large single doses, Cu uptake is more pronounced than the same soil level achieved by repeated incremental applications over time. Some deleterious crop response has been noted at soil solution levels of 10-40 ppm Cu, which translates into a variety of soil concentrations, depending on cation exchange capacity and organic content. A conservative estimate of the critical accumulation limits for land application of Cu is 125 kg/ha, 250 kg/ha and 500 kg/ha for soils of CEC values of 0-5, 5-15, and >15 meq/100 g, respectively.

Lead (Pb)

As a heavy metal, lead is toxic to human beings at certain levels and is nonessential to plants. Lead content of and uptake by plants should thus be kept at minimum to reduce entry into the food chain. Pb toxicity is almost completely reduced when applied at correct rates to soils by contrast to the demonstrated hazards of atmospheric Pb.

The behavior of lead in soils is complex, so only the major pathways for assimilation are known. Soluble lead added to the soil reacts with clays, phosphates, sulfates, carbonates, hydroxides, sesquioxides and organic matter such that the Pb solubility is greatly reduced.

Vlek and Lindsay (1974) indicated that in very acid soils $PbSO_4$ was a reaction product while Pb_3 $(PO_4)_2$ and various other lead phosphates were found in soils of intermediate pH. $PbCO_3$ was the principal species in soils above pH 7.0. Jurinak and Sontillan-Medrano (1974) concluded that lead is retained as the hydroxide or hydroxyphosphate in acid soils and as the carbonate in calcareous soils. Solubility studies with Mo showed that $PbMoO_4$ is a very important reaction product of Mo in soils that governs the solubility of $MoO_2^-{}^3$. This demonstrated the interdependence of the solubility of one metal on another.

Lead is more tightly held or fixed by humus soils than other cations (Hassler 1943). At pH values above 6, lead is either adsorbed on clay particles or forms lead carbonate. Lead in soil forms organic metal complexes and chelates with humus (Broadbent and Ott 1957). Lead can be mobilized as the redox status of a soil is changed. Prolonged flooding of soil or anaerobic incubation with plant materials can produce such changes, thus requiring industrial land application sites to be designed for predominantly aerobic waste assimilation (Lisk 1972). The influence of pH on lead solubility adsorption and equilibrium concentrations is depicted in Figure 10.5. Liming to raise the pH of soil in which clay is present would reduce the ionic level of Pb, thus reducing availability to plants. Lead would be concentrated in the surface zone of soils on a land treatment site in a similar manner as found surrounding air pollution sources such as smelters and highways (Page and Grange 1970). The lead sorption capacity in soils is quite high since insolubilization and organic binding occur in great magnitude.

Plants take up lead in the ionic form from soils. The amount of lead taken up from soil decreases as the pH, cation exchange capacity and available

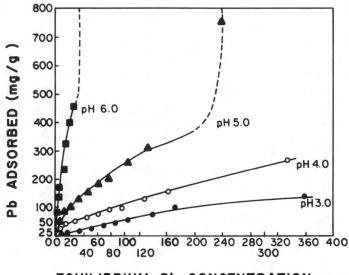

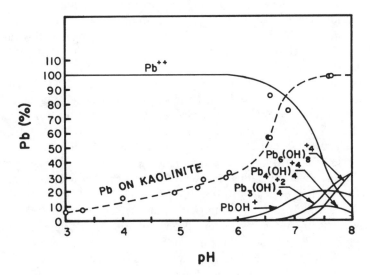

Figure 10.5 Lead adsorption at equilibrium with kaolinite (25°C) and pH-dependence of lead species and that on kaolinite (Griffin and Shimp 1976).

phosphorus of the soil increase. Basicially, the greater the Pb insolubilization or sorption the less the plant can remove from the soil system. Brewer (1966) found that if Pb were added in a single inorganic dose, then $Pb(No_3)_2$ was more available to plants than Pb CO_3. Thus, insolubilization in the soil requires a certain time. If forms of lead are added to soil in smaller increments over time, the availability to plants of a given soil level is reduced. However, very high concentrations of lead in soil may greatly restrict plant growth (Lag et al. 1969) and induce weak iron chlorosis in plants (Hewitt 1953). The stomatal route for Pb uptake exists for airborne lead, but when applied as a constituent in industrial waste, Pb uptake is primarily via the roots. Typical lead contents of many food plants and trees are in the range of 1-5 ppm. In fungi, algae and phytoplankton lead may be several fold higher in concentration than crop plants (Bowen 1966). In general, the lead content of roots is higher than that of plant tops with fruits and seeds showing the lowest content.

The crop response to land application Pb can be interpreted from a number of studies using sludges, inorganic Pb and areas of naturally occurring lead. Giordano et al. (1975) and Giordano and Mays (1976b) measured Pb uptake from sewage sludge on land containing approximately 100 kg/ha of Pb (CAST 1976). The fruit and foliar Pb concentrations (seven truck crops) were not significantly elevated over the control. One root crop, potato, also showed no Pb uptake at this level of waste application. Lead additions to soil at about 200 kg/ha of Pb did not produce plant toxicity (Swaine 1955). Clapp investigated a range of Pb applications as sewage sludge of 325, 650 and 1,275 kg/ha of Pb (CAST 1976). For corn grown on this sandy soil (low CEC) there was no elevation of leaf or grain tissue at any of the Pb application rates.

Inorganic lead was added to a soil growing barley (Brewer 1966). Application rates were 124-1,025 kg/ha of Pb as $Pb(NO_3)_2$ and 1,025 kg/ha of Pb as $PbCO_3$. The barley yields were unaffected. In the tops of the plant a safe food standard of 2 ppm was reached at 500 kg/ha of Pb as Pb $(NO_3)_2$, but only at 2,100-4,200 kg/ha of Pb as $PbCO_3$. At these critical levels the root content was 100-200 times that of the barley tops. Soluble Pb measured in the soil at the time of planting was about 2 ppm at these critical soil application rates and had decreased to 0.12-0.4 ppm after harvest. Thus, the form of the Pb added to soils, even inorganic, can affect crop response to lead. Comparing the previous results of Clapp and Brewer for approximately 1,000 kg/ha of Pb addition, the tops concentrations were about threefold lower for sludge-applied lead as compared to inorganic lead.

Land application of lead as a constituent of oil waste was investigated at a rate of about 600 kg/ha of Pb (Raymond et al. 1976). Turnip concentration of lead on a dry weight basis were compared for the control and the oil waste applied with fertilization to assure similar plant yields. The turnip tops and bottoms contained the same Pb level for the control vs the waste-applied plots. Lead carbonate applied up to 3,000 kg/ha on soils was found to have no yield effect on barley (Keaton 1937). Applications up to 3,200 kg/ha of soluble lead salts were tested for corn uptake (Baumhardt and Welch 1972). The leaf tissue was elevated in Pb concentration, but the grain was unchanged.

Areas surrounding ore-smelting operations were examined to determine heavy metal pollution and crop concentrations (Davies and Roberts 1975). In garden and field soils the total lead content of the surface zone was approximately 6,000 kg/ha of Pb, of which about 50-70% was extractable or available lead. Uncontaminated soils in the area contained only 37 ppm lead with about 40% being available. The lead content of the radish crops exceeded the safe food level of 2 ppm in 82% of all samples taken. These soil levels represent conditions that do not permit acceptable crop growth and thus have resulted in degradation of the terrestrial resources.

The response of humans and animals to lead poisoning has been widely documented. Substantial attenuation of lead seems to occur when land applications cation is practiced. From the crop response to applied Pb, expecially in wastes, it would appear that at levels of 1,000-3,000 kg/ha of Pb, conditions exist that might pose an adverse effect on future land use. Therefore, the Environmental Protection Agency has recommended that an upper limit of soil concentration of Pb be 500, 1,000 and 2,000 kg/ha of Pb for soil soil CEC values of 0-5, 5-15 and >15 meq/100 g soil, respectively.

Manganese (Mn)

The response of the plant-soil system and the total food chain indicate that Mn is essential for plants and animals, so the crop response will follow the general trends shown in Figure 10.4. The relation of Mn in the soil and the crop is similar to that described for previous metals, *i.e.,* as manganese is added to soils a major percentage is precipitated or sorbed, and the crop then responds to the solution phase or easily extractable concentration of Mn.

In a soil system the absolute level of Mn is not a useful measure of the assimilative capacity. Unfortunately, the majority of all data specify only total Mn soil levels and, therefore, only approximate assimilation capabilities are available. Two principal factors control the solution availability of Mn: (1) the soil pH, and (2) the redox status of the soil. The latter factor involves the reduction to divalent Mn under anaerobic conditions, thus solubilizing greater amounts of metal for possible plant uptake. Correction for redox effects on Mn mobility is simply to ensure that the industrial waste land application site is well aerated so that a class of microbial and chemical reactions will immobilize the applied manganese by conversion to insoluble tetravalent oxide of Mn.

At soil pH below 5.5, Mn is quite soluble. Below approximately pH = 5.7 and above pH = 7.5, the bacterial conversion of Mn^{+2} to Mn^{+4} is restricted, slowing the formation of the insoluble form (Figure 10.6). Field results corborate the Mn-phytotoxic effects in very acid or overlimed (alkaline) soils (Bromfield and David 1976). Observations of Mn deficiencies in peat soils

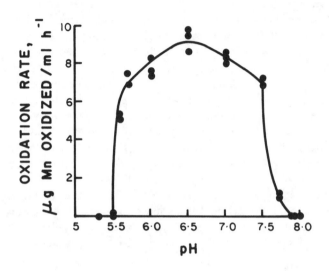

Figure 10.6 Dependence of Mn^{2+} oxidation rate on pH for bacterial suspensions (Bromfield and David 1976).

imply organic complexing, which reduces availability to plant uptake. Manganese added to soil at high levels can affect microbial reactions such as ammonification of land-applied organics and nitrification of ammonium (Brown and Minges 1916). Manganous chloride, $MnSO_4$ and MnO_2 applied above 2,200 kg/ha on a clay loam soil depressed the rates of ammonification and nitrification, while at between 110-220 kg/ha, these processes were stimulated. Manganous nitrate evidenced depressed microbial rates at 550 kg/ha of Mn. The authors concluded that the crop effects of Mn application were related to the stimulation or depression of nitrogen availability.

Crop effects of land application of Mn are related to the individual crop tolerances and the level of applied metal. Relatively tolerant crops are oats, rye, sugar beets, celery and broad beans. Intermediate tolerance is exhibited by barley, potatoes and red clover. Brassicas are in the most susceptible category.

The relationship of total soil Mn, available Mn and herbage content was determined (Wain 1938). For soils at pH $\geqslant$ 7, the total Mn was 260-780 ppm, with the ammonium acetate extractable Mn of 1 ppm. Herbage levels were 50-110 ppm on a dry weight basis. On soils having an acetate-soluble level of 6 ppm, the herbage content was 430 ppm. Average grass forage levels of Mn were 208, 78 and 47 ppm for orchard, Kentucky blue and alfalfa, respectively (Balin 1934). Citrus leaf contents of 3-5 ppm Mn resulted from manganese-deficient soils (as compared with normal levels of 14-26 ppm (Chapman et al. 1939). Thus, plant matter contents of 100-200 ppm Mn on a dry weight basis represent normal vegetative content.

With the soil addition of 20-40 kg/ha of Mn, tomato yields were increased (Skinner and Ruprecht 1913) as were oat yields (Bertrand 1905). Application of 10-110 kg/ha of Mn on sandy soils evidenced improved wheat growth, while on a nondeficient loam there was no increase (Skinner et al. 1914). At an application of 100 kg/ha of Mn, oats yielded a positive growth response (Clausen 1912). Under acid soil conditions the application of 280 kg/ha of Mn over 5 years gave depressed yields of wheat, rye, corn, cowpeas and potatoes (Skinner, et al. 1914), thus demonstrating Mn toxicity at low pH.

At somewhat higher rates, sugar beets achieved optimum growth yields at 200-400 kg/ha of Mn (Bartmann 1910). Wheat and kidney beans had improved yields when manganese applications of 145 kg/ha were used, but carrots and potatoes responded negatively (Andouard and Andouard 1911). With applications up to 600 kg/ha of Mn, sugar beets showed increased yield on a clay soil (Schulze 1915). Oats raised in a sandy soil gave increased straw yield at 1,000 kg/ha of Mn but depressed yeild at 2,000 kg/ha of Mn (Popp 1916). At 3,300 kg/ha of Mn, adverse effects on grass were highly pronounced (Agr. Gaz. 1911).

While a complete study using a number of manganese application rates has not been performed, individual studies indicate certain trends. In the range of 0-200 kg/ha, Mn applications produce improved yields for a variety of crops. In the range of 200-300 kg/ha under acid soil conditions, an adverse crop effect could occur. At up to 1,000 kg/ha of Mn, crop selection can eliminate adverse Mn response and often a positive yield response is found. In the range of 1,000-4,000 kg/ha, adverse yields and plant growth are more probable.

Several studies have been made of industrial wastes being land-applied as manganese slags (Popp 1916, Vlasyuk 1937, 1939). Manganese and Fe-Mn slimes were land applied on chernozem soil at about 100 kg/ha of Mn and an increased crop yield of sugar beets was measured. The response at higher applications was not reported (Vlasyuk 1937). Slags from the manufacture of ferromanganese and spiegeleisen were used in pot experiments of up to 2,000 kg/ha of Mn (Popp 1916). Oats grew with increased yield for application rates less than 1,000 kg/ha of Mn, but adverse crop yields occurred at

2,000 kg/ha of Mn. The response of manganese slags was similar to that of inorganic Mn salts. (Bromfield 1958).

Based on experimental results and the pathways for stabilization of Mn in soils, the approximate plant-soil assimilative capacity was developed (Bromfield and David 1976). Acceptable applications of Mn will vary with soil pH because of lower plant tolerance under acid conditions. That is, at pH less than 5.5 the solution-phase Mn is sufficiently increased that only lower total Mn loading can prevent adverse crop effects. As the pH is increased, greater insolubilization leads to greater assimilative capacity. As the pH exceeds 8.0, the effects of Mn toxicity again become evident. In the plant, a Mn concentration of 200-400 ppm is in the range of acceptable crops. At such plant concentrations the amount of Mn in a grass crop would be about 3-5 kg/ha/yr of Mn uptake. The assimilative capacity or upper soil level at which Mn might interfere with future productive land use was inferred from the available experiments as approximately 500-1,000 kg/ha, when soil pH is in the range of 6-7.5.

Mercury (Hg)

In an extensive survey of over 900 soil samples taken about 80 km apart throughout the U.S., Shacklette et al. (1971) reported a range of 0.01-3.4 and 0.01-4.6 ppm Hg for the eastern and western sections, respectively. Mercury occurs world-wide in belts that correspond to the earth's zones of dislocation. Mercury is sometimes associated with gold deposits and potential mines. The amount of mercury released every year world-wide by weathering processes and by combustion of coal is 230 and 3,000 metric tons, respectively. Other pollution-creating sources of Hg include crop fungicides, paint manufacturing, chloralkali plants, ore smelting, geothermal wastes and organic chemical manufacturing complexes. Mercury is a nonessential element. Toxic effects have been attributed to methyl, dimethyl and ethyl compounds of Hg, which binds to the -SH groups in body protein. The ionic or elemental forms of Hg are much less toxic.

The pathways for the assimilation of mercury in soils involve the biological and abiotic retention and volatilization losses. The atmospheric losses of Hg are inversely related to the soil capacity to insolubilize or sorb this metal. In comparing a soil with CEC = 25.5 to that of 8.4, the percentage of Hg volatilized was 5-8% and 25%, respectively, for an application of 1 ppm Hg as $Hg (NO_3)_2$ (Landa 1978). This rate of loss was stimulated by glucose addition to increase microbial populations. The pathway leading to the atmospheric loss is the microbial and chemical conversions of Hg^{+2} to elemental mercury, which is more volatile (vapor pressure of 0.007 mm Hg @ 24°C).

Mercury is predominantly held in the soil surface zone in a number of forms. Insolubilization forms include HgS and $HgCl_3^-$, thus waste containing both sulfur or chlorides along with Hg can enhance soil assimilative capacity. Organic-bound Hg absorbed in soils in addition to inorganic forms (Hassler 1943, CAST 1976). Humic acid can bind 18-80 mg Hg/g of humic acid (Bonner and Bustamonte 1976).

Microbial methylation of mercury to a toxic species in the terrestrial environment is a subject of debate. Rissanen et al. (1970) reported that no methylation has been observed in soils or sewage sludge. However, Rogers (1977) reported that by using the extract from soils, methylation could be achieved abiotically. This capacity was lost on exposure to UV radiation. At pH < 5 the reaction was more rapid, while under alkaline soils no conversion occurred. In addition, the sandy loam soil had lower methylation rates than the clay loam, so the significance of methyl mercury formation under conditions of land application of industrial waste is unclear.

Vegetation can take up Hg via the root (CAST 1976). Uptake of fungicides may also occur in the leaf area (Huisingh 1973). After absorption, Hg is transported easily throughout the plant (Ross and Stewart 1962, Araki et al. 1965 and Smart 1968). However, the relation between crop response and available land-applied mercury has not been established. Application of 25-37 kg/ha of Hg did not reduce yields of wheat, oats, barley, clover and timothy, nor did it reduce the activity of N-fixation in soils (MacLeod and Howatt 1934). Mercury levels in tomatoes were reported as high as 12 ppm after application of a high mercury sludge to an alkaline soil (Van Loon 1974).

These data are not conclusive with respect to the soil assimilative capacity for Hg. If one uses the average ratio of Hg:N for municipal sludges (Sommers 1977), the application of mercury would be 8-10 kg/ha/yr at 500 kg/ha/yr of N. Assuming that the existing municipal sludge systems are functioning acceptably, this range of mercury loadings would be the approximate assimilative capacity. The upper limit of Hg applications cannot yet be specified, and monitoring of the crop Hg content will be required. It should be noted that the biomagnification of Hg in a plant-soil system is usually less than 3- or 4-fold, while in aquatic systems, values of > 3,000 have been found (Huisingh 1973), which indicates that certain inherent advantages accrue with the use of land treatment systems.

Nickel (Ni)

The requirements for nickel in plants and animals are not completely understood. In plants, Ni is classified as nonessential; however, with a variety of vegetable crops, the foliar addition of nickel (0.01 N $Ni(NO_3)_2$) increased the citrin concentration while fertilizer did not (Lo and Wu 1943, Lo and Chen 1946). Scharrer and Schropp (1933) found that small additions of Ni to water culture or sand soils increased the yield of oats, wheat, rye, barley, maize and peas. Thus any essential nature of Ni is at very low concentrations. With regard to animals, nickel may also be essential, but at very small quantities (Elliot 1977).

In soils, Ni is adsorbed on reactive surfaces such as hydrous oxides of manganese and iron. The environmental impact of Ni is related to the exchangeable or solution levels, rather than to toal soil Ni concentrations. Mitchell (1945) measured the extractable Ni (2.5% acetic acid) on 8 soils in comparison to total Ni. The percentate exchangeable was 2.3-8.3% (average 4.2%) with the highest value for a soil containing the largest amount of natural nickel. This ratio is an approximate relation of acid solution phase to soil level of Ni. For a neutral extracting solution (pH 7.0, ammonium acetate) the range of extractable Ni was 0.09-1.65% (average 0.48%), which may better represent the solution Ni ratio for normal soils.

In addition to surface adsorption in soils, nickel can be chelated with organics, but less strongly than copper (CAST 1976). In the soil, nickel at sufficiently high levels can be toxic to microbial activity (Jackson and Watson 1977). Quantification of such toxic levels is not available, but Parker-Rhodes (1941) reported that in comparison of metal-sulfate salts, Ni = Zn = 2 x Cu for the fungicidal action on *Macrosporium* isolated from clover.

Plants are affected by land application of nickel from industrial wastes. Ni competes with iron in root uptake; thus, in the presence of sufficient iron, nickel phytotoxicity is reduced (Crooke 1955, Nicholas and Thomas 1954). The uptake mechanism and relative effect of organic-bound vs inorganic Ni is not well established. Dekock (1956) showed that mustard plants were protected from Ni toxicity when Ni-chelates were present in comparison to the inorganic species. However, Cunningham *et al.* (1975) concluded that unlike Cu and Zn, that Ni in sewage sludge was more readily taken into plants than as inorganic Ni. Nevertheless, at a certain soil Ni level, plant uptake can lead to phytotoxicity. The uptake of Ni was found to exceed that of Co for pasture crops (Mitchell 1945).

Using a water culture experiment, the solution concentration of Ni that was toxic to horse beans and corn was determined (Haselhoff 1890). All concentrations above 2.0 mg/l were found to be toxic. Using the approximate range of extractable to total Ni in neutral solution described previously, the toxic effect might then exist above about 400 mg Ni/kg dry soil. The relation between acid-extractable Ni in soil and the Ni content of pastures was examined (Mitchell 1945). On similar soils the dry matter content of the plant was 1.0-2.8 times the acid-extracted Ni/g dry soil. Again, if one uses the approximate acid extractable to total Ni, the total Ni soil level to plant dry matter content of Ni is about 25 to 70-fold for pasture species. In another study Mitchell (1945) determined that soils with 4-13 mg acid-extractable Ni/kg dry soil did not inhibit crop growth, while levels of 27-86 ppm were inhibitory. The above studies attempt to relate Ni phytotoxicity to soil available Ni while most studies relate total Ni applied to crop response.

The acceptable level of Ni in plants has not been firmly established. Efforts made to keep crops healthy would require Ni remain less than 50-100 ppm on a dry weight basis (Elliot 1977). Reported levels from agricultural lands are 0.1-5.0 ppm on dry weight basis (Pietz *et al.* 1978b, Souchelli 1969, Mitchell 1945, Mitchell 1967).

Sludge containing nickel was applied to a silty soil at 2.0 and 4.1 kg Ni/ha and the corresponding fescue and alfalfa uptakes measured (Stucky and Newman 1977). No difference in Ni content was observed. Giordano and Mays (1976) applied 4.5 kg Ni/ha as sewage sludge on a silt loam. In edible portions of brocoli, potato, tomato, cucumber, eggplant and string beans there was no increase in nickel concentration with some Ni levels decreasing. In their plant leaves, the Ni was lower on sludge areas than for the control. A range of Ni loadings as sewage sludge was evaluated on a silt loam and a sandy loam (Kelling *et al.* 1977). Up to a rate of 5.2 kg/ha of Ni, rye and sorghum-sudan contained about the same Ni concentrations as the control. Above this application rate sorghum sudan increased with the maximum Ni levels of 2-2.5 ppm for rye and 1.0-1.3 ppm for sorghum sudan at an application of 42 kg/ha of Ni. Corn stover and grain had only slightly elevated Ni contents at a rate of 42 kg/ha of Ni.

Clapp *et al.* (1976) conducted sludge applications in the range of 42-165 kg/ha of Ni. In the leaf tissue of corn there was a nearly linear increase in Ni over this range of application. At 165 kg/ha of Ni, the corn grain contained to about 4.0 ppm Ni dry basis. No significant plant toxicity was evidenced. In another sludge study, Cunningham (1975) found no decrease in corn or rye plant yeilds at 180 kg/ha. Rye tissue concentrations were 32 ppm dry basis while corn was 13.5 ppm.

Management and design of industrial waste land application sites can be used to reduce the food chain effects of Ni. Crop selection can alter the ratio of plant to soil Ni. On two soils (0.32 ppm and 0.60 ppm acetic acid extractable Ni), Timothy and rye grass Ni levels were the same or lower when comparing high level to low. However, red clover doubled over the same range of soil Ni. Thus, grass would be more favorable over clover in assimilatory capacity. Experiments in which acid soils have been limed and the relative Ni uptake measured under both circumstances, have not been performed. However, based on the pH relationships in soil absorption capacity and precipitation reactions, it can be inferred that Ni availability to plants and soil water transport will increase in acid soils. Toxicity of Ni has been reported on acid soils (CAST 1976). A direct measurement of pH effect is the ability to extract Ni from soils with solutions of varying acidity. Mitchell (1945) found the ratio of Ni extracted with pH 4.5 to that with pH 2.5 was

0.26-0.56 while the same ratio comparing pH of 7.0 to that of 2.5 was 0.06-0.2 again demonstrating the advantages of neutral pH.

Land application of industrial wastes containint nickel was practiced and observed by Haselhoff (1890). The waste was applied inadvertently to land as irrigation water and at rates exceeding the assimilative capacity leading to the studies of Haselhoff. Based on the available data described above, it appears that soil accumulations of applied material in the 0-10 kg/ha of Ni pose no hazards. Between 10 and about 200 kg/ha of Ni there are no plant yield effects, but the plant content of nickel continues to increase. However, all reported plant contents are in a presently acceptable Ni concentration range, and site and crop selection and pH control can be used to further reduce nickel burden. Soil levels of 400-800 ppm generally have Ni toxic effects. Thus, as a conservative approach, nickel levels in soils should be allowed to 50, 100 and 200 kg Ni/ha for soil CEC values of 0-5, 5-15 and > 15 meq/100 g, respectively (EPA 1976).

Palladium (Pd)

A study was made of crop response to Pd in which solution culture tests were conducted (Brenchley 1934). Concentrations of 0.033-0.13 mg/l of Pd did not affect the total dry matter yeild of barley and wheat. Oats were more sensitive, while peas grew well up to 0.26 mg/l of Pd and broad beans up to 1.1 mg/l of Pd. Above these levels germination and total dry matter yield were substantially reduced. Thus, crop selection can be used effectively to increase the plant-soil assimilative capacity for palladium. The actual assimilative capacity would have to be developed from laboratory tests in which soil solution levels were measured. Comparison to the work of Brenchley (1934) would bracket the assimilative capacity relative to the solution studies with other metals.

Platinum (Pt)

Experimental data are available for kidney beans and tomatoes grown in sand with applications of H_2PtCl_6 (Hamner 1942). For both vegetables there were no effects on plants grown or behavior when 1.4 ppm Pt or less were added to the soil. At the next level investigated, 14 ppm Pt, inhibition was observed. From substudies, the authors concluded that platinum influenced the intake of Ca and produced the plant effect.

Rubidium (Rb)

Interest in the application of rubidium to land arises from the similar atomic category with potassium and, hence, as a potential K-substitute.

It has been found that many European soils have received a prolonged application of Rb through the use of Chilean nitrate (Dieulafait 1884). The behavior of Rb in the soil has not been specifically addressed, so it can only be speculated that the pathways of K are also related to rubidium.

Response of plants to Rb has been more completely examined. Robinson *et al.* (1917) concluded that crop uptake was not consistent among species, ranging from zero to appreciable uptake. Rb uptake was found to occur primarily at the root tip in barley (Overstreet and Jacobson 1946). When potassium is deficient, Rb has a toxic effect above certain concentrations, which was attributed to blockage of Ca uptake (Richards 1941). Average Rb levels in phanerograms and cryptograms were 20 (1-98) and 120 (2-1,510) mg/kg dry matter (Bertrand and Bertrand 1946). In another study of *Gramineae, Legiminosae, Polygonaceae, Ranunculaceae, Compositae* and *Chenopodiaceae,* the Rb levels on an as-is basis were 1-10 ppm (Borovik-Romanova 1944). The Rb/K ratio in land plants was found to be roughly equal to that ratio in the medium of growth.

Experiments on Rb solution culture represent the behavior of crops in soils when the soil-water extractable concentrations reach similar levels. In potassium-deficient solutions, Rb could only substitute for K at very low levels. Sustained plant growth was observed with about 80 mg/l of Rb with a barley crop. Above this level Rb blocks the uptake of P and, hence, reduces the plant yield. When potassium was not deficient Rb levels up to 107 ppm Rb in solution did not affect the germination or growth of barley, peas or broad beans.

Rubidium added to soils at 0.2 ppm was reported to stimulate growth of cabbage, barley and spinach while little stimulation occurred at 1 ppm (Loew 1903). At a level of 80 ppm Rb a reduction in tobacco dry matter yield was observed (Alten and Goltwick 1933). A plant uptake by beets was determined as 0.25 kg/ha of Rb per growing season (Pfeiffer 1872).

In summary, rubidium, when added to soils with sufficient potassium for good plant growth, does not appear particularly toxic. Assimilative capacities might be estimated at 100 ppm Rb in soil. Specific tests are recommended above this level since greater assimilative rates may be possible.

Silver (Ag)

Because of the intrinsic value of silver, large quantities are not often found in industrial wastes. However, as a trace element, Ag appears in many materials, including superphosphate (10-15 ppm) and sewage sludge (5-150 ppm). Studies of silver in the plant-soil system have been comprehensive in relation to cloud seeding using silver compounds.

In soils, ionic silver is absorbed very strongly. Analysis of soils near cloud seeders used for a number of years have shown levels of 250 ppm (ash basis) in the surface 2 cm with only 0.8 ppm (ash basis) at 8-10 cm. Traditionally, studies of silver toxicity have been limited because of Ag adsorption to all manner of organics and clays (Cameron 1973). In addition, the solubility of the commonly used AgI compound is extremely low. Thus, leaching movement of Ag does not appear to be significant.

One of the primary concerns for Ag is the biological inhibition of ionic silver. Antimicrobial activity is related to inactivation of sulfhydryl groups of proteins (Snodgrass *et al.* 1960). In field tests, the total microbial population was unchanged from the control with additions up to 100 ppm Ag as the nitrate or iodide complex (Klein and Molise 1975). However, some decreased in microbial function (glucose mineralization) was observed at 0.5-0.8 ppm Ag. The authors concluded that the result of Ag addition to soil was to shift the organic matter toward the decomposer microorganism compartment of the ecosystem. By comparison, free silver concentrations of 0.001-0.01 ppm inhibit soil *Arthrobacter* species (Klein and Sokol 1976), again emphasizing the relationship between soil sorption and free Ag in the soil solution.

Crop response and uptake are also of concern with silver applications. Whether plants can take in Ag foliarly and/or via the root structure could not be determined from studies near cloud seeders; however, certain evidence from the Emerald mountain study indicates root absorption as the main uptake mechanism. About 30-45% of Ag found on surrounding vegetation could not be washed off (Teller *et al.* 1976). Background Ag levels in grass and Aspen foliage are 0.05-0.5 ppm (ash basis), while in pine foliage levels of 0.5-1.0 ppm occur.

No effect on yields of four crops (soybean, cron, wheat and sunflower) was measured when treatments of up to 460 ppm Ag (as AgI) were added to a loam and a sand soil (Weaver and Kalrich 1973). Herbage yield of pasture plots was unchanged at 100 ppm Ag (as $AgNO_3$ or AgI). In seedling vigor for lodgepole pine, the addition of 28 ppm Ag (as AgI) reduced emergence (White 1973). As the amount of applied Ag increases the content of the vegetative cover increases, representing an increase input to the food chain (Table 10.7). However, on a relative basis, the amount of silver in the crop compared to that in the soil decreased about 10-fold. Similar results were found in pasture plots receiving the same levels of silver.

The impact of silver in vegetation on animals does not appear to be large. Intake of about 10 ppm Ag (ash basis) by goats did not accumulate nor affect the rumen digestion (Roy and Bailey 1974). The abundance of chloride, proteins and other organic matter precluded the presence of Ag ions, thus lowering toxicity. Similar results were found for rabbits.

Silver has substantially different properties in the organic vs the inorganic form which, unlike Hg or Pb, appear to make similar biomagnification unlikely (Cooper and Jolly 1970, Gmelin 1972). However, quantitative levels

Table 10.7 Relationship of Soil Silver Level and Form to Plant Silver Content in an Aspen Community (Klein and Molise 1975)

Treatment	ppm	Soil[a]			Plant[b]			Ratio of Plant Silver / Soil Silver
		Percent Ash	Silver in Ash	Silver in Soil	Percent Ash	Silver in Ash	Silver in Plant	
Control		51	0.47	.23	10.2	2.00	0.20	0.87
AgNo3	1	37	15.60	5.77	10.8	0.97	0.11	0.02
	10	59	16.50	9.74	10.8	7.50	0.81	0.08
	100	55	200.00	110.00	10.0	60.00	6.00	0.05
AgI	1	46	6.00	2.76	12.6	0.78	0.09	0.03
	10	50	46.00	23.00	11.8	1.00	0.12	0.005
	100	68	181.60	123.49	10.8	6.40	0.69	0.006

[a] Average of data for four separate subplots.
[b] Grouped samples from four sub-plots used in analysis.

to which accumulation could occur are not well-specified. Based on available data, an assimilative capacity of 200 kg/ha of Ag appears reasonable.

Tellurium (Te)

Few data exist for the behavior of Te in plant-soil systems. In culture solutions of wheat with 0-32 ppm Te (K_2TeO_3) and selenium, the growth inhibition and dry weights were consistently lower than for Se additions (Martin 1937). At 1 ppm Te there was essentially no toxic effects while at 2 ppm Te growth inhibition occurred. In rat feeding trails the toxicity effect of Se was more pronounced relative to Te than that found in crop response. Thus, as a conservative estimate, the assimilative capacity for tellurium should be equal to that for Se.

Thallium (Tl)

As a metal, Tl would be expected to participate in such soil processes as cation exchange and precipitation. Detailed studies of Tl behavior in soils are not plentiful. Crafts (1936) determined the saturation of a Yolo clay occurred at 6,800 Tl/kg dry soil. Leaching with 200 m of distilled water had no effect on the initial Tl location nor the resultant toxicity. In separate experiments McCool (1933) concluded that leaching of Tl-treated soils with large quantities of water did not prevent the deleterious effects of Tl. Microbial nitrification in soils was not inhibitied by Tl addition until levels exceeding crop tolerance were reached (McCool 1933).

The crop response and uptake mechanisms have been studied for a number of species and thallium loading rates. Uptake of Tl appears to be via the root with translocation to the stems and cotyledons (Borzini 1935). At sufficient levels, Tl reduces germination and growth rate, and inhibits chlorophyll formation. Crop effects studies have been made in solution cultures as well as soils and these are presented separately.

Tobacco in 1 ppm Tl solution was either killed or slowed in growth with chlorosis and typical frenching symptoms of narrowed thickened leaves (McMurtrey 1932). Spencer (1937) found that $TlNO_3$ and Tl_2SO_4 had similar effects on tobacco in solution culture and that at 0.051 ppm, Tl chlorotic effects were found. At solution concentrations of <0.04 there was only a slight effect on tobacco response, while 0.04 and 0.1 ppm Tl were toxic (Bortner and Karraker 1940). Thus, if the soil-water equlibrium yields a thallium concentration less than 0.04 ppm, then growth response should be unimpaired.

Accumulation of thallium in soils would reach a critical level above which land will not be acceptable for other societal uses. Crafts (1936) reported that 0.003 kg/ha of Tl was very acceptable as an application to agricultural land. Investigations with growth of tobacco in pure sand determined that between 0.5 and 0.65 kg/ha of Tl produced an adverse growth effect (Bortner and Karraker 1940). Such sand had little cation exchange capacity. Field applications up to 9 kg/ha of Tl showed no injurious plant effects and some stimulation was observed (Horn et al. 1936). Bortner and Karraker (1940) also used farm soils with greater organic and exchange content and found that at about 14 kg/ha of Tl, slight chlorosis was detected in tobacco plants. A somewhat lower tolerance to thallium was reported by McCool (1933). At 3.4 kg/ha of Tl, slight retardation of ryegrass, soybean, wheat and alfalfa were reported, and at 14 kg/ha of Tl, serious injury resulted. In laboratory and field studies of vegetables and grasses on loam soils, Horn (1936) detected no vegetative response below 20 kg/ha of Tl. Application of thallous sulfate and 25 kg/ha of Tl reduced pasture grass yield by 50% (Crafts 1936). Complete soil sterilization occurred at 40-45 kg/ha of Tl (Brooks 1932) and at 90 kg/ha of Tl (Crafts 1936).

Management techniques can reduce Tl impact. Crop selection correlated to known resistance to frenching would lead to greater thallium tolerance (Spencer 1937), while the presence of high Ca, K or Al also reduced Tl toxicity. On the soils where Tl is applied in moderate amounts, lime and fertilizer reduced injury to several plants (McCool 1933). These researchers also observed that Tl toxicity was inversely proportional to base exchange capacity.

Assimilative capacity for thallium would appear to be in the range of 5-10 kg/ha of Tl, with some accommodation necessary for the soil CEC and base saturation. As soil moisture accumulates, solution concentrations in excess of 0.04 ppm would inhibit future uses of land application.

Tin (Sn)

For plants and animals tin is a nonessential element. In the soil Sn is largely insoluble, which implies that the impact on vegetation yield or uptake will be very small. The minimal impact of Sn due to the behavior in soils has been documented by researchers and summarized by Romney (1975). Experiments have been conducted on tin in solution culture simulating crop response to soil-water solution and on tin applied to soils.

Cohen (1940) experimented with corn, peas and sunflower and determined that below 5 ppm Sn there was no yield depression in solution culture. The response of $SnCl_2$ and $SnSO_4$ were similar. Recent studies with bush beans in solutions revealed that between 12 and 120 ppm Sn there was a reduction in plant yield, but if $CaCO_3$ were present no yield reduction was found when at 120 ppm Sn. The Sn accumulation occurred in the roots, stems and leaves when Ca was not present although the buildup relative to the control was less in the leaf portion. With Ca present, a condition likely with land application sites, little Sn buildup was found in the stem and leaves (below 120 ppm Sn) but root accumulation did occur.

In two soils of the loam classification, no yield reduction of barley or bush beans were detected with the application of approximately 1,000 kg/ha of Sn (Romney *et al.* 1975). The presence of tin at this level altered the uptake of other metals. In addition, lowering soil pH to 4.5 substantially reduced barley yields at 1,000 kg/ha of Sn as the soil-water solution availability of tin was presumably altered. For the bush beans grown, there was little accumulation of Sn in the stem or leaf. As a result, the authors concluded that despite yield response the Sn foliage content was not affected.

The assimilative capacity recommended for tin in industrial waste land treatment is 1,000 kg/ha of Sn under pH 6-8. As more acid soils are encountered a lower total accumulation should be utilized.

Titanium (Ti)

Throughout the earth crust Ti is the tenth most abundant element, with a tendency to be widely distributed instead of being in large, single deposits. As soil parent material weathers, titanium is found most abundantly in clays, followed by loams, and least in sandy and calcareous soils (Askew 1930). Ti does not appear to be essential (Pratt 1966b) in higher plants. There is no evidence for animals or any living organism that titanium is in any way essential (Underwood 1971).

In soils, Ti is very insoluble and thus would be expected to have little impact on plants. A characteristic of titanium group chemistry is the reactivity with oxidizing agents, hence, TiO_2 is the expected compound in land treatment systems. In the most important oxidation state, +4, TiO_2 is extremely inert. Two water or pure sand culture experiments have been reported. Warington (1946) found that at 0.1 ppm Ti, lettuce dry weight was increased, while Haas and Reed (1927) found a slight stimulation of orange plants at 0.2 ppm Ti.

Field studies of pasture grasses and soils have reported a range of Ti concentrations (Askew 1930). Soils ranged from 3,000-8,400 kg/ha of Ti with corresponding plant values of 7-50 ppm Ti (dry basis). One of the difficulties in comparing soil and plant Ti contents is that the universality of Ti in soil, which makes dust or soil contamination likely. Because plants contain 1,000-10,000 times less Ti than the soil, titanium is used as an index of soil contamination (Mitchell 1948, Barlow *et al.* 1960).

In summary, no quantitative data on adverse effects of high levels of Ti were found. Titanium appears to be very nontoxic. The observation of a very low ratio of plant to soil Ti implies that there is a large assimilative capacity of plant-soil systems for titanium constituents of industrial wastes.

Vanadium (V)

The role of V in plants and animals has not been completely ascertained. Recent evidence suggests that V is essential for chicks and, therefore, may be in other higher animals (Underwood 1971). Vanadium is essential for certain lower plants (green algae) but probably is nonessential for higher plants. Levels of V in soils and consequently in plants are highly variable. Typical grass concentrations are 0.03-0.16 ppm V dry basis (Mitchell 1957) while 62 diverse plant materials were found to contain 0.27-4.2 ppm V, dry basis (Bertrand 1941).

In water culture experiments, V solutions were used to evaluate stimulating or toxic effects on corn (Scharrer 1935). At 0.5×10^{-10} mg/l of V and above there was a 92% reduction in corn dry matter yield and 78% reduction in root weight after 25 days growth. This corn response was approximately the same up to 0.05 mg/l of V, above which severe growth retardation occurred.

The response of wheat, rye, barley, corn, and peas during an initial 15-day growth period was examined with the incorporation of 1.4×10^{-10} to 140 ppm V into pure sand (Scharrer and Schropp 1935). Between 0.14 and 1.4 ppm V, the most sensitive crops, rye and barley, had a substantial yield reductions. For the other species the lower yield was detected between 1.4 and 14 ppm V. Experiments with barley in sand determined that injury occurred at 100 ppm V with VCl_3 but not with NH_4VO_3 (Gericke and Rennenkampff 1939). When the barley was grown in soil at a neutral to alkaline pH there was not adverse growth response at 1,250 ppm V. However, the plant content was not measured with these vanadium applications and so the crop uptake is not known.

Vanadium is not a particularly toxic metal to humans and may play a role in inhibiting pulmonary tuberculosis lesions, depressing chlosterol synthesis and preventing dental caries (Underwood 1971). The main route of toxic contact has been through inhalation of vanadium in atmospheric dusts.

Some early reports on land application of V concerned basic industrial slags and the equivalent vanadium input from that source (Gericke and Rennenkompff 1940). In this form V was tolerated at 10 times greater levels than from the inorganic source. Based on available data it is difficult to determine an assimilative capacity for vanadium. The behavior of V is more toxic in solution, less toxic in pure sand and least toxic in a soil, parallels the land application response of other metals. In terms of neutral soils and tolerant crops, V may be satisfactory at levels of 10-1,000 kg/ha of V, although pilot tests are strongly encouraged for each soil type and crop prior to large-scale practice of vanadium land application.

Zinc (Zn)

In plants and animals, Zn is an essential element (Figure 10.4) and in comparison to most other metals in industrial waste, Zn is the most valuable with respect to crop nutrition. Zinc deficiency from continued cropping is widely identified, suggesting the benefit of zinc application (Viets 1966). Livestock and human deficiencies of Zn are also a prevalent problem (National Academy of Sciences 1974). The essential nature of Zn is based on a role in a number of enzyme systems.

Zinc has been applied to soils for years as a component of superphosphate (500-600 ppm Zn) in normal agricultural operations, thus field experience and data exist for assimilation (Walkley 1940). The divalent zinc form is that taken up by plants and is mobile, especially under acid conditions. In the soil structure, Zn is retained by sorption on clay or on iron oxide sites. Chelation with organic compounds is also an assimilative pathway (CAST 1976). From an analytical and a conceptual viewpoint, zinc exists in three forms in soils. First, Zn can be in water extractable form, *i.e.,* Zn in solution directly available for uptake. As a second form, Zn exists in chelated and easily extractable forms related to inorganic species. In acid soils much of this Zn is in solution while under neutral conditions the zinc is sorbed on exchange sites. The third form is nonreplaceable Zn as insoluble precipitates and tightly bound material. At low applications Zn is primarily in the solution and exchangeable forms while at higher amounts the Zn is more tightly bound. Near pH 7 the ratio of extractable to total Zn at a sludge application site was about 20% (while the control soil was about 3% (Hornick *et al.* 1976). The relationship of adsorbed to total Zn in a forest soils has been determined as a Freundlich and Langmuir expression (Table 10.8) (Sidle and Kardos 1977a). Cadmium

Table 10.8 Constants and Correlation Coefficients for the Freundlich and Langmuir Equilibrium Adsorption Equations (Sidle and Kardos 1977a)

Cation	Time of Equilibrium (hr)	Depth	Freundlich Equation K^a	1/n	r^d	Langmuir Equation K^b	b^c	r
Cu	0.3	0-7.5	386.4	1.00	0.989	-0.38	-0.67	0.411
	0.3	7.5-15	158.1	1.00	0.987	1.31	0.27	0.965
Zn	3.0	0-7.5	138.0	0.94	0.847	0.08	1.57	0.184
	3.0	7.5-15	30.9	0.69	0.991	0.32	0.15	0.976
Cd	3.0	0-7.5	32.4	0.82	0.955	47.19	0.0024	0.921
	3.0	7.5-15	7.4	0.71	0.977	29.96	0.0015	0.966

[a] μg metal adsorbed/g soil at an equilibrium concentration of 1 μg/ml.
[b] $C+ (mg/l)^{-1}$.
[c] mg of metal/g soil.
[d] Significant at the 0.05 and 0.01 levels, respectively.

and Cu results are also given. At an industrial waste application site with toxic plant effects the total soil Zn was about 1,200 kg/ha, while the exchangeable soil Zn was 550 kg/ha (pH 6.2 with 2,100 ppm exchangeable Ca). Another soil receiving this waste had about 1,200 kg/ha of total Zn, but only 14 kg/ha of exchangeable Zn (pH 7.3 with 5,800 ppm exchangeable Ca). The percent exchangeable Zn was 1% in one soil and 46% in the other. These approximate formulae and percentages can be used to interpret metals loading rates and to compare experiments using solution culture and soil media.

Water-soluble Zn was found to be easily leached (Gall 1936). Zinc was reported to be 10-fold more mobile than Cu in a clay loam (Sidle 1977a). For applications of 13 and 28 kg/ha of Zn in sludge, 3-3.2% of the Zn moved beyond 120 cm, although the percolate concentration was less than one-tenth the drinking water concentration standards. These low leach concentrations reflect the adsorption characteristics of Zn and thus are dependent on soil pH. Metals added to soils can influence the microbial population as well as the vegetative cover. The environment impact of this inhibition has not been evaluated, but a principle effect would be an increase in soil organic matter and an adaptation or evolution of microbial community. Chaney *et al.* (1978) found that an 380 ppm Zn, forest respiration was about 60% of control. However, when larger amounts of organics and N are present in a land application system, populations are elevated and the metal effect may be reduced.

Plants take up Zn in the divalent form via the root system and respond more directly to the solution concentration than the total soil concentration. Uptake of chelated Zn has also been documented (Benson *et al.* 1957, Wallace and Romney 1970). Because of the essential nature of Zn, deficiencies can occur with prolonged cropping or leaching. These sites can benefit from Zn addition in the form of industrial land application. Typical vegetation levels are 7-27 ppm Zn dry weight for plant parts deficient in chlorophyll, 40-95 ppm Zn in chlorophyll-rich materials, and 67-330 ppm Zn in cereal and legumes with legumes > cereals (Bertrand and Benzon 1928). Corn grain was reported to have on the average about 20 ppm Zn dry weight (Pietz 1978b). As the level of Zn is increased the plant content increases until toxicity occurs. At toxic levels the Zn content of plant tissue is in the range of 300-500 ppm Zn dry weight (Elliott 1977, CAST 1976). Knowles (1945) found Zn levels of 1,500-2,800 ppm in poorly surviving spring wheat, winter wheat and potato plant when toxic levels of Zn were land applied. However, the grain in this damaged wheat contained only 25 ppm Zn. Staker (1943) reported 900-1,300 ppm Zn dry basis in plants from high Zn peat identified as unproductive soils. Above a certain waste Zn concentration, a foliar burning can also occur due to zinc presence. Dufrenoy and Reed (1934) reported that solutions of 2,670 ppm Zn could burn tree leaves and that copper solutions were even more harsh. Thus, in addition to soil accumulation of Zn, the waste concentration may need to be considered in the assimilative capacity.

The response of plants to solution Zn concentration is often measured in water culture experiments rather than by assessing actual soil solution levels related to applied Zn and resultant crop response. Thus, solution culture data must be extrapolated with equilibrium relations to data available on soil applications of Zn. Researchers using oats (Lundegardh, 1927), peas (Reed 1942), and rice (Tokuoka and Gyo 1939) found that solutions less than 0.5 mg/l of Zn had reduced yields due to deficiency of zinc. At a level of 0.5 mg/l of Zn for rye, barley, corn and peas (Scharrer 1934), 1 mg/l of Zn for rice Tokuoka and Gyo 1939), 0.2-0.4 mg/l of Zn for barley and 2-4 mg/l of Zn for peas (Brenchley 1914) and 1 mg/l of Zn for roses there was a pronounced toxic effect on growth, root development, etc. The solution concentration found not to inhibit enzyme activity (takadiatase) was 15 mg/l of Zn, which is above that inhibitory to plant growth. At high levels, 400-2,000 mg/l of Zn, microbial *(Curvularia ramosa* and *Helminthosporium sativum)* growth was inhibited (Millikan 1938). If zinc applied as industrial wastes to soils is such that soil-water solution concentrations are lower than 0.2-1 mg/l of Zn then plant yield reduction should not occur.

The majority of impact data for zinc has been related to soil concentration or application amounts and crop response. Such information does not separate the effect of soil retention and the ability of the crop to take up or respond to solution levels of zinc. A summary of available data (Table 10.9) demonstrates that for certain soils (low exchange capacity) a toxic effect occurs in the range of 400-700 kg/ha of Zn. On other soils the zinc levels in excess of 2,000 kg/ha produce consistently toxic effects. Because of the large variation in experimental conditions, the assimilative capacity is conservatively estimated below these soil levels and is discussed below. At approximately 650 kg/ha of Zn, Liang and Tabatabai (1978) found the nitrification rate of ammonia to nitrate was 40% of control thus indicating an effect on microbial activity.

As the amount of Zn added to the soil increases there is an increase in crop uptake. Data for corn from severl research studies indicate a relationship at low to moderate application rates but widely differing results at higher application (Figure 10.7). Increases have also been found with vegetables (Giordano 1976b, Hornick 1976a). Thus Zn uptake is enhanced but yield is unaffected; however, the plant Zn content of healthy vegetation is well below any toxic effects to the animal or human food chain (CAST 1976). The relative plant uptake of zinc is reported to be $Zn > Cd > Pb$ (Lagerwerff *et al.* 1977). The daily uptake of zinc by fescue or alfalfa was found to increase until 180-200 days after growth initiation and then to decline (Stucky and Newman 1977).

Table 10.9 Crop Response to Zinc Additions in Soil

Application (kg Zn/ha)	Crop	Response	Reference	Comment
4.3-13	Corn, oats	Yield increase, earlier maturation	Barnette and Camp (1936)	Control soil was Zn deficient (ZnSO₄)
0.4-4	Corn	Beneficial to growth	Javillier and Camp (1912)	
4-8	Wheat	Decreased yield in acid soils	Teakle and Thomas (1939)	(ZnSO₄)
6-12	Wheat, oats	Superior growth relative to control	Millikan (1938)	Counteracted root fungi (ZnSO₄)
62-110	Rye	Little yield reduction relative to control	Lagerwerff et al. (1977)	Sewage sludge limed to pH 6.8 rye grown from seed, immediately after spreading
62-530	Rye	Little yield recution with respect to control	Lagerwerff et al. (1977)	Sewage sludge limed to pH 6.8 rye grown from seed, 7 weeks delay prior to planting
12	–	Reduced Zn deficiency dieback	Millikan (1946)	Highly alkaline soils (ZnSO₄)
40	Tung trees	Increased tung oil content	Bahrt et al. (1944)	(ZnSO₄)
200	Wheat	No affect on yield	Voelcker (1913)	As $ZnPO_4$, $Zn(NO_3)_2$, $Zn(CO_3)_2$
313	Alfalfa, fescue	Yield increase due to additional macronutrients	Stucky and Newman (1977)	Sewage sludge
350-700	Oats	Good yields relative to control when crop nutrients added	Lundegardth (1927)	Zn from ore roasting stack gases
400	Mustard	Growth ceased	Ghedroiz (1914)	$ZnSO_4$, $ZnCl_2$
400	Barley	Relatively good yield	Ghedroiz (1914)	$ZnSO_4$, $ZnCl_2$
400 (acid extractable)	–	Decreased plant growth	Jones et al. (1936)	Norfolk fine sand, $CaCO_3$ 1,100 kg/ha alleviated toxicity
500	Cow peas	Toxic effect above this level	Gall (1936)	Norfolk fine sand (ZnSO₄)
700	Corn	Toxic effect above this level	Gall (1936)	Norfolk fine sand (ZnSO₄)
555-2,175	Corn	No yield effect	Clapp et al. (1976)	Sewage sludge
400	Wheat	Promoted growth	Tokuoka and Gyo (1940)	Loamy soil pH 6.7 (ZnSO₄)
600-800 (acid extractable)	Corn	Toxic effect	Barnette (1937)	Orangeburg, Greenville and Norfolk soils

Table 10.9 Continued.

Application (kg Zn/ha)	Crop	Response	Reference	Comment
1,200 (14 exchangeable)	Wheat	Good yields	Knowles (1945)	Foundry waste (pH 7.3)
1,200 (550 exchangeable)	Wheat	Severe growth inhibition	Knowles (1945)	Foundry waste (pH 6.2)
1,390	Corn	No effect	Chesnin (1967)	Acid and alkaline soils
2,000	Rice, wheat	Toxic action evident	Tokuoka and Gyo (1940)	–
2,600	Grass	Toxic response	Mejier and Goldewaagon (1940)	Galvanized metal contamination (ZnO)
4,800-8,000	Oats	No adverse effect	Lott (1938)	(ZnO) silt loam neutral pH
8,600	Vegetable crops	Nonproductive soil	Staker (1942)	Naturally occurring high Zn peat

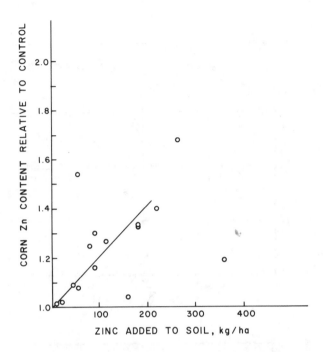

Figure 10.7   Corn uptake of Zn from sludge applications to land (Baker *et al.* 1976, Decker *et al.* 1976, Giordano *et al.* 1975, Kelling *et al.* 1976 and Singh *et al.* 1976.

Management of an industrial waste land application site can substantially increase the plant:soil assimilative capacity for zinc. Control of pH from acid to neutral greatly affects Zn availability in comparison to Cd, Cu or Ni. Addition of various calcium compounds to raise the pH has been shown to overcome the toxic effects of moderate levels of exchangeable Zn (Gall 1936, Lott, 1938, Knowles 1945). The pH effect on increasing Zn availability for plant uptake and hence plant toxicity is evident from Figure 10.8. At pH 6.7 and greater, the plant yields and Zn plant concentrations were very similar to the control. Below this level adverse responses were noted (Lott 1938).

Crop selection can also reduce the adverse effects of zinc by using more tolerant species. A partial listing by tolerance is as follows: mustard, buckwheat, sugar beet < peas, beans < grasses (Knowles 1945, Edgerton *et al.* 1975, Palazzo and Duell 1974).

Industrial wastes containing substantial Zn have been land-applied, although often without regard for the plant-soil assimilative capacity. When compared to $ZnSO_4$ or ZnO, use of stripping acid residue ($Zn_3(PO_4)_2$) or

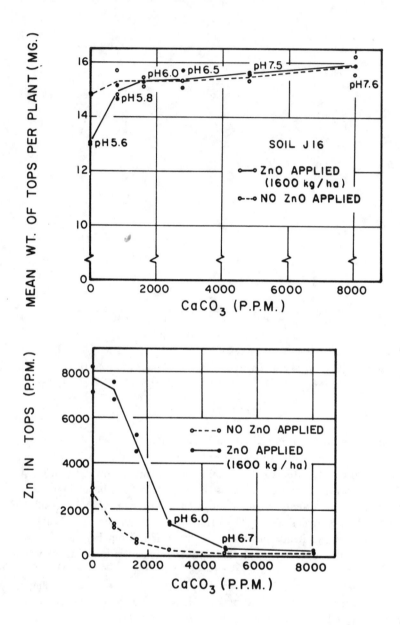

Figure 10.8 Effect of pH adjustment on weight of oat tops and Zn content of oats (Lott 1938).

zinc granules gave similar crop results; while blast furnace slag or zinc silicates were largely unavailable to plants (Lipman and Sommers 1928, Bown *et al.* 1957). Excessive application of calamine spoils (Griffith 1919), zinc ore roasting airborne fallout (Ludengardh 1927), munition foundry dross (Knowles 1945) and mine tailings (Davies 1941) have led to long-term zinc contamination of former agricultural soils. Thus, experience shows to utilize zinc application only up to the soil assimilative capacity.

The assimilative capacity for zinc is based on soil capacity to fix this metal, thus reducing availability to the plant. As proposed by the EPA in view of USDA research guildance, these are 250, 500 and 1,000 kg/ha of Zn for soils with a CEC of 0-5, 5-15 and >15 meq/100 g soil, respectively.

ENGINEERING UTILIZATION OF METAL ASSIMILATIVE CAPACITY IN INDUSTRIAL LAND APPLICATION

In all wastes and even in irrigation water there are metals and, hence, the differences are minor with respect to land application of a variety of materials. Industrial wastes represent a continuum in metal loadings so that the presence of metals does not rule out a particular waste from land application. Municipal sludge is one point on this continuum. Metals must be considered individually with respect to both the waste generation rate (kg/yr) and the assimilative capacity at a specific site or plant-soil system (kg/ha/yr). These two factors determine the area requirements for an environmentally acceptable land receiver, since the basic constraint of nondegradation is already built directly into the design process. If a metal, when compared to the other constituents present (Chapters 4-11), is the LLC, only then are metal loads the controlling factor. The required land area for a particularly high metal waste may be large and possibly uneconomical, but until compared on a cost basis to other waste management alternatives, land application should not be excluded. The statement "Waste X contains heavy metals and therefore is unsuitable for land application" does not recognize the continuum in assimilative capacity nor the potential for land treatment technology.

The information for the specific metals must be used in the context of the overall design consideration of waste and site-specific characteristics to complete the LLC analysis. Typical information for each metal is the amount which can be land applied to accumulate to a critical soil level, which will still allow use of the same area for certain reasonable societal purposes. However, allowances for all possible land use are not made.

The critical soil level is expressed as kg/ha of metal, which is actually kg/ha of metal—15 cm depth or mass per unit volume of soil. A management decision is necessary to determine the desired lifetime of the land treatment area. Apportioned over this lifetime one arrives at the kg/ha/yr for metal assimilative capacity. It is usually better to apply these metals on numerous occasions over the area lifetime rather than exhausting the plant-soil capacity in a single application. In this manner crop effects, soil pH shifts, anaerobic tendency, etc. are substantially reduced, enhancing the environmental benefits of land application.

Several design and operational options can further improve the plant-soil assimilative capacity. The design metal loading rates are expressed as kg/ha— 15 cm/yr, which is mass per unit soil volume per unit time. If, as is quite feasible, one can incorporate the industrial waste in the upper 30, 45 or 60 cm, then the allowable capacity is increased 2, 3 or 4 times the preliminary design criteria. This will reduce the land area requirements for the metal constituent of a waste, but only if metals are the LLC will the total waste area requirements be reduced. Use of deep incorporation is most feasible with solid or slurry wastes since incorporation or plowing is frequently a required step.

Another option in industrial waste land application systems is monitoring to allow feedback correction of design criteria. The long-term test of metals assimilation is the tolerance and low accumulation (within FDA guidelines) of the vegetative cover. Thus, the more tolerant plant species should be used and monitoring must be practiced, because present design criteria are imprecise and conservative. As one approaches the design lifetime of a waste application site, the vegetative cover may allow further metals application; the crop metals content analysis will indicate whether further loading is feasible. Ultimately, because of the conservative and accumulative nature of metals, these waste constituents will require a change of site; although the time frame may range from 10 to 1,000+ years, depending on the waste and soil properties.

RESEARCH NEEDS

Within the agencies and private firms concerned with land application of waste, certain research needs exist relative to the assimilation of metals in plant-soil systems. The principal focus of work is the basic mechanisms involved in the interaction of a larger variety of metals with different soil and crop types. Present emphasis on five principal metals, because of prevalence or impact in municipal sludge, falls short of the real needs. The present approach does not allow much predictive capacity, since it is oriented toward a specific waste-municipal sludge, and not at the constituents common to many wastes.

More information is needed on the micro-level for such phenomena as a ratio of solution metal to total soil metal, on the crop tolerance and the uptake mechanism, and on long-term behavior. Such data are preferred over experiments in which with various masses of waste material per unit area are applied to land, since the former approach avoids duplicating land-based treatment experiments for every waste and soil type.

Brown (1975) raised following specific questions, which certainly warrant additional experiments:

1. What is the mechanism that has prevented or delayed metal toxicity conditions in some sludge-amended soil, even though the accumulated metal in the soils is greater than the quantity of metal salt that would be expected to cause toxicity conditions?
2. Will the protective mechanism function regardless of the metal accumulation in the soil?
3. Will the protective mechanism cease when sludge applications to soils are terminated, or is there a residual effect of the sludge that will continue to protect plants from metal toxicity until the time when the applied metal has reverted to chemical forms unavailable for plant uptake?
4. What is the tolerance of various crops and of various crop varieties to metal accumulations in the soil?

REFERENCES

Agr. Gaz. N. S., Wales. "Manganese Causing Bare Patches in the Soil," *Agr. Gaz. N.S., Wales, V.* 22(1):70 (1911).

Allaway, W. H. "Agronomic Controls over the Environmental Cycling of Trace Elements," *Adv. Agron.* 20:235-271 (1968).

Alten, F., and R. Gottwick. "A Contribution to the Question of the Substitution of Rubidium and Caesium for Potassium in Plant Nutrition," *Ernahr. Pflanze.* 29:393-399 (1933).

Andersson, A. "Heavy Metals in Swedish Soils: On their Retention, Distribution, and Amounts," *Swed. J. Agric. Res.* 7:7-20 (1977a).

Andersson, A. "Some Aspects on the Significance of Heavy Metals in Sewage Sludge and Related Products used as Fertilizers," *Swed. J. Agric. Res.* 7:1-5 (1977b).

Andersson, A. "The Distribution of Heavy Metals in Soils and Soil Materials as Influenced by the Ionic Radius," *Swed. J. Agric. Res.* 7:79-83 (1977c).

Andouard, A., and P. Andouard. "The Fertilizing Action of Manganese. *Engrais, V.* 26(33):915-916 (1911); *Chem. Abs.* 6(3):403 (1912).

Antonovics, V. A., A. D. Breadshaw and R. G. Turner. "Heavy Metal Tolerance in Plants," *Adv. Ecol. Res.* 7:1-85 (1971).

Araki, T., S. Toyota, Y. Mizusawa and N. Suzuki. "Penetration and Translocation of Mercury in Rice Plants Sprayed with 203 Hg Labeled Phenyl Mercuric Acetale and the Accumulation of 203 Hg in Various Organs of Rats Fed with 203 Hg Containing Rice Grains," in *Residue Analysis of Organo-Mercuric Fungicides Sprayed on Rice and Fruits,* Japan Plant Protection Assoc., Tokyo (1965).

Arnon, D. I. "Ammonium and Nitrate Nitrogen Nutrition of Barley at Different Seasons in Relation to Hydrogen Ion Concentration, Manganese, Copper, and Oxygen Supply," *Soil Sci.* 44:91-121 (1937).

Askew, H. O. "Titanium in some New Zealand Soils and Pastures," *New Zealand J. Sci. Technol.* 12:173-179 (1930).

Aubert, H., and M. Pinta. "Trace Elements in Soils," *Develop. Soil Sci.* 7:135 (1977).

Bahrt, G. M., R. Jones, E. Angelo, A. F. Freeman, F. C. Pack and R. S. McKinney. "The Effects of Zinc and Other Trace Elements on Oil Content of Tung Fruits," *Proc. 10th Ann. Conv. Am. Tung Oil Assoc. and United Tung Growers' Assoc.* (1944), pp. 98-101.

Baker, D. E., M. C. Amacher and W. T. Doty. "Monitoring Sewage Sludges, Soils, and Crops for Zinc and Cadmium," paper presented at the 8th Annual Cornell Conference, Rochester, NY (1976).

Balin, D. W. "The Manganese Contents of Grasses and Alfalfa from Grazed Plots," *J. Agri. Res.* 48:657-63 (1934).

Barber, S. A. "Influence of the Plant Root on Ion Movement in Soil," in *The Plant Root and Its Environment,* Vol. 1, E. W. Carson, Ed., (Charlottesville, VA: University Press of Virginia, 1974), pp. 525-554.

Barlow, R. M., D. Purves, E. J. Butler and I. V. McIntyre. "Swayback in South-East Scotland," *J. Comp. Pathol. Ther.* 70:396-410 (1960).

Barnette, R. M. "The Occurrence and Behavior of Less Abundant Elements in Soils," *Fla. Agric. Exp. Sta., Ann. Rep.* 61 (1936).

Barnette, R. M., and J. P. Camp. "Chlorosis" in Corn Plants and Other Field Crop Plants," *Fla. Agric. Exp. Sta., Ann Rep.* 45 (1936).

Bartmann, H. "Manganese in Field Experiments," *J. Agric. Prat., n. Ser.* 20(47):666-667 (1910).

Baumhardt, G. R., and L. F. Welch. "Lead Uptake and Corn Growth with Soil-Applied Lead," *J. Environ. Qual.* 1:92-94 (1972).

Beavington, F. "Some Aspects of Contamination of Herbage with Copper, Zinc, and Iron," *Environ. Poll.* 8:65-71 (1975).

Benson, N. R., L. P. Batjer and I. C. Chmelir. "Response of Some Deciduous Fruit Trees to Zinc Chelate," *Soil Sci.* 84:63-75 (1957).

Bertrand, G. "On the Successful Use of Manganese as a Fertilizer," *Comp. Rend. Acad. Sci. (Paris)* 141(26):1255-1257 (1905).

Bertrand, D. "Distribution of Vanadium in Plants," *Comp. Rend. Acad. Sci.* 212:1170-1172 (1941), *Chem. Znetr.* I:1891 (1942).

Bertrand, G., and D. Bertrand. "Rubidium in the Cryptogams," *Comp. Rend. Acad. Sci.* 222:572-574.

Bertrand, G., and B. Benzon. "The Content of Zinc in the Principal Foods of Vegetable Origin," *Bull. Soc. Sci. Hyg. Ailment* 16(10):457-463 (1928).

Bertrand, D., and A. DeWolf. "Requirement for the Trace Element Chromium in the Growth of Potatoes," *Comp. Rend. Acad. Sci. Ser. D.* 266:1494-1495 (1968).

Bonner, W. P., and R. P. Bustamonte. "Behavior of Mercury in Suspended Solids and Bottom Sediments," Water Resources Research Center, Report 50, University of Tennessee, Knoxville, TN (1976).

Borovik-Romanova, T. F. "The Content of Rb in Plants. II. *Dokl. Akad. Nauk. S.S.S.R.* 44:313-316; *Comp. Rend. Acad. Sci. U.R.S.S.* 44:258-288 (1944).

Bortner, C. E., and P. E. Karraker. "Frenching of Tobacco, with Particular Reference to Thallium Toxicity," *J. Am. Soc. Agron.* 32:195-203 (1940).

Borzini, G. "The Influence of Thallium Ions on the Germination of Several Seeds, and on the Initial Growth of the Young Plants," *Boll. Staz. Patol. Vegetale, (N.S.)* 15:200-231 (1935).

Bouwer, H., J. C. Lance and M. S. Riggs. "High-Rate Land Treatment II. Water Quality and Economic Aspects of the Flushing Meadows Project," *J. Water Poll. Control Fed.* 46:844-849 (1974).

Bowen, H. G. M. *Trace Elements in Biochemistry* (New York: Academic Press, Inc., 1966).

Bown, L. C., F. Viets, Jr. and C. L. Crawford. "Plant Utilization of Zinc from Various Types of Zinc Compounds and Fertilizer Materials," *Soil Sci.* 83:219-229 (1957).

Brady, N. C. *Nature and Properties of Soils,* 8th Ed. (New York: MacMillan Publishing Co., 1974).

Breeze, V. G. "Land Reclamation and River Pollution Problems in the Croal Valley Caused by Waste from Chromate Manufacture," *J. Appl. Ecol.* 10(2):513-525 (1973).

Brenchley, W. E. "The Action of Certain Compounds of Zinc, Arsenic, and Boron on the Growth of Plants," *Ann. Bot. (London)* 28(110):283-301 (1914).

Brenchley, W. E. "The Effect of Rubidium Sulfate and Palladium Chloride on the Growth of Plants," *Ann. Appl. Biol.* 21:398-417 (1934).

Brewer, R. F. "Lead," in *Diagnostic Criteria for Plants and Soils,* H. D. Chapman, Ed., University of California, Division of Agricultural Science, (1966), pp. 213-217.

Bringmann, G., and R. Kuhn. "Water Toxicological Studies with Protozoans as Test Organisms," *Gesundheits-Ing.* 80:239 (1959).

Broadbeut, F. E., and J. B. Ott. "Soil Organic Matter-Metal Complexes. I. Factors Affecting Retention of Various cations," *Soil Sci.* 83:419-427 (1957).

Bromfield, S. M. "The Solution of γ-MnO_2 by Substances Released from Soil and from the Roots of Oats and Vetch in Relation to Manganese Availability," *Plant Soil* 10(2):147-159 (1958).

Bromfield, S. M., and D. J. David. "Sorption and Oxidation of Manganous Ions and Reduction of Manganese Oxide by Cell Suspensions of a Manganese Oxidizing Bacterium," *Soil Biol. Biochem.* 8(1):37-43 (1976).

Brooks, S. C. "Thallium Poisoning and Soil Fertility," *Science* 75:105-106 (1932).

Brown, R. E. "Significance of Trace Metals and Nitrates in Sludge Soils," *J. Water Poll. Control Fed.* 47(12):2863-2875 (1975).

Brown, P. E., and G. A. Minges. "The Effect of Some Manganese Salts on Ammonification and Nitrification," *Soil Sci.* 2(1):67-85 (1916).

Cameron, D. R. "Modeling Silver Transport in the Soil," Ph.D. Thesis, Colorado State University, Fort Collins (1973), p. 116.

CAST (Council for Agricultural Science and Technology). "Application of Sewage Sludge to Cropland: Appraisal of Potential Hazards of the Heavy Metals to Plants and Animals," Report No. 64, prepared for Office of Water Program Operations, EPA-430/9-76-013, Washington, D.C. (1976).

Chaney, R. L. "Crop and Food Chain Effects of Toxic Elements in Sludges and Effluents," in *Recycling Municipal Sludges and Effluents on Land*, EPΛ, ORD (1973) pp. 129-143.

Chaney, R. L. National Program Staff; Soil, Water, and Air Sciences, Beltsville, Md., Unpublished results (1974).

Chaney, R. L. "Metals in Plant-Absorption Mechanisms, Accumulation and Tolerance," in *Metals in the Biosphere*, Procedure of a Symposium by the Department of Land Resource Science, University of Guelph, Ontario, Canada (1975).

Chaney, W. R., J. M. Kelley and R. C. Strickland. "Influence of Cadmium and Zinc on Carbon Dioxide Evolution from Litter and Soil from a Black Oak Forest," *J. Environ. Qual.* 7(1):115-118 (1978).

Chapman, H. D., G. F. Liebig, Jr. and E. R. Parkes, "Manganese Studies: California Soils and Citrus Leaf Symptoms of Deficiency," *Calif. Citrograph, V.* 24:427, 454; 24:11, 15 (1939).

Chesnin, L. "Corn, Soybeans and other Great Plains Crops," *Micronutr. Manual Farm Tech.* 23(6) (1967).

Clapp, C. E., R. H. Dowdy and W. E. Larson. Unpublished data. Agricultural Research Service, U. S. Department of Agriculture, St. Paul, MN (1976).

Clausen, H. "Manganese and Stimulation in the Soil," *Deut. Landw. Presse.* 39(97):113-1132 (1912).

Cohen, B. B. "Some effects of Stannous Sulphate and Stannic Chloride on Several Herbaceous Plants," *Plant Physiol.* 15(4):755-760 (1940).

Cooper, C. F., and W. C. Jolly. "Ecological Effects of Silver Iodide and Other Weather Modification Agents; A Review. *Water Resources Res.* 6(1):88-98 (1970).

Crafts, A. S. "Some Effects of Thallium Sulfate Upon Soils," *Hilgardia* 10:377-398. (1936).

Crecelius, E. A., C. J. Johnson and G. C. Hofer. "Contamination of Soils Near a Copper Smelter by Arsenic, Antimony and Lead," *Water, Air, Soil Poll.* 3:337-342 (1974).

Crooke, W. M. "Further Aspects of the Relationship Between Nickel Toxicity and Iron Supply," *Ann. Appl. Biol.* 43(3):465-476 (1955).

Cunningham, J. D., D. R. Keeney and J. A. Ryan. "Phytoxicity and Uptake of Metals Added to Soils as Inorganic Salts or in Sewage Sludge," *J. Environ. Qual.* 4:460-462 (1975).

Davies, G. N. "Effect of Zinc Sulfate on Plants," *Ann. Appl. Biol.* 28:81-84 (1941).

Davies, B. E., and L. J. Roberts. "Heavy Metals in Soils and Radish in a Mineralized Limestone Area of Wales, Great Britain," *Sci. Total Environ.* 4:249-261 (1975).

Decker, A. M., R. L. Chaney and D. C. Wolf. "Effects of Sewage Sludge and Fertilizer Applications on Yields and Chemical Composition of Corn and Soybeans," in *Crop and Soils Report 1975* (College Park: University of Maryland, Department of Agronomy, 1976).

Dekock, P. C. "Heavy-Metal Toxicity and Iron Chlorosis," *Ann. Bot. (London) (N.S.)* 20:133-141 (1956).

deKoning, H. W. "Lead and Cadmium Contamination in the Area Immediately Surrounding a Lead Smelter," *Water, Air, Soil Poll.* 3:63-70 (1974).

Dieulafait. *"Agricultural Chemistry," Comp. Rend. Acad. Sci., (Paris)* 98 (1884).

Drucker, H., R. E. Wildung, R. T. Garland and M. P. Fujihara. "Influence of Seventeen Metals on Microbial Populations and Metabolism in Soil," *Agron. Abst., Div.* S-3:90 (1974).

Dufrenoy, J., and H. S. Reed. "Pathological Effects of the Deficiency or Excess of Certain Ions on the Leaves of Citrus Plants," *Ann. Agron. (N.S.)* 4:637-653 (1934).

Edgerton, B. R., W. E. Sopper and L. T. Kardos. "Revegetating Bituminous Strip Mine Spoils with Municipal Wastewater," *Compost Sci.* 16(4):20-25 (1975).

Elliot, F. F., Ed. "Soils for Management of Organic Wastes and Waste Waters," (Madison, WI: Soil Science Society of America, ASA and C. SSA, 1977), p. 635.

Fiskel, J. G. A., and P. J. Westgate. "Copper Availability in High Copper Soils," *Fla. Agric. Expt. Stn. J. Ser.* 424:5 (1955).

Gall, O. E. "Zinc Sulfate Studies in the Soil," *Citrus Ind.* 17(1):20-21 (1936).

Gemmell, R. P. "Use of Waste Materials for Revegetation of Chromate Smelter Waste," *Nature (London)* 240:569-571 (1972).

Gericke, S. "The Effect of the Trace Element Cr on Plant Growth," *Bodenk. Pflanzenernahr.* 33:114-129 (1943).

Gericke, S., and E. V. Rennenkampff. "Effect of the Trace Element Vanadium on Plant Growth," *Prakt. Blat. Pflanzenbau Pfanzenchutz.* 17:1722 (1939).

Gericke, S., and E. V. Rennenkampff. "The Effect of Vanadium on Plant Growth," *Bodenk. Pflazenernahr.* 18:305-315 (1940).

Ghedroiz, K. "The Influence of Zinc Vessels in Culture Experiments," *Selsk. Khoz. i. Liesov.* 345:625-627 (1914).

Giordano, P., and D. A. Mays. "Effect of Land Disposal Applications of Municipal Wastes on Crop Yields and Heavy Metal Uptake," EPA Document (1976a).

Giordano, P., and D. A. Mays. "Yield and Heavy Metal Content of Several Vegetable Species Grown in Soil Amended with Sewage Sludge," in *Biological Implications of Metals in the Environment,* 15th Ann. Hanford Life Sci. Symp., Richland, WA (1976b).

Giordano, P. M., J. J. Mortvedt and D. A. Mays. "Effect of Municipal Wastes on Crop Yields and Uptake of Heavy Metals," *J. Environ. Qual.* 4:394-399 (1975).

Gmelin, L. "Handbuch der Anorganischen Chemi," 8th ed. Weinheim/Bergstrasse, West Germany: Verlag Chemi GMbH., (1972), System Nr. Bl.

Griffin, R. A., and N. F. Shimp. "Effect of pH on Exchange Adsorption or Precipitation of Lead from Landfill Leachates by Clay Minerals," *Environ. Sci. Technol.* 10(3):1256-1261 (1976).

Griffith, J. J. "Influence of Mines upon Land and Livestock in Cardiganshire," *J. Agric. Sci.* 9:366-395 (1919).

Haas, A. R. C., and H. S. Reed. "Significance of Traces of Elements not Ordinarily Added to Culture Solutions, for Growth of Young Orange Trees," *Bot. Gaz.* 83:77-84 (1927).

Hamence, J. H., and G. Taylor. "Experiments on the Toxicity of Chrome-Bearing Fertilizer to Plants Including a Brief Survey of the Literature on Chromium Toxicity," *Fert. Feed. Stuffs J.* 34:449-453 (1948).

Hamner, C. L. "Effects of Platinum Chloride on Bean and Tomato," *Bot. Gaz.* 104(1):161-166 (1942).

Haselhoff, E. "Experiments on the Injurious Effect on Vegetation of Water Containing Nickel," *Landw. Jahrb.* 22(6):862-868 (1890).

Hassler, A. "Retention of Copper in Soil," *Mitt. Lebensm. Hyg. (Bern)* 34(1-2):79-90 (1943).

Hewitt, E. J. "Metal Interrelationships in Plant Nutrition. I. Effects of Some Metal Toxicities on Sugar Beet, Tomato, Oat, Potato, and Marrow Stem Kale Grown in Sand Culture," *J. Exp. Bot.* 4:59-64 (1953).

Hinesly, T. D. Unpublished results.

Hodgson, J. F. "Chemistry of Micronutrient Elements in Soils," *Adv. Agron.* 15:120-159 (1963).

Hodgson, J. F. "Chemistry of Trace Elements in Soils with Reference to Trace Element Concentration in Plants," in *Trace Substances in Environmental Health,* Vol. III, D. D. Hemphill, Ed. (University of Missouri Press, 1970), pp. 45-58.

Horn, E. E., J. C. Ward, J. C. Munch and F. E. Garlough. "The Effect of Thallium on Plant Growth," *U.S. Dept. Agric. Cir.* 409:8 (1936).

Hornick, S. B., R. L. Chaney and P. W. Simon. Unpublished results (1976).

Huisingh, D. "Agri-Ecology of Heavy Metals in Plant Disease Relationships," paper presented at the 2nd International Congress Plant Pathology, Minneapolis, MN (1973).

Jackson, D. R., and A. P. Watson. "Disruption of Nutrient Pools and Transport of Heavy Metals in a Forested Watershed Near a Lead Smelter," *J. Environ. Qual.* 6(4):331-339 (1977).

Javillier, M. "The Use of Zinc as a Catalytic Fertilizer, *Orig. Commun.* 8th Internatl. Congress Applied Chemical, Washington and New York, 15 (Sec. Vii):145-146 (1912).

Jenne, E. A. "Controls on Mn, Fe, Co, Ni, and Zn Concentrations in Soils and Water: The Significant Role of Hydrous Mn and Fe Oxides," in *Trace Inorganics in Water,* R. A. Baker Ed. *Adv. Chem. Ser.* 73:337-387 (1968).

Jenny, H., and R. Overstreet. "Surface Migration of Ions and Contact Exchange," *J. Phys. Chem.* 43:1185-1196 (1939).

Jones, H. W., O. E. Gall and R. M. Barnette. "Reaction of Zinc Sulfate with the Soil," *Fla. Agric. Expt. Sta., Bull.* 298:5-43 (1936).

Jones, R. L., T. D. Hinesty and E. L. Ziegler. "Cadmium Content of Soybeans Grown in Sewage Sludge Amended Soil," *J. Environ. Qual.* 2:351-353 (1973).

Jurinak, J. J., and J. Sontillan-Medrano. "The Chemistry and Transport of Lead and Cadmium in Soils," *Research Report* #18, Agricultural Experimental Station, Utah State University, Logan, UT (1974).

Keaton, C. M. "Influence of Lead Compounds on the Growth of Barley," *Soil Sci.* 43:401-411 (1937).

Keeney, D. R. "Environmental Impact of Cadmium and Other Heavy Metals from Land Applied Sewage Sludge," Water Resources Center, University of Wisconsin, Madison, WI, *Tech. Rep. WIS,* WRC 76-03 (1976).

Kelling, K. A., D. R. Keeney, L. M. Walsh and J. A. Ryan. "A Field Study of the Agricultural Use of Sewage Sludge: III. Effect on Uptake and Extractability of Sludge Borne Metals," *J. Environ. Qual.* 6(4):352-358 (1977).

Klein, D. A., and E. M. Molise. "Ecological Ramifications of Silver Iodide Nucleating Agent Accumulation in a Semiarid Grassland Environment," *J. Appl. Meteorol.* 15(4):673-680 (1975).

Klein, D., and R. A. Sokol. "Evaluation of Ecological Effects of Silver Iodide Seeding Agent," in *Ecological Impacts of Snowpack Augmentation in the San Juan Mountains,* H. W. Steinhoff and J. D. Ives, Eds., Final Report, Colorado State University Publication, Fort Collins, CO (1976).

Knowles, F. "The Poisoning of Plants by Zinc," *Agric. Prog.* 20:16-19 (1945).

Koenig, P. "Stimulative and Depressive Effects of Chromium Compounds on Plants," *Chem. Zeit.* 35(49):442-443; 35(51):462-463 (1911).

Lag, J., O. O. Hvatum and B. Bolviken. *Nor. Geol. Unders. (Skr.)* 266:141-159 (1969).

Lagerwerff, J. V. "Lead, Mercury and Cadmium as Environmental Contaminants," in *Micronutrients in Agriculture,* (Madison, WI: Soil Science Society of America, Inc., 1972), pp. 593-636.

Lagerwerff, J. V., G. T. Biersdorf, R. P. Milberg and D. L. Brower. "Effects of Incubation and Limig on Yield and Heavy Metal Uptake by Rye from Sewage-Sludged Soil," *J. Environ. Qual.* 6(4):427-430 (1977).

Landa, E. R. "Microbial Aspects of the Volatile Loss of Applied Mercury (II) from Soils," *J. Environ. Qual.* 7(1):84-86 (1978).

Leeper, G. W. "Reactions of Heavy Metals with Soils with Special Regard to Their Application in Sewage Wastes," Contract No. DACW 73-73-C-0026. Department of Army Corps of Engineers, Washington, DC (1972).

Liang, C. N., and M. A. Tabatabai. "Effects of Trace Elements on Nitrification in Soils," *J. Environ. Qual.* 7(2):291-293 (1978).

Lindsay, W. L. "The Chemistry of Metals in Soils," in *Metals in the Biosphere,* Procedure of a Symposium by the Department of Land Resource Scientists, University of Guelph, Guelph, Ontario, Canada (1975).

Lipman, C. B., and A. L. Sommers. "Zinc and Boron for Plants," *Science* 67:12 (1928).

Lisk, D. J. "Trace Metals in Soils, Plants, and Animals," *Adv. Agron.* 24:267-325 (1972).

Lo, T-Y., and C-H. Wu. "The Effect of Various Elements on the Formation of Vitamin P in Germinated Mung Beans," *J. Chinese Chem. Soc.* 10:58 (1943).

Lo, T-Y, and S-M Chen. "Vitamin P Content of Vegetables as Influenced by Chemical Treatment," *Food Res. V.* 11:159-162 (1946).

Loew, C. "The Physiological Effect of Rubidium Chloride on Plants," *Bull. Coll. Agric., Tokyo Imp. Univ.* 5(4):461-465 (1903).

Lott, W. L. "The Relation of Hydrogen-Ion Concentration to the Availability of Zinc in Soil," *Soil Sci. Soc. Am. Proc.* 3:115-121 (1938).

Lundegardh, H. "The Importance in the Development of plants of the Quantities of Zinc and Lead Added to the Soil from Smoke Gases," *Meddel. Centralanst. Forsoksv. Jordbruksomradet (Sweden)* 326:14 (1927).

MacLeod, D. J., and J. L. Howatt. "Soil Treatment in the Control of Certain Soil Borne Diseases of Potatoes," *Am. Potato J.* 11:60-61 (1934).

Martin, A. L. "A Comparison of the Effects of Tellurium and Selenium on Plants and Animals," *Am. J. Bot.* 24:198-203 (1937).

McCool, M. M. "Effect of Thallium Sulfate on the Growth of Several Plants and on Nitrification in Soils," *Contrib. Boyce Thompson Inst.* 5(3):289-296 (1933).

McMurtrey, J. E., Jr. "Effect of Thallium on Growth of Tobacco Plants," *Science,* 76(1960):86 (1932).

MacLeod, D
Centralanst.

Meijer, C., and M. A. J. Goldewaagen. "A Case of Zinc Poisoning due to Galvanized-Iron Sheeting," *Landbouwk. Tkjdschr.* 52:17-19 (1940).

Melsted, S. W. "Soil-Plant Relationships," in *Proc. Recycling Municipal Sludges and Effluents on Land,* EPA, Washington, DC (1973) pp. 121-129.

Mertz, W., E. W. Toepfer, E. E. Roginsky and M. M. Polansky. "Present Knowledge of the Role of Chromium," *Fed. Proc.* 33:2275-2280 (1974).

Millikan, C. R. "A Preliminary Note on the Relation of Zinc to Disease in Cereals," *J. Dept. Agric. Victoria* 36:409-416 (1938).

Millikan, C. R. "Zinc Deficiency in Flax," *J. Dept. Agric. Victoria* 44:69-73, 88 (1946).

Mitchell, R. L. "Cobalt and Nickel in Soils and Plants," *Soil Sci.* 60(1):63-70 (1945).

Mitchell, R. L. "The Spectrochemical Analysis of Soils, Plants and Related Materials," *Commonwealth Bur. Soils, Tech. Commun.* No. 44 A:223 (1948).

Mitchell, R. L. "Trace Element Contents of Plants," *Research (London)* 10:357-362 (1957).

Mitchell, R. L. "Trace Elements in Soils," in *Chemistry of the Soil,* F. E. Bear, Ed. (New York: Van Nostrand Reinhold Company, 1967), pp. 320-366.

Moore, D. P. "Mechanisms of Micronutrient Uptake by Plants," in *Micronutrients in Agriculture* J. J. Mortvedt *et al.,* Eds. (Madison, WI: Soil Science Society of Agronomy, 1972).

Moitis, H. E., and D. R. Swingle. "Injury to Growing Crops Caused by Application of Arsenical Compounds to the Soil," *J. Agric. Res.* 34(1):59-78 (1927).

Murphy, L. S., and L. M. Walsh. "Correction of Micronutrient Deficiencies with Fertilizers," in *Micronutrients in Agriculture,* R. C. Dinaver, Ed. (Madison, WI: American Society of Agronomy, 1972).

Murrmann, R. P., and F. R. Koutz. "Role of Soil Chemistry Processes in Reclamation of Wastewater Applied to Land," in *Wastewater Management by Disposal on the Land,* Cold Reg. Res. and Eng. Lab., Report 171, Hanover, NH (1972).

Myttenaere, C., and F. M. Mousny. "The Distribution of Chromium−51 in Low Land Rice in Relation to Chemical Form and to the Amount of Stable Chromium in the Nutrient Solution," *Plant Soil* 41(1):65-72 (1974).

National Academy of Sciences. "Geochemistry and the Environment. I. The Relation of Selected Trace Elements to Health and Disease," Washington, DC, (1974).

Nicholas, D. J. D., and W. D. E. Thomas. "Some Effects of Heavy Metals on Plants Grown in Soil Culture," Part II. *Plant Soil* 5:182-193 (1954).

Overcash, M. R., and F. J. Humenik. "Concepts for Pretreatment Land Application for Small Municipal Systems," paper presented at the Second National Conference on Complete Water Use, American Institute of Chemical Engineers, Chicago, IL (1975), p. 16.

Overcash, M. R., J. W. Gilliam, F. J. Humenik and P. W. Westerman. "Lagoon Pretreatment: Heavy Metal and Cation Removals," *J. Water Poll. Control Fed.* 50(8):2029-2036 (1978).

Overstreet, R., and L. Jacobson. "The Absorption by Roots of Rubidium and Phosphate Ions at Extremely Small Concentrations as Revealed by Experiments with Rb[86] and P[32] Prepared Without Inert Carrier," *Am. J. Bot.* 33(2):107-112 (1946).

Page, A. L. "Fate and Effects of Trace Elements in Sewage Sludge when Applied to Agricultural Lands," A Literature Review Study, Environmental Protection Technology Series, EPA-670/2-74-005, Cincinnati, OH (1973).

Page, A. L. "Fate and Effects of Trace Elements in Sewage Sludge when Applied to Agricultural Lands," EPA-670/2-74-005, Cincinnati, OH (1974).

Page, A. L., and F. T. Bingham. "Cadmium Residues in the Environment," Residue Rev. 48:1-44 (1973).

Page, A. L., and T. J. Gange. "Accumulation of Lead in Soils for Regions of High and Low Motor Vehicle Density," *Environ. Sci. Technol.* 4:140-142 (1970).

Palazzo, A. J., and R. W. Duell. "Responses of Grasses and Legumes to Soil pH." *Agron. J.* 66:678-682 (1974).

Parker-Rhodes, A. F. "Mechanisms of Fungicidal Action, I. Preliminary Investigation of Nickel, Copper, Zinc, Silver and Mercury," *Ann. Appl. Biol.* 28:389-405 (1941).

Patel, P. M., and A. Wallace. "Some Interactions in Plants Among Cadmium, Other Heavy Metals and Chelating Agents," *Ann. Meeting Am. Soc. Agron., Abst.* 31 (1975).

Patterson, J. B. E. "Metal Toxicities Arising from Industry," in *Trace Elements in Soils and Crops. Min. Agric. Fish. Food, Tech. Bull.* 12:193-207 (1971).

Pfeiffer, E. "The Occurrence of Rubidium in Beets," *Arch. Pharmaz.* 200:101 (1872).

Pietz, J. R., C. Peterson, C. Lue-Hing and L. F. Welch. "Variability in the Concentration of Twelve Elements in Corn Grain," *J. Environ. Qual.* 7(1):106-110 (1978a).

Pietz, R. I., R. J. Vetter, D. Masarik and W. W. McFee. "Zinc and Cadmium Contents of Agricultural Soils and Corn in Northwestern Indiana," *J. Environ. Qual.* 7(3):381-385 (1978b).

Popp, M. "Manurial Experiments with Manganese Slag," *Fühling's Landw. Zeit.* 65(15-16):354-360 (1916).

Porter, J. R. *Bacterial Chemistry and Physiology* (New York: John Wiley and Sons, Inc., 1946).

Pratt, P. F. "Chromium," in *Diagnostic Criteria for Plants and Soils*, H. D. Chapman, Ed. University of California, Divison of Agricultural Science (1966b), pp. 478-479.

Purves, D. "Trace Elements Contamination of Soils in Urban Areas," *Trans. 9th Int. Congr. Soil Sci.* 2:351-355 (1968).

Randhawa, N. S., and F. E. Broadbent. "Soil Organic Matter—Metal Complexes. 5. Reactions of Zinc with Model Compounds and Humid Acid," *Soil Sci.* 99(5):295-300 (1965).

Raymond, R. L., J. P. Hudson and V. W. Jamison. "Oil Degradation in Soil," *Appl. Environ. Microbiol.* 31(4):522-535 (1976).

Reed, H. S. "Relation of an Essential Micro-element to Seed Production in Peas," *Growth* 6(4):391-398 (1942).

Reinhold, J., and E. Hausrath. "Experiments with Trace Element Fertilization of Cucumbers," *Gartenbauwiss.* 15:147-158 (1940).

Reuther, W., and C. K. Labanauskas. "Copper," in *Diagnostic Criteria for Plants and Soils*, H. D. Chapman, Ed., University of California, Division of Agricultural Science (1966), pp. 157-179.

Richards, F. J. "Physiological Studies in Plant Nutrition. II. The Effect of Growth on Rubidium with Low Potassium Supply and Modification of This Effect by Other Nutrients. I. The Effect on Total Dry Weight," *Ann. Bot. (N.S.)* 5:263-296 (1941).

Rissanen, K., J. Erkama and J. K. Miettenen. "Experiments on Microbiological Methylation of Mercury (2+) Ion by Mud and Sludge Anaerobic Conditions," Conference Marine Pollutants, Rome, F.A.O.-FIR WP/70/E-61 (1970).

Robinson, W. O., L. A. Steinkoenig and C. F. Miller. "The Relation of Some of the Rarer Elements on Soils and Plants," *U.S. Dept. of Agric. Bull.* No. 600:25; *Abst. Exp. Stat. Rec.* XXXVIII:409 (1917).

Rogers, R. D. "Abiological Methylation of Mercury in Soil," USEPA, ORP, Environmental Monitoring and Support Laboratory, Las Vegas, NE, EPA-600/3-77-007 (1977).

Romney, E. M., A. Wallace and G. V. Alexander. "Response of Bush Bean and Barley to Tin Applied to Soil and to Solution Culture," *Plant Soil* 42:585-589 (1975).

Ross, R. G., and D. K. R. Stewart. "Movement and Accumulation of Mercury in Apple Trees and Soil," *Can. J. Plant Sci.* 42:280-285 (1962).

Roy, D. R., and J. A. Bailey. "Effect of Silver from Cloud Seeding on Rumen Microbial Function," *Water, Air, Soil Poll.* 3:343-351 (1974).

Scharrer, K., and W. Schropp. "Sand- and Water- Culture Experiments with Nickel and Cobalt," *Zeit. Pflanzernähr., Dung., Bodenk.* A 31:94-113 (1933).

Scharrer, K., and W. Schropp. "Sand and Water Culture Experiments on the Effect of Zinc and Cadmium Ions," *Z. Pfanzenernahr., Dunung Bodenk* 34A:14-29 (1934).

Scharrer, K., and W. Schropp. "The Action of Vanadium upon Cultivated Plants," *Z. Pflanzenernahr., Dungung Bodenk.* 37:196-202 (1935).

Schulze, B. "Contribution to the Question of the Action of Stimulants on Plant Development," *Landw. Vers. Stat.* 87(1):1-24 (1915).

Schweneman, T. J. "Plant Response to and Soil Immobilization of Increasing Levels of Zn^{2+} and Cr^{3+} Applied to a Catera of Sandy Soils," Ph.D. Thesis, Michigan State University, East Lansing (1974).

Shacklette, H. T., J. C. Hamilton, J. G. Boerngen and J. M. Bowles. "Elemental Composition of Surficial Materials in the Conterminous United States," *U.S. Geol. Survey, Prof. pap.* 574-D:D71 (1971).

Sidle, R. C., and L. T. Kardos. "Adsorption of Copper, Zinc, and Cadmium by a Forest Soil," *J. Environ. Qual.* 6(3):313-317 (1977a).

Sidle, R. C., and L. T. Kardos. "Transport of Heavy Metals in a Sludge-Treated Forested Area," *J. Environ. Qual.* 6(4):431-437 (1977b).

Sillanpaa, M. "On the Effect of Some Soil Factors on the Solubility of Trace Elements," *Agrogeol. Publ.* 81:1-24 (1962).

Singh, R. N., R. F. Keefer, D. J. Horvath and A. R. Khawaja. "Sewage Sludge Application to Soils: 1. Yield and Chemical Composition of Corn and Soybeans," *Am. Soc. Agron. Abstr.* 33 (1976).

Skinner, J. J., and R. W. Ruprecht. "Some Results of Soil Fertility and Fertilizer Experiments with Tomatoes on Glade Soils of Dade County, Florida," Florida Dept. of Agric. Soils and Ferts., New Series No. 31:148-153 (1913).

Skinner, J. J., *et al.* "The Action of Manganese in Soils," *U.S. Dept. Agric. Bull.* 42:32 (1914).

Smart, N. A. "Use and Residues of Mercury Compounds in Agriculture," *Residue Rev.* 23:1-36 (1968).

Smith, P. F. "Mineral Analysis of Plant Tissue," *Ann. Rev. Plant Physiol.* 13:81-108 (1962).

Snodgrass, P. J., B. L. Valee and F. L. Hock. "Effects of Silver and Mercury on Yeast Alcohol Dehydrogenase," *J. Biol. Chem.* 235:504-508 (1960).

Soane, B. D., and D. H. Saunder. "Nickel and Chromium Toxicity of Serpentine Soils in Souther Rhodesia," *Soil Sci.* 88:322-330 (1959).

Sommers, L. E. "Chemical Composition of Sewage Sludges and Analysis of Their Potential Use as Fertilizers," *J. Environ. Qual.* 6(2):225-232 (1977).

Souchelli, V. *Trace Elements in Agriculture,* (New York: Van Nostrand Reinholt Co., 1969).

Spencer, E. L. "Renching of Tobacco and Thallium Toxicity," *Am. J. Bot.* 24:16-24 (1937).

Staker, E. V. "Progress Report on the Control of Zn Toxicity in Peat Soils," *Soil Sci. Soc. Am. Proc.* 7:387-92 (1942).

Staker, E. V. "Sulfur-Zinc Relationships in Some New York Peat Soils," *Soil Sci. Soc. Am. Proc.* 8:345 (1943).

Stucky, D. J., and T. S. Newman. "Effect of Dried Anaerobically Digested Sewage Sludge on Yield and Element Accumulation in Tall Fescue and Alfalfa," *J. Environ. Qual.* 6(3):271-273 (1977).

Swaine, D. J. "The Trace Element Content of Soils," *Common. Bur. Soil Sci. Tech. Comm.* No. 48 (1955).

Teakle, L. J. H., and I. Thomas. "Recent Experiments with 'Minor' Elements in Western Australia. IV. The Effect of 'Minor' Elements on Growth of Wheat in Other Parts of the State," *J. Dept. Agric. W. Australia* 16:143-147 (1939).

Teller, H. L., D. R. Cameron and D. A. Klein. "Disposition of Silver Iodide used as a Seeding Agent," in *Ecological Impacts of Snowpack Augmentation in the San Juan Mountains,* H. W. Steinhoff and J. D. Ives, Eds. Colorado, Final Report, Colorado State University Publication, Fort Collins, CO (1976).

Tiffin, L. O. "Translocation of Micronutrients in Plants," in *Micronutrients in Agriculture,* J. J. Mortvedt, *et al.,* Eds. (Madison, WI: Soil Science Society of America, Inc., 1972).

Tiffin, L. A., V. V. Lagerwerff and A. W. Taylor. "Heavy Metal and Radionuclide Behavior in Soils and Crops," a review, AEC Res. Contract AT (49-70-1) 182 (1973).

Turner, R. G. "Heavy Metal Tolerance in Plants," in *Ecological Aspects of Mineral Nutrition of Plants,* I. H. Rorison, Ed. (Oxford, England: Glackwell, 1969), p. 484.

Turner, M. A., and R. H. Rust. "Effects of Chromium on Growth and Mineral Nutrition of Soybeans," *Soil Sci. Soc. Amer. Proc.* 35:755-758 (1971).

Tokuoka, M., and O. R. Gyo. "The Effect of Zinc on the Growth of Rice," *J. Sci. Soil Manure, Japan.* 13:211-216 (1939).

Tokuoka, M., and S. Gyo. "The Effect of Zinc on the Growth of Wheat," *J. Sci. Soil Manure, Nippon* 14:587-596 (1940).

Underwood, E. J. *Trace Elements in Human and Animal Nutrition* (New York: Academic Press, 1971).

U.S. Environmental Protection Agency. "EPA Technical Bulletin on Municipal Sludge Management," Section 2-4.2. (1976).

U.S. Congress. *Federal Register* "Proposed Rules for Solid Waste Disposal Facilities," 43(25):4942-4955 (February 6, 1978).

Van Loon, J. C. "Mercury Contamination of Vegetation Due to the Application of Sewage Sludge as a Fertilizer," *Environ. Lett.* 6:211-218 (1974).

Vanselow, A. P. "Microelement Research with Citrus," *Calif. Citrograph* 37:77-80 (1951).

Vanselow, A. P. "Cobalt," in *Diagnostic Criteria for Plants and Soils*, H. D. Chapman, Ed., University of California, Division of Agricultural Science (1966), pp. 142-156.

Viets, F. J. "Zinc Deficiency in the Soil-Plant System," in *Zinc Metabolism*, A. S. Prasad, Ed. (Springfield, IL: C. C. Thomas Publishers, 1966).

Vlasyuk, P. A. "Utilization of the Waste Products of the Manganese Ore Industry for Fertilizing Sugar Beets," *Osnovnye Vyvody Nauch.—Issledovatel. Rabot VNIS* 30:161-173 (1937, 1939), *Khim, Referat. Zhur.* 8:55-56 (1940).

Vlasyuk, P. A. "Industrial Wastes as New Fertilizers," *Sveklovichnoe Polevodstvo* 9:50-52 (1939).

Vlek, P. L., and W. L. Lindsay. "Molybdenum Solubility Relationships in Soils Irrigated with High Mo Water," *Agron. Abs.* 126 (1974).

Voelcker, J. A. "Pot Culture Experiments," *J. Roy. Agric. Soc. (England)* 74:411-422 (1913).

Volecker, J. A. "Pot Culture Experiments," *J. Roy. Agric. Soc. England* 82:287-297 (1921).

Wain, R. L. "Manganese in Soils and Herbage at Wye," *J. Southeast. Agric. Coll., Wye, Kent* 42:146-153 (1938).

Walkley, A. "The Zinc Content of Some Australian Fertilizers," *J. Council Sci. Ind. Res.* 13:255-60 (1940).

Wallace, A., and E. M. Romney. "The Effect of Zinc Sources on Micronutrient Contents of Golden Bantamcorn," *Soil Sci.* 109:66-67 (1970).

Wallace, A., R. T. Mueller. "Effects of Chelated and Non-Chelated Cobalt and Copper on Yields and Microelement Composition of Bush Beans Grown on Calcareous Soil in a Glasshouse," *Soil Sci. Soc. Amer. Proc.* 37(6):907-908 (1973).

Wallace, A., S. M. Soufi, J. W. Cha and E. M. Romney. "Some Effects of Chromium Toxicity on Bush Bean Plants Grown in Soil," *Plant Soil* 44(2):471-473 (1976).

Warington, K. "Molybdenum as a Factor in the Nutrition of Lettuce," *Ann. Appl. Biol.* 33:249-254 (1946).

Weaver, T. W., and D. Klarich. "Impact of Induced Rainfall on the Great Plains of Montana," Final Report to Bureau of Reclamation, Dept. of Botany, Montana State University (1973), p. 19.

Webber, J. "Effects of Toxic Metals in Sewage on Crops." Water Pollut. Contr. 71:404-413 (1972).

White, R. W. "Silver Content of Soil and Plants on the Medicine Bow National Forest and its Influence on Plant Growth," Ph.D. Thesis, University of Wyoming, Laramie, WY. (1973).

Wober, A. "The Toxic Action of Arsenic, Antimony, and Fluorin Compounds on Some Cultivated Plants," *Angew Bot.* 2(6):161-178 (1920).

CHAPTER 11

NITROGEN

INTRODUCTION

Domestic and municipal wastewaters contain nitrogen in the average range of 15-25 mg/l (Barth and Dean 1971). Industrial wastes vary widely with many containing as little as 10% of that in municipal wastes. Industrial wastes can also contain large nitrogenous loads, such as in:

1. the manufacture of nitrogen-based explosives;
2. the organic chemicals production process employing nitrogenous compounds;
3. the manufacture and use of nitric acid;
4. ammonia and ammonium salts, and nitrogen fertilizers;
5. nitrophenols, aniline, nitroaniline pesticides, herbicides, nitrification inhibitors (N-serve), etc.;
6. coke ovens, steel works and coal mining industries; and
7. cyanide and thiocyanate manufacture and use.

Nitrogen is also present in the wastes from slaughterhouses, vegetable processing, swine, poultry, dairy and animal-based industry, and wine-making operations, as well as in the enzyme, antibiotic, pharmaceutical, and paper and pulp industries.

Nitrogen in the waste effluents is present in inorganic and organic forms. Inorganic forms include NH_4^+, NH_3, NO_2^-, NO_3^-, whereas examples of organic forms are amino acids, proteins (enzymes), nucleic acids, antibiotics, biomass, nitroglycerine, lignoprotein, nitroprotein, nitrocellulose, nitrophenol and nitroaniline complexes. The assimilative capacity for nitrogen is thus an important part of the overall design of an industrial land treatment system (Figure 11.1).

459

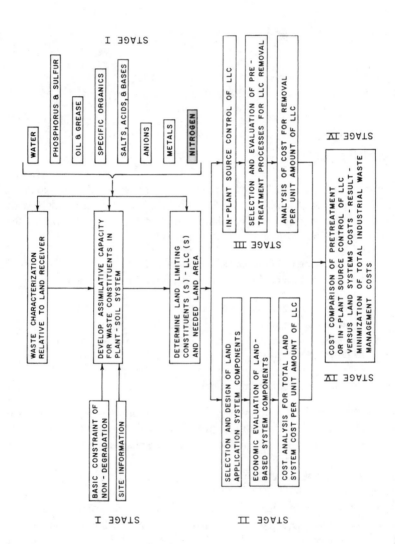

Figure 11.1 Relationship of Chapter 11 to overall design methodology for industrial pretreatment–land application systems.

NITROGEN IN SOIL-PLANT SYSTEMS

Nitrogenous organic compounds in soils are derived from plant and animal residues, organic composts, manures, biological N fixation, root exudates and microbial cells.

In most mineral soils the organic matter ranges from 1-15%, of which only about 5% is nitrogen. Thus, on a total soil basis, 0.05-0.75% is in the nitrogen pool. The division between organic N and inorganic N is shifted toward the organic-based nitrogen (Figure 11.2) with only 2-3% as inorganic. About 24-37% of the organically bound nitrogen is comprised of bound amino acids; 5-10% bound amino surgars; and 1-10% nucleic acids. The remainder of the organically bound nitrogen is in the humic acid fraction as lignin-ammonia, quinone-ammonia, etc., complexes. Thus, the pool of available N for such processes as plant uptake, microbial growth and leaching is quite small in comparison to total soil nitrogen. The depth distribution of nitrogen reflects the accumulation in organic compounds near the active soil surface zone and the leaching movement of NO_3-N, Figure 11.3.

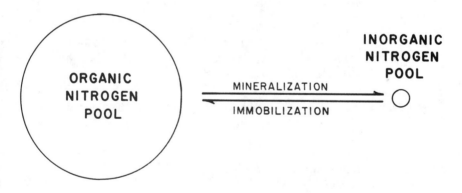

Figure 11.2 Relative proportion of forms of nitrogen in soil.

The overall soil nitrogen cycle, including the interaction with industrial wastes containing nitrogen compounds is depicted in Figure 11.4 Although not a large fraction in soils, nitrogen is essential to microbial and plant life and has been exhaustively studied by Bartholomew and Clark (1965), Black (1968), and others.

As a general class of constituents in industrial wastes, the nitrogenous compounds represent species that are assimilated by plant and microbial uptake. That is, to achieve the basic constraint of nondegradation (Chapter 1), the N assimilative capacity of a plant-soil system is in the first category of waste constituents, *i.e.*, the rate is controlled by soil decomposition or uptake by plants. To establish the assimilative capacity and ensure an understanding of the mechanisms of land-based treatment, this chapter reviews briefly the pathways for N assimilation. Following these sections, the assimilative capacity is presented to allow the land limiting constituent (LLC) analysis for an industrial waste containing nitrogen.

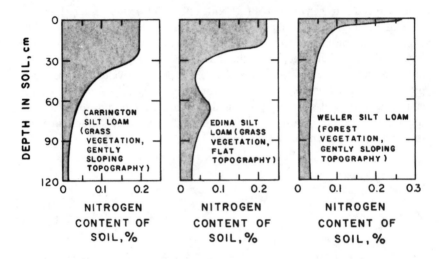

Figure 11.3   Distribution of nitrogen with depth below soil surface (Pearson and Simonson 1939).

FATE OF NITROGENOUS COMPOUNDS APPLIED TO PLANT-SOIL SYSTEMS

Nitrogen Mineralization

The conversion of organic nitrogen to inorganic forms is known as nitrogen mineralization. The microbial mineralization of nitrogen occurs in a sequence of steps (Figure 11.5). Except for some specific compounds (such as urea), which degrade very rapidly, or for conditions when NO_3-N formation is inhibited, the rate-limiting mineralization step is that of organic N to NH_4^+. Thus, little accumulation of NH_4^+ occurs.

In the steady-state soil conditions the mineralization of organic nitrogen and organic carbon are related. In an unamended soil, the two elements are mineralized at parallel rates, and the ratio of carbon dioxide: carbon

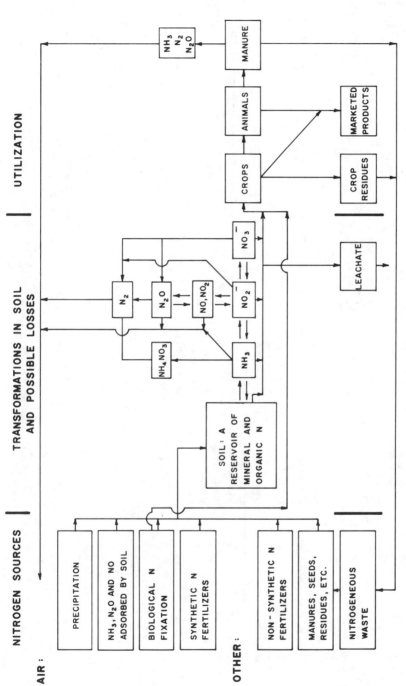

Figure 11.4 Soil nitrogen sources, transformations, and fate of the end products (Allison 1965).

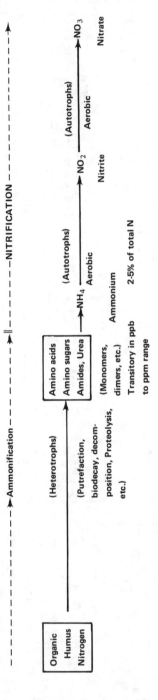

Figure 11.5 Schematic of nitrogen mineralization in soils.

that evolves and inorganic nitrogen that is released is essentially between 7:1 and 15:1. This is not necessarily true for organics from industrial wastes when applied to land, because of the variable C:N ratio, specific waste compound or the influence of other waste constituents.

The transformation of organic nitrogen compounds to ammonia or ammonium is termed ammonification. Where the applied waste contains proteins, hydrolysis results in long-chain amino acid fragments of simple peptides, which, on continued hydrolysis, yield free amino acids. Amino acid may then be:

1. metabolized by microorganisms;
2. transformed by microbial enzymes with the formation of ammonia;
3. adsorbed by clay minerals or incorporated into humus fraction; and
4. utilized by plants.

The reactions of protein breakdown are shown below where enzyme proteases hydrolyze the peptide bonds:

$$\ldots \underset{\underset{R}{|}}{\overset{\overset{H}{|}}{N}}HCCON\underset{\underset{R'}{|}}{\overset{\overset{H}{|}}{H}}CCON\underset{\underset{R''}{|}}{\overset{\overset{H}{|}}{H}}CCO \ldots \xrightarrow[\text{enzyme}]{H_2O} \ldots \underset{\underset{R}{|}}{\overset{\overset{H}{|}}{N}}HCCOH + H_2NCCON\underset{\underset{R'}{|}}{\overset{\overset{H}{|}}{H}}CCON\underset{\underset{R''}{|}}{\overset{\overset{H}{|}}{H}}CCO \ldots$$

Other more resistant N compounds, such as purines or pyrimidines are mineralized to ammonia and simple organic compounds (Figure 11.6).

Anaerobic decomposition of proteins may release foul-smelling compounds during putrefaction. The final products of complete anaerobic decay are NH_4^+, CO_2, CH_4, H_2, amines, organic acids, indole, skatole, mercaptans, hydrogen sulfide, phosphene, etc. Because of the nuisance factor with these anaerobic end products, and for a variety of reasons discussed in relation to other waste constituents, anaerobic conditions in a land treatment system are to be avoided.

Ammonification rates of industrial nitrogenous compounds depend on a number of environmental factors. The rate of mineralization, however, appears maximum for a soil near 70% of water-holding capacity. Acidity tends to depress but does not eliminate mineralization. Liming of acid soils stimulates to mineralization as the pH is brought closer to the optimum for the active microflora. Temperature influences the mineralization sequence as each biochemical step is catalyzed by temperature-sensitive enzymes produced by microorganisms, whose growth is also conditioned by temperature. Optimum temperature for ammonification is between 40 and 60°C, demonstrating the activity of thermophiles. However, conversion of ammonium to nitrate essentially ceases at 45°C. These high temperatures are seldom encountered under field conditions. Mineralization ceases when frozen conditions occur. Between 0 and 35°C, a range of temperatures at land application sites, the mineralization rate increases with temperature (Alexander 1977, Stanford *et al.* 1973). As an approximation, the first-order mineralization rate constant can be corrected for temperature as follows (Reddy *et al* 1977):

$$K_{T_1} = K_{T_2}\, \theta^{T_1 - T_2} \tag{11.1}$$

where K = rate constants
T = temperature, °C
θ = 1.07

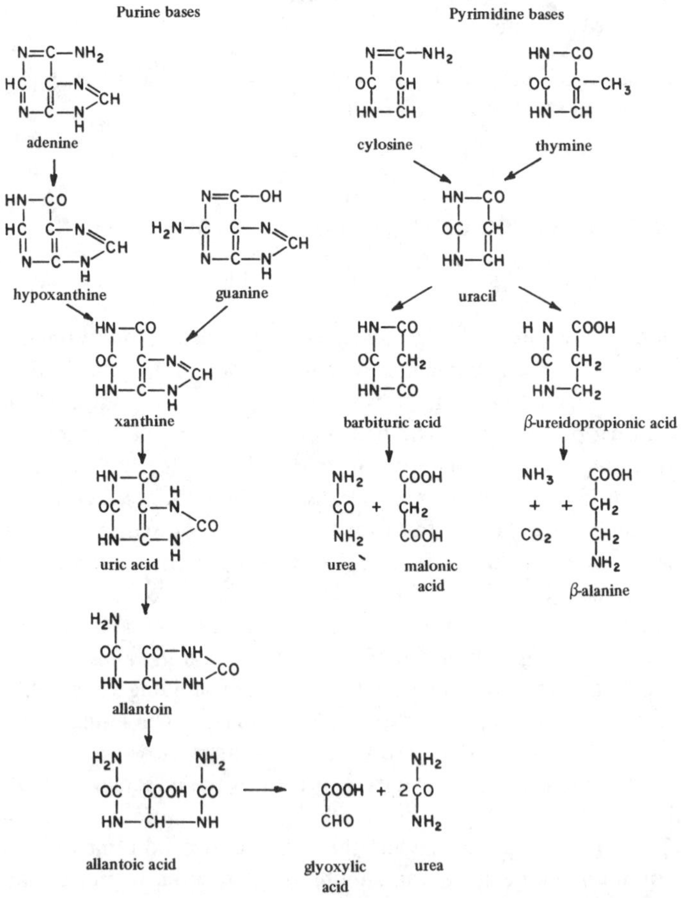

Purine bases   Pyrimidine bases

adenine   cylosine   thymine

hypoxanthine   guanine   uracil

xanthine   barbituric acid   β-ureidopropionic acid

uric acid   urea   malonic acid   β-alanine

allantoin

allantoic acid   glyoxylic acid   urea

Figure 11.6 Biochemistry of purine and pyrimidine decomposition (Alexander 1977).

Biological formation of nitrites and nitrates from ammonium is called nitrification. This process occurs in two steps by two distinctly different groups of autotrophic aerobic bacteria. The first step of nitrification is:

$$2NH_3 \xrightarrow{\;O_2\;} 2HO - NH_2 \xrightarrow{\;-4H\;} HO - = N - OH \xrightarrow{\;O_2\;} 2HO - N = 0$$

Oxidation State

-3	-1	$+1$	$+3$
Ammonia	Hydroxylamine	Hyponitrite	Nitrite

Several intermediates are formed between ammonium and nitrite. The principal microorganism involved is *Nitrosomonas*, although *Nitrococcus*, *Nitrosospira*, *Nitrosocystis* and *Nitrosogloea* also are involved. The second step of nitrification involves:

$$HO - N = 0 \xrightarrow{\;+H_2O\;} HO - N \big\langle \substack{OH \\ OH} \xrightarrow{\;-2H\;} HO - N \big\langle \substack{0 \\ 0}$$

Oxidation state $+3$ $+5$

Nitrite   Nitrate

The bacteria that oxidize nitrite to nitrate belong to genera *Nitrobacter* and *Nitrocystis*. This step involves a change in oxidation state of the nitrogen atom from $+3$ to $+5$ by a biochemical mechanism. The nitrifying chemoautotrophs are limited in their oxidative capacity to nitrogen compounds for their energy, while carbon for cell synthesis is derived from CO_2, carbonates or bicarbonates.

Biochemical efficiency expressed as the ratio of inorganic nitrogen oxidized to carbon assimilated (C:N) varies 14 to 70:1 for *Nitrosomonas* and 76 to 135:1 for *Nitrobacter* (Alexander 1977). The activities of nitrifiers are affected by a number of environmental conditions, particularly those influencing pH and aeration. The optimum pH for growth of nitrifying organisms tends to be related to the reaction of the soil from which they are isolated. Strains isolated from low-pH soils are more acid tolerant than those from an alkaline environment. Although nitrification will proceed in soils with a pH as low as 4.5, the rate is reduced under such conditions. The addition of lime frequently improves the rate of nitrification after initial adoption. Soil pH below 4.5 greatly impedes nitrification, although ammonification may still proceed. With the application of acid wastes exceeding the assimilative capacity of the soil, the biological processes of nitrate production and, hence, plant growth, may be impeded.

As obligate aerobes, nitrifiers are particularly sensitive to excess moisture or soil conditions that may limit oxygen supply. Thus, the design for the assimilation of the hydraulic component of an industrial waste must allow for unsaturated conditions of sufficient length that reaeration is complete. The presence of organics, which exert an oxygen demand during decomposition, must also be accounted for in assessing soil type to ensure a predominantly aerobic soil environment (see Chapter 7).

Nitrogen Immobilization

Microbial assimilation and transformations of inorganic nitrogen to less available organic or biological forms is referred to as immobilization. These less available forms represent the majority of the soil nitrogen pool (Figure 11.2). A considerable portion of the added nitrogen in soil is not recovered in the crop because of biological immobilization and chemical stabilization, in addition to losses through volatilization and leaching. The conversion of fertilizer inorganic nitrogen to organic forms, which become progressively less available with time, is termed nitrogen reversion (Broadbent and Nakashima 1967). Nitrogen reversion is shown below, in which the relative rates of reaction are proportional to length of arrow:

$$\text{Microbial-N} \underset{\text{(biological)}}{\overset{\text{(biological)}}{\rightleftarrows}} NH_3 \underset{\text{(biological)}}{\overset{\text{(nonbiological)}}{\rightleftarrows}} \text{Stable-N}$$

The process of nitrogen reversion is expected to hold for industrial wastes containing inorganic and organic nitrogen. The extent of nitrogen immobilization is less for a bacterial flora and greater for actinomycetes.

The aerobic flora that immobilize inorganic nitrogen consume between 20 and 25 parts of carbon per unit of nitrogen, C:N = 20-25. Industrial wastes with C:N $>$ 23 would be expected to generate little inorganic N, and the organic decomposition process is nitrogen limited. In this instance net immobilization is occurring (Figure 11.7). For wastes with a ratio of C:N $<$ 23, the decomposition would proceed with adequate N, and inorganic N for plant usage would be generated. This is a net N mineralization. From this critical C:N ratio of 23 an approximate ratio of 10:1 is ultimately put into the biological and humus fraction of the soil (Figure 11.8). The remainder of the carbon is lost as carbon dioxide in microbial growth. Although not designed for land application, anaerobic systems require less N (C:N $\cong$ 75).

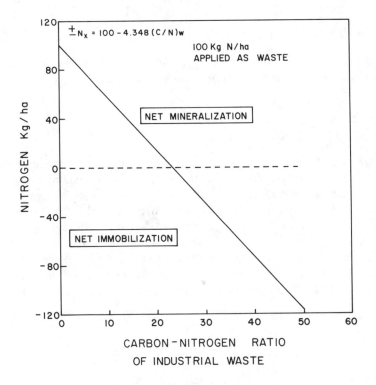

Figure 11.7 Predictive tool for net mineralization of wastes with varying C/N ratio (Reddy *et al.* 1977).

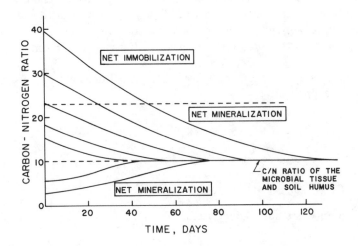

Figure 11.8 Sequence of carbon and nitrogen changes for various waste materials having different initial C/N ratios (Reddy *et al.* 1977).

Ammonia Volatilization

When industrial wastes containing ammonium are applied to land, the magnitude of the volatilization losses is largely determined by:

1. method of application,
2. temperature, and
3. soil CEC (Reddy *et al.* 1978).

The losses of ammonium through ammonia volatilization occur in a short period—1-3 weeks after an application event. If an industrial waste is applied at the plant-soil surface, there is little to inhibit ammonia movement, and thus a large fraction of NH_3 is expected to be lost (61-99%) (Lauer *et al.* 1976, Lemon 1977). When wastes are incorporated into the soil by plowing or direct injection, the loss of NH_3 is substantially reduced, with the magnitude of retention being proportional to the soil CEC. Assuming a first-order kinetic equation for NH_3 volatilization, the rate constant for incorporated waste can be corrected for soil CEC (Adriano *et al.* 1971).

$$K_2 = K_1 F_{CEC_2}/F_{CEC_1} \qquad (11.2)$$

where

$$F_{CEC} = 1.00 - 0.038 (CEC) \qquad (11.3)$$

in which cation exchange capacity (CEC) is expressed in meq/100 g soil. After 1-3 weeks the volatilization is exceeded by mineralization, and the ammonia lost is thus fixed in the soil. When wastes are irrigated and > 1.25 cm of liquid applied per event, the ammonia loss is similar to that for incorporated wastes because of the physical transport of NH_4 into the soil profile.

Temperature corrections have been calculated from field data using the following expression:

$$K_2 = K_1 \theta^{T_2 - T_1} \qquad (11.4)$$

where K = first-order rate constant for ammonia volatilization
  T = temperature, $^{\circ}C$
  θ = constant $\cong 1.08$ (Reddy *et al.* 1978)

For a given waste application system the NH_3 losses are depicted in Figure 11.9.

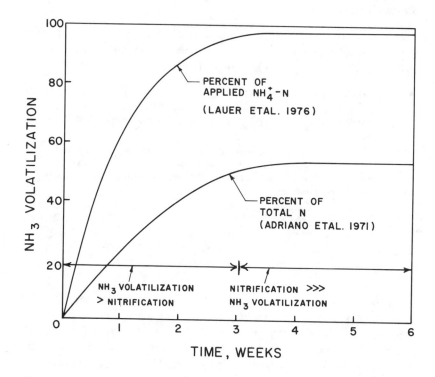

Figure 11.9 Pattern of ammonia volatilization or conversion from land applied waste.

Denitrification

The microbial reduction of nitrates and nitrites to diatomic nitrogen and/ or nitrous oxide is considered denitrification. The denitrifiers are hetero-trophic organisms utilizing organic sources of carbon for energy and growth. These include species of *Pseudomonas, Achromobacter, Bacillus* and *Micrococcus.* The facultative aerobic and heterotrophic microbes use nitrates in place of oxygen as an alternate terminal electron acceptor in the absence of free oxygen or under partially anaerobic conditions.

Denitrification represents a net loss of N from the plant-soil system, but specific quantitative inclusion in design for assimilation of nitrogeneous constituents is not included. The inability to ensure that all conditions for denitrification in a field situation are favorable and to prevent denitrification are the reasons for combining these losses with the others in land treatment design.

Nitrogen Fixation

Diatomic nitrogen from the atmosphere can be utilized by a number of free-living microorganisms, as well as by those in symbiotic association with higher plants. The nonsymbiotic nitrogen fixers include:

1. heterotrophic bacteria that require a source of organic carbon and energy for growth and effective fixation of N_2, *e.g.*, *Azotomonas, Bacillus polymyxa, Beijerinckia, Clostridium, Pseudomonas*, and *Aerobacter;*
2. photosynthetic bacteria, which use sunlight as an energy source to fix both CO_2 and N_2 as C and N sources, respectively, for their growth, *e.g.*, *Chlorobium, Chromatium, Rhodomicrobium, Rhodopseudomonas* and *Rhodospirillum;*
3. Chemoautotrophic bacteria, which use the energy derived from the oxidation of inorganic compounds to fix the N_2, *e.g.*, *Methanobacillus omelianskii;* and
4. blue-green algae, which are mostly photosynthetic and reduce N_2 for their use, *e.g.*, *Anabaena, Calothrix, Cylindrospermum, Nostoc* and *Tolypothrix.*

Microbes that assimilate N_2 have the ability to utilize NH_4^+. In fact, ammonium is used preferentially and often at a greater rate than diatomic nitrogen. The presence of ammonium inhibits the fixation of diatomic N_2 gas. For wastes with a high C:N ratio, such as oil sludges, those organisms that can fix N have an advantage, so appreciable N can be received from the atmosphere.

Symbiotic nitrogen fixation also occurs in an association between *Rhizobia* and legumes at nodules on the roots. The resting cells of Rhizobia depend on energy and carbon supply from the host plant and cannot fix atmospheric nitrogen in the absence of host plants. Leguminous plants benefit markedly from their symbiotic relationships with *Rhizobia*; consequently, the soils in which they are grown also benefit when these crops are incorporated back into the soil. Conservative estimates of the annual fixation of all legumes average from 65-125 kg/ha/yr. Much higher values have been recorded frequently for alfalfa, sometimes in excess of 325 kg/ha/yr of N. Besides the leguminous family, 8 general of Angiosperms, comprising 190 species of widely distributed trees and shrubs, also produce nodules where the atmospheric nitrogen is fixed but the microsymbionts have not been well characterized or isolated.

Nitrogen-Plant Relations

Nitrogen not immobilized in the soil system is predominantly inorganic and, as an essential element, is utilized by plants. The assimilative capacity for nitrogen constituents of industrial wastes is predominantly controlled by the plant uptake rates. Nitrogen enters into the structure of amino acids, amides, nucleic acids, alkaloids and chlorophyll in plants. Plants cannot make use of the organic nitrogen efficiently. The assimilation of nitrates by plant and microbes involves biological reduction of nitrates to ammonia or the amino level, which finally ends up in cell proteins. These assimilatory pathways are controlled genetically in each species of living organisms.

The response of plants to N is shown schematically in Figure 11.10. As amount of applied nitrogen increases the plant takes up more N, until an asymptotic region is reached. In this range, the majority of the nitrogen is available for leaching, and not plant uptake. The N application rate at which asymptotic behavior occurs and the magnitude of the plant uptake are very dependent on crop type. The range in plant response is broad, thus allowing crop selection to substantially increase the design assimilative capacity.

Leaching and Runoff Losses

The movement of nitrate through soils, leaching, or the transport of organic and inorganic N species in surface runoff, are undesirable losses of the nitrogenous constituent of land-applied wastes. Mechanisms and pathways for these losses have been developed and reviewed extensively (Khaleel *et al.* 1977, Davidson and Rao 1971, ARM 1977). These losses represent nonpoint source contributions from land application sites. The control of such nonpoint source inputs is by use of "best management practices," instead of complete collection and treatment (EPA 1972). Best management practices (BMP) include a variety of agricultural practices that: (1) reduce runoff and, hence, potential pollutant transport; (2) prevent entry of runoff sediments into surface wastes; or (3) reduce availability or amount of nutrients for runoff or leaching (Sietz 1978). A fairly complete list of BMPs has been assenbled by Walter *et al.* (1977).

On an industrial waste land application site, efforts must be made to control the runoff and leaching losses of nitrogen compounds. With leaching, the basic approach is to have a viable crop to retain the N and not to greatly exceed the assimilative pathways for nitrogen. This will prevent excess nitrate formation and potential leaching. Concerning runoff, the basic approach is to implement the best BMPs at a particular site. It must be recognized that runoff from pristine areas contains nitrogen, so the objective is not 100% control, but to reduce runoff levels to background conditions.

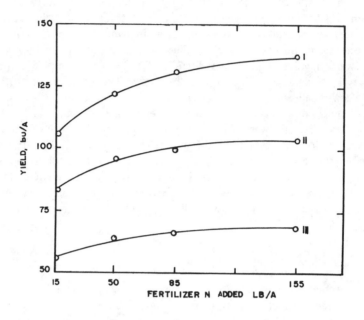

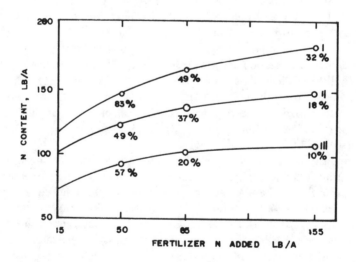

Figure 11.10 Corn yield and nitrogen content as a function of fertilizer N application
rate (Lathwell *et al.* 1970).

ASSIMILATIVE CAPACITY

Within the basic land application design constraint of minimal impact on the receiving waters and nondegradation of the plant-soil system, the nitrogenous constituent assimilative capacity can be developed for each site and waste situation. Assimilation via plant uptake, immobilization, volatilization and denitrification is favored, while leaching and runoff losses are reduced (Figure 11.11).

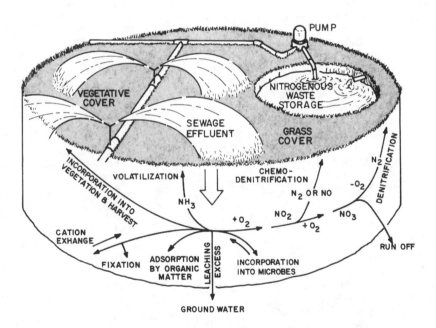

Figure 11.11 Schematic of nitrogen pathways in spray irrigation of industrial waste (Sewage Effluent Disposal at Fort Ord 1974).

An initial estimate of the assimilative capacity is based directly on the plant uptake of N (Carlile and Phillips 1976). The capacity depends on geographical location, in the same manner as the agricultural response of a certain crop varies. Crop uptakes are listed on Table 11.1 for species considered for land treatment. Carlile and Phillips (1976) concluded that the immobilization and denitrification losses of N, when combined with crop uptake, could be estimated as 150% of the crop N uptake. This estimated technique was based on minimal impact in nitrate leaching from numerous field-scale, land application sites. Thus, the designer would refer to Table 11.1 and multiply by 1.5 to arrive at the N assimilative capacity, kg/ha/yr of N.

Table 11.1 Expected Nutrient Removal by Forage and Field Crops, and Forest Trees

Vegetative Cover (yield goals)	Nitrogen Uptake (kg/ha/yr)
Forage and Field Crops	
Coastal Bermudagrass with rye overseed	570 + 205 = 775
Coastal Bermudagrass	480 - 600
Reed canary grass	226 - 359
Fescue	275
Alfalfa	155 - 220
Sweet clover	158
Red clover	77 - 126
Lespedeza hay	130
Johnson Grass, 27 metric ton/ha	890
Peanuts, 7.5 metric ton/ha	140
Corn (7.6-12.9 m^3/ha	155
Soybeans (5.2 m^3/ha	94 -113
Irish potatoes	108
Cotton	66 - 100
Milo maize	81
Wheat	50-76
Sweet potatoes	75
Sugar beets	73
Barley	63
Oats	53
Tobacco (flue cured, 3,300 kg/ha)	85
Forest Trees	
Mixed hardwoods	200
Red pine	160
White spruce (old field vegetation)	250
Pioneer succession vegetation	250

When dealing with more complex industrial wastes, different operational and management alternatives, and the need to include site-specific characteristics, the design assimilative capacity can be refined. The procedural approach described below is based on the fundamental processes described earlier in this chapter.

Waste characterization for the percentage of ammonium, nitrate and organic N is performed on a reliable composite sample of the waste under consideration for land application. If the waste is to be surface applied, 75-90% of the ammonium present is assumed to be lost. For irrigation or incorporation of wastes, a first-order rate expression can be used:

$$\frac{TAN}{TAN} = \exp(-Kt) \tag{11.5}$$

where TAN = total ammonium, NH_4-N, in the soil-waste system
       TAN = total ammonium, NH_4-N, in the soil-waste system at application $t = 0$
       K   = volatilization rate constant, day^{-1}
       t   = 10-20 days, the period of volatilization

The value for K is not well established, but from available field data, a value of 0.4 day^{-1} is a reliable first estimate (Reddy et al. 1977). Correction of K for temperature and soil absorption properties (CEC), are given in Equations 11.2 and 11.4. The solution for Equation 11.5 indicates the approximate fraction of waste ammonium lost. The remaining ammonium is assumed to be converted to nitrate or immobilized. Such an approach is quite simplistic and reflects the present limited availability of field data.

Organic N will mineralize to yield inorganic compounds, which must be accounted for in the assimilative capacity. The rate of mineralization varies over a tremendous range depending on both the nitrogenous compound and the presence of adequate microbial populations to decompose such compounds. Rate of mineralization is important for matching crop needs to inorganic N availability. These kinetic rates are generally not known or are imprecise. Thus, an assumption is made that, in the steady-state situation of repeated waste application and predominantly aerobic conditions, a certain constant fraction of the annual applied organics will become inorganic.

The fraction of nitrogenous waste constituents that becomes inorganic depends on the C:N ratio and the immobilization pathways. Utilizing a C:N = 23 as a balance point, Equation 11.6 can be used:

$$NX = NW - 0.043 (NW) (C/N)_w \tag{11.6}$$

where NX   = amount of waste N that is nitrified, kg/ha
       NW   = total amount of waste N applied, kg/ha
      $(C/N)_w$ = ratio of carbon to total N in the waste (excluding volatilized N)

The amount not converted to nitrate (NW-NX) is immobilized and added to the soil organic N pool:

$$(NW-NX) + ON_S = TON \tag{11.7}$$

where ON_S = soil organic N
      TON = total N in the soil-waste system

The TON is also assumed to be mineralizable, but at a much lower rate—more characteristic of nitrate formation in background or undisturbed soils.

Total inorganic N from an industrial waste is thus the ammonium not lost by volatilization, the waste nitrate, plus NX. Therefore, crop selection must be made to match this fraction of the total waste nitrogen.

Selection of a crop should be compatible with recommendations of agricultural specialists for a particular site. That is, not all crops are grown successfully in every climatological region. Weather factors such as temperature and length of growing season are important, while drought response may not be a factor when liquid wastes are being land spread. Thus, waste application can overcome certain climatological limitations associated with conventional crop production. The soils of an area may restrict growth of certain species because of excess or poor water movement, or soil acidity. Input from agricultural specialists is useful here. Also, the waste properties can improve soil characteristics, thus recovering marginal agricultural lands, e.g., organic wastes providing enhanced moisture-holding capacity of soils.

Another factor in crop selection is crop value. Consideration of the economic return from crops of a land application site should be studied, especially in the long-term operation, to improve the economics of land-based systems.

Certain combinations of crops in Table 11.1 can be utilized together within a yearly cycle, since growing periods do not overlap. This is referred to as double cropping and, although it involves greater management, will certainly improve the nitrogen assimilative capacity of the soil system. A classic example is using Coastal Bermuda from May to October, and winter rye from October to April, or using a part of a field for Coastal Bermuda and another part for fescue, and alternating applications between fields. Maximum nitrogen utilization of both crops can be used in waste application calculations. Typically, a maximum rate of utilization is about 600 kg/ha/yr of N for a double crop Coastal Bermudagrass and winter rye, which, using the 150% design factor (Carlile and Phillips 1976), could then accept rates of 800-900 kg/ha/yr of N of waste nitrogen. Most other grass systems operate slightly below this rate (400-600 kg/ha/yr of N).

REFERENCES

Adriano, D. C., P. F. Pratt and S. E. Bishop. "Fate of Inorganic Forms of N and Salt from Land-Disposed Manure from Dairies," *Proc. Int. Nat. Symp. on Liverstock Wastes*, ASAE, St. Joseph, MI (1971), pp. 243-246.

Agricultural Runoff Management (ARM). "Model Version II: Refinement and Testing. EPA-600/3-77-089, Research Grant No. R803772-01 to A.S. Donigian, D. C. Beyerlein, H. H. Davis and N. H. Crawford (1977).

Alexander, M. *Introduction to Soil Microbiology*, 2nd ed. (New York: John Wiley & Sons, Inc, 1977).

Allison, F. E. "Evaluation of Incoming and Outgoing Processes that Affect Soil Nitrogen," in *Soil Nitrogen,* W. Bartholomew and F. W. Clark, Eds. ASA Monograph No. 10 in the Series of Agronomy (1965), pp. 573-606.

Barth, E. F., and R. B. Dean. "Nitrogen Removal from Wastewater: Statement of the Problem," *Symp. on Adv. Waste Treatment,* Dallas, TX (1971).

Bartholomew, W. V., and F. E. Clark, Eds. "Soil Nitrogen," Monograph No. 10 in the Series of Agronomy (1965).

Black, C. A. *Soil-Plant Relationships,* 2nd ed. (New York: John Wiley & Sons, Inc., 1968).

Broadbent, F. E., and T. Nakashima. "Reversion of Fertilizer Nitrogen," *Soil Sci. Soc. Am. Proc.* 31(5):648-651 (1967).

Carlile, B. L., and J. A. Phillips. "Evaluation of Soil Systems for Land Disposal of Industrial and Municipal Effluents," Report 118, Water Resources Research Institute, University of North Carolina, Raleigh, NC (1976).

Davidson, J. M., and P. S. C. Rao. Personal communication (1978).

Khaleel, R., G. R. Foster, K. R. Reddy, M. R. Overcash and P. W. Westerman. "Predicting Sediment Transport from Land Areas Receiving Animal Wastes," EOS, *Am. Geophys. Union Trans.* 59(4):283 (1977).

Lathwell, D. J., D. R. Bouldin and W. S. Reid. "Effects of Nitrogen Fertilizer Applications in Agriculture," in *Relationship of Agriculture to Soil and Water Pollution,* Cornell Agric. Waste Mgmt. Conf., Rochester, NY (1970), pp. 192-206.

Lauer, R. C., S. D. Klausmer and T. W. Scott. "Applications of Agricultural Wastes to Land," in *Land Treatment and Disposal of Municipal and Industrial Wastewater,* R. L. Sanks and T. Asano, Eds. (Ann Arbor, MI: Ann Arbor Science Publishers, Inc., 1976).

Lemon, E. "Gaseous Ammonia Implications to Agriculture and Ecology," Unpublished Results (1977).

Pearson, R. W., and R. W. Simonson. "Organic Phosphorus in Seven Iowa Soil Profiles: Distribution and Amounts as Compared to Organic Carbon and Nitrogen," *Soil Sci. Soc. Am. Proc.* 4:162-167 (1939).

Reddy, K. R., R. Khaleel, M. R. Overcash and P. W. Westerman. "Conceptual Modeling of Nonpoint Source Pollution from Land Areas Receiving Animal Wastes: I. Nitrogen Transformations," ASAE Paper No. 77-4046, presented at 1977 Annual Summer Meeting, ASAE held at North Carolina State University, Raleigh, NC (1977).

Reddy, K. R., R. Khaleel, M. R. Overcash and P. W. Westerman. "A Nonpoint Source Model for Land Areas Receiving Animal Wastes. II. Ammonia Volatilization. *J. Environ. Qual. Trans. ASAE* (In press).

Sewage Effluent Disposal at Ford Ord, California, CH2M Hill, Inc., Sacramento, CA (1974).

Seitz, W. D. "Alternative Policies for Controlling Nonpoint Agricultural Sources of Water Pollution," EPA 600/5-78/005, Athens, GA, (1978), pp. 314.

Stanford, G., M. H. Frere and D. H. Schwaninger. "Temperature Coefficient of Soil Nitrogen Mineralization," *Soil Sci.* 115:321-323 (1973).

U.S. Environmental Protection Agency. "Federal Waste Pollution Control Act Amendments of 1972," Section 208 (1972).

Walter, M. F., T. S. Steenhuif and D. A. Haith. "Soil and Water Conservation Practices for Pollution Control," presented at 1977 meeting, ASAE, Chicago, 1977.

CHAPTER 12

LAND TREATMENT SYSTEM DESIGN
AND ECONOMIC ANALYSIS–STAGE II

INTRODUCTION

In the context of the overall industrial waste design methodology for pretreatment-land application systems, stage II is related in a manner shown in Figure 12.1. Stage II represents the complete engineering design and agricultural practices necessary to transport an industrial waste from the generation point to final application on land. That is, the designer has already completed stage I; a detailed examination, based on site data, has been made; and the assimilative capacities have been developed from Chapters 4-11. The land limiting constituent (LLC) has been identified and the corresponding land area requirement determined (Chapter 2).

To this point (completion of stage I) the work has been related primarily to quantifying the pathways by which waste constituents are assimilated in the plant-soil system. All calculations have been based on the fundamental constraint of nondegradation so that waste constituents are assimilated in an environmentally acceptable manner. Stage II is based on the results of stage I and involves engineering design of the components of a land system. "Components" refers to the individual parts of the system that are necessary to actually implement and operate a land treatment project. Table 12.1 is a summary of the components necessary or usually found in the land treatment of industrial wastes.

The purpose of stage II is to design and specify each of the components in Table 12.1, so that accurate economic evaluations of land treatment can be made. Obviously, a site must have been selected to allow for stage II design. Where flexibility has allowed preliminary examination of four or five sites and detailed stage I evaluation of two sites, the actual component design is usually undertaken for only one site. Specific site data are needed, such as distances for transmission, topography and usable land area. These site

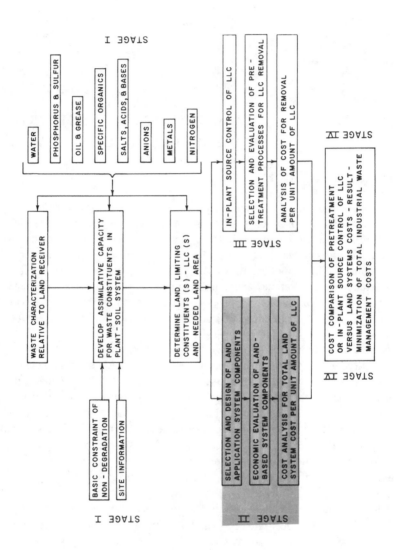

Figure 12.1 Relationship of Chapter 12 to overall design methodology for industrial pretreatment-land application systems.

Table 12.1 Typical Land System Components of Industrial Pretreatment-Land
Application System

Application System (spreader, irrigation, subsurface injection, etc.)
Storage
Transmission or Conveyance
Land Acquisition
Buffer Zone
Diversion Structures
Beautification and Security
Monitoring
Operational Control Systems
Vegetative Cover and Associated Agricultural Equipment

data will be discussed more completely in relation to each land system component.

Design of the land system is first based on the LLC and required land area for assimilation of 100% of the LLC generated. However, closer examination indicates that certain land system component costs are fixed, while others vary in some manner with the amount of LLC to be applied to land. Therefore, a series of costs can be developed in stage II that represents the land system cost per unit amount of LLC applied. That is, if the LLC is reduced in the waste stream, then what are the corresponding cost savings in terms of the land system. The relationship between the investment or total annual cost for the components needed for land application and the amount of LLC to be land applied is expressed as a land system cost curve (Figure 12.2). Reduction of the LLC from 100% or the raw waste stream level to 80% would then result in the corresponding cost savings, as shown in Figure 12.2. The development of Figure 12.2 is based on land area as modified by the specific site characteristics, and is the focus of Chapter 12. The evaluation of the land system cost curve is based on the design and economics of the individual components comprising a total land treatment system.

In the stage I analysis of assimilative capacity, waste generation or the LLC(s), there was no reference or dependence on waste source or physical characteristics. Thus, sludges, solids and liquids were treated alike. However, in stage II, specific allowance must be made for the nature of the waste stream to be land applied. In general, two main categories are used throughout Chapter 12:

1. liquids and certain slurries, and
2. sludges, solids and certain slurries.

This differentiation is a result of the various engineering and equipment requirements for the respective waste types.

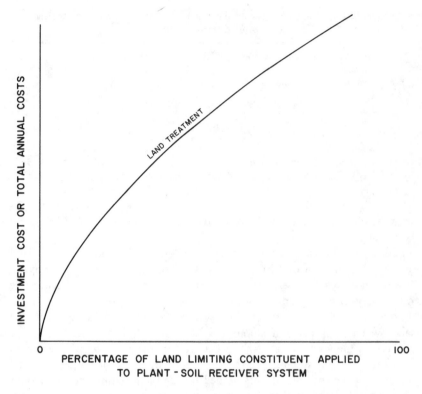

Figure 12.2 Representative cost curve for a land system in which varying amounts of raw waste land limiting constituent are applied to plant-soil receiver system.

In summary, stage II emphasizes the design factors and economics for the individual components necessary to deliver waste from the industrial source to the final application on land and maintenance of that land receiver. For each component an effort is made to discuss (1) design, (2) operation and maintenance requirements, and (3) economics. Then, based on the site under consideration, the components can be assembled as needed, and the land system design and costs can be developed in relation to the amount of LLC applied to the plant-soil system (Figure 12.2). Parallel to stage II and the material in Chapter 12, the designer must also complete stage III (Chapter 13) to assess pretreatment and in-plant source control options for the LLC.

WASTE APPLICATION SYSTEMS

Industrial wastes may be applied to land using two broad categories of techniques:

1. liquid and slurry distribution, involving pumping or gravity flow, and
2. sludges and solids distribution, involving tanks, vehicles and spreading devices.

The physical characteristics of the waste comprise one factor in selecting the application technique. Wastes can be pumped and handled as liquids if the solids content is less than 8% and if the paticles are less than 2.5 cm in diameter. As these upper limits of liquid handling are reached, specialized equipment is needed. Wastes containing greater than 8% solids and/or particles greater than 2.5 cm are better handled as solids or sludges. Corresponding to these two categories are different technologies for land application as a part of the overall land system design. This chapter section discusses these application categories separately.

In addition to the physical characteristics of an industrial waste, the LLC and required land area may control the land application technique chosen. This is particularly true for industrial wastes in which water is not the LLC, and in fact, very low volumes of water are land applied. For example, in an LLC analysis completed for a manufacturing facility, nitrogen was the LLC, with the hydraulic application on the order of 0.2 cm/wk. Although the waste was a liquid and could have been irrigated, the cost of installing a system on a large number of hectares exceeded the cost of using tank truck hauling and application. Techniques common for handling sludges were used successfully as the cost-effective application system. Therefore, with the complexities of industrial wastes and site characteristics, selection of the land application technique can vary substantially, depending on both technical and economic factors.

Liquid Wastes

There are two application methodologies for liquid wastes:

1. surface irrigation
2. sprinkler irrigation

These techniques differ substantially with respect to energy requirements, labor, initial investment, control of application rate, etc. In terms of irrigation of all liquids, including water for crop production, surface irrigation is the most common (more than 75% of U.S. systems). However, in a recent summary of 80 industrial land treatment systems, only about 25% were surface irrigation (Sullivan *et al.* 1973). This shift in application technique reflects the general trends to sprinkler irrigation (Powell *et al.* 1972). The two most important questions to answer in determining which of these techniques to choose: (1) do the site conditions and amount of waste to be applied suit surface irrigation criteria; and (2) what are the relative investment and operating costs? If conditions are not suitable for surface systems, then a variety of operational problems and system failure can result at installation (Powell 1972). However, within site constraints, surface irrigation can generally achieve similar uniformity of water coverage at lower investment and operating costs. In less developed countries, surface irrigation is more widely used for waste application than in the United States.

Surface Irrigation

  Methods. Two different concepts are used for surface irrigation:

  1. the flooding of level basins and subsequent infiltration and percolation of wastewater, and
  2. the introduction of wastewater to sloped areas with resultant overland flow and infiltration.

The first concept is referred to as basin or check irrigation and can be used for a variety of liquid industrial wastes, regardless of the LLC (Figure 12.3). Design for basin irrigation is determined predominantly by the LLC assimilative mechanisms and time period between application, as derived in the stage I analysis. The principal engineering design would focus on the volume of

Figure 12.3 Basin irrigation (Nutter, 1978).

water, area and depth of surrounding levees to retain the waste. Basin irrigation requires nearly level ground and, by the very nature of construction, prevents runoff from waste and rainfall inputs.

Surface irrigation on sloped land is designed as (1) border, (2) ridge and furrow and (3) corrugations. The border method of irrigation directs a flow of water over and down an area and confines it with dikes at either side. The area between dikes is referred to as the border strip (Figure 12.4). The strips are 10 to 20 m wide. Water is turned into the strips from a head ditch or supply pipe.

Furrow irrigation is a method by which water is applied to a soil in a series of well defined parallel channels so spaced as to saturate the root zone of the plant (Figure 12.4). The channels or furrows may vary greatly in size and shape from a small furrow between crop rows used in flat cultivation to deep narrow furrows used in ridge planting.

Corrugation irrigation is used with close growing crops such as wheat and pastures. The corrugations are small furrows (Figure 12.4) installed to supply water for soils which crust heavily when flooded. The disadvantage of corrugations is field roughness when machinery moves perpendicular to the furrows.

In terms of industrial waste, surface irrigation on sloped areas can only be used if water is the land limiting constituent (LLC). Wastewater application and percolation into the soil occur at maximum rates allowable by soil texture and field conditions. Thus if another parameter were the LLC, the waste would still be entering the plant-soil system at the maximum hydraulic rate. Excess loading of the LLC would occur and the basic design constraint for land treatment would be violated.

The design of surface irrigation for sloped areas centers on the volume of water flow and the velocity as determined by slope. The objective is to control the length of flow and velocity such that nonerosive conditions exist (Israelson and Hansen 1962). For these surface flow irrigation techniques, the design basis is the unit stream flow, Q_u, or the volume of flow per 0.3-m width of wastewater flow per 30-m of land application length. In Table 12.2, the maximum Q_u as a function land slope for bare soil is given along with the recommended total flow length and spacing, as a function of soil type. If the land area is in sod, then higher nonerosive velocities can be used. Slopes of 0.2-0.4% are preferable for surface irrigation, although from 0.1-7.5% have been used successfully. Land preparation is necessary to make the water flow perpendicular to the slope as uniform as possible, so that channeling does not occur. A single slope should be used in the direction of flow. If several slopes occur, the wastewater levees and distribution dikes should be constructed for each sloped area. Extensive work in further optimizing these surface irrigation techniques has been performed by Powell (1972), Fok and Bishop (1965) and Christiansen et al. (1966).

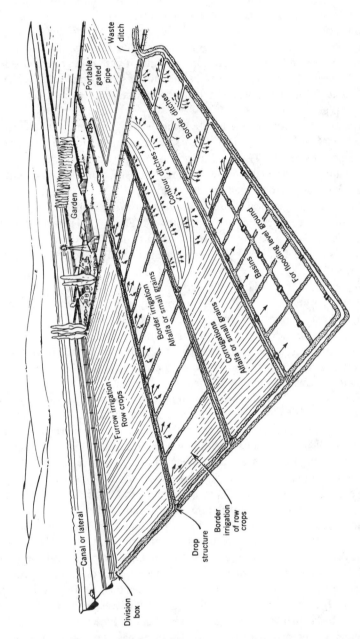

Figure 12.4 Schematic of furrow, border and corrugation methods of surface irrigation (U.S. Soil Conservation Service, 1947).

Table 12.2 Design Criteria for Surface Irrigation—Allowable Erosion Free Flows
per 0.3 meter of Border Width and 30 m of Border Length (SCS, 1957)

Slope	Maximum Stream $(Q_u)^a$ (m^3/min)	Slope	Maximum Stream $(Q_u)^b$ (m^3/min)
0.3	0.25	1.0	0.10
0.4	0.20	1.5	0.073
0.5	0.17	2.0	0.060
0.6	0.15	2.5	0.051
0.7	0.13	3.0	0.044
0.8	0.12	4.0	0.036
0.9	0.11	5.0	0.031

[a]For border strips without sod protection. Larger streams may be used with sod cover.

Lengths of Run for Furrows and Corrugations
(U.S. Bureau Reclamation 1951)—Length of Furrows or Corrugations in meters

Slope (%)	Loamy Sand and Coarse Sandy Loams	Sandy Loams	Silt Loams	Clay Loams
0.2	75-120	90-200	200-400	270-400
2-5	60-90	60-90	90-200	120-270
5-8	45-60	45-60	60-90	75-120
8-15	30-45	30-60	30-60	60-90

Optimum Furrow or Corrugation Spacing (McCulloch *et al.* 1973)

Soil Condition	Optimum Spacing (cm)
Coarse sands—uniform profile	30
Coarse sands—over compact subsoils	45
Fine sands to sandy loams—uniform	60
Fine sands to sandy loams—over more compact subsoils	75
Medium sandy—silt loam—uniform	90
Medium sandy—silt loam—over more compact subsoils	100
Silty clay loam—uniform	120
Very heavy clay soils—uniform	90

Surface irrigation is advantageous where wind velocities are excessive, the soil system has a low infiltration capacity, or a lower initial investment and availability of inexpensive labor are factors. Since certain areas are periodically submerged, the design should ensure adequate drainage and aeration time, so that the top soil zone remains aerobic and viable. Successful operation can be achieved for wastewater systems with surface irrigation techniques.

Economics Evaluation. Investment costs for surface irrigation are very site specific. If constant slope or level areas exist with the correct field dimensions, then land grading is not necessary and a substantial savings can be realized. Generally, earth moving in site preparation is needed; average estimates are given in Table 12.3a. Once graded, the actual furrows or levees can be constructed and a distribution system installed. Estimated investment and operations/maintenance costs are given in Table 12.3a and 12.3b. Lifetime factors for amortization are also approximated. These costs and economic analysis factors represent the surface irrigation component of a complete land system design. Manipulations to obtain annual costs for surface irrigation can be performed using economic criteria relevant to each industry.

In summary surface irrigation by flooding of basins is appropriate for any industrial waste with liquid characteristics while surface irrigation by ridge and furrow, border or corrugation techniques is only appropriate for industrial wastes in which water is the LLC. Design for surface irrigation involves construction of retaining levees or channels, and maintaining non-erosive wastewater flow velocities. However, site characteristics are the

Table 12.3a Surface Irrigation Investment Costs

Surface Irrigation System Component	Cost ($/ha)	Lifetime (yr)
Land Leveling[a]		
900 m³/ha, moderate or average	920-1,230	
1,400 m³/ha, substantial or large	1,400-1,850	
Furrow or Border Construction		
2.5-5 hr/ha	110	1
Distribution		
Gated pipe[b] (12 m lengths)	440	10
Concrete trapezoidal ditch[a]	420	10
Representative Total Cost Range (1978)	1,450-2,400	
Representative Total Cost Without Land Leveling (1978)	500-590	

[a]Pound et al. (1975)
[b]Sneed (1978), ditch or border reconstruction yearly.

Table 12.3b Operating and Maintenance Costs for Surface Irrigation

Site	Operation and Maintenance ($/ha/yr)
Grean Giant Co. (NM, 1953)	125
Citrus Plant (CA, 1950)	92
Bakersfield (CA)	220
General Design (Pound et al. 1975)	340-880

predominant factors, since earth moving to achieve a site suitable for standard design can be very expensive. The advantages of surface irrigation can be achieved by site selection. Economic analysis is based on the costs in Table 12.3 when adapted to the proposed sites for such factors as land leveling, dimensions of fields and nature of distribution. Surface irrigation is characterized by low-to-moderate investment costs and moderate operation and maintenance costs. Conditions of availability and low cost labor further enhance the cost-effectiveness of surface irrigation, e.g., in developing countries. Finally, the use of surface irrigation eliminates the need to provide buffer zones, another component of a land system. The nuisance effects for industrial waste or the aerosolized effects of wastes containing potentially hazardous microorganisms are virtually eliminated with this method of waste application.

Sprinkler Irrigation

Using rotating impact sprinklers (Figure 12.5) with main and lateral piping systems for industrial waste, in place of surface irrigation, represents

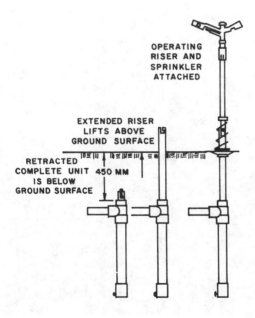

Figure 12.5 Telescoping riser and rotating impact sprinkler (Pair, *et al.*, 1969)

an attempt to better control liquid application and increase the number and type of sites that can receive waste. Sprinkler irrigation is thus an evolution of general irrigation practice, which has had a substantial impact on industrial and municipal land treatment. In a 1973 survey, more than 75% of industrial land application was by means of sprinkler systems (Sullivan *et al.* 1973). The evolution of sprinkler technology assumed low energy costs and relatively low materials costs. As the energy picture changes, there are modifications toward more efficient operational techniques for spray irrigation, as well as consideration of surface irrigations. Sprinkler techniques for industrial waste application to land allow greater control, so that the nature of the LLC does not affect the design and more sites are amenable for use in land treatment.

Methods. There are a number of options within the range of sprinkler or spray systems. The three principal techniques are:

1. fixed position (solid set) with rotating impact sprinklers (Figure 12.6).
2. traveling pipe and sprinkler (Figure 12.7), and
3. center pivot with rotating pipe and sprinkler (Figure 12.8).

These systems will be discussed separately then compared with respect to design limitations and economics.

Figure 12.6 Solid set sprinkler irrigation.

Figure 12.7 Traveller device for sprinkler irrigation.

Figure 12.8 Center pivot irrigation.

Solid set irrigation is the most common system for industrial land treatment, although the economic or technical justification for this choice is not well established. Experience or mimicry of existing land application systems appears to dictate this choice at this stage of industrial land treatment technology. Solid set arrangements may be either permanent or movable, with the construction being similar (Figure 12.9). Mainlines or headers convey waste throughout the site, and laterals with sprinklers distribute the waste over the land area. Movable systems are primarily of aluminum construction with a wide diversity of connection techniques. However, with most waste systems, worker acceptance of handling and moving pipe has been unfavorable. Thus, movable piping and lateral systems are not widely used at present, although with proper education and training the economic advantages of such transportable modular systems could be realized.

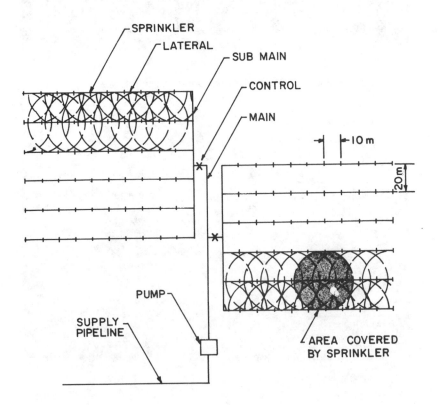

Figure 12.9 Schematic of design for solid set sprinkler irrigation (Norum, 1976).

The dominant piping-sprinkler system is designed to completely cover the land application area, with either aboveground aluminum pipe or belowground PVC pipe. Specific site conditions, such as vegetative cover, climate, topography and management options, dictate the selection of type of laterals and mainline. For above- and belowground piping, the risers containing the sprinkler (Figure 12.5) must be supported firmly perpendicular to the slope of the land. The primary factors to achieve the design application rate are the size and flow characteristics of the sprinkler, and the time of pumping/irrigation.

Traveler irrigation equipment is a second option for sprinkler systems (Figure 12.7). The development of traveling devices was aimed at reducing the substantial investment cost per hectare associated with solid set systems. When compared with a modular piping, multisprinkler system, which is transported from zone to zone at a land application site, the traveling units also require less labor per unit area. The principal traveling system used in industrial land treatment is a self-propelled vehicle with a single high-capacity sprinkler (Figure 12.7). The vehicle speed is controlled by a winch, powered by an internal combustion engine or a water-powered system. The water-powered winch can be driven by a water turbine or by a ratcheting cylinder drive. The unit moves itself along a travel lane at a predetermined speed, pulling a flexible hose for its wastewater supply. The hoses are usually 10.2-'1.4 cm (4-4.5 in.) in diameter, in continuous lengths of up to 402 m (1,320 ft). The winch is supplied with 402 m (1,320 ft) of cable, thus controlling the length of run on either side of a main wastewater supply hydrant. In addition to the self-propelled traveler, there are several other traveling systems that reduce the investment cost per unit area, but these are less commonly used for industrial land treatment systems. These alternative systems are described in detail by Woodward (1969).

The center pivot irrigation device represents a further refinement toward reduced investment cost per land treatment area. Center pivot machines cover up to 73 ha and consist of a central wastewater hydrant and a radial distribution boom, which moves circumferentially to cover a circular land area. While center pivot can be used on areas of 4 ha, the real cost savings are economies of scale associated with large land areas. Since the center pivot machine can be transported between main waste distribution hydrants, the investment cost per land area can be further reduced.

The three sprinkler irrigation options described above represent the large majority of delivery techniques for land treatment of liquid industrial wastes. Actual selection depends on site conditions and the objectives of reducing investment costs or utilizing available labor in land application. Subsequent sections of this chapter will discuss the design, selection considerations and economics of these three sprinkler application options.

Design. The design of land application equipment for industrial waste liquids is complex. Considerable field experience, repetitive design and related engineering judgment are necessary, especially when compared to the magnitude of the investment involved, the implications of an unreliable waste management system and the need for a functioning, long-term delivery capability. These considerations, and the authors' experience with less-than-well-designed systems, argue strongly in favor of having industrial waste irrigation experts do the actual engineering specifications and plans.

The purpose of the design section of this book is to acquaint the industrial environmental engineer or coordinator to a sufficient level that a perceptive review and definition of irrigation needs is possible. That is, the responsible industry personnel or consultant must be able to describe thoroughly the needed application system to the industrial waste irrigation specialist, as well as review the design to assure reliable performance relative to land treatment.

Solid set, multisprinkler systems can be understood by using a design step technique developed by Myers (1974) (Figure 12.10). The approach illustrates well the concepts from which the industrial waste application system can be designed or reviewed. These principal variables are: (1) weekly industrial waste liquid flow; (2) weekly assimilative capacity or loading rate, expressed as depth of liquid (equivalent to volume of liquid per unit area); (3) hourly application rate; (4) sprinkler spacing; and (5) nozzle operating pressure. From a conceptual viewpoint these variables are considered independent; however they are actually interrelated, as would be evident in an actual engineering design of such systems (Klose 1978a).

The waste liquid generation rate (volume per unit time) from an individual facility, in combination with the weekly loading depth, determine the overall

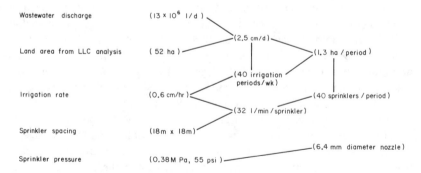

Figure 12.10 Design flowchart for solid set sprinkler irrigation system for industrial waste (Myers, 1974).

size of the land area and irrigation distribution system. The loading depth is usually expressed in cm of wastewater applied each week over the entire disposal area. For an industrial waste, the LLC analysis (Chapter 2) specifies the land area requirements to assimilate the controlling waste constituent. The volume of waste (ha·cm/wk or ac·in./wk) divided by the needed land area (ha or ac) yields the depth of liquid to be applied. Thus, for 13 million l/wk and an area of 52 ha, the application rate is 2.5 cm/wk, while an area requirement of 104 ha would lead to an application of 1.25 cm/wk.

Data collected at the land treatment site using drill rigs, undistributed soil core sections and topographical analysis, are used to determine the allowable infiltration rate or hourly application of an industrial waste, cm/hr. The infiltration rate is the instantaneous inflow of wastewater beyond the surface layer of a plant-soil system. Both the soils and vegetation must be accounted for in specifying the infiltration. The objective is to prevent any runoff of applied waste liquid, hence the need for site-specific information.

The application rate, expressed in cm/hr, in conjunction with the loading depth per week (cm/wk), establishes the number of irrigation periods per day or per week. Thus, for a 2.5 cm/wk loading depth applied on one day (and an application rate is 0.6 cm/hr), each irrigation period would be 4.2 hours, and there could be approximately 6 irrigation periods per day or 40 periods per week, (Figure 12.10). That is, 40 separate zones could be irrigated each week to cover the entire industrial land treatment site. When using an application rate of 0.4 cm/hr, each irrigation period would be 6.25 hours, and there could be about 4 irrigation zones per day or 26-27 periods per week.

Having followed the design from total waste load to number of irrigation periods per day or week, the physical dimensions and sprinkler flows can be determined. As a first estimate, a spacing of 18 m between laterals and 18 m between lateral risers is used for forest land. In open fields, 30 m between laterals and 24 m between laterals risers is appropriate. Using a 18 m x 18 m spacing and the infiltration rate, the flow per sprinkler, l/min, is determined (Figure 12.10). The reader is cautioned that the above spacings do not represent the least-cost dimensions, since the specific site and waste conditions are the determining criteria. The industrial waste irrigation specialist would adapt the design to the individual circumstances for the most economic design.

Selection of a rotating impact sprinkler is based on the needed flowrate per sprinkler and the operating pressure economically available from the pump transmission system. These two criteria and sprinkler head design (which varies with manufacturer) allow specification of a nozzle size (Figure 12.10).

With the waste application liquid depth on a weekly basis, the maximum number of application periods per week, and the flowrate of each sprinkler, further outline of the sprinkler system can be made. Management choice, climatic patterns and economic design factors are used to determine the optimal number of irrigation zones and periods per week or month. Then with the waste application liquid depth per irrigation and periods per week, the wetted area per period or the zone size can be determined (Figure 12.10). Finally, the number of sprinklers per zone is determined, knowing the flow-rate per sprinkler. By following the conceptual approach (Myers 1974) and adapting for industrial land application, the reader can accurately understand a detailed solid set sprinkler design.

As a practical matter, developed from field experience, there are important additional considerations in a spray system for industrial waste. In low rainfall areas the problem of scheduling industrial waste applications is simplified. Irrigation can occur as frequently as desired, with little chance of simultaneous rainfall. Thus, with slight adjustments for infrequent rain events, there is little need for storage, and even daily irrigation is possible. Because of these frequent spray conditions, the waste liquid can be applied with smaller pumps, lower-diameter transmission lines, and in a larger number of sprinkler zones. Substantial cost savings can thus be realized in investment expenditures.

In higher rainfall regions, greater attention is needed for the interaction of precipitation events, which occur frequently (at least during large portions of the year), and the application of an industrial waste. The climatological constraint involves (1) the ability to have sufficient soil storage to receive a waste in which water is the LLC, or (2) the elimination of irrigation for impending or occurring rainfall, as a means reducing rainfall-runoff transport when other than water is the LLC. Because of the random nature and higher frequency of rainfall, there is a good chance of waste application conflicting with rain events when irrigation is required 4-7 times per week. This interaction requires substantial storage, since successive rainfall events continue to conflict with at least a part of the weekly irrigation. Thus, greater success is achieved with larger pump and transmission capacity, so that the entire weekly waste load can be applied in one or two days. Flexibility of irrigation can be maintained and, based on ongoing field observation, the application can be conveniently completed at varying irrigation days within a given week. This consequence of rainfall region means a greater expense, but is necessary to assure reliable performance. The design, climatological simulation and cost optimization are strong justification that industrial waste irrigation expertise is essential for a workable land application system. Furthermore, fixed or uniform design criteria, as contemplated by regulatory agencies, will prove invalid because of the critical, site-specific characteristics of land-based treatment.

Winter operation poses additional design complexities, which need to be addressed for any year-round industrial waste discharge. Process heat present in the liquid waste should be considered to determine whether any icing is likely. For many industrial wastes a design/economic analysis must be constructed to balance storage vs the cost of specialized winter operation equipment. Since these analyses are highly site and industry specific, only the major factors for consideration are covered in this chapter.

Myers (1966) has reported on equipment for winter operation. Solid set mainline or laterals must be buried below the potential frost line, or if aboveground, must be equipped with self-draining pipe connectors. In this manner, no water stands in the riser or pipes for freezing. The rotating impact sprinkler ices substantially during freezing conditions. A rotating deflector spray head was found to be more reliable, but required closer spacing and still did not give uniform distribution (Kardos *et al.* 1974). The extra lateral and riser cost and mixed performance lead to the conclusion that a reasonable performance can be obtained with the regular sprinkler head. A ridge and furrow or a gated pipe trickle system were found to work very well under freezing conditions in Pennsylvania and Minnesota; thus, these represent a workable alternative to achieve year-round operation. The low hydraulic load associated with most industrial wastes (water is not LLC) improves the opportunity for year-round operation because irrigation events can be varied to match favorable temperature conditions. However, in areas of high snow and prolonged subfreezing temperatures, often the only option is storage.

In addition to the preceding solid set design considerations, a series of basic performance factors must be included in industrial sprinkler irrigation systems. Flow variations at the sprinkler can not deviate by more than 10% throughout the entire irrigation zone, especially under conditions in which the LLC is not hydraulic load. For wastes with hydraulic LLC, the flow through soil tends to equilibrate nonuniformities; but for other constituents, which do not migrate, design for uniformity is critical. Elevation changes, economic factors favoring pipe size changes, operating sprinkler pressure under prolonged use and pipe system drainage for freeze protection represent complicating factors that must be included in an industrial sprinkler delivery system design.

Traveler systems are designed with more attention toward a specific piece of equipment. Although there are more than 10 types of moving or traveling systems, the self-propelled device (Figure 12.7) is the most widely used traveler for industrial wastes. A traveling gun design requires that several variables be specified. These are (1) the waste volume to be applied per hectare of land (from LLC analysis), (2) the wetted diameter of the available sprinkler gun, (3) the rate of movement or travel, and (4) the power package to be used for machine propulsion. With the needed land area and resultant waste volume per unit area for a given industrial waste, the designer can decide on the amount to be sprayed per application, or the number of applications per year. Constraints from the plant-soil assimilative capacity (Chapters 4-11), such as no more than x kg of waste constituent are allowed per application, may set the pattern for industrial irrigation frequency. The rainfall factors described previously under *solid set* design can dictate frequency of application, as can the economics of pump and transmission line sizes. Whatever the deciding factors, the industrial waste irrigation frequency must be established, from which the total waste volume per application per unit area can be determined.

The area covered by one rotation of the sprinkler gun and the rate of traveler speed determine the instantaneous application rate, cm/hr. The lowest application rate presently feasible is 0.8 cm/hr, which exceeds the infiltration capacity of some soils, particularly those without forests or grass sods. Thus, the soil infiltration, as measured by field testing in the Stage I process, controls the upper limit of allowable application rate. If the maximum feasible application rate is less than the waste volume per unit area to be applied, then more than one travel cycle must be included. Thus, from the soils and the traveler equipment specifications, the traveler velocity and wetted area are determined. Spacing between traveler lanes are then set to provide the overlap and application uniformity critical to many industrial wastes.

As a final design stage, the propulsion unit is selected, based on energy use, engineering and economic considerations. Three basic alternatives exist: (1) engine powered, (2) hydraulic piston driven, and (3) the radial inflow turbine movement. Natural gas, diesel or other fuels are used in engine-driven travelers with reliable results, particularly during the first three years. For industrial wastes, these units are preferable because the liquid, often containing large solids, is not used to operate the propulsion engine. Corrosion and clogging of the traveler are thus reduced. Engine-driven units are the most expensive traveler.

The piston-driven units are conceptually simpler in design, although mechanical stress on operating parts requires heavy duty construction. Thus, these units are larger in size. Additionally, the waste used to drive the pistons must be discharged separately from the sprinkler nozzle. By contrast, the inflow turbine propulsion unit is smaller and less expensive to purchase. All waste liquid exits through the nozzle. However, problems of constant traveler speed still remain.

Traveler devices operate from a series of hydrants (Figure 12.11). Hydrants are usually supplied by buried mainline pipe at an industrial land treatment site. Operating pressure needed at the traveler averages 620 kPa (90 psi), depending on gun size. There is a high degree of spray distortion with wind velocities above 8 km/hr. There is good design flexibility for irregular field shapes, making the traveler the most portable of the sprinkler alternatives. The design of travel lanes can easily accommodate power poles and other

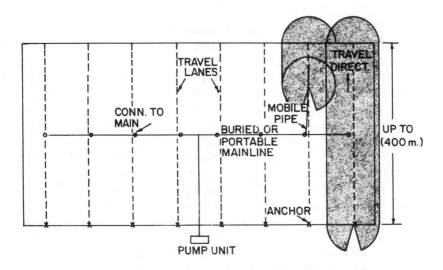

Figure 12.11 Schematic of design for traveler device for sprinkler irrigation (Norum, 1976).

obstructions present in land application fields. Also, the traveling gun is the only sprinkler option capable of handling substantial solids present in an industrial waste.

Under winter operations, traveling gun machines can operate under brief, moderate freezing conditions. Ice buildup on the gun and rear structural areas can exceed the power or strength conditions; however, with an engine winch traveler, the waste heat can be utilized, along with other techniques, to keep ice buildup to a minimum. Again, the added expense of specialized operating equipment should be evaluated vs storage. In comparison to solid set systems, winter performance of certain travelers can be superior. Under prolonged freezing or snow pack conditions, storage is the only option.

Center pivot equipment is designed based on two factors: (1) required land area (from LLC analysis), and (2) the waste volume per unit land area. The first factor determines the size of the machine, while the second factor dictates the pump rate and ground speed. Field topographical data and shape are needed to determine the feasibility of center pivot use and, hence, the necessary design. The entire center pivot machine must be drained to prevent freezing. For many industrial wastes, the limitation of aluminum as the center pivot material may be such that corrosion replacement costs must be considered in the sprinkler system selection and design. Drainage for freeze or corrosion protection requires a sump or other facility to prevent runoff of the industrial waste liquid.

Under winter conditions, a center pivot device can be adapted to allow irrigation during prolonged freezing periods. Top-located rotating sprinklers are replaced with low-pressure nozzles attached to downward risers. In this manner, ice does not build up on the superstructure, and a rotating sprinkler is not used (Dorset 1978). Thus, for extensive winter operation, center pivot devices are more feasible than the solid set or traveler options. For irrigation of solids the center pivot and solid set system have the same limitations associated with sprinkler head nozzle sizes.

A sprinkler system comparison of design factors is presented in Table 12.4. This description is brief and the reader can refer to the previous detailed sections, as needed. The sprinkler system options are purposefully limited to four out of more than 15 possible sprinkler irrigation techniques, because these four have evolved as most widely used for industrial wastes. Obviously there are certain advantages and limitations for each sprinkler system option, so the designer must review the particular system under consideration. For example, if a forested area is the receiver system, then a traveler or center pivot device are eliminated. If winter operation is required, then the solid set or traveler options are less desirable. Thus, site characteristics impose a first level of design constraint. However, it is also obvious that for many conditions, all of these alternative sprinkler systems are, or can be designed to work successfully. Adaptation of design variables to deliver the required industrial waste volume to a specified land area at, or during, certain time periods can be made by specialists with direct consideration of the relevant characteristics of the waste. Economics of the respective sprinkler options, as designed for each site, then controls the selection of a system.

Some general system comparison is useful for the industrial designer to better understand the detailed equipment design by an industrial waste irrigation specialist. The multisprinkler solid set system, whether portable or not, has the greatest flexibility and control of the application for most liquid wastes. Highly irregular field shapes, typical of several regions of the United States, and topographies, can be easily accommodated. There are no limitations on vegetative cover, even including forest irrigation. However, the solid set systems require a very large piping system, since water is transmitted to each section of land to be wetted. Limitations also exist for winter operation and for more difficult-to-handle industrial liquids. In total, most industrial waste liquid application systems are solid set, thus indicating the emphasis on this option.

The traveler and center pivot evolved to replace the large piping system and cost of solid set systems, by using a moving application device. The traveler is more versatile in terms of field shape, terrain and crop selection, although neither can function in forested areas. Both of these movable systems have a definite lower instantaneous application rate, which exceeds acceptable

Table 12.4 Design and Performance Comparison of Principal Sprinkler Irrigation Options for Industrial Land Treatment Systems

	Permanent Solid Set	Hand-Move Solid Set	Self Propelled Traveler	Center Pivot
Prolonged Freezing Operational Capabilities	Poor to moderate	Poor to moderate	Moderate, with adaptions	Good, with adaptions
Solids Handling Capacity	Poor	Poor	Good	Poor to moderate
Wind Drift	Moderate	Moderate	Substantial	Low, with adaptions
Allowance for Obstructions	Excellent	Excellent	Good to excellent	Poor
Field Coverage for Single System (ha)	Unlimited, controlled by technical and economic factors	Unlimited, controlled by labor supply	2-16	4-75
Instantaneous Application Rate (cm/hr)				
Min.	0.13	0.13	0.75	1.2
Max.	5	5	5	3
Averaged Application Rate (cm/hr)				
Min.	0.13	0.13	0.8	0.8
Max.	5	5	5	5
Maximum Field Slope (%)	Unlimited, controlled by soil and ground cover to prevent runoff	Unlimited, controlled by soil and ground cover to prevent runoff	Unlimited, controlled by soil and ground cover to prevent runoff	<5
Shape Limitations	No limit	No limit	Rectangular	Square
Unusable Crop Area Within Wetted Area	None	None	<4%	<1%
Maximum Crop Height (m)	Unlimited	Unlimited	Any field crop height can be accommodated	3.5
Average Nozzle Pressure kPa (psi)	200-400 (30-60)	200-400 (30-60)	350-700 (50-100)	100-400 (15-60)

Table 12.4 Continued

	Permanent Solid Set	Hand-Move Solid Set	Self Propelled Traveler	Center Pivot
Portability	None	Disconnect and reassamble	Tractor or truck hitch and tow	Rotate wheels and tractor tow
Energy Consumption	Least, since no system movement and low pressure are needed	—	Highest, due to movement energy and high nozzle pressure	Intermediate, with low pressure nozzle but pipe and sprinkler movement
Comments on Use	Aboveground version well suited to forest, below-ground moderately suited, low labor and versatile to design	Labor intensive, amenable to forest systems, versatile	Extremely portable, restricted by water infiltration, well suited for industrial wastes with solids, open fields only	Economics to large scale, restricted by water infiltration and field conditions, open fields only, advantageous for winter operation

limits for significant land areas, particularly in the eastern U.S. However, the advantages of being easily portable are very substantial in reducing the size, and therefore cost, of equipment needed to cover a specified land area. This is an economic factor covered in a later section of this chapter. In addition, the traveler device is the best sprinkler option for solids-containing industrial waste, while center pivot is superior for prolonged freezing operations.

Economics. The cost of irrigation systems has an apparently wide range when a literature compilation of existing or projected systems is made. However, in-depth analysis reveals that often these systems include different costs or conditions, so are not expressed on a common denominator. The economic values for sprinkler irrigation systems *per se*, as well as the transmission to the site, pump and other land system components, can be established within reasonable reliability.

The costs described below are to be used by the reader as benchmarks for evaluating economic estimates for land treatment. Costs can vary because of increasing prices of goods and materials, degree of difficulty at land application site or depending on the proper use of competitive bidding. In an industrial waste system, the components of the land application, including the sprinkler or surface irrigation portion, should be investigated and bid separately. This approach identifies potentially high-cost items, which can be reviewed for accuracy and potential substitution. In addition, the very high cost systems (those exceeding rule-of-thumb costs given below) are generally excessively overdesigned or represent some error in judgment about the system.

For the principal sprinkler irrigation options, as one component in an actual land application system for industrial waste, investment costs are presented in Table 12.5. Costs from a number of sources indicate the variation in design and installation expenses, as well as the increasing costs with inflation. There is some increased cost per hectare associated with smaller systems (<20 ha) for the traveler and center pivot devices. For the solid set option there is little economy of scale for larger systems. Solid set system investment costs on an installed basis are approximately the same for aboveground aluminum as for buried PVC pipe.

The economic estimates derived from Table 12.5 represent minimum cost values. As with many items, unless sound competitive bidding is used, the actual costs can be higher. Another factor that has evolved from field experience is the inflationary effect of complex specifications as a trade-off for experience in installing irrigation systems. The bid costs in the former are usually much higher than those obtained from using experienced industrial waste irrigation supervision. As a representative investment cost, using 1978 values, the solid set, traveler and center pivot would be $2,500-3,000/ha, $300-1,000/ha and $800-1,000/ha, respectively. Allowances for size factors

Table 12.5   Investment Cost for Sprinkler Irrigation Component
of Land Treatment System

Source	Investment Cost, $/ha		
	Solid Set	Traveler	Center Pivot
Klose (1978[a]), based on recent industrial installations	2,100-3,100	320-1,000 (6-16 ha)	–
Kardos (1974), 1967 dollars	1,300	–	–
Pound et al. (1975) (16-3,250 ha) (194[a])	1,800[b]	–	–
(30-3,250 ha)	–	–	520
Allender (1972) (119.4[a])	520[b]	–	–
Pierce and Bramerel (1974)	1,980[b]	–	–
Norum (1975) (16 ha)	–	–	440
Pound and Crites (1973) (52 ha) (192[a])	1,200[b]	–	–
Norum (1976)	1,000-2,500[c]	300-600	440-860
Sneed (1972)	1,000-2,200	300-740	300-600
Hill and Keller (1978)	2,100[b]	–	–
Dorset (1978) (4-30 ha)	–	320-800	1,100-2,600
(30-75 ha)	–	–	890
Value judged as representative of 1978	2,500-3,000	300-1,000	800-1,000
Lifetime for industrial land treatment system	10 yr	2-5 yr	5-7 yr

[a]EPA Sewer Construction Cost Index.
[b]Aboveground aluminum pipe.
[c]Buried PVC pipe.

can be judged from Table 12.5. To determine annual depreciation, the sprinkler irrigation system lifetime must be estimated. For industrial waste liquid irrigation, the conventional agricultural lifetimes were reduced by 50% as a conservative estimate (Table 12.5). The nature of the industrial waste being irrigated would substantially control these depreciation lifetimes.

Operation and maintenance costs are important in assessing the annual financial requirements of an ongoing system. Labor needs of the several sprinkler irrigation options are given in Table 12.6 to be evaluated with the required annual number of application events and hourly labor costs. In general, the materials replacement costs on an annual basis are 10-20% of the labor expenditures (Pound et al. 1975). The traveler and center pivot options require separate propulsion power requirements, and these are included in Table 12.6. Thus, on a given industrial system, the sprinkler irrigation operation and maintenance cost can be estimated from labor expenditures, materials replacement costs and system power costs, if appropriate.

Total annual costs for a sprinkler irrigation system are determined as the sum of amortized investment costs and the annual operational expenses.

Table 12.6 Operational Factors for Economic Analyses of Sprinkler
Irrigation Systems for Industrial Wastes

Source	Operation Labor Requirements (hr/ha/application event)			
	Multisprinkler			
	Solid Set	Hand-Move	Traveler	Center Pivot
Sneed (1972)	0-12-0.5	2.0-3.7	0.25-0.75	0.12-0.37
Norum (1976)	0.12-0.25	0.5-1.2	0.25-0.75	0.12-0.37
Pound (1975)	0.5-1.2	–	–	0.6-1.2
Allender (1972)	0.04-0.08	–	–	–
Value Judged Representative	0.12-0.5	2.0-3.7	0.25-0.75	0.12-0.6
	Power Costs in Addition to Delivery Pumping Costs to Site (kWh/hr of operation)			
Pound et al. (1975)	None	None	–	0.42-0.68

From Table 12.5 and with a depreciation rule (straight line, decreasing balance, etc.), the annual amortized investment is determined and added to the operation and maintenance costs and computed for a year, to yield the total annual costs. Typically, the O&M expenses are 40-60% of the total annual costs.

Sludge and Solid Wastes

Industrial wastes containing less than 92% moisture, or particles in excess of 2.5 cm in size, are applied to the plant-soil system with a variety of techniques and equipment. This cateogry of materials is referred to in this chapter as sludge, partial fluids, thickened slurries, semiliquids or solid waste, without regard to the industrial origin of the waste, because of the commonality of application technique. As an industrial waste category, the application of these materials is labor intensive and very dependent on the amount of material (percent water or liquid) to be transported. Solids or sludge applications are characteristically infrequent on a given area of soil with rotation of spreading areas, although with some industrial sites, frequent spreading of small waste amounts is required and practiced.

There are a large number and diversity of application devices, handling equipment and options for land spreading; but in overview, two principal methods can be used to summarize the available alternatives:

1. semiliquid material, and
2. solid or low-moisture materials.

Semiliquids include materials between 3 and 15% solids, thus including a range of physical properties. Low-moisture materials are those in excess of 15% solids. The design of industrial land application systems for sludge and solid materials will be discussed separately for these two principal handling methods. The economic factors will be discussed as one section. The reader should note that by the very nature of a sludge hauling/application system, the distribution on land and the transmission to the site are usually together, by contrast to the surface or sprinkler application, in which transmission to the site is considered in a later section as a separate component of the total land system. This differentiation is important in constructing the overall economic evaluation of the land treatment system.

Semiliquid Systems

Equipment and techniques in this land spreading category consist of a movable tank in which mixing or agitation can be attained and from which the waste is applied to land by a spreader mechanism. The movement or propulsion of the tank can be by tractor hauling (Figure 12.12) or with a tank mounted on a truck frame (Figure 12.13).

Figure 12.12 Tractor-hauled liquid spreading or injection equipment (Badger Northland, Inc., Kaukauna, WI).

Figure 12.13 Truck-mounted hauling equipment for semiliquids and sludges (Ag-Chem. Equipment Co., Minneapolis, MN).

Truck-mounted tanks range from 5,000-22,500 ℓ, with a typical size of 9,500 ℓ. Tank material should be a minimum of carbon steel, with stainless or coated steel for appropriate industrial waste. Safety valve, top access hatch and mechanism for complete draining should be included. Agitation and field application are provided by (1) a pump, centrifugal or positive displacement depending on usage, (2) pump-drive shaft with auger vanes, or (3) a pressurized system with compressors to create vacuum for loading waste, and pressure displacement for unloading or spreading. Agitation in the first option is by pump-tank recycling, in the second option by shaft rotation, and in the third option by forced air sparging. Specifications are based on a prescribed solids content and flow properties, as related to the industry-particular waste.

As characteristic of many industrial wastes, the application to land is a year-round operation. Therefore, special provisions are needed for truck movement across wet fields. Flotation tires (Figure 12.13) allow for a low compaction factor of approximately 100 kPa (15 psi), thus permitting increased field access. Special consideration in design or specification must be directed toward the steering, braking and maneuverability of the truck, on which high flotation tires are used. The increased cost of flotation tires is more than offset in lower storage requirements, lower field compaction and decreased field damage.

An alternative to use of flotation tires and hauling to a treatment site is the pump-injector system (Smith *et al.* 1975). Wastes are pumped in headers to a series of hydrant points located in each 7.5-ha land section. A track or wheel-type tractor with the injection sweeps is connected to a hydrant by a 10-cm-diameter flexible hose. The operator then drives and injects throughout the 7.5-ha section, discharging the waste under the wings of the high-lift plow blades, and the soil folds back on top of the waste. The principal advantages are: (1) little weight of the waste is present at any one time on the field, reducing compaction, and (2) pump and pipe are substituted for driving time, to get the waste to the land treatment site. As described in a later section, the alternative of flotation tires and hauling may be less expensive than the continuous injection system with pump and transmission pipe, depending on the waste and site used.

Tractor-haul tank wagons are generally smaller in size (1,900-9,500 ℓ), but are constructed and operated in the same manner as the truck-mounted tanks. The same options and specificatons are available for both methods of tank movement. Flotation tires must be specified as an extra item as justified for the truck tank waste applicators. Tractor modification is also required to have an efficient waste spreading facility. Rear flotation tractor tires are recommended as are front weights for pulling and traction capacity. An enclosed, heated cab is necessary for year-round operation.

Liquid spreading devices vary in degree of sophistication, amount of control and land spreading objective. Similar devices are found on tractor haul or truck tanks. The simplest is a single pipe or "T" pipe, from which the waste flows onto the ground. Since overlap is needed to cover the entire ground, the vehicle is driven on freshly applied material, and traction can become a problem. A second or improved level of distribution is the pressurized spray, or a rotating spinner disc, which provides a wide coverage (up to 12 meters), thus eliminating the need for vehicle overlap. The most sophisticated option is a subsurface injection device, in which the waste is applied at 15-25 cm deep and automatically buried. For industrial wastes, at least the second level of distribution is needed.

In addition to the instantaneous distribution of wastes, control must be maintained throughout the application cycle. This is referred to as positive ground speed control of waste application rate. Design of land spreading must eliminate the hydraulic head effect on waste flow, since with a full tank, greater material would reach the ground. Without positive ground speed control, less material would be applied traveling downhill than on an upgrade. The needed control is offered by several tank-truck suppliers, and presumably could be adapted to tractor-haul equipment. For industrial wastes the emphasis on assimilative capacity requires this higher degree of waste application uniformity.

Figure 12.14 Low-moisture solids spreading by tractor-spreader wagon (Badger Northland, Inc., Kaukauna, WI) and truck-mounted hopper spreader with flotation tires (Ag Chem. Equipment Company, Minneapolis, MN).

the ultimate responsibility for waste; therefore, authority and control must parallel this responsibility. With industry application of waste, the control of uniformity, maintenance of vehicles and maintenance of the plant-soil system will be in line with the desire for effective, long-term system performance. The goal of assimilation will meet the RCRA responsibilities. Contract hauling and spreading separates the operation and long-term effects from the industry responsibility, thus making the industry vulnerable.

Economic Evaluation

The overall cost for either semiliquid or low-moisture solids systems is centered on the round-trip time requirement, as developed in the detailed design section. Figure 12.15 reflects the increased total cost per ton of dry solids with increasing round trip time (Anderson 1977). A breakdown of these costs on a percentage basis is given in Table 12.7. The largest expense is for labor (56% for salary and fringe benefits) and vehicle maintenance (19%).

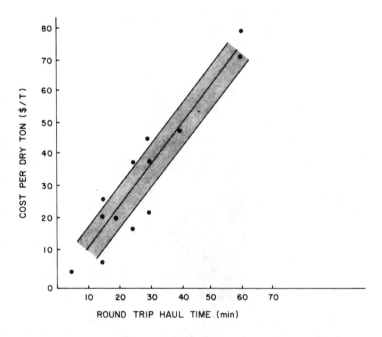

Figure 12.15 Total operational, maintenance and depreciation costs per dry ton of semiliquid or sludge materials as a function of round-trip time requirements (Anderson, 1977).

Table 12.7 Percentage Analysis of Total Annual Cost for Land Spreading
Semiliquid or Sludge Material Based on 15 Systems (Anderson 1977)

Item	%
Personnel Salaries	42
Fringe Benefits	14
Vehicle Operation and Maintenance	19
Investment Depreciation	16
Utilities and Miscellaneous Costs	9

For some industrial wastes the LLC is often sufficiently restrictive that when land applied, less material is put on a unit area than is the case with municipal sludge. Installing an irrigation system over an entire land treatment system is not as preferable as a hauling spreader system for medium to large systems. However, generalization is difficult, and the designer is simply cautioned to review all application alternatives when evaluating industrial land treatment of semiliquid or sludge materials.

The land spreading costs for low-moisture materials are not as well documented because of the lower number and wider diversity in such systems. An economic analysis of parallel handling of a sludge material as a 4.5% solid versus dewatering and application as a 16% solid material was performed (Anderson 1977). The land spreading cost for the 16% solid material was approximately three times the cost for the more liquid wastes (4.5% solids). While this percentage difference will vary with the actual site, the elements in the respective land speading analysis must be evaluated in a manner described for semiliquid wastes to determine whether dewatering is cost-effective.

The costs in Figure 12.15 and Table 12.7, and the discussion of low-moisture solids provide a sound evaluation based on a number of operating systems and based on round trip time as the principal variable. In addition to these costs, the following section provides estimates of the investment levels necessary for hauling/spreading of industrial wastes. Equipment costs depend on the size of the tank or hopper/conveyor, so the decreased number of round trips is balanced against the increased investment. Table 12.8 compiles the investment and operation costs typical of equipment used. In an economic analysis, the equipment lifetime is used in amortizing the investment costs. Hauling/spreading equipment should be depreciated over an approximate 7 to 9-year period.

In summary, the land treatment of semiliquid materials, sludges or low-moisture solids represents a substantial alternative for industrial wastes. Design of these systems involves a number of equipment and field experience factors to ensure that the required assimilation of waste constituents by the plant-soil system is achieved. Specialists in industrial land treatment

Table 12.8 Investment Cost Estimates for Hauling/Spreading Equipment (1978)

Equipment	Cost ($)
Tank-truck spreader, 4,500 l tank	43,000
6,000 l	46,000
8,250 l	50,000
14,250 l	82,000
Injection option	3,500
Tractor with heated cab and flotation tires	15,000
Tank wagon	
3,000 l	3,700
5,600 l	6,600
9,800 l	9,600
13,500 l	13,200
Low-moisture solids truck spreader	
5.5 m^3 hopper	45,000
7.7 m^3 hopper	49,000
9.7 m^3 hopper	86,500
Low-moisture solids wagon spreader	
5.6 m^3	2,700
7.2 m^3	3,300

should be involved to ensure the successful, long-term implementation of a waste distribution system as developed from the LLC analysis in stage I. The costs for such systems are developed from the time requirement for one cycle of the land spreading, including amortized investment and operation expenditures.

STORAGE

Overall Volume

In the overall land system, the storage component is a buffer device between the daily generation of an industrial waste and the intermittent application existent in the plant-soil receiver. The magnitude of storage is generally dictated by (1) the climatological factors controlling the ability to have wastes applied and assimilated by the land treatment system, and (2) the management preference for convenient frequency of application and (3) contingency protection for unknown factors related to either in-plant or land system components.

There are seven situations or elements, each of which must be quantified to develop the required storage volume. All seven may not be necessary for any given industrial land treatment system, but only site and facility conditions can determine the magnitude of each element. These situations or elements are:

1. periods of nonapplication
2. crop cycle requirements
3. repair and emergency allowances
4. a single 25-year, 24-hour rainfall event
5. freeboard
6. economics of startup and shutdown
7. periods of restricted application

The first five are determined by site, industry and equipment, and must be included in storage specification. The last two are flexible and determined by economics. If it is less expensive to store an industrial waste than to apply waste frequently or over enlarged areas (resulting from periods of the year when the assimilative capacity is reduced), then storage is justified and should be included. A detailed description of each storage element is presented in the following sections.

One conclusion can be drawn before proceeding with the storage element analysis: for any given industrial land treatment system, arbitrary storage requirements, such as 30 days, usually are either too small or too large. If it is too large, this storage component is not cost-effective and should be reduced. If it is too small, reliable performance year-to-year cannot be achieved. These considerations strongly suggest that a procedural approach, with site- and waste-specific data, is necessary to specify storage volume; and that fixed storage criteria are not environmentally or economically satisfactory.

Periods of Nonapplication

The first storage element is based on nonapplication periods at the land treatment site. Nonapplication periods are those days in the annual cycle in which adverse pollution potential or system failure (nonassimilation) would be probable due to climatic conditions at the land site. Two climatic conditions are important:

1. certain periods when rainfall inputs considerably exceed evaporation plus drainage, such that saturated field conditions occur, and
2. cold periods of extended freezing or snow cover.

The mere existence of saturated or frozen conditions does not necessarily require storage. The nature of the LLC and the critical assimilative pathways are also essential in determining the need for nonapplication storage. If the assimilative pathways for the LLC are functioning, this storage element is not needed. For example, an industrial waste of low-moisture solid containing principally heavy metals (Cu is the LLC) is assimilated by soil cation exchange, precipitation and adsorption sites. These are operable under a wide variety of low-temperature and short-term water saturated conditions, so no storage is needed during these normally nonapplication periods. The variety of constituents that are the LLC in different industrial wastes is what differentiates these materials from municipal wastes, and thus the storage requirements definitely are not the same. For an industrial waste with water as the LLC, the assimilative pathways are similar to municipal effluent; hence, similar storage requirements exist for nonapplication periods.

The occurrence of saturated soil conditions is determined from climatic data and soil conditions, using the techniques in Chapter 4 on hydraulic assimilative capacity. As with the assimilative capacity, the determination of saturated soil periods is highly dependent on soil and site characteristics. Simulation over weather records will permit determination of the time periods for excessive moisture inputs, which occur at some return frequency. The recommended return frequency is once in 10 years, and the resultant excess moisture periods are used for the saturated nonapplication period in storage volume determinations.

One refinement of the above analysis is necessary for industrial sludges or low-moisture solids, which are applied to the plant-soil system by a hauling/ spreading system. In this instance, not only saturated conditions, but those prohibiting traction or vehicle support, must be quantified. These traction determinations are not widely appreciated in storage design, but from the authors' experience are essential with such industrial land treatment systems for hauling/spreading. Traction and support determinations involve geotechnical calculations, and experts in this field (as related to land treatment) should be involved in specifying this storage time element.

A minimum storage for saturated nonapplication periods can be determined from climatic data developed by Natural Oceanic and Atmospheric Administration (NOAA) (1976). Saturated condition storage is needed when rain conditions are sufficiently large or extended that, regardless of soil type, the plant-soil system cannot accept any further waste volume. This is a minimum estimate, since the volume of waste, size of land treatment site, and magnitude of lateral and vertical movement are not included. These factors would restrict, rather than increase, permissible waste application volumes.

Precipitation of greater than 1.25 cm was used to define an unfavorable day in which storage was required. Then the stored material was kept until a favorable day and applied at an incremental percentage above the normal application. Thus, several days would be needed to provide drawdown of stored material. Using these criteria, simulations were run over years of weather record. The maximum periods of saturated conditions storage are shown in Table 12.9. Results are expressed for the southeastern and northwestern United States as the only areas with significant rain-related storage requirements. Improvements on these methods to realistically appraise the site conditions have often increased these storage requirements substantially and even required such storage in other regions of the United States (Nutter 1978).

The second climatic condition in determining nonapplication periods for industrial land treatment is extended cold periods. If the assimilative capacity for waste constituents is affected by temperature (Chapters 4-11), or if the method of waste distribution is affected, then cold conditions are a factor in determining waste storage volume. As a practical matter, snow depths limit the mechanics and assimilation in land treatment systems. Frozen ground prohibits injection. Thus, climatic data must be used to assess cold and snow periods at an industrial site.

Whiting (NOAA 1976) established criteria of snow depth greater than 2.5 cm and/or daily maximum temperatures of 4.4°C as unfavorable for land application and, hence, needing storage. Waste storage and drawdown on favorable days were determined in a manner similar to the saturated soil case. Simulation over available weather records at reporting stations was used to determine the maximum cold condition storage requirements (Figure 12.16). These storage days times the daily waste generation yield this portion of the nonapplication period storage element at an industrial land treatment site.

Crop Cycle Requirements

Crop cycles may also restrict the application of industrial waste onto a land treatment site. This would then represent a volume of waste that would be included in the storage determinations. Crop-related constraints fit into two quantifiable categories:

1. those related to growth or crop quality, and
2. those related to harvesting the vegetative cover.

The first constraint is related to the type of crop and waste constituents. For food chain crops, a certain period prior to harvest is required to allow levels of critical waste constituents to return to acceptable plant concentrations. For example, caprolactam in a corn vegetative cover would be decomposed in plant tissue within 16 days (Kapauna 1975), so a delay period would exist with requisite storage needs. For some crops, such as grasses, field application rotation can be used to provide the necessary delay prior to harvest. Forested areas do not require such crop constraints of storage. In addition to food chain considerations, the growth pattern of some crops restricts addition of N or

Table 12.9 Contingency Storage Days for Land Application of Liquid Wastes, in Which Water is the LLC (NOAA, 1977)

Station	Years	Max	Station	Years	Max
Alabama			Greenwood	1949-74	16
Bay Minette	1949-73	13	Jackson	1943-61	12
Brewton	1949-73	17	Meridian	1948-73	14
Clanton	1949-73	24	Pontotoc	1949-71	20
Mobile	1949-73	15	Poplarville	1945-73	25
Selma	1949-73	20	Stoneville	1951-72	17
Thomasville	1953-74	25	Vicksburg	1942-62	17
Arkansas			North Carolina		
Dumas	1951-71	19	Charlotte	1949-73	12
Little Rock	1949-72	12	Raleigh	1949-73	14
California			Weldon	1926-50	11
Los Angeles	1952-69	5	Wilmington	1949-73	11
San Francisco	1949-72	13	Oregon		
Florida			Eugene	1948-73	35
Avon Park	1945-72	13	Medford	1948-72	21
Belle Glade	1951-73	11	Roseburg	1935-64	22
Daytona Beach	1955-73	8	Salem	1948-73	36
Tampa	1953-73	16	South Carolina		
Georgia			Charleston	1948-73	24
Augusta	1949-72	10	Columbia	1949-73	15
Macon	1949-72	12	Conway	1945-73	9
Savannah	1949-72	17	Tennessee		
Louisiana			Crossville	1953-73	24
Houma	1950-73	17	Texas		
Lafayette	1948-73	12	Abilene	1950-70	6
Lake Providence	1951-72	19	Amarillo	1949-71	11
Leesville	1950-73	35	Brownsville	1953-72	12
Monroe	1949-72	12	Corpus Christi	1951-72	13
New Orleans	1954-73	16	Dallas	1948-72	15
Shreveport	1949-73	11	Houston	1948-72	14
Schriever	1948-73	16	Wichita Falls	1950-72	8
Winnfield	1950-73	16	Washington		
Mississippi			Longview	1948-73	60
Aberdeen	1951-72	24	Olympia	1949-73	65
Biloxi	1951-73	14	Seattle	1949-73	47
Canton	1948-73	16	Sequim	1948-73	6
Clarksdale	1949-72	18	Vancouver	1924-50	31
Columbia			Walla Walla	1950-73	14

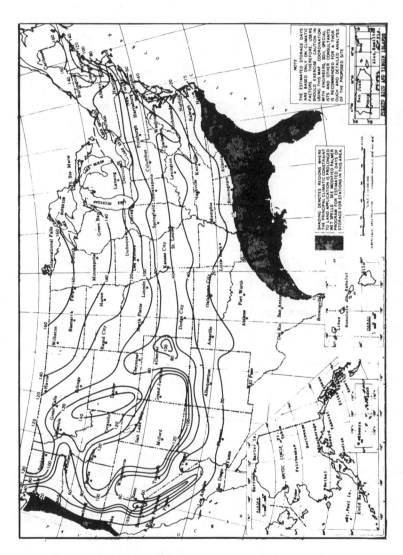

Figure 12.16 Estimation of maximum storage requirement (days) due to cold conditions.

other nutrients late in the growing cycle, to assure good yield of the commodity grown. For example, doses of N without potassium balance near the end of fescue growing period increases the likelihood of winter kill and loss of grass stand. Similar restrictions and consideration must be addressed for other crops in storage calculations. Another common crop restriction for hauling/spreading systems is when crop height is sufficient to restrict field access, in which case storage or field rotation are required.

The second type of crop-related constraint is due to harvesting. Conventional agricultural equipment, such as mower-conditioners, combines, balers and tractors cannot operate with wet field conditions. Thus, for industrial wastes in which the LLC permits large doses of liquid per unit area, storage must be sufficient to allow conditions favorable for harvesting. The frequency and duration of harvest varies from 30 years in forested areas to 3-4 weeks for grass forage crops. Thus, this storage element is site specific.

Repair and Emergency Allowances

The storage element attributable to emergency conditions is a requirement in industrial land treatment. Pump failure, irrigation malfunction, engine repair on spreading vehicles, power blackout and damage repair are examples of contingencies that might be considered in determining this storage element. Based on experience, it is usually less expensive to stock spare equipment or parts, than to provide substantial increased storage. This rule of thumb is most valid for large waste volumes, and a more critical cost trade-off calculation would be necessary for low volume industrial streams. Approximately 3-5 working days are a minimum for contingency storage, with increasing days due to the nature of possible malfunction and subsequent repair.

Single 25-Year, 24-Hour Rainfall Event and Freeboard

The design of a storage pond must include contingencies for severe weather conditions. A standard used by many regulatory agencies and the Soil Conservation Service is a 25-year recurrence interval storm lasting 24 hours (25 year, 24-hour storm). Weather data have to be tabulated and mapped to allow geographical estimation in the U.S. of the magnitude of a 25-year, 24-hour event (Figure 12.17). In addition, freeboard is usually added. These two volume elements depend on the surface area of the pond, since the values are in cm or m, i.e., depth of pond. Advantage in construction cost is derived by using deeper storage units and reducing the 25-year, 24-hour and freeboard elements relative to the total pond volume.

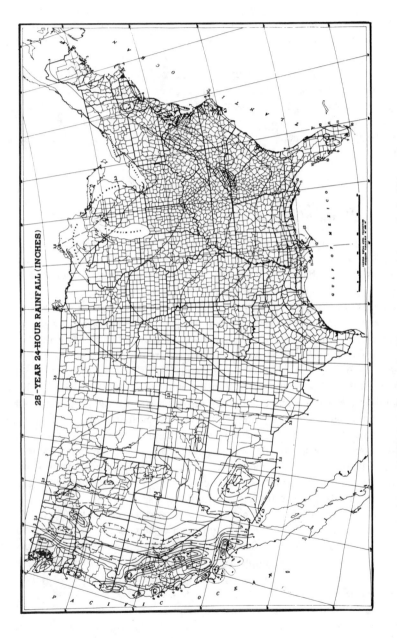

Figure 12.17 Rain amounts for 25 years recurrence interval of a 24-hr event.

The above five storage elements are, to a large degree, fixed by the site, climate and system factors for a given industrial land treatment system. The sixth element is related to the incremental cost associated with startup and shutdown. For example, if a sprinkler system must be drained after each application, spraying in small increments would require frequent draining and labor cost. Detailed equipment inspection is needed on startup of any sprinkler system, as well as establishing siphon tubes or gate opening for surface irrigation. Therefore, startup costs are reduced if storage and large, less frequent, irrigations are used. All applications amounts must comply with the assimilative capacity constraints, but the timing and storage can be offset to achieve a more cost-effective system.

Periods of Restricted Application

The final element in storage calculations is related to the periods in an annual cycle when application rates are restricted to less than the average rate but greater than zero (nonapplication). This seventh storage element is referred to as volume for periods of restricted application and depends heavily on the LLC analysis, site characteristics and the relative economics of storage vs land application. Since industrial waste is generated at a more or less constant rate, and since the application or assimilative rate (for the LLC) may vary within an annual cycle, the land area requirements will vary, unless storage can buffer the situation. For example, if the LLC were oil, the climate-temperature variations would dictate the land areas based on the monthly assimilative capacities (Figure 12.18). Oil, as a part of the industrial waste, is generated at a constant rate. Thus, for any given month, the land area requirements varied from 3-27 ha. To provide land treatment year-round, a system would be needed for 27 ha without storage. The use of storage in the low assimilative periods would reduce land area needs, and the amount of storage would be determined by balancing land system vs storage costs. Thus, economics control this restrictive application storage element.

Of the industrial waste constituent categories used in this book, an industrial material with water, oil/grease, nitrogen or organics as the LLCs would need to determine the cost-effectiveness of storage during restricted or reduced application periods. A technique for determining the balance between storage and land treatment costs was developed by Overcash and Pal (1976) and is presented below.

For a detailed example, consider an industry waste in which water is the LLC. The detailed hydraulic assimilative capacity was developed at the selected site according to the techniques in Chapter 4. In this instance, the

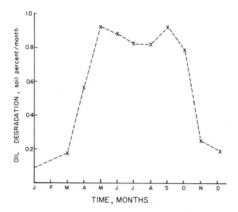

Figure 12.18 Annual variation due to temperature dependence of the oil decomposition rate in soils.

soil type is Cecil, slope 4-5%, 30% lateral movement, and the area of design is variable. The industrial manager was seeking an optimal, least-cost storage design. To simplify the calculations for this example, assume there are no fixed storage elements from nonapplication, crop cycle, emergency or startup/shutdown cycles. The reader can easily include a fixed storage and repeat these calculations. The hydraulic assimilative capacity (water = LLC) is given in Table 12.10. The waste volume is 10,000 m³/mo or 100 ha·cm/mo.

If no storage is provided, then the land area and application system required for January is 22.2 ha. This area and application system are purchased and made operational; then in each month except January there exists an excess assimilative capacity (EAC). That is, for any month i,

EAC_i = assimilative capacity for month i minus the restrictive application rate in the annual cycle used to set system size

For February, EAC_2 = 5.2 cm - 4.5 cm = 0.7 cm.

The summation over the year is the annual unused assimilative capacity, $\sum_{i=n}^{12}$ $(EAC)_i$, where n is the number of months of storage. For no storage, this annual value is 60.8 cm. The land area needed is the ratio of the monthly waste volume, Q_w, and the most restrictive monthly assimilative capacity, AC_n. This is 22.2 ha. Thus, the volume of additional waste that could be assimilated is shown in Table 12.10 as

$$A_n \cdot \sum_{i=n}^{12} (EAC)_i \qquad (12.1)$$

Table 12.10 Industrial Waste Land Treatment Demonstration of Calculational Process for Cost-Effective Storage Volume

Month	Assimilative Capacity (cm/mo)	Excess Monthly Assimilative Capacity for a given Application Rate, cm				
		4.5 cm/mo (n=0)	5.2 cm/mo (n=1)	6.5 cm/mo (n=2)	7.2 cm/mo (n=3)	8.6 cm/mo (n=4)
January	4.5	0	0	0	0	0
February	5.2	0.7	0	0	0	0
March	6.5	2.0	1.3	0	0	0
April	9.7	5.2	4.5	3.2	2.5	1.1
May	10.8	6.3	5.6	4.3	3.6	2.2
June	12.3	7.8	7.1	5.8	5.1	3.7
July	13.1	8.6	7.9	6.6	5.9	4.5
August	13.2	8.7	8.0	6.7	6.0	4.6
September	13.5	9.0	8.3	7.0	6.3	4.9
October	10.2	5.7	5.0	3.7	3.0	1.6
November	8.6	4.1	3.4	2.1	1.4	0
December	7.2	2.7	2.0	0.7	0	0
$\sum\limits_{i=n}^{12} (EAC)_i =$		60.cm	53.1 cm	40.1 cm	33.8 cm	22.6 cm
Monthly Waste Volume, Q_W		100 ha cm	100 ha cm	100 ha cm	100 ha cm	100 ha cm
$A_n = Q_W/AC_n =$		22 ha	19.2 ha	15.4 ha	13.9 ha	11.6 ha
$A_n \sum\limits_{i=n}^{12} (EAC)_i =$		1,350 ha cm	1,020 ha cm	618 ha cm	470 ha cm	262 ha cm
Storage Months, n		0	1	2	3	4
$nQ_W =$		0 ha cm	100 ha cm	200 ha cm	300 ha cm	400 ha cm
Net Annual Unused Assimilative Capacity $Y_n =$		1,350 ha cm	920 ha cm	418 ha cm	170 ha cm	-138 ha cm

The waste volume stored (zero for this first iteration) is the number of months of storage times the monthly waste volume:

$$nQ_W = (0)(100 \text{ ha cm/mo}) \tag{12.2}$$
$$= 0 \text{ (for the no storage iteration)}$$

Since any waste volume stored must be dispersed during the remainder of the year, we define the unused assimilative capacity, when n months of waste are stored, as

$$Y_n = \sum_{i=n}^{12} (EAC)_i - nQ_W \tag{12.3}$$

For positive Y_n, the capacity of the system for the LLC exceeds the annual application.

Providing storage for the waste generated during the month with the lowest assimilative capacity, the required system size is dictated by the second lowest assimilative capacity (52. cm/mo for February in this example). The same procedure is repeated (Table 12.10) and another Y_n is generated. The iteration process is continued until $Y_n < 0$, in which case the stored material cannot be applied throughout the remainder of the year. In other words, even at the higher assimilative capacity associated with the more favorable periods of the year, the stored material and the monthly waste generated cannot be applied in sufficient amount such that no net material remains in storage after one year.

This storage period, from n = 0, to n, at which $Y < 0$ (Figure 12.19), is not necessarily the most cost-effective storage volume. The relative cost of storage and land system cost must be considered. With each increment of storage, 100 ha·cm, there is a cost for a storage pond. This may or may not be a linearly increasing cost. In this instance, the storage cost curve developed for a location at this industrial site is shown in Figure 12.20. Associated with each increment of storage (0, 1, 2. . . months of waste volume) is the land area for the month with the lowest assimilative capacity, A_n. The cost of a land system of that size is developed from the various components listed at the beginning of this chapter. The cost curves for the land system is shown in Figure 12.20. Costs are only calculated as a function of months of storage up to the point of $Y_n < 0$, since technical feasibility does not exist for greater storage.

Three situations are depicted in Figure 12.20, in which the total cost is the summation of storage plus land system costs. The storage cost curve was fixed because of site considerations, but three alternative waste distribution systems were considered—solid set, hauling/spreading and center pivot. For the solid set system, the total cost continued to decrease as storage increased,

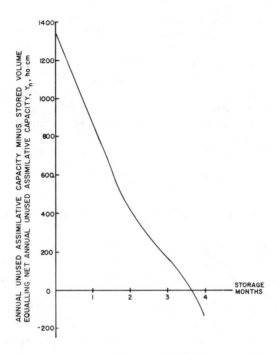

Figure 12.19 Balance between irrigatible liquid amount and volume buildup due to storage.

so that the maximum storage (~ 3.5 months, $Y_3 > 0$, $Y_4 < 0$) was the most cost-effective. A different cost relationship exists for hauling/spreading, reflecting the higher fixed costs of equipment. The total cost exhibited a minimum at a storage volume slightly less than the maximum permissible (~ 3.15 months). In this instance, neither maximum storage nor zero storage were the most cost-effective, but rather some intermediate amount of volume. With a center pivot device, the distribution costs are lower, which shifts the optimum balance toward less storage as the least-cost total system (~ 2.15 months) (Figure 12.20).

Costs used in the storage-land application optimization (Figure 12.20) are investment expenditures. Industry may prefer to use total annual cost (amortized investment plus O&M) as the cost index for optimization. For storage, O&M costs are very low, and total annual cost will be primarily investment depreciation. In the land system, the O&M costs are significant, and the total annual cost would be significantly above debt retirement expenditures. These factors can shift the optimal balance toward increased storage. Thus, a choice exists as to the most appropriate cost index to use in determining the storage element for periods of restricted land treatment.

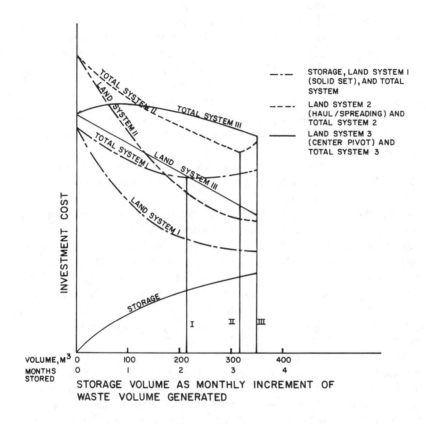

Figure 12.20 Least-cost combination of liquid storage facility and required land application system for industrial waste with water as LLC, three typical land system cost relationships.

The storage requirements for periods of reduced land treatment capacity, as well as those for incremental costs of startup and shutdown, are determined by an economic analysis. The industrial waste LLC is very important in determining whether (and how much) storage is needed, depending on the assimilative capacity. Site characteristics are critical factors in determining storage, since both construction cost and variation in assimilative rates are site depndent. As a general rule, as the variation in LLC assimilative capacity becomes greater in an annual cycle, the economic incentive for storage increases. That is, if the allowable application rates are constant throughout the year, there is no advantage in storing waste for periods of higher assimilative capacity.

The analysis presented above was done on a monthly increment but is not restricted to such large time blocks. The above analysis was based on average monthly assimilative capacities for illustrative purposes. A year-by-year evaluation over available years of weather records near the industrial site can be used. In this manner, more attention can be directed to conditions that occur 1 year in 10 or 25, as a more conservative evaluation of the variation in assimilative capacity. Finally, the site climate and economic dependence of the seven storage volume elements further emphasizes the need to evaluate storage requirements on an individual system basis. A design procedure, such as that developed herein for industrial waste, is then essential, while fixed storage criteria are unworkable.

Although it is not commonly considered, there is one other climatic condition that is relevant under certain circumstances in determining industrial waste storage. For waste in which the LLC depends on decomposition or plant uptake for meeting the basic nondegradation constraint, prolonged dry periods should be evaluated. Under very low moisture conditions (due to low rain and waste volume), the soil microbial processes are essentially halted. Further, waste application might exceed soil levels of constituents, so that when adequate moisture occurs the assimilative pathways are overloaded. Thus, storage would be necessary. Industrial land treatment specialists with experience in low-moisture conditions should be consulted for advice in this matter. An example of such dry climatic effects and needed systems was presented by Brown and Deuel (1977).

Storage must be provided in each industrial land treatment system, and the magnitude of storage is controlled by: (1) the waste characteristics, and (2) the site characteristics. The procedure necessary to determine storage volume requirements is based on calculation of each of the seven volume elements previously described. Generally, not every volume element is needed, so that particular volume can be equal to zero. For example, with an enclosed storage for a thick waste material, no freeboard or 25-year, 24-hour storm elements are needed. In forested areas, no crop cycle volume element is needed. Thus, the land treatment site and waste information are very important in establishing the criteria for industrial storage volume. A volume calculation expression for storage is as follows:

$$
\begin{aligned}
\text{Total industrial land treatment storage volume} ={}& V_{\text{nonapplication}} + V_{\text{crop cycle}} + V_{\text{repair and emergency}} \\
& + (D_{\text{25-year, 24-hour storm}} + D_{\text{freeboard}}) \cdot \text{Area of storage pond} \quad (12.4) \\
& + (V_{\text{startup and shutdown}} + V_{\text{restricted application}}) \Big|_{\substack{\leftarrow \text{ determined by} \\ \leftarrow \text{ economic balance}}} + V_{\text{management decision}}
\end{aligned}
$$

where V_x are expressed in m^3, D_x in m, and area in m^2.

The $V_{management\ decision}$ is a discretionary volume, which may be included for a variety of reasons peculiar to the industrial facility being studied. Such reasons as future expansion, extra conservative factor, ease of scheduling waste application with plant functions, aesthetics or "would make for better fishing," are included in this discretionary element. Having established the total storage facility volume, the design and economic analysis can then be completed.

Design

Storage facilities vary in design depending on the nature of the industrial waste. The principal options are:

1. earth ponds
2. concrete tanks
3. tanks of special material for corrosive or other special waste properties
4. low-moisture solids stacking

The volume of these storage units is determined by adding the elements or volumes determined in the previous section on storage (Equation 12.4).

Earth Ponds

The most common storage unit associated with industrial land treatment, particularly of liquid wastes, is the earth pond. Design and construction of land treatment storage ponds can utilize the site topography to considerable advantage. The cost of such storage units is determined by the amount of earth moved in constructing the dam, so desirable areas are those in which a small dam can impound a large volume of waste liquid. At the other extreme is a pond that must be constructed on a flat area, for which all the storage volume must be obtained by excavation. In this instance, the excavated material may be used for the dam construction, thus reducing the earth moving to the range of 50% of the total storage volume needed. Geotechnical assistance can be used to substantially reduce pond cost by location selection and determination of best soil materials (Smith 1978b).

All industrial waste storage ponds must be constructed with a clay dam core (Figure 12.21), usually of trapezoidal shape. The core is constructed down through the upper soil horizons to a subsurface clay layer, or to 3 m. Core construction is needed to reduce seepage. The infrequent use of the pond (only for short-term, adverse land treatment periods) further reduces seepage. Waste flow to the pond must include an inlet structure to support the pipe and a concrete splash pad down to the pond floor. Pad construction is necessary to eliminate dam erosion, especially when the storage facility is empty. Immediately after construction, all berm, interior and exterior dam

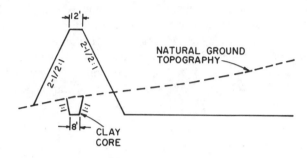

Figure 12.21 Antiseepage core for earth dam of storage pond.

surfaces should be seeded, fertilized and mulched to stabilize the structure. On larger structures, embankment protection with rip rap is recommended.

Earth storage ponds can be modified with liners or soil additive material to reduce seepage of waste constituents. These modifications include liners of butyl neoprene or PVC, and soil additives or sealers of bentonite, soil cement and petromat. Installation should follow manufacturers specifications and a careful geotechnical review of which seepage control option will work at the pond site. The relative cost of workable options must be determined.

Odor control is another important factor in the design of industrial waste storage ponds. Whether a waste liquid will produce an offensive odor when held under pond conditions (likely to be anaerobic) must be determined first. For those wastes that will have odor, several options must be evaluated. The first is situating the pond in the most isolated location, to reduce the odor potential at neighborhood locations where nuisance complaints might exist.

Additional odor control can be achieved by: (1) constructing a sufficiently large pond that surface aerobic conditions exist, or (2) installing mechanical aerators to control odor. The first alternative is based on observations of animal waste lagoons, in which as larger pond volume per mass of waste chemical oxygen demand (COD) was provided, the odor potential and daily odor probability were decreased. Between 4 and 9 m^3 of lagoon per kg/wk of COD was a threshold below which odor potential increased substantially (Humenik and Overcash 1976). These limits appear to be somewhat waste dependent, as a laboratory study of pharmaceutical waste revealed a volume of 18 m^3/kg/wk of COD was needed for odor control.

Use of mechanical aeration for odor control is designed for less oxygen or power inputs than that needed for waste organic stabilization. The emphasis is on surface zone aeration, with the remainder of the storage being anaerobic. Since most aeration equipment is not aimed at limited surface oxygen control, there are few data for design purposes. Using studies of animal waste in field-scale lagoons, the area influenced for odor control was about 150 m^2/kW of aerator power (Overcash and Humenik 1976). Greater areas/kW resulted in odor-producing surface zones. Again, these criteria are very approximate when dealing with a variety of industrial wastes.

Concrete Tanks

Concrete or metal tanks are utilized for short-term storage (1-3 days) of industrial waste liquids or for longer-term storage of more concentrated materials. For liquids, the objective is to equalize, allow agitation to prevent solids settling, and to serve as a sump for direct land application. Standard construction specifications, facility for positive pump inlet suction, and underdraining to prevent buoyant displacement, should be used (Figure 12.22). More concentrated waste is often stored to facilitate hauling/spreading or irrigation on an infrequent basis, and to serve as a waste collection unit outside the actual industrial plant. The size of the concrete tank storage for liquid and concentrated waste is dictated by the factors previously discussed for storage volume, as well as management decisions based on operational ease.

Specialized Tanks

In the variety of industrial materials that can be land treated successfully, there are a number of particular wastes that require specialized storage facilities. These requirements are usually for corrosive materials, aseptic wastes, high viscosity or nonNewtonian liquids, powders, etc. In one industry, heated and insulated tanks are used to keep a waste suspension sterile, so that prolonged nonodorous storage can be successful. At another site, a corrosive material is kept in coated, covered tanks to eliminate transfer to the air prior to a hauling, soil injection system for land application.

Low-Moisture Solids Stacking

Low-moisture solids can be stored in relatively unconfined structures, similar to dairy manure storage. The principal design considerations are volume of waste, roof coverage, access for loading and unloading, and allowance for waste drainage or seepage. Conveyor equipment can be used effectively to stack low-moisture waste (Figure 12.23). In many areas, a roof structure is essential to prevent rainfall additions. Walls may be constructed of wood, concrete or earth. Most waste material will continue to drain liquids after stacked in storage, so the floor should slope to one location where the liquid can be removed periodically. Removal of waste can be designed for front end loaders, scraping conveyors or ramp construction. Each of these options would empty into a hauling vehicle for transport to the

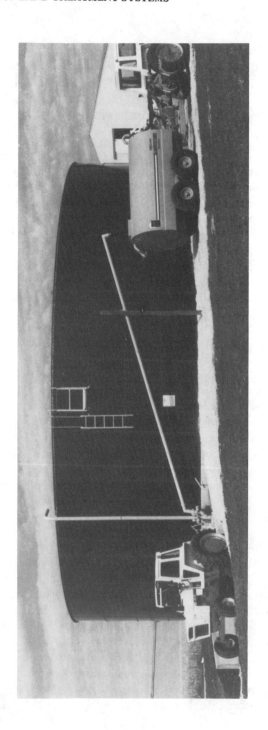

Figure 12.22 Above ground storage tank for semi-liquid or sludge wastes (Smith, 1978).

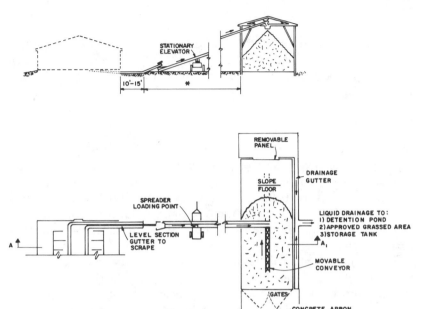

Figure 12.23 Storage and handling facility for low-moisture solids.

industrial land treatment site. The factors involved in storing low-moisture solids should be considered in detail as a part of the design process. Such storage is necessary to buffer the constant waste generation with the capacity of the land system.

Economic Estimates

The annual cost of a storage facility if predominantly that of investment depreciation, although some very definite labor and materials inputs are necessary. Earth storage pond construction has been evaluated on a cost basis in several reports (Patterson and Banker 1971, Chamblee 1978, Pound *et al.* 1975). There are wide variations in construction costs, so some interpretation is necessary relative to storage facilities for industrial land treatment systems.

The economic values in Figure 12.24 represent land treatment storage units. However, the volume axis must be corrected to coincide with the volume from Equation 12.4 and corresponding calculations. Figure 12.24 assumes 0.9 m for freeboard, plus 25-year, 24-hour storm contingency, so the total volume in Equation 12.4 must be reduced by these two volume elements to obtain storage volume plotted in Figure 12.24. Recent industrial pond construction costs (LANDTECH, 1978) have been in agreement with the values in Figure 12.24.

The corresponding costs for pond construction as presented in reports of Black and Veatch (Patterson and Banker 1971) and by Dames and Moore (Chamblee 1978) are two-four times higher for an equivalent size pond. These results reflect valid data, but also manifest the municipal approach to design, procurement and construction of conventional treatment facilities. Design groups specializing in industrial land treatment and related components of land systems have generally been able to tailor the storage facility needs so that costs are kept a minimum.

Construction costs range from $1.30-1.60/m³ of storage volume in the range of 200-4,000 m³. As pond size approaches 4 million m³, the expenditures are reduced to the order of $0.15-0.20 m³. The use of embankment protection is dependent on the size and operation mode of the facility. For many small storage units typical of industrial waste needs, grass is used for both interior and exterior banks. Since liquids are usually kept below the pond spillway, the grass functions effectively to protect the banks. During drawdown, the grass can spread to cover areas left exposed. On certain, easily erodible soils, and for larger storage ponds, the protection of embankment must include rock material on the inside slope of the dams. The cost of embankment protection is 50-160% of the earth pond construction outlay (Figure 12.24).

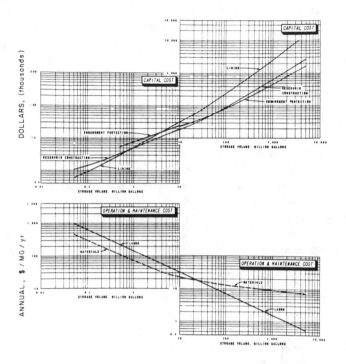

Figure 12.24 Investment and operational costs for storage ponds (Pound, 1975).

Storage ponds for certain industrial wastes or in difficult seepage control areas require special practices to prevent concentrated waste constituent migration. Soil additives or liners are then necessary, and for an asphalt-based material the cost values are shown in Figure 12.24. Adjustment factors for other seepage control materials are as follows (Pound *et al.* 1975):

Material	Adjustment Factor
Bentonite	0.86
PVC (10 mil) with Soil Blanket	1.21
Soil Cement	1.21
Petromat	1.24
Butyl Neoprene (30 mil)	1.97

Earth pond operation and maintenance costs are relatively low. The grass on the pond berms must be kept mowed. The spillway is maintained in cleared condition. Once a year all trees or shrubs growing on the dam slopes must be removed to maintain structural integrity. Labor estimates for storage ponds are given in Figure 12.24. The materials costs in Figure 12.24 are related to the presence of a liner, which must be scraped and patched an average of once every 10 years. To these material costs should be added fertilizer and lime, to maintain the vegetative cover used on the embankments. Annual soil testing should be used to establish quantities to be added.

Concrete or coated steel storage tanks are frequently used as sumps or to contain semiliquid or sludge materials. Because of the high cost per unit volume associated with this storage option, only concentrated or low-volume waste streams can be stored in this manner. Concrete and steel tanks can be prefabricated or constructed onsite. An approximate 1978 installed cost for metal aboveground or concrete tank units is $19-27/m^3$, with capacity up to 4,000 m^3 in size (Smith 1978a, Midwest Plan Service 1977). Negligible annual operation costs are required for the storage tank. Costs for specialized metal tanks, insulation, agitation or other requirements, must be evaluated on an individual system basis.

Cost estimation for low-moisture solids stacking and storage is based on the expenses of constructing a floor and/or roofed structure. As a minimum, a concrete pad with ramp for loading waste onto hauling vehicles can be estimated at $10-15/m^2$. Construction of a roof and wall structure with concrete floor would require $85-120/m^2$ of area.

Total annual cost for storage facilities can be determined from the investment and operation costs presented above. Earth ponds have an indefinite lifetime, so individual industrial policy should be used to depreciate the initial investment. Concrete and coated steel tanks would be projected for a lifetime of 20 years. Liner material survival must be obtained from manufacturers or suppliers for the individual waste and site conditions. Annual operation costs are given in Figure 12.24 for earth ponds, but for concrete or steel tanks, or stacking areas, the costs would have to be established separately. Addition of depreciated investment costs and annual operational expenses is the total annual cost of storage as an element in the overall land system.

MISCELLANEOUS LAND TREATMENT SYSTEM ELEMENTS

Pump and Pump Station

Within the land system for industrial waste is a component related to the transmission or pumping of the waste. For liquid materials, pumps transfer waste either from the plant to the irrigation system, from plant to storage, or from storage to an irrigation system. For semiliquids, sludges or low-moisture solids, the pumping, if any, is generally included in the hauling/spreader system, so is not included in this section.

Predominant Pump Types

Two principal pumping systems are used in connection with industrial land treatment: (1) centrifugal or volute, or (2) turbine, primarily vertical. The centrifugal pump is usually specified as horizontal acting, in which an impeller with casing is used to introduce liquid and to discharge the liquid at a higher pressure. The impeller is connected to a horizontal shaft, which is turned by a motor power unit (Figure 12.25). A number of impeller types are available for different liquids or to provide different pressure-liquid velocity relationships.

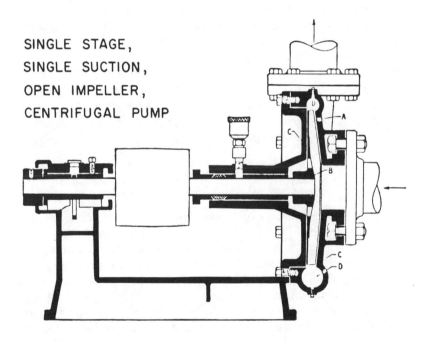

SINGLE STAGE, SINGLE SUCTION, OPEN IMPELLER, CENTRIFUGAL PUMP

A. IMPELLER   C. CASING

B. IMPELLER HUB   D. DISCHARGE VOLUTE

Figure 12.25 Internal schematic of horizontal centrifugal pump (Badger, 1936).

The vertical turbine pump also uses centrifugal action to impart energy and develop pressure head. Impellers with angled vanes rotate so that the incoming and exit liquid streams move parallel to each other. The diverging passages formed by the rotating vanes or propellers direct the liquid vertically, so that the exit liquid is increased in velocity and pressure (Figure 12.26). The impeller and vanes are placed beneath the liquid surface in a section referred to as the bowls. Several pumping stages can be placed in series to develop about 30 m of head per stage. Power is provided by a motor at the upper end of the impeller shaft. Depending on usage conditions, the impeller type, shaft bearings, lubrication and material of construction can be varied to give optimal performance.

Design

Certain basic factors are necessary to specify a pump unit for the several possible transmission options in an industrial land treatment system. These are reviewed here to permit those responsible for the waste management design to discuss and evaluate the equipment recommended. Actual specification of pumps should rely on engineering specialists for land application systems, because of the potential cost savings, improved reliability through experience and improved, long-term materials wear that can be obtained.

Pump specification begins with establishing the total dynamic head or pressure that must be developed to discharge to a storage pond or operate a sprinkler irrigation system. In pump design, head and pressure are used interchangeably by means of the following equations:

$$\begin{aligned}
\text{pressure (psi)} \quad \times \quad 2.31 \quad &= \quad \text{head (ft)} \\
\times \quad 0.705 \quad &= \quad \text{head (m)} \\
\text{pressure (kPa)} \quad \div \quad 9.8 \quad &= \quad \text{head (m)}
\end{aligned} \qquad (12.5)$$

Total dynamic head consists of several different pressure heads, which are summed for the total. The first is static suction lift or head (h_s), which is the distance between the center line of the pump and the water level (>0 if water level is below the pump). A second head is the static discharge pressure (h_d), which quantifies the height of water when the pipes are full between the pump and the discharge. This pressure is included if the discharge is higher or lower than the pump. If the pump is used to operate a sprinkler irrigation system, then a certain pressure must be maintained at the sprinkler. Table 12.4 details the pressure head (h_p) required by the principal irrigation options. More precise, h_p requirements would be based on the actual sprinkler used.

As liquid moves through mainline pipes, energy is expended, overcoming fluid drag forces along the pipe wall. This energy loss must be compensated for by a friction pressure head (h_f). Pipe material varies in roughness or

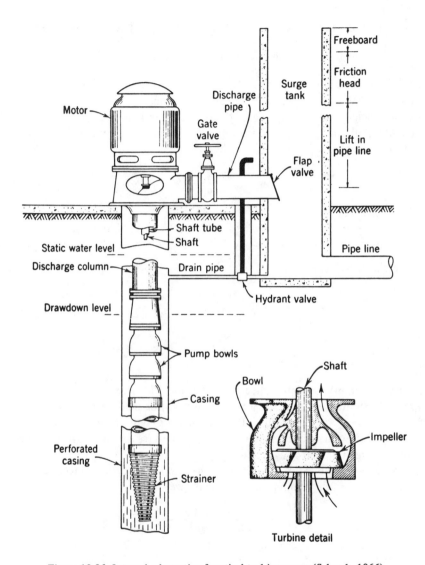

Figure 12.26 Internal schematic of vertical turbine pump (Schwab, 1966).

degree of wall drag, so must be considered in determining h_f. As the waste liquid moves, the velocity is changed from a minimum of stationary at the pump suction to the pump discharge or pipeline velocity. This Bernoulli energy (h_v), although actually low, should be included to ensure complete design under a wide variety of possible circumstances.

Summation of all pressure elements yields the total dynamic head (tdh):

$$tdh = h_s + h_d + h_p + h_f + h_v \qquad (12.6)$$

Several of these pressure components are very site dependent, because overall distance, elevations, piping and irrigation equipment can vary widely. The pump selected must be capable of delivering waste liquid (often with solids present) at the tdh. Power requirements are based on the tdh as well as the flowrate, and can be determined as follows:

$$whp = \frac{gpm \times tdh}{3960} \qquad (12.7)$$

where whp (=) horsepower
      gpm (=) gallons per minute
      tdh (=) ft of water

In metric units,

$$wp = \frac{lpm \times tdh}{2560} \qquad (12.8)$$

where wp (=) kW
      tdh (=) m
      lpm (=) liter per min

Flowrates (l/min) are determined based on the LLC, area required and the waste distribution system. The water power (wp) or theoretical power needs must be corrected for the actual pump efficiency to determine the pump size. Efficiency is evaluated from pump operating curves, so varies considerably with the large number of available equipment. Advice of a specialist is necessary to obtain the greatest efficiency at the flowrate required. Having determined efficiency, the pump power needs are given by:

$$wp = \frac{l/min \times tdh}{(2560)\,(efficiency)} \qquad (12.9)$$

$$0 < efficiency < 1$$

The source of power for pumping can be electricity, gasoline or diesel engines, gas engines or power takeoff from tractors. Electricity is the predominant source for industrial land treatment systems. Electricity-driven pumps/motors must be enclosed or protected from the environment by construction of a pumphouse. These structures must have drains or dry mounting for the pump, good insulation from cold and ventilation during high temperatures, protection from rodents, and safety shutoffs for overloading, low voltage or overheating. Typically, three-phase power is needed.

The intake structure is very important to ensure reliable startup and liquid delivery to the spray system. Centrifugal pumps are best installed below the liquid level to create a flooded suction line. This requires installation through the dam wall during pond construction, or near a sump at the waste source. For an existing waste-containing structure, a self-prime centrifugal pump can be used with a floating intake and flexible intake hose. Such an arrangement allows response to storage pond volume fluctuations. Vertical turbine pumps can be installed during storage pond construction or on a platform for existing ponds.

A pumphouse serves to protect the irrigation or transmission pump from weather conditions and vandals. A block or wooden structure is usually constructed over a concrete pad or floor. Certain accessories should be included to facilitate pump and system operation. Insulation for winter conditions and ventilation for summer temperatures are used to maintain a more uniform environment for operation. Supplemental heating is needed for colder regions. The pumphouse floor should have a drain to maintain dry conditions during repairs or malfunctions. Electric lights, power outlets and pump hour meters are usually recommended.

As a part of the pump station, pressure guages are used to monitor operation. An alarm for sudden line pressure changes can shut down the system when it malfunctions. The alarm should sound in the industrial plant to alert attention. Sufficient floor and head space are needed to facilitate repairs. Manual start-stop control is recommended to facilitate responsible management. The pumphouse should be secured and locked when not in use.

Economics

As described earlier, centrifugal and vertical turbine pumps are the prevalent devices for transferring industrial liquid wastes to storage directly, to irrigation or from storage to irrigation. Exceptions necessary for substantial concentrations of solids are usually selections of intake, outlet and impeller orifices, to accommodate the industrial waste particle size. Open impellers can be used for waste transmission but do not generally develop sufficient pressure for irrigation.

The initial investment costs vary between the two basic pump constructions. Figure 12.27 details the investment, installation and pumphouse construction costs for vertical turbine and centrifugal pumps. Pump flowrate is determined by the flow variations of the particular industrial processes generating waste. Pumping of daily waste from an industrial plant to storage, or directly to irrigation, requires a backup or reserve pump. Dual pump capacity requires almost twice the size pumphouse and two pumps with installation, so double the values in Figure 12.27 should be used. When the

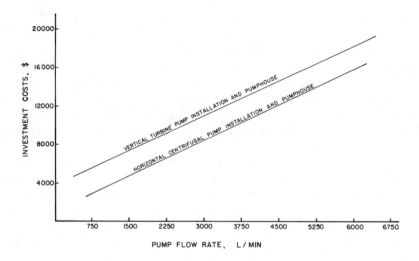

Figure 12.27   Capital costs of various-size vertical turbine and horizontal centrifugal pumps.

pumps can be located in-plant, these costs can be reduced by approximately one-third by using the plant as the pumphouse structure.

Pump stations located at storage ponds have the storage pond capacity to offset the consequences of pump failure. Depending on the storage provided, pump motor repairs can be made in 3-5 days, bowl assemblies (most common malfunction with vertical turbines) should be purchased for standby, and contract arrangements made for emergency repairs with appropriate service companies. In this manner only a single pump and pumphouse are needed. The required pump size is based on the average flow, since the pond serves as a sump-equilization unit and on the irrigation system requirements.

Municipal pump stations are sized and built with different conditions of construction and bidding (often as a part of a larger project). The redundancy with the investment costs presented by Pound *et al.* (1975) indicates a 10-fold lower cost for industrial pump stations, reflecting the differences alluded to above.

However, the operation and maintenance costs for either municipal or industrial pump systems are similar since the major factors are electricity and labor for wear repair (Figure 12.28). The total annual cost contains an amortized investment cost plus the operation and maintenance expenses (Figure 12.28). Pump and pump component lifetimes are given in Table 12.11, assuming the same lifetime as for irrigation operation. The desired industrial depreciation formula can thus be used to establish the total annual cost.

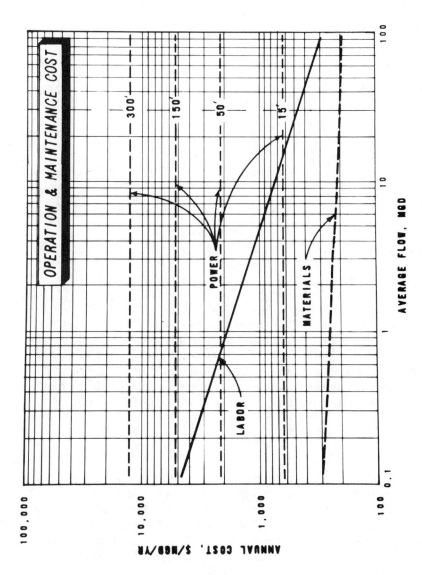

Figure 12.28 Operation and maintenance costs of electrical irrigation pumps (Pound, 1975).

Table 12.11 Suggested Service Life for Components of an Irrigation System Pump

Pump Component	Service Life (hr)
Pump, Vertical Turbine	
Bowl (about 50% of cost of pump unit)	16,000
Column, etc.	32,000
Pump, Centrifugal	32,000
Power Transmission	
Gear head	30,000
V-belt	6,000
Flat belt, rubber and fabric	10,000
Flat belt, leather	20,000
Power Units	
Electric motor	50,000
Diesel engine	28,000
Gasoline or distillate	
Air-cooled	8,000
Water-cooled	18,000
Propane engine	28,000

Transmission

Alternative Methods

Pumping conveyance of liquids from plant or pretreament unit to storage or from storage to the irrigation system is termed transmission. The pipe size is determined by pressure requirements and flowrate. Peak flows must be accommodated or, more commonly, are equalized, and the pipe size selected accordingly. Transmission can be belowground PVC or other pipe used as force mains, or aboveground aluminum pipe. The aboveground pipe must drain to the equalization sump or storage pond, or into gravel pits if freezing can occur.

Economic Evaluation

Economic estimates of transmission for aboveground aluminum or buried PVC (According to ASAE standard S376) are given in Figure 12.29 for industrial installations. As pipe size exceeds 102 mm (4 in.), the aboveground aluminum costs about the same as the installed PVC. These estimates are for straight lines of pipe, typical of transmission layouts. Conditions of rock or installation below 1 m would require additional costs. Again, the municipal price structure is somewhat higher. Prices in Figure 12.29 can vary ± 30% depending on the competitive bid situation.

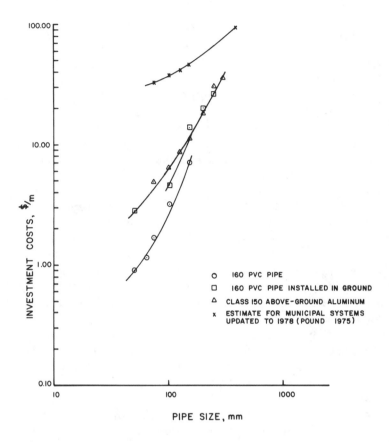

Figure 12.29 Capital costs for transmission pipe.

The projected lifetimes for transmission pipe are needed to determine amortized investment costs. Using a conservative assumption for industrial waste usage, the lifetime is estimated at 50% of that found for agricultural irrigation. Aboveground aluminum, buried coated steel and buried PVC have estimated lifetimes of 6-8 years, 10 years and indefinite, respectively. For more corrosive waste materials, the aluminum and steel service length is reduced.

Diversions and Runoff Control

Land application systems must be designed and operated to have no direct surface runoff of applied effluent (so-called application runoff). This constraint can be met by not exceeding the infiltration-percolation capacity of the soil, cm/hr, with the application system. Site data on infiltration must be taken to determine the capacity for waste application. No application waste runoff implies that the operation must not allow application during saturated or near-saturated conditions.

The question of rainfall-runoff impact is not answered for land application sites, but it cannot be denied that such runoff will occur. The relative quality of such runoff, especially in comparison to runoff from surrounding areas in agriculture, silviculture or urban use, is the critical determinant in the environmental impact of land treatment sites. Limited data from a land application area for animal waste indicate that, for rates of manure application approximating the assimilatory capacity of crop uptake, the runoff quality is similar to that from surrounding runoff (Overcash 1974). Data such as these and other regulatory input have led to the rejection of requirements to collect and reapply rainfall runoff as a control mechanism, especially for nontoxic industrial wastes. For designated or toxic pollutants applied at the low land treatment rates necessary for assimiliation, the runoff would appear to contain possibly only lower than detection limit concentrations. The very low mass of those constituents in relation to sediments and biological organics transported in runoff and the large volume of runoff liquid would further reduce the nonpoint source impact of these parameters. At this stage of land treatment knowledge the rainfall runoff transport of specific materials of concern appears to be unknown. The use of assimilation of waste constituents as the design criteria may minimize rainfall runoff to a practical lower limit.

As a problem of environmental protection, the transport of constituents in rainfall-runoff has been designated as a nonpoint source of pollution. This incudes the runoff from many types of land areas, such as industrial land treatment sites. The solution legislated by the U.S. Congress and implemented by the U.S. Environmental Protection Agency (EPA) for nonpoint source control has been to utilize best management practices (BMP). That is, collection of rainfall runoff and treatment prior to stream discharge is not to be used. Instead, sound agricultural practices are to be implemented, so that sediment and chemical constituents are reduced in the runoff liquid. BMPs are of course site specific, and effectiveness of control is not well established; however, several researchers are actively evaluating such practices (Horney 1978, Walters et al. 1977, Romkens et al. 1973, Johnson and Moore 1978). Diversions, grass buffer strips, sod cover and erosion control are examples of nonpoint source control practices that can be used on industrial land treatment sites. Thus, the objective of land system design should be to minimize runoff by implementing BMPs.

A diversion structure is an earth construction designed to intercept runoff from a designated land area. The most functional diversion for industrial waste land application sites is to prevent entry of rainfall-runoff from adjacent lands onto the land application site. These diversions prevent large-scale flows, associated with runoff events in a watershed, from transporting soil or residual material from an application site. Use of these diversions reduces stream impact from runoff and serves as a substantial BMP.

Design of diversion structures follows established agricultural practices for controlling erosion by storm runoff management. The specific land site dictates the actual diversion layout. Basically, diversion ditches (Figure 12.30) are cut on the uphill side of the application site, directing water around the site and sloping toward a natural surface drainage outlet. A typical layout is shown in Figure 12.30. These diversions and the decisions regarding the need for such facilities should be evaluated by the land application designer or consultant.

A second type of diversion is developed to contain all surface runoff from the land application site and can be used in conjunction with the diversions described above for keeping water from the site. These treatment site containment diversions are of similar design, but are located to direct all surface drainage from the site to a collection pond. In general, containment diversions are not required at properly designed industrial waste sites, since the nonpoint source contribution is controlled more realistically by BMP. Potential need for such containment would exist where infrequent, heavy land applications, such as sludge, might contribute substantially to stream pollution.

Investment costs for diversion structures depend on the equipment used to construct the channels and the site topography. Under average site conditions and with a trained bulldozer operator, diversion construction would cost $2.50-3.50/m of length. After construction, these channels should be fertilized, seeded and mulched, and stabilization netting used as needed. Depreciation lifetimes are indefinite, so individual industrial policy should be used. Operation costs are those for mowing and annual maintenance fertilization, as needed.

Acceptance

In the designing of a land treatment system it is sometimes necessary to consider the potential neighbor problems in implementing this waste management option. No industrial plant likes to be perceived as an offensive member of the community. While the use of land treatment or any other treatment alternative is an industry or private sector decision, adverse reaction can heighten regulatory concern.

The first step is to decide whether a nuisance problem is likely to occur. This centers around the nature of the waste and the proximity of nearby residences. If no obvious problem exists, further interaction should only be

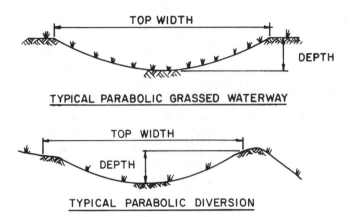

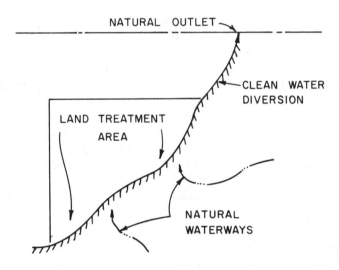

Figure 12.30 Schematic of field diversions for control of surface runoff.

to provide general information about the nature and benefits of the intended industrial facility.

Should a nuisance problem be obvious, the involvement of neighbors becomes necessary to help them understand the overall picture. Specific information should be provided so that the proposed design can be viewed as a reasonable choice. It has been helpful to point out, usually in an invited meeting, the environmental basis for the design, *i.e.*, the concept of nonde-gradation. Emphasis on the agricultural nature and the analogy to manureing practices are useful in relating to the system. The concept of improving stream quality by nondischarge can be used to demonstrate the improvement in environmental quality associated with using the land alternative.

Presentation of this information should allow questions and discussions, so that a one-sided situation does not arise. The firm scientific basis, the environmental concern of the industry, and the desire to work together for the betterment of the community should be conveyed throughout such a neighborhood or community meeting. This effort to inform and attain a certain consensus among those impacted cannot eliminate all opposition. Many nuisance complaints are without basis. However, this process of accept-ance, where nuisance problems are likely, can greatly reduce negative attitudes through a program of information dissemination.

Buffer Zones

The concept and subsequent regulations of buffer zones for land applica-tion sites evolved with municipal land treatment systems. The concerns were primarily public health oriented. However, the purposes for buffer zones include two different requireemnts:

1. buffers for nuisance protection, and
2. buffers for public health protection.

The width of a buffer zone for control of aersolized microorganisms or odors is typically established in a state guideline and referred to as a "standard distance." The assumptions regarding the land application site, for which the standard buffer distance is related, are that:

1. municipal effluent with secondary treatment, followed by standard chlorine disinfection, is to be applied;
2. a solid set sprinkler irrigation system is to be used;
3. an open field exists between the application area and the property line or nearest residence.

Variations from these assumptions must result in different requirements for buffer widths—either an increase or a decrease, depending on the conditions. The standard or required distances in various states have been reviewed by Morris and Jewell (1976).

The alterations in buffer zone width are developed from two factors:

1. waste generation or pretreatment changes, and
2. engineering design techniques that reduce the airborne transport of waste constituents of concern.

Quantification of the percent decrease or increase in buffer zone size is difficult, but sound engineering judgment suggests that such alterations are quite prudent.

Concept of Buffer Zone

The concept of buffer zone decrease or increase relative to variations in the pathogenic or odor potential of a waste is based on the mechanism for airborne transport (Hickey and Reist 1975). Health risk or nuisance probability are directly related to the number of microorganisms or quantity of material passing beyond the *outer* edge of the buffer zone. A sprinkler device puts out a constant number of droplets capable of being transported (under a given set of operational and climatological conditions). A removal or reduction of microorganisms and other materials of concern occurs through the buffer zone reaching an acceptable limit at the outer edge of the standard buffer zone. Therefore, if the number of organisms in the waste to be spray irrigated is reduced, then less buffer width is needed to achieve the acceptable limits previously achieved over the entire standard buffer width. This is illustrated in the limit of no harmful constituents present in the spray irrigated waste, for which there is no buffer zone, *i.e.*, agricultural irrigation of water requires no buffer zone. The width of a buffer zone must thus be designated to meet the conditions of the type of material to be sprayed on land.

In decreasing the buffer zone width for industrial waste, it would appear that a minimum distance is reached below which it would not be in the best interest of an industry to approach the property limits. This is a general nuisance or neighborly width, of the range of 5-10 m, in which a screen or beautification program could be used.

Buffer zones should be reduced to the minimum value for industrial wastes that do not contain pathogenic organisms or substantial adverse odors (Table 12.12). Instances in which domestic waste sludge is irrigated, the fecal organism content should be compared to that of secondary treated effluent with chlorine disinfection. For each order of magnitude increase in indicator fecal organism, the buffer might be increased by 50% of the standard state guideline (Table 12.12).

Pretreatment to enhance disinfection beyond normal chlorination, such as extended ozonation, ultraviolet radiation and greater chlorination, should be reviewed with respect to the fecal organism concentration. For each order of magnitude reduction, below levels characteristic in the standard assumptions the buffer width might be reduced 20-30% of the standard distance. Thus, a pretreatment that reduced the fecal indicator levels by 1,000 would require a buffer of 10-40% of the standard (60-90% reduction), but not less than the minimum width of 5-10 m.

Table 12.12 Modification Possibilities of Land Treatment Buffer Zone Width Based on Waste Characteristics, Pretreatment Options or Land System Design

System Change	Buffer Zone Width (% of standard requirement)
Secondary-treated, chlorine-disinfected municipal effluent; applied with solid set irrigation in open field.	100% (standard requirement)
Industrial waste without potentially pathogenic microorganisms or substantial adverse odor.	Minimum value[a]
Municipal sludge.	150% (for one order of magnitude increase in fecal indicators) 200% for two orders of magnitude
Pretreatment of domestic waste beyond normal chlorination, as by extended ozonation, ultraviolet radiation, greater chlorination, etc.	70-80% (for one order of magnitude decrease in fecal indicator) 40-60% (for two orders of magnitude) 10-40% (for three orders of magnitude)
Hybrid of 1% domestic, treated, chlorinated effluent with 99% industrial process waste not containing potentially pathogenic microorganisms or substantial adverse odor.	40-60% (twofold dilution)
Traveler gun system with secondary-treated, chlorine-disinfected municipal effluent, applied in an open field.	200%
Traveler gun system with industrial waste not containing potentially pathogenic microorganisms or substantial adverse odor.	Minimum value[a]
Forested buffer with secondary-treated, chlorine-disinfected municipal effluent, applied by solid set system.	50%
Special spray design for maximum trajectory height of 0.6-0.7 m aboveground, irrigating secondary-treated, chlorine-disinfected municipal effluent in wooded area.	30-50%
Gated pipe for application of secondary-treated, chlorine-disinfected municipal effluent in open field.	Minimum value[a]

[a]5-10 m

In industrial cases where domestic waste is included with the process waste to be spray irrigated, there is a hybrid situation. Consider a typical case in which the domestic component is treated, disinfected and is 1% of the total material to be sprayed. This is a 100-fold reduction in fecal indicator concentrations, so a buffer zone of 40-60% of the standard width would be appropriate

(Table 12.12). If the combined waste stream undergoes further treatment, such as storage prior to spray irrigation, then the relative concentrations of constituents of concern should be determined to judge whether further buffer zone refinement is warranted.

Design and Pretreatment to Reduce Buffer Area

The concept of a buffer zone decrease or increase based on engineering design techniques associated with the spray system or the site conditions is also based on the mechanisms for airborne transport. Land system techniques that reduce the transport potential relative to a solid set spray irrigation system with an open field between the wetted spray distance and the property line, would thus require less buffer area. The enhanced removal through a buffer, such as forest, or the reduced generation rate by a special application technique, result in fewer fecal indicators reaching the outer buffer edge than for the standard assumed spray system. Thus, a shorter buffer width is needed to achieve the same indicator level as that achieved under the standard conditions. For example, if the waste is applied at ground level, as in surface irrigation, or if injected in the soil, these techniques obviate the need for any buffer larger than the minimum distance previously described (Table 12.12).

Use of a traveling gun device with a domestic effluent would require a larger buffer zone, because of the high pressure and high trajectory of the spray pattern. The degree of buffer increase has not been established by experiment, but a 100% increase might be considered. For a traveler device with an industrial waste not containing fecal indicators, only the minimum width is needed, since the aerosolized problem has been solved prior to spray application.

Forested buffer zones through which wind can penetrate are usually judged to require only 50% of the standard width for open fields (Table 12.12). Another technique for reducing buffer width is related to the solid set spray liquid trajectory, which under standard conditions reaches a height of 3-5 m for a typical sprinkler spacing. For certain small systems with land treatment of disinfected domestic effluent, the trajectory can be reduced to 0.6-0.7 m above the ground. This technique reduces greatly the droplets transported, since the wind velocity is significantly reduced at this height (Figure 12.31). Furthermore, surrounding such a spray system with a shrub-tree line would continue to reduce the wind velocity and, hence, aerosol transport. Use of low-level spray systems may allow a 30-50% reduction of the standard buffer width (Table 12.12). Obviously, in the limit of gated pipe with openings at ground level, the buffer would be continually reduced to the minimum 5-10 m.

The reader should note that this discussion of buffer zone is directed at the concept of matching site, waste and application characteristics to the

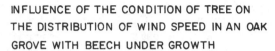

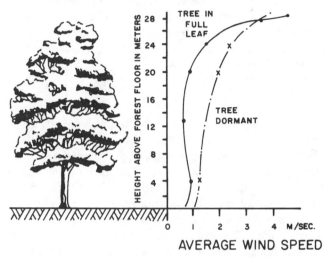

Figure 12.31 Buffer effect on wind velocity of the presence of trees (Geiger, 1969).

need for a buffer zone to control airborne microorganisms of concern or constituents likely to create odor. There are few, if any, quantitative data to
establish the exact increases or decreases necessary to accomplish the desired
buffer control. Recommendations in Table 12.12 are thus only preliminary
judgments. It is clear that a fixed regulatory buffer width will be too small
under certain conditions and too large in other situations, hence the need for
adaptive guidelines.

The cost of buffer zones is primarily the initial investment as established
by the purchase value per hectare. Consideration should be given to siting, so
that unusable or uncleared land can serve as the buffer area, thus reducing the
percentage of unusable land. An annual operation cost may be necessary to
mow or otherwise manage the buffer zone.

Beautification

At an industrial land treatment site it is important to gain the public
acceptance necessary to minimize nuisance considerations. For many persons,
the only appreciation of the industrial environmental efforts is from the roads
viewing storage, pretreatment units, irrigation or spreading operations. A

staggered or triangular shrub or tree spacing along highways will provide an effective beautification measure. The preferable species planted should be rapid growing, have year-round foliage, be flowering and tolerable of climatic and soil conditions at the industrial site. Additional shrubs or trees are recommended surrounding storage ponds to reduce any odor impact. Maintenance of beautification areas involve mowing to assure a pleasant appearance. An estimated beautification cost based on recent installation (1978) is $900-1,000/100 m of shrub or tree line.

Security

Security of an industrial land treatment site involves both the safety of unauthorized persons and of the system equipment and facilities. Unless there are large numbers of neighbors or special problems related to the waste, the first security phase should be posting of information and warning signs forbidding entry to the site. All access roads into the site should be secured with locked gates and fit into the beautification zones. Industry personnel should maintain surveillance of the site to identify any persons not heeding the posted instructions. Personal communications can usually solve these problems. If conditions cannot be maintained secure, a fence would be considered, but only after the information approach had been tried. The use of posting restrictions is usually successfully, thus eliminating the need for fencing.

Weather Restriction and Operational
Control Systems

In any industrial land treatment system, certain control devices or techniques must be included as an additional design and economic component. On-off controls are needed to account for weather conditions. This control may be via the land system operator or by some automatic device. The most common control is to eliminate any irrigation during a rainfall event, regardless of the type of irrigation option used. The goal is to eliminate the potential for direct surface runoff of applied waste. Often, with the center pivot, traveler or hauling/spreader devices, no application can occur for 1-2 days following a large rain event, because of soil traction and support strength.

Following rain restriction or control on the application cycles, the control restrictions for freezing or snow conditions are most common. In regions where cold periods are not extensive, there is generally no application during freezing conditions, thereby reducing problems of ice formation and pipe breakage. However, in regions where winter snow conditions exceed 1-2 weeks, application may be necessary during these adverse periods.

Available Options

Storage should first be considered, but for certain wastes, low-rate application under snow conditions is feasible even with the more costly equipment. For systems that must operate in winter environments the design engineer should consult the design factors presented earlier in this chapter.

Sludge applications can be undertaken under moderate snow conditions. To prevent snow melt runoff, sludges must be incorporated into the soil, rather than surface applied. Again, extra precautions are needed to prevent adverse environmental impact or system damage when operation under freezing conditions is utilized.

A third control device is the regulation of spray irrigation (solid set, center pivot and travelers) under adverse wind conditions. During windy periods, uniform application is not possible, and nuisance drift problems can become acute. Generally, irrigation during the early morning hours will reduce the problems of daily wind conditions. For winds in excess of rates given in Table 12.4, irrigation should cease. Such control can be made automatic with a wind gauge.

Another type of overall management control, minimizing rainfall-runoff potential, can be obtained by selecting the application periods of the year. This is particularly important for sludge applications that tend to be infrequent and heavier. Basically, this control technique interprets local weather data to establish the weeks within an annual cycle in which the rainfall-runoff potential is minimized. A plot is made of the average streamflows and the intensity of rainfall events vs the months of the year. Periods in which rainfall intensity is high, or stream flow is low, should be avoided in scheduling sludge applications. Such an approach can minimize the environmental impact from an industrial waste site.

When an irrigation system is utilized to apply industrial waste to land, additional control devices are necessary to ensure reliable performance. A valve system is needed for operation of a spray field by sections. It may be either manual or automatic. The manual are gate valves installed below the ground surface in the pipeline. A protective casing (0.2-0.3 m in diameter) is installed around each valve to allow operator access. The major advantages of these valves are the reliability and the need to have the operator go into the field to change irrigating sections. The frequent presence of the operator facilitates surveillance for malfunctions.

Automatic control valves may be either pressure- or electric-operated. The electric operational valve systems are more reliable, although slightly higher in initial investment and installation costs. Pressure operation may use either water or air as the working fluid. Special precautions are needed to ensure no particles are in the system, as clogging of ports can be extensive.

Both water- and air-based systems have been made to work in industrial waste usage, with air being easier to keep particle-free.

In all pipes, valves and equipment transporting wastewaters, large orifices should be used to eliminate any possibility of buildup or blockage. Another useful device is a pressure cutoff to detect when there is a sudden drop in line pressure. These devices detect a break in the irrigation lines and shut off the irrigation pump when effluent is being discharged in a very concentrated area. Prevention of this flooding is especially important where sodium or toxic substances are involved because the soil can be severely impacted and rendered unusable by heavy doses.

Irrigation is controlled in zones, *e.g.*, 10 ha of solid set, 2 basins, or 1 hydrant or pass of traveler. Each zone is accessed and controlled by a valve, either manual-operated or automatic. The cost of zone control valves is given in Table 12.13. The attention needed for manual valve operation is recommended to oversee for malfunctions. However, some prefer an automated system, in which case valve costs (for electrical and hydraulic) are estimated in Table 12.13. In addition, the automated timer/controller must be included at a price of $180-300, depending on the number of zones. The wire or hydraulic tubing cost are inconsequential when compared to the transmission pipe, Thus, the automated system has a greater investment cost, consisting of valves and a controller. Labor savings are small where proper supervision of irrigation performance is done.

Table 12.13 Control Valve Investment Cost Estimates (1978)

Valve	Size (cm)	Representative Cost Estimate ($)
Electrical Activated	7.5	160-170
	10	600-700
	15	1,000-1,100
Hydraulic Activated	7.5	90-110
	10	600-650
	15	1,000-1,100
Manual (Globe and Butterfly)	7.5	160-170
	10	225-245
	15	300-320

Artificial Drainage

For industrial wastes in which water is the LLC or for which predominantly aerobic conditions are difficult to maintain due to an elevated water table, an underdrain system is sometimes suggested. An artificial drainage system is a grid arrangement of low-strength plastic pipe, with slots to allow water flow. The pipe is buried 1-2 m deep and at a spacing dictated by soil type and water inputs. Design of such systems should be undertaken with knowledgeable personnel and using new, improved techniques (Skaggs 1979). In addition to plastic drain systems, open-trench grids of relatively wide spacing can also be used for drainage.

The primary disadvantages of using artificial drains are associated with the potential regulatory response. A tile drainage system is viewed more as a point source, which may be subject to a zero effluent criterion. A land treatment system functioning properly without artificial drainage will have these same subsurface water flow and discharges, as discussed in Chapter 4. In addition, so will all adjacent land (forests, agriculture, etc.), and the water quality from these nontreatment areas will not have a zero concentration of various chemical parameters. However, quality data on subsurface flow are not widely available, so the land application site, particularly if tile drained, is not subject to judgment in relation to natural or uncontrollable conditions characteristic of adjacent land areas. A monitoring system and correctly designed system can alleviate some of the potential problems in this regard. Another tile drainage disadvantage is that it often reduces the treatment zone for wastewaters, because little material moves below the drain lines or over the long distances associated with undrained fields and natural drainage outlets. Still, the typical subsurface flow distances are 10-30 m, which is very substantial with respect to the assimilative pathways.

Artificial drain systems are often necessary when an industrial site has already been chosen without regard for the potential for land application. Unfortunately, there are ample examples of this situation. In other cases, factors besides waste management are particularly favorable at a given site (highway access, railroads, labor force, etc.), so that a given area seems highly advantageous. In these instances a total economic analysis would provide insight into site advantages vs costs of land application. In most cases, a compromise site in which the land application potential is improved can result from waste management input early in the decision process.

The cost of underdrains is substantial with respect to land purchase or irrigation systems. Estimates of investment for 30- and 120-m spacing, buried 1.5-2 m deep, are $950/ha and $250/ha, respectively (Skaggs 1979). These cost values are based on recent installed values, reflecting realistic design and competitive bid prices. When compared to underdrain system costs developed under municipal conditons, the industrial system is 2.5- to 3.5-fold less expensive. Realistic underdrain lifetimes are 20-25 years for tile drains. Labor requirements for tile inspection and maintenance of the outlets are given in Figure 12.32 (Pound *et al.* 1975). In general, no cleaning of actual lateral or main drain lines is likely over the 25-year period. The total

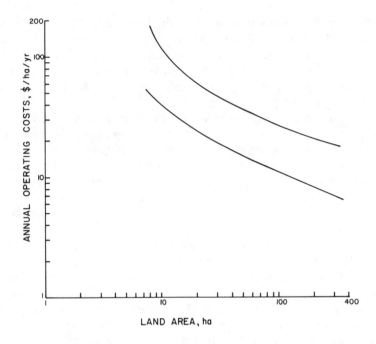

Figure 12.32 Operational costs of underdrain system (Pound, 1975).

annual cost can be determined using industry depreciation rate guidelines plus the annual operation expenses.

VEGETATIVE COVER AND ASSOCIATED AGRICULTURAL EQUIPMENT

Introduction

The critical need for a vegetative cover at an industrial land treatment site is well established from field experience and based on a fundamental non-degradation objective. The presence of a vegetative cover provides enhanced rainfall and irrigation infiltration over noncropped areas, due to plant roots, vegetative canopy and senescent plant matter. Increased infiltration decreases runoff, which, because of this characteristic nonpoint source impact, is the recommended method for areawide pollution control, the so-called "best management practice." The vegetative cover is one of the most effective monitoring systems of the satisfactory waste assimilation by a soil-plant system. The ability to maintain viable crop or forest production is roughly equated to acceptable application rates of waste.

Under certain conditions the vegetative cover can be used to recover plant nutrients in an industrial waste, thus providing an economic return. The value of such crops can often equal the annual operation and maintenance costs, providing a strong incentive for land application. In terms of management and operation of a land application system, the requirement to maintain agricultural yields stimulates the responsible operation of the waste management system, another advantage of a vegetative cover. Responsible operation is a large part of a successful waste management system, so factors that encourage such operation should be recommended. Finally, an important contribution of a vegetative cover is to promote an aesthetic system. With a crop and more natural conditions, neighbors, passersby, manufacturing personnel and customers are more favorably impressed, overcoming many potential nuisance or sociological conflicts. Whether the crop selected is perennial, annual or used on a rotating field basis, it is an essential component of the land system required for plant-soil treatment of industrial wastes.

Purpose of Vegetation

The primary principle in selection of vegetative species is to reduce the total annual cost of the overall land system. Secondary principles are the unknowns and concerns by industries regarding substantial involvement in agricultural operations. This latter concern often runs counter to the proven need for responsible management for purposes of land treatment and avoiding surface landfilling situations.

Major Types of Vegetation

Reviewing over 60 industrial land treatment systems it is obvious that there are only two or possibly three predominant vegetation types used. Thus, to simplify the design considerations, the crops to be discussed are:

1. grass species,
2. forest, and
3. other fiber ornamental or nonfood crops

In the grass category are a number of annual and perennial varieties. These represent the major, direct food chain link, since nearly the whole plant is consumed in grazing or as hay. The major grass species in the U.S. are given in Table 12.14. This listing is only representative of the more widely used, tolerant species. Other grass may provide better waste management capacity, but specific local agricultural recommendations should be sought in those instances. Additionally, double cropping can be elected using a perennial grass that grows in one temperature regime (warm or cool), and an annual variety that grows in the opposite season.

Table 12.14 Major Grasses and Legumes in the U.S. (Heath *et al.* 1973)

Region	Legume or Grass Most Commonly Produced
Northeast	Timothy
	Alfalfa
	Red clover
	Birdsfoot trefoil
Central and Lake States	Kentucky bluegrass
	Timothy
	Alfalfa
	Red clover
Southeast (upper)	Kentucky bluegrass—white clover
	Orchardgrass—ladino clover
	Tall fescue—ladino clover
	Orchardgrass—alfalfa
(lower)	Hybrid bermudagrass
	Tall fescue—white clover
	Bahiagrass
	Common bermudagrass
Northern Great Plains	Bromegrasses
	Wheat
	Bluegrass
	Bluestems
	Blue grama
Intermountain	Alfalfa
	Smooth bromegrass
	Tall oatgrass
	Orchardgrass
	Red clover
Southern Great Plains and Southwest	Bluestem
	Switchgrass
	Grama
	Alfalfa
Pacific Coast	Alfalfa
	Clover
	Timothy

Forest alternatives consist of a number of deciduous and evergreen trees. The principal species characterizing land treatment sites are given in Table 12.15. Existing forest stands may be combinations of these tree types. While the predominant end use of forests is not for food, there is some wild animal usage of tree portions for food. This is a very small part of the overall tree, but does indicate the considerations necessary in use of forests in land treatment.

Table 12.15 Major Forest Species in the United States

Region	Species Most Commonly Produced
Western	Douglas fir
	Poplar
Southern	Pine
	Mixed pine—hardwood
Lake States	Poplar
Eastern	Red pine
	White spruce
	Mixed hardwoods

Nonfood or fiber crops are important land treatment system alternatives consisting mainly of: (1) cotton, (2) jute or hemp, and (3) crops grown for fermentation production of fuels (*e.g.*, corn and tapioca fermented for ethanol to be used as a petroleum substitute). These crops are annuals grown under managed conditions, so pose the least threat to a food chain.

Regulatory attention is often directed toward the crop or vegetation present at an industrial land treatment site. Concern is expressed regarding toxic, hazardous, priority and other pollutants. These concerns must be tempered on the basis of a thorough understanding of the assimilative pathways present in the plant-soil system. These concerns are not valid if based on the behavior of constituents in an aquatic system. The logic that these materials will ultimately end up in a water reservoir is simply an aquatic centrical theory, which does not focus on the predominant terrestrial impact. Therefore, a balance is needed for evaluating land treatment vs other technological alternatives, including economic factors. Some of the relationships between industrial waste constituents and the vegetative cover are discussed briefly below. More detailed information may be obtained from the chapters on assimilative capacity.

Bioaccumulation or magnification is an area of environmental concern with respect to many chemicals. For example, in aquatic systems, bioaccumulation ratios for PCB have been reported as 300-800 (Herbst *et al.* 1976). By comparison, in a soil-carrot system, the bioconcentration was only two, while the metabolites were not biomagnified (Moza *et al.* 1976). It is readily acknowledged that there are far too few data on terrestrial bioaccumulation, but as a preliminary judgment of toxicologists from the limited data, it appears that less potential hazard exists with land treatment pathways. That is, there may be an inherent advantage in the low application rate, assimilative mechanism of land treatment, when compared to any potential stream discharge. This certainly would not be true of landfill operations.

A second factor in considering crop implications is the alternative of using nonfood vegetation. Two of the three vegetation alternatives recommended for industrial land treatment are nonfood options. However, the nonfood option does not allow prolonged or permanent soil conditions in which crop concentrations might be unfit for use in the food chain. That is, nondecomposable or nonmobile constituents cannot be disposed on the plant-soil system such that the potential for growing some food chain crops or the edible (by nondomestic animals) portions of fiber-use crops is adversely affected. Thus, the nonfood crop can be useful in solving short-term (0.5-5 yr), phytotoxic or plant uptake conditions, but the long-term plant-soil conditions must not be degraded.

A third vegetation factor is the behavior of organic compounds taken up by the plant. This facet is just now being evaluated with respect to chemicals not used as pesticides but present in industrial wastes. The present information is preliminary, but it appears that the uptake by plants is not in itself an indication of the transmission of toxic chemicals into the food chain. The plant appears to have a demonstrated capacity to decompose or detoxify organic compounds. For conservative elements such as metals, no disappearance occurs, so control must be exercised in the amounts added to the soil. The assimilative capacities in Chapter 10 are developed on this control of soil level. The pathway of plant uptake and subsequent neutralization by plant metabolism has been demonstrated for 23 complex and toxic organic compounds (Kapauna 1975, Dressel 1973). These results must be reviewed with caution, but it does appear that with certain management schemes, the short-term uptake, prior to soil and plant stabilization of organics, may not pose a severe limitation or food chain impact. More information is needed regarding this crop facet of land treatment systems.

The three factors described above allow the conclusion that proper management and design can reduce to acceptable levels the impact of organic chemicals in industrial wastes, when applied to a plant-soil system. The assimilative capacity loading criteria are such that a vegetative cover can, and therefore should be, used at an industrial land treatment site. Crop selection and management would be used to assure satisfactory performance.

Crop Selection

Three factors are used in reaching a decision regarding the type of vegetative cover at an industrial land treatment site. These are, in sequential order:

1. the natural range of crop species,
2. waste factors, and
3. economic return.

The land system designer must undertake a review of site- and waste-specific inputs in relation to this crop selection process, so that the needs of the particular industrial facility are met.

Natural Range

In every region, state and even at a county level, the annual cycle of temperature and moisture are the most important conditions controlling what can be grown. Soil factors play a lesser role in vegetation potential. For a given crop type, these temperature and moisture cycles define a geographic range in which successful cultivation is probable. Therefore, in evaluating species of trees, grasses or nonfood crops at a given industrial site, those crop types should be eliminated that are not viable because of natural range. For example, in the northeast U.S., Coastal Bermudagrass is outside the range of growth due to cold conditions creating winter-kill, and therefore would be ruled out of further evaluation.

These natural ranges for crop species have been developed from extensive forestry/agricultural experiments and observations under routine environmental conditions. One exception to the natural range considerations is the limitation due to moisture, in which land application of industrial liquids may extend the areas in which a crop can be grown. This is similar to the potential for crop growth under irrigated agriculture. The designer must consider the nature of the restrictions due to natural crop range to determine if the waste itself can eliminate the normal restrictions.

Waste Factors

Throughout the analysis of site characteristics and assimilative capacity, the restrictions and crop response variations have been addressed (Chapters 4-11). The LLC analysis considers each constituent in the waste and plant response as one part in each of these considerations. Based on the calculated LLC, the crop restrictions for these constituents must be addressed in crop selection. Improving the assimilative capacity of the LLC(s) by using more tolerant or more uptake capacity crops can then reduce the land area or improve the operational considerations.

See Figures 7.17, 7.35, 8.5, 9.2 and 10.6; and Tables 5.2, 5.6, 6.27, 7.15, 7.23, 8.1, 8.4, 8.6, 9.4, 9.6, 10.7 and 10.9; and discussion of each constituent, for the tolerance of various crops to waste application. Most, but not all, crop considerations focus on immediate phytotoxicity. That is, after a brief initial period, the microbial, adsorptive, reaction and other pathways have reduced the soil level of constituent to allow satisfactory crop growth. In other instances, the phytotoxicity is related to foliage contact, while the same material incorporated directly in the soil has little crop effect.

In cases of this brief phytotoxicity, crop rotation as well as selection can be used to considerable advantage. This technique involves selecting the more tolerant species in the appropriate grass, tree or fiber crop category. The total land area is planted, except for one section on which the waste is first applied. After a brief time the immediate phytotoxic effects are reduced and the crop is planted on that section. Waste is next applied on another section, killing some or all of the crop. Then this section is reseeded and the process continues. In this manner the immediate phytotoxic effects are dealt with by sacrificing one yield cycle. However, the replanting and growth until next waste application provide all the critical benefits of a vegetative cover discussed previously in this section.

Economic Return

The relative value to be derived from among the alternative vegetation types remaining at this point in the selection process is the final factor in evaluating crop species. Among the three main classes of vegetation, the fiber, or nonfood, crop is likely to result in the largest economic gain. The grass species and then the trees have succeedingly lower economic gain. Estimated revenues of crop alternatives are discussed in the economics section related to the vegetative cover component of the land system. Often, crops of lower economic value are selected because of waste considerations related to toxic materials, *e.g.*, selection of forests in place of grasses. Within the category of trees, there is a variation in economic value, hence some management choice is available.

Crop Establishment

The annual temperature and moisture cycles in the locale of an industrial land treatment site establish the time or times of year that the vegetative species can be planted. This is a fixed site characteristic which, if not included in the planning schedule, can cause a delay of up to a year in the implementation of the land system. For the perennial grass and tree options, establishment is a one time activity, while nonfood crops or annual grasses require yearly planting. In each case there is a relatively narrow time period for planting, so that all site preparation should be scheduled for completion just prior to the optimal planting dates.

Evergreen or deciduous trees are established as seedlings within a 2-4 month time period in an annual cycle. After the seedlings are planted, the remaining uncovered areas are planted in a convenient perennial grass, such as pine seedlings with fescue. This is necessary to eliminate erosion at the site. As the trees grow and close at the crown level, the grass dies back due to lack of sunlight. If the waste is applied via spray irrigation, the solid set system can be made adaptable to the tree growth stages (Reed 1978). The window for seedling planting is in later winter.

The wide variety and growth patterns of grasses makes generalization of planting recommendations very difficult. The designer must establish which grass is to be planted, then refer to agricultural forage bulletins for guidelines on seeding or sprigging periods. Generally, planting is done in the fall or spring. Cool season grasses are planted in the early fall to allow growth prior to winter. Seeding or sprigging must be done on the contour if there is a reasonable probability of rain in the early growth periods.

If land area must be converted from forest to grass, steps must be taken to clear and prepare the land system. The first stage is tree harvesting, with stumps left for removal. The clearing contractor must remove all trees greater than 8 cm, regardless of market value, so that the clearing and grading can be done efficiently. Next, all stumps, roots and other debris are collected and burned. Finally, thorough root raking, hand labor and some earth moving are used to finish the site. It is very important to make the site reasonably free of debris, which will interfere with grass harvesting and waste application. Finally, fertilizers are added according to soil tests, seeding or sprigging occurs, and mulch may be applied if immediate erosion is a potential problem. The supervision of this phase of crop establishment is very critical, and qualified assistance cognizant of the objectives of land treatment should be utilized (Klose 1978a). Cost of conversion from forested to grasses or establishment of grasses are given under the economics section for the vegetative cover.

Cotton is planted in the spring and harvested in late summer. A rotation to a winter forage is recommended to control erosion. This crop may be harvested or disked-in, depending on the waste constituent influences. Other crops grown for nonfood purposes should be planted annually, according to agricultural guidelines for the site locale.

If sufficient attention is not directed to planting at the optimal times, the vegetation stand will be sparse. Under these conditions, erosion and gully formation occurs. These make subsequent agricultural operations difficult because of surface irregularities. The bare areas promote surface runoff and thus increase substantially the nonpoint source impact from a land treatment site.

Crop Management

Little vegetation management is needed throughout the tree growth cycle. At the start, the grass grown among the trees must be mowed to maintain the stand. Hence, the tree spacings must be sufficient to permit mowing equipment. After 10-20 years the tree stand is usually thinned by cutting and hauling, to allow further growth. Throughout the overall growth period, downed or diseased trees must be removed. Assistance of a trained forester should be used to keep the tree crop viable (Nutter 1978). Final clearing should remove all trees regardless of condition, with harvesting in sections preferred. Replanting of grass and seedlings are recommended shortly after cutting.

The management of grasses or nonfood annuals must be integrated with the waste application cycles. If the waste creates wet soil conditions or the waste constituents are to be limited in application prior to harvest, then harvesting must occur within the waste cycle, or storage must be provided during harvesting (see previous storage section). The objective in crop management is to maintain reasonable yields and to remain efficient in terms of operating costs.

For grasses and a single set of farm equipment, about 30 ha can be effectively managed. This would require a sequence of cutting (possibly with conditioning), drying, raking, baling and bale removal (Figure 12.33) in one section (7.5 ha) of the total area. After one cycle the next section is harvested

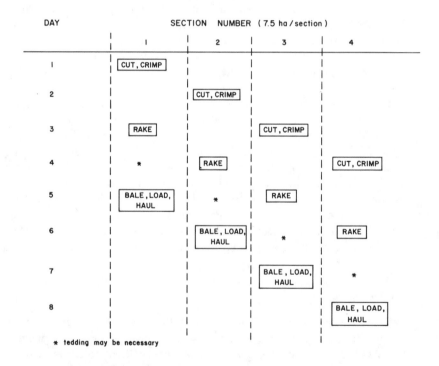

Figure 12.33 Operational cycle for harvesting grass at industrial land treatment system.

and the process continued. Bales should be removed from the field and stacked. Covered hay storage of from 50-100% of the annual cuttings usually is cost-effective when compared to losses due to exposure and fluctuations in hay prices. When grass rotation is used to offset initial phytotoxic effects, the crop management must also include soil disking and replanting. Areas larger than 30 ha would require additional increments of farm equipment and labor.

Cotton is harvested with special equipment and labor. These should be investigated for the site conditions during the land system design process. Seeding should occur in late spring (April-May). Cotton provides advantages of a nonfood chain crop, but field rotation is needed, so the management plan can be quite complex.

Economic Evaluation

The cost for planting tree seedlings and establishing grass for erosion control between tree seedlings varies between $570/ha and $870/ha. The tree sale value depends on the tree type, size and management input. Harvested value thus ranges from $250-6,000/ha from forest production. This value includes the cost of harvesting and, when spread over a 25-year growth cycle, is about $10-240/ha/yr.

Clearing forested areas to establish grass species on recent industrial land treatment sites costs approximately $3,000-3,500/ha. Establishment of grass on existing agricultural land requires about $500-750/ha. From the production of hay a sale revenue of $350-400/ha can be expected on an annual basis. Typically, this revenue will equal the labor input for the entire waste management system, but not the depreciation of investment.

Farm implement investment costs for the harvesting of grasses are given as estimated values in Table 12.16. Optional equipment needed for annual crops or for rotational schemes is also given.

Table 12.16 Equipment Costs (1978) for Grass Harvesting and Field Rotation

Farm Implement	Investment Costs ($)
Tractor, 45 kW	13,000
Mower/Conditioner, 2.7 m	5,900
Rake, 2.7 m	1,600
Baler	5,900
Stack Retriever	12,100
Hayloader plus Wagon	3,500
Seeder	700
Disc Harrow	2,500

Cotton planting and harvesting are done annually. Detailed costs for cotton production are presented by Reed (1974). Direct sales revenue is on the order of $1,000/ha. Thus, as greater operational expense is needed, *i.e.*, forest to grass to cotton, the revenue also increases.

MONITORING PRETREATMENT-LAND APPLICATION SYSTEMS

Basis of Monitoring Plan

From a strict constructionist viewpoint, monitoring requirements should be discussed in each stage of the total industrial waste management program. However, to coordinate the understanding and to present the reader with the complete concept of system monitoring, the overall structure is presented in this chapter. The land system, which is the subject of this chapter, is the ultimate basis and central rationale for monitoring, so the following discussion is appropriate in covering the entire industrial pretreatment-land application system.

Previous references to land treatment monitoring have been heavily oriented toward municipal effluent and sludge systems (Bauer 1978, Melsted 1973, Keeney 1978). A common description of monitoring was "no firm recommendations or basis could be given." However, if one interprets carefully the LLC analysis developed in Chapter 2, a rational basis for monitoring *any* land treatment system appears clear. Such a basis is essential when faced with the diversity of wastes and site conditions commonly present in industrial systems.

The LLC analysis weighs the proportion and amount of waste constituents being evaluated for industrial land treatment vs the plant-soil assimilative capacity for these parameters, as developed from actual site information. Waste complexity, site characteristics and the basic design constraint of non-degradation are thus considered in detail as a part of the LLC analysis. The result of these inputs and calculations is a subset of parameters that require the greatest land areas. These are, by virtue of the LLC analysis technique, the constituents for which the waste application rate and assimilative capacities are nearly equal. That is, these parameters are "closest to the firewall" of land treatment performance capability. It is therefore the LLC and subset of parameters requiring the largest land area that should be the basis for monitoring. That is, the LLC analysis, in addition to the basis for land system design and pretreatment selection, is the rationale for the monitoring program development.

The presence and degree of pretreatment does not alter the monitoring strategy. Pretreatment reduces the level of the LLC or possibly changes the

the particular constituents considered limiting, but there is still a subset of parameters in the pretreated waste that is limiting or dictates land areas. These parameters are determined by the same process of calculations, so the LLC results continue to serve as the monitoring basis. Pursuing this logic for monitoring selection, the parameters chosen are those to be evaluated most critically from the very start, in-plant and as the raw waste; through the pretreatment, if used; and in the various portions of the plant-soil system and adjacent waters. Use of the LLC approach in justifying a monitoring program helps eliminate the tendency to sample and analyze many parameters simply because in some circumstances the constituents could be a concern in land systems. Thus, cost savings can be achieved by eliminating nonessential monitoring.

The overall dual objectives of land treatment monitoring have been agreed upon by most researchers and consultants. These objectives are:

1. to corroborate the performance as a waste treatment system subject to the nondegradation type of constraint and to verify long-term plant-soil viability; and
2. to maintain the needed essential elements for crop growth.

Sampling occurs throughout the total industrial waste management system with the constituents monitored being similar, as determined by the LLC analysis. The frequency of sampling varies according to the component in the total system being considered. In a schematic form, the critical monitoring stages and frequencies are presented in Figure 12.34. Two tiers or levels of sampling and analysis are used at each monitoring point. The first is the critical parameters selected from the LLC procedure. First tier data frequency is directly related to the section under consideration. For example, the waste generation section is sampled more frequently than the groundwater module of a pretreatment-land application system. The second tier is established to assess other parameters that are less likely to be of concern, either because of the lower level in the waste or because of the high assimilative capacity. However, as a conservative approach and to obtain a better confirmation of system performance, these parameters are included, but at a much lower frequency of sampling. As related to the LLC analysis, the parameters in this second tier are those of interest to the designer or those with intermediate land area requirements when a ranking is made from the largest area (LLC) to the least needs. In the following discussion, each of the system sections subject to monitoring (Figure 12.34) are reviewed relative to industrial pretreatment-land application systems.

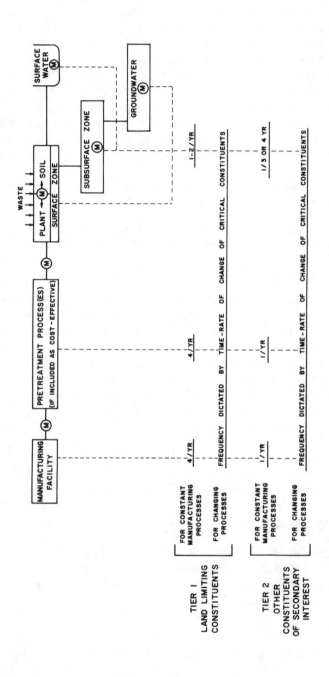

Figure 12.34 Diagram of monitoring system utilized for industrial pretreatment-land application system.

Industrial Facility

Industrial facility is the first component of the total system that must be considered in a monitoring program. As a part of the stage I portion of industrial land treatment design, a complete and composite waste analysis has been performed (Chapter 2). However, monitoring cannot routinely include such a comprehensive testing. Instead, the several parameters requiring the greatest land area (from the LLC procedure) must be evaluated on a regular basis. The objective is to determine with some confidence the total mass of these compounds generated at the plant. In addition, tier one includes additional waste characteristics that affect the pretreatment process efficiency. For example, one industry had determined that nitrogen was the LLC and that ammonia stripping was cost-effective. Although flow and pH were not critical land system factors, the pretreatment depended strongly on these factors. Hence, pH and hourly flow were tested in the first tier or level of monitoring.

The critical, first-level parameters must be monitored on a cycle that reflects anticipated variations in waste streams. If a single process is being operated on a routine basis, then waste monitoring frequency can be quarterly. In this case, the volume of waste or the manufacturing level can be related to waste amount of the critical parameters, and this correlation is used to determine waste generation based on annual product manufactured. Duplicate composite samples can be obtained over the short-term plant cycles (8 hour, one day, one week, etc.) to obtain this quarterly sample of raw waste.

In the case of continually changing industrial processes, such as with new products, new reactants and altered process variables, the monitoring frequency must keep pace with such changes. Within the time period of similar process operation, the sampling can be done on a less frequent basis; but with likely significant changes, the concentrations and mass of constituents, based on the product volume or process duration, must be updated frequently. The basic objective is simple—to determine the mass and typical concentration of the critical constituents to be pretreated, if cost-effective, or to be land applied.

The basic structure and parameters determined for monitoring, from the LLC procedure, can be used to project or predict proposed manufacturing process changes. That is, what are the waste management implications of plant changes manifested in the waste? Are the needed land areas, storage, monitoring, etc. beyond the technical or economic feasibility of the available land treatment system? This is a very important management capability, associated with a well-developed waste monitoring system.

The second level or tier used to monitor the industrial waste generation must also reflect probable shifts in manufacturing processes. In comparison to level one sampling, the second tier is monitored less frequently (Figure

12.34). Constituents selected are those in the middle range of land area requirement, as determined by the LLC analysis.

Pretreatment Process Performance

This monitoring in the overall industrial waste management system is based on the parameters critical to the ultimate plant-soil receiver. The constituents identified from the LLC analysis are determined in samples collected from the pretreatment effluent streams. Frequency of sampling depends on the factors discussed previously for variations in manufacturing waste characteristics. Often, pretreatment attenuates short-term changes in the waste generated, thus reducing the frequency of required sampling. Substantial manufacturing changes will be reflected in altered pretreatment process performance, and hence necessitates a monitoring frequency responsive to such changes. The selection of level one and two constituents is based on the same rationale presented for the raw waste monitoring.

Land System Operation

Evaluation of the behavior and requirements of the land system focuses on the performance with respect to waste treatment (Figure 12.34) and the agronomic needs for a viable vegetative cover (Table 12.17). These two types

Table 12.17 Agronomic Assessment Parameters for Land Treatment Systems

Phosphorus
Potassium
Calcium
Magnesium
Salinity
pH
Sulfur
Essential Micronutrients

of monitoring are important and will be discussed separately in this section. The plant-soil monitoring program developed below can be justified to: (1) assist the industry environmental manager in system operation; (2) demonstrate water quality protection; (3) provide public health protection; (4) meet regulatory concerns; and (5) establish industry and public confidence in land treatment systems.

Monitoring to evaluate land treatment of industrial waste constituents is categorized in four phases related to the behavior of the critical constituents in the plant-soil system and to the methods used to collect samples. These categories are:

1. plant and soil-surface zone
2. soil and soil-water subsurface zone
3. groundwater
4. surface waters

Plant and Soil-Surface and Subsurface Zones

Plant or vegetation monitoring is necessary in industrial land treatment. Often the plant itself, growth and chemical analysis, serve as a biological monitor of the entire soil-plant-waste system. The degree and type of plant testing depend on the LLC analysis. For example, at one industrial site the LLCs were several mobile anions and cations. Plant uptake and growth pattern were not a factor, so the vegetative cover was not included in the level one monitoring program. Rarely, if ever, is crop monitoring not included in at least tier two sampling. For LLCs in the categories of specific organics, nitrogen, metals and hydraulic, the plant analysis is an essential part of level one monitoring. Plant monitoring includes chemical composition of particular compounds and growth or yield as a measure of overall acceptable soil conditions.

A vegetation sampling procedure is established to account for the substantial field variability of growth and chemical composition. Random sampling techniques have been developed for industrial land treatment sites, so that a vegetation sample consists of 5-15 subsamples (Overcash 1978). Duplicate samples are taken of each different principal vegetative cover. When the crop is to be harvested and used in the food chain, the entire harvested portion is sampled, *e.g.*, grasses, corn, soybeans. When the crop is not to be used purposefully in the food chain, only those portions edible by wildlife are sampled, *e.g.*, forest crops, cotton crop. Once the subsamples of the appropriate crop portion have been taken and composited to a single sample and the process repeated to obtain a duplicate sample, then preservation is needed.

For specific organics as the LLC, the plant samples would be frozen, and, when analyzed, are defrosted, ground and extracted on a fresh weight basis. Extraction procedures and subsequent analyses are compound specific. Methods follow those developed for pesticides (Santelmann 1977, Weber 1975). For nitrogen, metals and specific inorganics the plant sample is dried as soon as possible at 65°C, ground and stored. Analysis procedures then follow standard methods for plant materials (Jones and Steyn 1973).

To assess vegetative growth as a measure of overall acceptable soil waste conditions, the annual crop yield per ha from the site must be measured (bushels/ha, bales/ha, etc). For perennial tree crops another measure of growth is used, *e.g.*, height increase, tree diameter change, etc. Results of the annual crop performance are compared against control areas (described in later section) to ensure reasonable growth.

Soil monitoring is used at level one intensity for almost every industrial site. That is, all LLC categories are related to acceptable soil concentrations of those compounds, hence soil testing is needed to verify land treatment performance. There are variations in whether the surface soil zone or deeper layers are to be sampled, depending on the movement potential of the LLC species. In both instances the sampling is done on a random basis, with 5-15 subsamples or locations being composited to obtain a sample. Duplicate samples are to be taken within the land treatment area for each appropriate sampling unit. A sampling unit is commonly taken to be each soil type. This can be reduced, if two or more soil types are similar in the characteristics necessary for the LLC assimilative mechanisms. For example, if metals are the LLC, then soil types with similar CEC, pH and texture in the surface zone can be sampled as one unit. Again, the sampling unit is defined by the assimilative pathways necessary for those industrial constituents requiring the greatest land areas—the LLCs.

The subsamples are thoroughly mixed, which requires patience and energy, especially with moist soil material. After mixing, the required analytical volume is taken and air-dried as soon as possible for the inorganic analyses. A representative sample is better obtained if the entire volume of subsamples are air dried and then mixed. High-temperature drying ($>39°C$) can alter the analyses and should be avoided. After drying, the material for analysis is ground (<2 mm size) and stored until testing in airtight containers. Soil analyses for inorganic compounds follow standard procedures (Walsh and Beaton 1973, Black 1965).

Soil volume is also taken of the freshly collected soil and extracted with appropriate solvents when organic species are to be analyzed. Extraction techniques are compound specific. Representative extractions have been presented by Cifrulak (1969), Medvedev and Davidov (1972) and Barik *et al.* (1976). The extract is then analyzed by the appropriate technique. Specific care must be taken to avoid contamination throughout the inorganic and organic sampling, preparation and analysis scheme.

Groundwater

Monitoring is performed at tier 1 intensity on those systems in which the critical or limiting constituents from the system design (stage I) are likely to migrate with the soil-water movement. Sampling points for groundwater include springs, spring-fed ponds, in which surface runoff is diverted, and on-site wells (either existing or newly constructed). These points, when located in the land treatment area, represent the maximum or first impact of the applied industrial waste. In addition, an offsite or perimeter groundwater sample should be taken in the direction of hydrologic flow, where this is clearly established. Extensive perimeter wells are not generally needed if one or two wells are put in the most prevalent flow direction.

Permanent groundwater wells should be constructed to give an unbiased sample of the groundwater. A typical well schematic is given in Figure 12.35. Well size is determined by the sample method (submersible pump, hand pump, weighted sampling bottle, etc.). The locked well cap is essential to ensure an uncontaminated sample. Well installation should be by means of coring or auger installation, not by cable-tool rig. This will provide a detailed drill log of the well to ensure a more representative interception pattern of groundwater flow and quality changes.

Well sampling usually involves pumping out a certain volume of the existing water. Seepage from the surrounding groundwater supply refills the well, and then a sample is taken. These steps are accomplished with a submersible pump made portable with sufficient electrical cable and powered by an engine-driven electrical generator. The pump is connected to flexible hosing and discharges to the ground level. Since the concentrations in groundwater are low, great care with respect to contamination is needed if pump, pipe and wire are transferred from well to well. Instead, it is recommended to install the pump, pipe and wire permanently, then with a portable power supply to pump out and collect the needed sample. Groundwater samples should be preserved to prevent chemical and biological changes until the appropriate analyses can be performed. These analyses follow those described in *Standard Methods* (APHA 1975).

Surface Waters

Surface waters surrounding an industrial land treatment site should be a part of a monitoring program, where lateral movement of waste constituents through soils is likely. The sample collection frequency is determined as tier 1 or 2, Figure 12.34, depending on the LLC analysis. Preservation and analysis proceed in a similar manner to groundwater samples.

In all of the four land system sampling locations the industrial waste and site characteristics are used to assess the more critical waste constituents. Then, depending on the three categories of calculational techniques used to determine the assimilative capacity, the appropriate portion of the land system to be monitored can be determined. Those portions (surface zone, subsurface soil, groundwater, surface waters) that are essential for the LLC parameters are then selected for tier 1 frequency. Not all four portion are necessarily included in level one. As shown in Figure 12.34, tier 1 samples are

WATER QUALITY MONITORING FACILITY

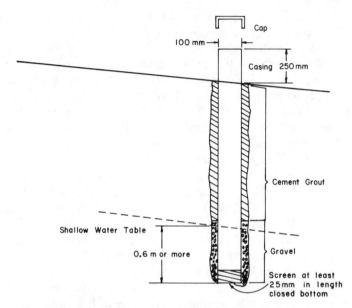

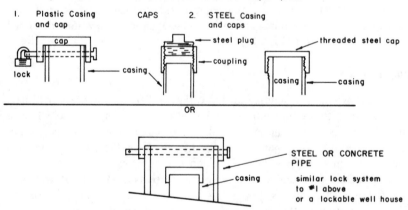

Figure 12.35 Water quality monitoring well.

taken once or twice per year, usually in the fall and/or spring. For those parameters and points in the land system that are in tier 2, fewer samples are collected.

Monitoring of the land treatment system operation also includes agronomic sampling and testing to maintain a viable vegetative cover (Table 12.17). The objective is to provide a balanced fertilization program and to correct soil conditions to enable reasonable crop growth. These agronomic soil samples are collected and analyzed to determine needed corrective action on the land treatment site. In some cases, the agronomic and waste treatment sampling can be combined to avoid duplicate analyses of the sample parameters. The agronomic-related analyses are performed once per year, usually in the autumn. Analyses can usually be obtained from the state Department of Agriculture.

Land System—Background and Baseline

The assessment of background levels of various critical constituents that are in a particular industrial waste as well as in the natural environment is very important. Several major questions of operational success and regulatory concern must be based on knowledge of background conditions. Background or baseline evaluation involves two testing schemes:

1. determination of conditions at a site prior to any waste application, and
2. determination of conditions and changes over time of surrounding areas unaffected by the waste application.

Determination of prior conditions is usually covered in the site data collection associated with the stage I LLC analysis. Emphasis is placed on the tier 1 parameters. Sampling should include the soil surface and subsurface zone, vegetative cover, groundwater and receiving waters. It is recommended that monthly samples for one year be collected of surface and groundwater supplies. Sufficient soils and plant samples should be collected for present analyses, and preserved for future use and reference. These archival materials can then be used to respond to future questions.

Determination of baseline data reflects the conditions at a site, if the land treatment activity is not present; but the surrounding land use practices continue to impact the area. In other words, the conditions in the terrestrial environment are not static, but will change over time, so the criteria for background or baseline conditions may change. For the plant-soil portion, duplicate control areas (approximately 5 m x 5 m) are established and protected from waste application. The same vegetative cover is grown, but on the control plot only fertilization and natural inputs occur. Where several different soil associations are present on a land treatment site, control plots for each should be established. For groundwater or receiving water, sampling stations must be located up the hydraulic gradient from the land treatment site. Sampling of the baseline stations should be on an infrequent, but regular, basis, e.g., once every two or three years. Parameters to be analyzed should be the level one constituents as defined previously.

Economic Evaluation

Both initial investment and ongoing operational costs are required for a land treatment monitoring system. The raw waste and pretreatment streams for nearly uniform industrial processes, lagoon or storage pond effluent, adjacent surface waters and some wells, can be sampled with a grab sampler capable of collecting liquid from any prescribed depth (Figure 12.36). Costs for such samplers are $120-150. In industrial situations for which greater daily and weekly variations are likely, a flow proportional sampler is necessary (Figure 12.37) (N-CON Surveyor II). Typical expenditure for these devices are $550-750. Specialized materials of construction for corrosive or other peculiar waste conditions can be obtained at additional cost.

The soil-plant sampling involves certain equipment items. These have been assembled in a kit, which therefore represents the typical cost for such equipment separately (N-CON 1979). Depending on the flexibility needed, the initial investment for plant-soil sampling is $500-1,500. Preservation, described previously with standard methods of crop and soil analysis, must be done shortly after sample collection, and thus is usually undertaken at the industrial facility. Investment costs for weighing scales and drying are in the range of $650-800, depending on the size and sophistication selected. Soil extraction for organic constituents is relatively complex and costly, needing weighing scales, solvent and a solvent extraction apparatus (approximately $800-1,000). Groundwater monitoring requires specialized well installation. Typical cost of installation at a site for wells of 3-, 15-, 30- and 100-m depths are $130, $400, $750, and $2,100 respectively (Taylor 1978). Cost estimates must be increased if rock or other difficult soil conditions are encountered. A submersible pump and attachments for 30-m deep well would cost approximately $250-350 per well. A portable generator of sufficient size requires $750-850 and is usable on the total number of wells. It is recommended that each well contain a permanent submersible pump, discharge line and electrical hookup, and that a portable, gasoline-driven electrical generator be used to operate all the pumps.

Labor requirements for a complete monitoring system are determined by the frequency of sampling and the land area. For the tier 1 sampling with relatively uniform processes generating the waste (Figure 12.34), approximately 8 days/yr would reflect the sample collection and the review of results

Figure 12.36 Grab sampling device (Wheaton Instruments, Millville, NJ).

Figure 12.37 Flow Proportional Sampler (N-CON Systems Co., New Rochelle, NY).

for an entire pretreatment-land application system. This input would cover the entire monitoring requirements from raw waste to the final receiver system. The chemical analyses needed will depend on the results of the LLC procedure, as previously described. Approximate analytical costs on a unit sample basis for more commonly used parameters are presented in Table 12.18.

The complete monitoring system cost can be developed from the above components. An industrial system example is as follows: from the waste characteristics and site data four parameters are determined as needing level one sampling, while an additional eight are examined at level two. The industrial processes are somewhat variable, and a lagoon was used as pretreatment. The land treatment site size was 22 ha, resulting in 5 monitoring wells. Table 12.19 describes the necessary monitoring system costs for a typical pretreatment-land application system. The reader should note the strong dependence of monitoring system cost on the specific waste and the actual land treatment site characteristics.

In summary, the selection of a total monitoring system can be made cost-effective by using the site and waste as the basis for sampling and analysis. Excessive monitoring leads to excessive costs and loss of perception of the

Table 12.18 Estimated Waste, Soil and Plant Analytical Costs

Analysis	Price per Constituent and Per Sample ($)
Routine Waste Constituents	2-5
Forms of N, Forms of P, Ca, Mg, Na, Cl, SO$_4$, conductivity pH, K, Cu, Cr (wet chemistry and colorimetry)	
Metals and Aggregate Organics	10-20
Zn, Cd, Co, Ni, Se, COD, TOC, BOD$_5$ (atomic adsorption, wet chemistry)	
Microbial	15-25
FS, FC, total count	
Oil and Grease	8-10
Multiple Screening of Industrial Chemical	20-30
Specific Organic Analyses	40-60
Estimated Soil and Plant Tissue Analysis Costs	
Routine Essential Elements	2-3
(P, K, Mn, Ca, Mg)	
Nonessential Elements Commonly of Concern	4-8
(Na, Cl, Mo, EC)	
Plant Tissue	2-3
Routine essential elements	
Less Routine Analyses	5-8

critical constituents for land treatment. The LLC procedure (Chapter 2) can be used effectively to establish a tiered approach, which focuses monitoring attention on a relatively small number of parameters, with less frequent determinations of noncritical parameters. This constituent selection is valid throughout the waste system, from manufacturing process generation to the final plant-soil receiver.

TOTAL LAND SYSTEM

The land system, as designed for industrial land treatment, contains some or all of the individual components discussed in this chapter. The precise components and design size are determined directly by the specific waste characteristics and site selected. Therefore, the economic evaluation of the land treatment portion of a total pretreatment-land application system can only be developed with any accuracy after a specific site has been identified. This site-specific characteristic was also shown to be critical in the stage I

Table 12.19 Example of Industrial Land Treatment Monitoring System Costs

	$
Industrial waste—composite sampler	550-750
Lagoon pretreatment and surface waters	
—grab sampler	120-150
Soil and plant preservation (for inorganic and organic species)	1,450-1,800
Well installations (based on 250 miles travel distance)—	
5 wells, 15-m deep	1,000-1,500
Well pump installation	2,000-2,500
Total Investment	5,120-6,700
Annual Labor—8 Days	510-640
Chemical Analyses	
Tier 1: 4 Constituents (1 metal, 2 inorganic, and 1 specific organic); raw waste, (12),[a] lagoon effluent (4), soil (1), plant (1), 5 wells (1) and two streams (1)	1,300-2,200
Tier 2: 6 Constituents (3 metals, 2 inorganics and oil); raw waste, (2),[a] lagoon effluent (1), soil (0.5), plant (0.5), 5 wells (0.5) and two streams (0.5)	300-100
Total	2,100-3,460

[a]Number in parentheses is sample per year.

portion of the overall design methodology, which determined the LLC and the land area requirements.

Combining the necessary land system components one can determine:

1. investment cost,
2. operation and maintenance cost, or
3. total annual cost.

The results of this cost analysis is a primary objective of stage II in the overall methodology developed in this book for pretreatment-land application systems for industrial wastes. Any of the above cost indices can be selected for comparison to pretreatment expenditures (stage III, Chapter 13) to arrive at the most cost-effective balance (stage IV, Chapter 14) between pretreatment and land treatment.

To determine the sensitivity of the land treatment system to reductions in the LLC (by in-plant source control or by pretreatment), the cost index selected must be evaluated for several land areas. That is, if the LLC is 90%, 80%, 60%, 30%, or 10% of what is projected for the design raw wastes, how much is the land system cost correspondingly reduced? Figure 12.2 depicts this type of relationship. The components of the land system vary in different amounts with the same reduction in LLC, hence a straight line relationship does not usually describe this overall land system cost curve.

From the design description sections for all needed components covered in this chapter, the reader can obtain the component cost estimate based on design variables. If the cost index to be used is total annual cost, the designer must also select:

1. a depreciation interest rate
2. a method of depreciation (straight line, declining balance, etc.)
3. a component or total system lifetime (the industrial planning period)

The component lifetime is given in the respective sections of this chapter. The other amortization factors are dictated by individual corporate policy.

To illustrate stage II of the overall pretreatment-land application methodology, an example of industrial site design for a case study is presented as the summary of this chapter.

An industry in the middle Atlantic States was constructing a new manufacturing facility. It desired to assess land treatment as one alternative for meeting federal and state environmental regulations. The complete waste was combined in a liquid containing 1% solids.

At the completion of the LLC analysis it was determined that a land treatment area of 20 ha was required for the complete manufacturing plant waste. The LLC was a substituted aromatic, which appeared as an intermediate product in the manufacturing process.

Two sites had been investigated with respect to assimilative capacity in stage I and found to be nearly equal. In acquiring land, one parcel was more favorable economically, so the detailed design and economic analysis was performed at that site. The land area purchased was 27 ha, since certain sections were unusable (creeks, field surface waterways and poorly drained land) (Table 12.20). Transmission to the site is 0.6 km.

A storage analysis was performed to obtain the inputs for Equation 4. No data existed to establish the need for storage due to variations in the assimilative capacity throughout the year. Volume for nonapplication was determined by freezing conditions as 45 d. No crop cycle requirements existed. Repair and emergency needs were estimated at 5 d. The 25-year, 24-hour storm was 15 cm, with a freeboard of 30 cm. One week of waste was added to the volume for use of pond as a sump. These were combined, the storage unit (earthen pond) was sited, and based on the volume needed, the storage investment cost was determined.

Pump selection for irrigation was a horizontal centrifugal unit, since installation could be readily made during storage pond construction. An in-plant centrifugal pump was used to convey the material to the storage pond at the land treatment site. A solid set irrigation system was selected for waste application. In the land treatment area the crop to be used was fescue, so establishment and harvesting equipment expenses were needed. To reduce

Table 12.20 Total Land System Ecomonics Developed from Individual Components

Land System Component	Investment Cost ($ 1978)
Purchase (27 ha)	73,000
Pumping Transmission to site—0.6 km 600 l/m pump without pumphouse	1,200
Storage	34,000
Irrigation pump and pumphouse	6,300
Solid set irrigation (20 ha)	50,000
Establish grass on existing farm fields (20 ha)	10,000
Farm Equipment	38,000
Diversions to remove adjacent field runoff (500 m)	1,500
Monitoring	6,000
Buffer zone (6 m), available within 27 ha tract purchased	–
Beautification (1,300 m, remainder already wooded) and security posting	12,000
Weather equipment	500
Total	233,000

rainfall-runoff impact, diversions were constructed to eliminate runoff from adjacent areas.

The monitoring system is that presented in Table 12.19. A buffer zone of 6 m was selected because no pathogenic organisms were in the waste. This area (about 1 ha) was available within the total 27-ha tract purchased.

The total system cost for the land treatment of the complete industrial plant waste was then determined as the summation of the individual components (Table 12.20). Investment cost savings for the land system were evaluated for reduction in the amount of LLC. For this waste, the removal of the LLC did not substantially affect the volume to be handled, although the actual land area was reduced in direct proportion to the LLC removal. The results for 90%, 75%, 50% and 30% of the LLC, but with the same site and other conditions, are presented in Figure 12.38. A nonlinear reduction in cost was found because some land system components represent nearly fixed costs, while others are strongly area dependent.

Information displayed in Figure 12.38 includes the results of the stage II portion of the overall pretreatment-land application methodology. These results are very site specific, as the reader should note from the inputs used in these calculations. There are some waste-specific inputs, principally in the method of handling and application of the industrial materials. Thus, at this step in the overall design process, the land system costs for the complete

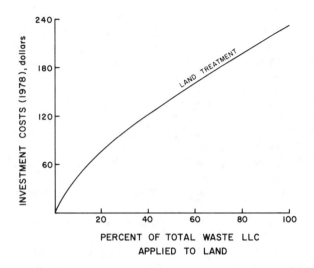

Figure 12.38   Industrial land treatment system investment cost as a function of the amount of LLC applied and, hence, land area required.

waste and the cost sensitivity to reductions in the LLC are known and can be used in the remainder of the overall design.

REFERENCES

Allender, G. C. "The Cost of a Spray Irrigation System for the Renovation of Treated Municipal Wastewater," M.S. Thesis, Pennsylvania State University, University Park, PA (1972).

American Public Health Association. *Standard Methods for the Examination of Water and Wastewater* (1975).

Anderson, R. K. "Cost of Land Spreading and Hauling Sludge from Municipal Wastewater Treatment Plants," EPA/530/SW-619, Environmental Protection Agency, Cincinnati, OH (1977).

ASAE Standard S 376. "Design, Installation and Performance of Underground Thermoplastic Irrigation Pipe."

Badger, W. L., and W. L. McCabe. *Elements of Chemical Engineering* (New York: McGraw-Hill Book Company, 1936).

Barik, S., R. Siddaramppa and N. Sethunathan. "Metabolism of Nitrophenols by Bacteria Isolated from Parathion–Amended Flooded Soil," *Antonie Van Leeuweenhoek* 42:461-470 (1976).

Bauer, J. W. "A Prototype Program for Monitoring Domestic Wastewater Land Treatment Systems," *Int. Symp. on Land Treatment of Wastewater*, Vol. 2, U.S. Army Corps of Engineers (1978), pp. 317-324.

Black, C. A. *Methods of Soil Analysis*. Agronomy #9 in the Series of Agronomy (Madison, WI: American Society of Agronomy, Inc., 1965).

Brown, K. W., and L. E. Deuel. "Soil Disposal of Petrochemical Waste Effluents," Soil Science Dept., Texas A & M University, College Station, TX. Presented at AIChE meeting, New York, 1977.

Chamblee, J. A. (Project Director). "Construction Costs for Municipal Wastewater Treatment Plants: 1973-1977," EPA 430/9-77-013 (1978).

Christiansen, J. E., A. A. Bishop, F. W. Kiefer, Jr. and Yu-Si Fok. "Evaluation of Intake Rate Constants as Related to Advance of Water in Surface Irrigation," *Trans ASAE* 9(5):671-674 (1966).

Cifrulak, S. D. "Spectroscopic Evidence of Phthalates in Soil Organic Matter," *Soil Sci.* 107:63 (1969).

Dorset, W., "Valmont Industries, Inc., Valley, Nebraska. Personal communication (1978).

Dressel, J. "Uber das Vorkommen von Nitrosaminen in pflänzlichem Material," *Landwirtforsch. Sanderh.* 28:273-279 (1973).

Fok, Yu-Si, and A. A. Bishop. "Analysis of Water Advance in Surface Irrigation," *Irrigation Drainage Div. Proc. ASCE* (March 1965), pp. 99-117.

Geiger, R. *The Climate Near the Ground*, Milroy N. Stewart *et al.*, translators (Cambridge: Harvard University Press, 1969).

Heath, M. E., D. S. Metcalfe and R. E. Barnes. *Forages* (Ames, IA: Iowa State University Press, 1973).

Herbst, E., I. Weisgerber, W. Klein and F. Korte. "Beiträge zur okolgischen Chemie- CXVIII, Bilanz Biokkumulierung and Umwandlung von 2, 2-dichlorobiphenyl 14-C in Goldfischen" (1976).

Hickey, J. L. S., and P. C. Reist. "Health Significance of Airborne Microorganisms from Wastewater Treatment Processes. Part I: Summary of Investigations," *J. Water Poll. Control Fed.* 47(12):2741-2757 (1975).

Hill, R. W., and J. Keller. "Irrigation System Selection and Management for Maximum Crop Profit," ASAE Paper 78-5024, presented at ASAE, Logan, UT (1978).

Horney, L. F. "Best Management Practices for Nutrient Control from Agricultural Nonpoint Sources," Dept. Biological and Agricultural Eng., North Carolina State University, Statewide North Carolina 208 plan (1978).

Humenik, F. J., and M. R. Overcash. "Design Criteria for Swine Waste Treatment Systems," EPA-600/2-76-233 (1976), p. 292.

Israelson, O. W., and V. E. Hansen. *Irrigation Principles and Practices* (New York: John Wiley & Sons, Inc., 1962).

Johnson, G. S. and J. A. Moore. "The Effects of Conservation Practices on Nutrient Losses," Agr. Eng. Dept., University of Minnesota, St. Paul, MN (1978).

Jones, J. B., and W. J. Steyn. "Sampling, Handling and Analyzing Plant Tissue Samples," in *Soil Testing and Plant Analysis*, L. M. Walsh and J. D. Beaton, Eds. (Madison, WI: Soil Science Society of America, Inc., 1973), pp. 249-270.

Kapavna. "Neutralization of Organic Substances in Wastewater by Plants," All Union Scientific Research Institute for Utilization of Wastewater in Agriculture (1975).

Kardos, L. T., W. E. Sopper, E. A. Myers, R. R. Parizek and J. B. Nesbitt. "Renovation of Secondary Effluent for Reuse as a Water Resource," EPA 660/2-74-016, Environmental Protection Agency, Ada, OK (1974).

Keeney, D. R. "Design of Soil-Plant Monitoring Procedures for Land Treatment Systems," *Int. Symp. on Land Treatment of Wastewater*, Vol. 1, U.S. Army Corps of Engineers (1978), pp. 365-375.

Klose, H. C., Soil Systems, Inc., 4900 Waters Edge Drive, Raleigh, NC. Personal communication (1978a).

Klose, H. C. "Land Application of Industrial Wastewaters," *Proc. 15th N.C. Irrigation Soc. Conf.*, November 8 (1978b).

LANDTECH. "Selected Reports on Industrial Land Treatment Systems," Cary, NC (1978).

McCulloch, A. W., *et al. Lockwood–Ames Irrigation Handbook*. W. R. Ames Co., Gering (1973).

Medvedov, A. A., and V. D. Davidov. "The Transformation of Various Coke Industry Products in Chernozem Soil," *Pochvovedenie* 22-28 (1972).

Melsted, S. W. "Soil-Plant Relationships (some practical considerations in waste management)," in *Recycling Municipal Sludge and Effluents on Land*, National Association of State Universities and Land Grant Colleges, Washington, DC (1973), pp. 121-128.

Midwest Plan Service. *Selecting Dairy Manure Handling Systems*, AED-18 (Ames, IA: Iowa State University Press, 1977).

Morris, C. E., and W. J. Jewell. "Regulations and Guidelines for Land Application of Wastes—a 50 State Overview," presented at 8th Annual Cornell University Agricultural Waste Management Conference, Rochester, NY (1976).

Moza, P., I. Weisgerber and W. Klein. "Fate of 2,2′dichlorobiphenyl–^{14}C in Carrots, Sugarbeets and Soil under Outdoor Conditions," *J. Agric. Food Chem.* 24(4):881-885 (1976).

Myers, E. A. "Engineering Problems in Year-Round Distribution of Wastewater," in *Management of Farm Animal Wastes*. ASAE, St. Joseph, MI (1966), pp. 38-41.

Myers, E. A. "Sprinkler Irrigation Systems: Design and Operation Criteria," in *Proc. Conf. on Recycling Treated Municipal Wastewater through Forest and Cropland*, EPA 660/2-74-003 (1974), pp. 299-309.

National Oceanic and Atmospheric Administration. "Use of Climatic Data in Estimating Storage Days for Soils Treatment Systems," EPA 600/2-76-250 (1976).

N-CON. Land Treatment Monitoring Kit. N-CON Systems Co., New Rochelle, NY (1979).

Norum, E. M. "Land Application of Potato Processing Wastes Through Spray Irrigation," ASAE Paper 75-2514, presented at ASAE meeting, Chicago, IL (1975).

Norum, E. M. "Design and Operation of Spray Irrigation Facilities," in *Land Treatment and Disposal of Municipal and Industrial Wastewater*, R. L. Sanks and T. Asano, Eds. (Ann Arbor, MI: Ann Arbor Science Publishers, Inc., 1976).

Nutter, W. Department of Forestry, University of Georgia, Athens, GA. Personal communication (1978).

Overcash, M.R. "Mechanism and Control of Rainfall Runoff Impact from Land Application Sites," Preliminary report of EPA grant, North Carolina State University, Raleigh, NC (1974).

Overcash, M. R. "Sampling Manual and Kit for Land Treatment Systems," (1978).

Overcash, M. R., and F. J. Humenik. "State of the Art: Swine Waste Production and Pretreatment Processes," EPA-600/2-76-290 (1976), p. 170.

Overcash, M. R., and D. Pal. "Industrial Waste Land Application," AIChE Today Series, AIChE, New York (1976).

Pair, C. H., W. W. Hinz, C. Reid and K. R. Frost, Eds. *Sprinkler Irrigation* (Washington, DC: Sprinkler Irrigation Assoc., 1969).

Patterson, W. L., and R. F. Banker. "Estimating Costs and Manpower Requirements for Conventional Wastewater Treatment Facilities," U.S. Environmental Protection Agency, Water Pollution Control Research Series 17090 DAN 10/71 (1971).

Pierce, J. C., and R. Bramerel. "Land Application of Wastewaters," *Proc. 11th N.A. Irrigation Soc. Conf.* (December 1974).

Pound, C. E., and R. W. Crites. "Wastewater Treatment and Reuse by Land Application," Vol. VII, EPA-660/2-73-00066, R. S. Kerr Environmental Research Laboratory, Ada, OK (1973).

Pound, C. E., R. W. Crites and D. A. Griffes. "Costs of Wastewater Treatment by Land Application," EPA-430/9-75-003, Office of Water Progress Operation EPA, Washington, DC (1975).

Powell, G. M., M. E. Jensen and L. G. King. "Optimizing Surface Irrigation Uniformity by Nonuniform Slopes," Winter ASAE meeting, Chicago, IL (1972), p. 15.

Reed, A. D. "Sample Costs to Produce Crops," University of California Cooperative Extension Circular MA-4 (July 1974).

Reed, H. T. "Apple Processing Land Treatment System," Soil Systems, Inc. Marietta, GA (1978).

Romkens, M. J. M., D. W. Nelson and J. V. Mannering. "Nitrogen and Phosphorus Composition of Surface Runoff as Affected by Tillage Method," *J. Environ. Qual.* 2(2):292-295 (1973).

Santelmann, P. W. "Herbicide Bioassay," in *Research Methods in Weed Science*, B. Truelove, Ed., Southern Weed Science Soc. (Auburn, AL: Auburn Printing, Inc., 1977), pp. 79-87.

Schwab, G. O., R. K. Frevert, T. W. Edminster and K. K. Barnes. *Soil and Water Conversation Engineering* (New York: John Wiley & Sons, Inc., 1966).

Skaggs, W. "Water Management Model for Shallow Water Table," Report No. 134, Water Resources Research Institute of the University of North Carolina, Raleigh, NC (1979).

Smith, A. O. Harvestone Products, Surrystore Systems (1978a).

Smith, W. Soil Systems, Inc., 6040 Old Pineville Rd, Charlotte, NC. Personal communication (1978b).

Smith, J. L., D. B. McWhorter and R. C. Ward. "On Land Disposal of Liquid Organic Wastes Through Continuous Subsurface Injection," in *Managing Livestock Wastes. Proc. 3rd Int. Symp. Livestock Wastes*, ASAE Proc. 275, St. Joseph, MI (1975), pp. 606-610.

Sneed, R. E. "The Economics of a Total Irrigation System," *Proc. 9th NC Irrigation Soc. Conf.* (November 1972).

Sullivan, R. H., M. N. Cohn and S. S. Baxter. "Survey of Facilities Using Land Application of Wastewaters," Office of Water Program Operation, EPA, Washington, DC (1973).

Taylor, G. Soil Systems, Inc., 525 Webb Industrial Dr., Marietta, GA. Personal communication, based on 20 industrial and municipal installations (1978).

U.S. Bureau of Reclamation. "Irrigation Advisor's Guide," U.S. Dept. of the Interior, Washington, DC (1951).

U.S. Soil Conservation Service. "First Aid for the Irrigator," U.S. Department of Agriculture, Misc. Publ. 124 (1947).

U.S. Soil Conservation Service. *Instructions and Criteria for Preparation of Irrigation Guides,* Engineer Handbook, Sect. 15, Pt. 1, Engineer and Watershed Planning Unit, Portland, OR (1957).

Walsh, L. M., and J. D. Beaton. *Soil Testing and Plant Analysis* (Madison, WI: Soil Science Society of America, Inc., 1973).

Walters, M. R., P. D. Robillard, R. Hexem and R. Gilmour. "Development of Best Management Practices for Agriculture—New York State Strategy," The 10th Annual Cornell University Conference (1977).

Weber, J. B. "Agricultural Chemicals and Their Importance as a Nonpoint Source of Water Pollution," *Proc. S. E. Regional Conf., Nonpoint Source of Water Pollution,* VPI & State University, Blacksburg, VA, May 1-2 (1975), pp. 115-129.

Woodward, G. O., Ed. *Sprinkler Irrigaion,* 3rd ed. (Washington, DC: Sprinkler Irrigation Association, 1969).

CHAPTER 13

PRETREATMENT AND IN-PLANT SOURCE CONTROL– STAGE III

INTRODUCTION

This chapter investigates the pretreatment or nonland treatment alternatives for reducing the total costs associated with an industrial pretreatment-land application system. That is, what options are available to reduce the land area requirements and still achieve an environmentally acceptable solution for the entire waste generated by an industrial facility? In perspective of the overall methodology developed in this book, Chapter 13 is delineated in Figure 13.1.

Evaluation of Total Streams from Pretreatment Processes

Examining a variety of unit processes, a similar flowchart can be used to describe many of these alternatives (Figure 13.2). Evaluation of all of the output streams when compared to the input material represents a total system analysis. The discussion which follows describes the ramifications of Figure 13.2 and a total system approach to pretreatment processes prior to a plant-soil receiver. The unit process or facility in Figure 13.2 can be an extramural treatment or an in-plant process that alters the total raw waste generated.

A large number of treatment processes utilized in conventional and advanced wastewater treatment (AWT) are really only separation processes. That is, an effluent that is relatively pure clean water is produced, but so is a sludge or highly concentrated stream. Often there is an inordinate focus on the clean stream and the subsequent discharge to a stream or lake. The sludge is often neglected with a small arrow on the design sheets.

→ SLUDGE TO ULTIMATE DISPOSAL

593

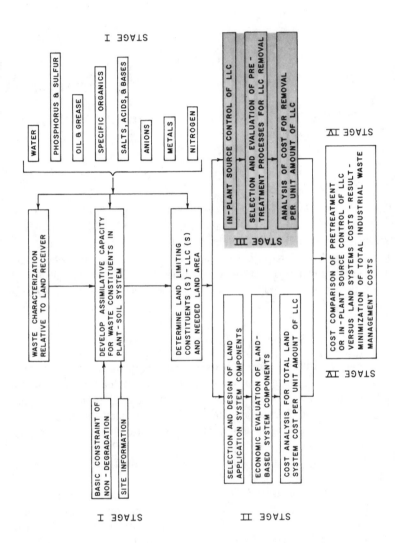

Figure 13.1 Relationship of Chapter 13 to overall design methodology for industrial pretreatment-land application systems.

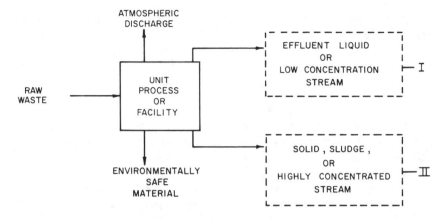

Figure 13.2 Pretreatment schematic with conventional separation to liquid and solid waste streams. Total waste management encompases all streams.

Activated sludge, reverse osmosis, coagulation and activated carbon are examples of such separation processes. In point of fact, the combined effluent and sludge are often not greatly dissimilar to the raw waste. Thus, the total system does not evidence a large amount of treatment or conversion to an environmentally acceptable material, only a highly effective separation.

The approach recommended in this book is that, where land is the ultimate receiver system, the entire waste generated (raw waste or effluent plus sludge) be included in the analysis. Thus, the land area requirements are dictated by assimilative capacity for the entire waste. In this approach, separation into several waste streams is not effective because both sludge and effluent must be applied to land. Instead, the objective is to convert waste constituents into environmentally acceptable materials, so that discharge to terrestrial, aquatic or atmospheric receivers creates few problems. For example, the nitrification-denitrification cycle creates N_2 gas, which is readily discharged to the air. Chemical reactions that convert toxic chemicals to nontoxic compounds are also viable pretreatments. This chapter will discuss the various constituent categories to evaluate such pretreatment options. For those materials not economically converted to an environmentally acceptable material, the plant-soil assimilative capacity must be established (Chapters 2-11). That is, land treatment capability is needed as the ultimate receiver of the pretreated industrial waste.

One exception to the lack of efficacy of separation processes is the case in which the separated stream can be used in an environmentally acceptable manner. For example, if the solids removed from an industrial waste could be used as a feed source or as an input to structural material, then the separation would be advantageous. In one instance a screen filter system recovered a predominantly cellulosic material that had a use in making accoustical tile. These waste constituents were thus effectively used after separation, and land treatment was needed only for the remainder.

Pretreatment–A Definition

Pretreatment, as defined in this chapter, is any process or scheme that alters the chemical or physical nature of the waste generated during the course of manufacturing or processing raw materials into a desirable product. Included are in-plant options such as process change, alteration in management or operation, and substitution of raw materials. Also included are extramural processes that remove or convert specific waste constituents. Pretreatment can range from simply turning off running water to the sophistication of vitrification processes or ozone treatment. Thus, both specific waste pretreatment processes as well as in-plant source control are included in the concept of reducing land area requirements by "pretreatment."

DETERMINATION OF PRETREATMENT OPTIONS

What criteria control the selection of "pretreatment" processes for industrial wastes for which the plant-soil system is the ultimate receiver? The answer is the LLC analysis, stage I of the overall design methodology. In stage I, the characteristic of the particular industry and wastes under investigation are taken into consideration by waste characterization. The specific land area with all soil and vegetation characteristics is directly taken into account. Then a balance is made between land assimilative capacity and the waste generation for each of the constituents under investigation. This balance produces a subset of the overall list of constituents requiring the greatest land areas, *i.e.,* the land limiting constituents (LLC).

Having identified the LLC(s) using the stage I methodology, the designer then knows the parameter(s), which, if reduced, will allow a reduction in needed land area. Hence, pretreatment, or in-plant source control, is dictated by the LLC analysis, or the plant-soil system. Therefore, the LLC analysis

1. establishes the needed land area,
2. determines the monitoring system, and
3. dictates the performance criteria for selection of pretreatment options.

The reader should note that especially where selective pretreatment removal is accomplished, the LLC analysis must be reconsidered continually. That is, as the amount of the LLC, derived on the basis of the untreated waste, is reduced, one of the other parameters may become the limiting constituent. At that point, all further pretreatment must result in the simultaneous reduction of two or more constituents. For example, at a location in North Carolina, the soil assimilative capacity and the waste generated from a poultry processing plant indicated that nitrogen content was the LLC. However, after the removal of 30% of this nitrogen, the hydraulic loading became a limiting constituent. Therefore, investment to reduce nitrogen beyond 30% was not justified without also reducing the overall waste volume or water loading. Further, the evaluation of effective pretreatment is an iterative process to ensure that the LLC(s) are being removed and that there is an effect on the land requirements.

In addition to the principal pretreatment criterion (reduction of the LLC), there are sometimes secondary factors that are also used in the selection process. These are primarily nuisance related. For example, it may be decided to reduce the oxygen demand to control odor when another constituent such as metals or water may be the LLC. Management may decide to reduce the water content of a waste so that it can be handled as a solid, even though the land area is controlled by nitrogen and not water content. These secondary factors are not discussed further in this chapter since the decision basis is very dependent on the objectives of a particular industrial facility.

While the LLC analysis dictates which parameters to remove by pretreatment or in-plant control, the extent of removal is controlled by economic factors. Specifically, the cost of removal relative to the cost of land treatment (as developed in Chapter 12) establishes the degree of pretreatment. A typical pretreatment economic analysis for industrial waste is presented in Figure 13.3. As a greater amount of the LLC is treated through the plant-soil system, the pretreatment cost curve continually decreases.

The degree of pretreatment is very high for conventional industrial waste treatment prior to stream discharge. Levels of 95%, 99% and 99[+]% are not uncommon. From Figure 13.3, the costs for very high removals of almost any constituent increase exponentially with increasing percent pretreatment. Thus, utilizing lower removals (<90%) results in very large savings. Pretreatment-land application systems typically utilize 30-50% removal as a cost-effective level. In that instance, the cost savings relative to 99% removal are very large, so the savings more than pay for land treatment of the partially pretreated industrial raw waste.

OVERALL DESIGN APPROACH FOR PRETREATMENT

Two broad categories of in-plant source control or pretreatment process design are commonly used with industrial wastes prior to land treatment. These are:

1. alteration of unit process variables to achieve differing removal or conversion performances, and

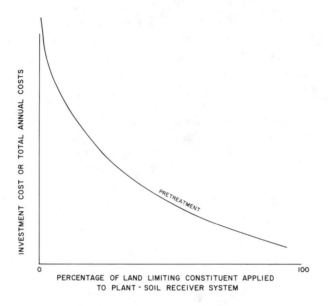

Figure 13.3 Generalized pretreatment or in-plant source control cost relationship to removal of LLC constituent from industrial waste.

2. selection among a variety of standard design processes to achieve differing removal or conversion performances.

Both categories produce a range of costs associated with removal of a particular industrial waste constituent. Using processes relevant to the industrial waste LLC, one can generate a part or all of Figure 13.3. This range of costs is necessary to compare with a companion curve for the land treatment system (Figure 12.2).

For the first pretreatment category, the designer must evaluate the effect of a number of possible design parameters. As an example, consider a nitrification-denitrification process conducted in two fixed tanks. The pharmaceutical waste has nitrogen as the LLC. As the flowrate to the pretreatment process is increased, the percentage of the waste ammonium and organic nitrogen that is nitrified is reduced. Since the denitrification reaction is much more rapid, it was assumed that the denitrification rates remained constant with increasing flow (Figure 13.4). The pretreatment costs per unit amount of nitrogen converted to N_2 gas decrease as the amount of nitrogen converted decreases. Thus, loss of 25% nitrogen was substantially less than loss of 40% N (Figure 13.4). The curve generated in Figure 13.4 can then be used to determine the most cost-effective level of pretreatment vs land application.

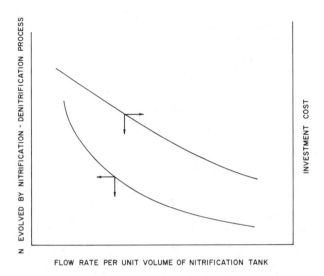

Figure 13.4 Process modification and effluent N performance for nitrification-denitrification sequence.

For the second category, the designer must establish the removal of the LLC as typically found in all available pretreatment processes. These are ranked according to percentage removal and estimated costs (investment or total annual). Such an analysis was performed for the removal of a compound in the class of a substituted aromatic, which had been determined to be the LLC. The results in terms of percentage removal for five different processes are shown in Figure 13.5. As greater removals were achieved, the costs were increased accordingly. This curve then allows selection of the appropriate pretreatment option, when matched with the land system cost curve, to yield a total cost estimate.

The remainder of this chapter is devoted to evaluating pretreatment options for the parameter(s) designated as the LLC. Since inclusive categories for industrial constituents were utilized in the waste characterization and the assimilative capacity development, these will also serve as the basis for Chapter 13. Eight categories (Table 13.1) will be examined, since within each category similar alternatives are available for the range of particular compounds likely to be present.

Many of the options considered as pretreatment may appear to be unconventional or highly advanced. This is due to the approach that a net conversion to an environmentally acceptable material must be accounted for in the land treatment design. Monitoring of pretreatment performance is not covered in Chapter 13. The reader should utilize the monitoring section of Chapter 12 for guidance in this area.

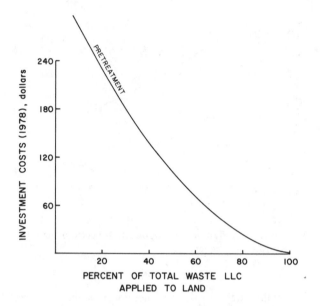

Figure 13.5 Alternative process costs leading to successively greater removal of substituted aromatic (the LLC), smoothed curve of many processes.

Table 13.1 Inclusive Categories for Industrial Waste Constituents

Water
Phosphorus and Sulfur
Oil and Grease
Specific Organics
Salts, Acids and Bases
Anions
Metals
Nitrogen

PRETREATMENT PERFORMANCE RELATIVE TO INDUSTRIAL WASTE CONSTITUENTS

Water

A number of industrial categories characteristically generate large volumes of water as production raw waste. All industries are increasingly aware of wastewater volume, as the regulatory criteria focus on total discharge and impact on streams, rivers, impoundments and estuaries. Thus, there is a strong trend toward water conservation and recycling. A substantial part of the U.S. Environmental Protection Agency (EPA) research budget is directed to water reduction.

For industrial wastes that will ultimately be applied to land, the "pretreatment" for reduction in water must first be examined relative to the plant-soil assimilative capacity. That is, unless the stage I analysis has identified hydraulic loading as the LLC, the fetish for low volume or water content is misapplied. Investment in water conservation or reuse techniques and requisite process changes provides relatively little savings in land system requirements and cost if water is not the LLC. First, the LLC analysis must identify water amount as the controlling land factor.

Once identified as the LLC, water can be reduced using two broad techniques:

1. reduction of volume at point of use, and
2. reuse of water to reduce total volume per unit of product.

These two options will be discussed in terms of typical concepts that would then have to be applied by knowledgeable personnel to a particular industrial situation under investigation. A third option of evaporation will not be covered as a viable alternative. The likely industrial waste organic contaminants in the water vapor released to the atmosphere implies that such discharge is not environmentally compatible.

Reduction

One consequence of reducing the water used at a process location is that other constituents are often not reduced proportionally. Thus, with water reduction comes increasing waste concentration. For most constituents, the land assimilative capacity is predominantly controlled by total mass of constituent, so water reduction has little effect on parameter generation. However, for salts, some anions and sodium, the increased waste concentration can limit land treatment. That is, the assimilative capacity depends on constituent concentration, so that as water volume is lowered, the LLC analysis needs to be continually reevaluated. These considerations must be kept in mind when determining the alternative of in-plant water minimization.

To reduce water at the point of use, the designer must determine what is wasted water. Often this is equated to negligent operation, such as allowing water hoses or taps to flow continuously, whether in use or not. Incorrect operation of valves, water flowmeters and other process-related water uses, also represent wasted water. The solution is management. Supervision of operations and careful instruction of proper procedures can be used effectively to stop water waste.

In addition to management options, the design of water usage equipment can be reviewed to determine what is the task performed by the water. Some techniques have evolved to successfully curtail water use in a variety of industrial situations. Substitution of high-pressure, low-volume nozzles and pumps have repeatedly allowed water washing with reduced water use. Substitution of pneumatic transfer systems for solids and chemical wastes can be used to minimize water consumption. Gravity or mechanical conveyance devices also eliminate the need for water. In cleanup of process equipment, an initial removal of material by mechanical or manual techniques followed by water reuse has often been found to decrease total water in the final waste.

Of course there are many water reduction techniques at the point of use. No book could be all-inclusive. The methods for evaluating how and what water can be conserved were described above from a conceptual viewpoint. Successful implementation has been demonstrated repeatedly for water reduction, so that if water is the LLC, then in-plant techniques are available.

Reuse

The basic question that exists with water reuse is what level of waste constituents will cause the manufacturing or processing product to be completely unacceptable? Reuse of waters containing higher levels would then not be feasible. The answer to this question is completely determined by the specific industry. The degree of possible reuse of wastewater appears to be linked to the quality or sensitivity of product. For example, Springer and Marshall (1975) reported that with the less-refined product of paperboard, the best mills could achieve complete recycle. The more quality-oriented fine paper mills could only achieve a level of 20.6 m^3 of effluent/metric ton of production. Certain industrial processes and products are thus more amenable to recycling of water. The widespread use of recycle water in cooling towers is another example of water reduction through reuse. Fleischman (1975) has reported the constituent limits for effective recycle through cooling towers, and hence a wastewater can be evaluated relative to this reuse potential.

Recycle or reuse effectively reduces water volume in proportion to the number of cycles used. At a certain time, the buildup of other constituents is sufficient that removal of the wastewater is required. The mass of other constituents is often the same with or without recycle, so water is the only parameter effectively removed. Examining the total plant wastewater for reuse potential is often not encouraging. However, experience has shown that the total waste consists of a number of separate streams of differing quality.

Thus, identification and segregation of less-contaminated streams can allow reuse of these. This is partial reuse.

Examples of the procedures and results of water reduction ans reuse have been reported increasingly in journals and reports. Nonintegrated fine paper and board manufacturing (Springer and Marshall 1975), plywood (Allison *et al.* 1975) and rayon production (Waggener and North 1975) are typical. The cost of water reduction at point of use or by means of reuse must be balanced against the savings in land system cost for each level of water volume possible with such techniques. This economic balance determines the degree of water source control that is cost-effective.

Phosphorus and Sulfur

These two constituents of industrial wastes are essentially conservative. Regardless of the form, the application to a plant-soil system will lead to mineralization and availability to the assimilative pathways present. The design analysis is based on the total quantity of phosphorus or sulfur applied to the land treatment site.

The pretreatment options are thus limited to:

1. recovery as a usable product, or
2. in-plant source control.

The decision to pretreat or use in-plant control must be based on the fact that phosphorus or sulfur is the LLC, and thus removal of these would result in land system savings. The first option has had considerable research effort with reports and patents describing the industrial waste recovery process. Phosphorus removal is usually by means of precipitation, such as calcium phosphate. Then a separation process is used to remove the phosphate, usually clarification. If acceptable with respect to trace contaminants, this material can be recycled as an animal supplement, acid neutralizer or as a raw material in certain industrial processes. Use as a fertilizer would not be cost-effective unless the commercial value or benefit on other land were sufficient to warrant the separation. Normally, separation into a lime material and an effluent and then applying both to land is more expensive than direct land treatment of the initial waste. Probable separation processes for industrial waste are reviewed by Sittig (1973, 1977).

Sulfur recovery from industrial waste has been developed for sulfides, sulfur dioxide, sulfates and other forms of sulfur (Sittig 1973). Recoverable forms include elemental sulfur, sulfuric acid and sulfate precipitates. These materials are of use in industrial process as reactants or acidifying agents. One pretreatment often performed is sulfide conversion. The reason is not based on land area, but on nuisance odor factors associated with the application to land of sulfides.

A detailed process for sulfide pretreatment was developed by Sayers and Langlais (1977) for use on industrial tanning wastes. The schematic is given in Figure 13.6. Sulfide is stripped as H_2S gas from the tanning waste at low pH. Recovery of the H_2S occurs in a sodium hydroxide absorber. The sodium sulfide is then substituted into the dehairing process of the plant to reduce the raw material costs. The recoverable sulfur was equal to $85/1,000 hides.

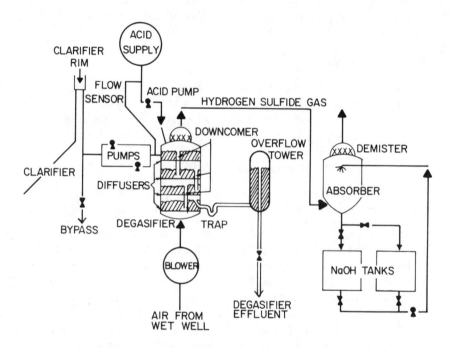

Figure 13.6 Sulfide reclamation system (Sayers, 1977).

In-plant source control of P and S centers on eliminating amounts in the waste or substituting other materials. Reducing the waste load involves detailed examination of sources and process usage of phosphorus or sulfur. Management supervision and design of process equipment can be used effectively to reduce losses of these constituents. Substitution is often possible to reduce waste generation of phosphorus or sulfur. Decisions of the role of these compounds in the manufacturing processes must be made; then acceptable substitution can be implemented. The degree of in-plant control to remove P or S is determined by the stage at which another parameter becomes the LLC and on the relative cost of in-plant vs land treatment.

Oil and Grease

The predominance of pretreatment options for this industrial waste constituent category involve the physical separation of oil and grease with a usable oil product generated. This product is thus utilized elsewhere as a substitute for oil (petroleum, animal or vegetable). For petroleum-based oils, the uses are as an energy source or as cooling/lubricating materials. As the supply of waste oil is increased, other uses also will evolve. Vegetable and animal fats are recovered and then heat-treated to be recycled as an animal feed. This process is widely established.

The plant-soil system typically has a large assimilative capacity for oils. Thus, only certain industrial wastes will have oil as the LLC. It is these materials (effluents or sludges) that may be considered as candidates for oil recovery pretreatment. That is, the presence of oil, even at high levels, does not require oil pretreatment, but when oil/grease are the LLC, such recovery may be cost-effective. Oil removal or recovery that does not have a *bona fide* end use should not be implemented, since the resulting concentrated oil stream will still have to be considered for land treatment.

Various unit processes have been used to effect separation of the oil fraction of an industrial waste. Holding tanks provide the first separation of the most easily removed material. The lower specific gravity of oil causes it to rise through the aqueous phase and be removed with surface skimming. Residence time in the holding tank can be used to control the percentage removal (Wallace *et al.* 1965). The removal efficiency range for normal operation of an API separator is shown in Table 13.2.

Table 13.2 Efficiencies of Oil Separations Process (Sittig 1973)

Treatment	Source of Influent	Percent Removal	
		Floating Oil (%)	Emulsified Oil (%)
Skimming Holding Tanks	—	30-60	—
API Separator	Raw waste	60-99	Not applicable
Air Flotation, Without Chemicals	API Effluent	70-95	10-40
Air Flotation, With Chemicals	API Effluent	75-95	50-90
Chemical Coagulation and Sedimentation	API Effluent	60-95	50-90

To achieve a higher level of oil removal, various combinations of dissolved air flotation and chemical addition are used. With air flotation, additional emulsified oil can be removed. Chemical addition, *e.g.,* aluminum sulfate, destabilizes the oil/water emulsion, thus allowing easier flotation and froth removal (Quigley and Hoffman 1966).

Centrifugation has been used for concentrated sludges to recover the oil fraction (Sittig 1973). For wastes that are predominantly oil, electrical processes have been used to concentrate the oil emulsions. Having separated the oil fraction from the industrial waste, a concentration process is sometimes needed to achieve a usable product. Less than 5% water is needed to permit reasonable combustion of recovered oil, while greater percentages can be tolerated for cooling and lubricating uses.

Most of the oil separation research has involved petroleum-based oils. The oils and greases from animal and vegetable sources are separated using very similar equipment and procedures. The end use of animal feed is the predominant difference (Seng 1973). Investment and operating costs can be determined on an individual case basis. Then the justification for use of oil removal can be made on an economic basis by comparison to the savings in land treatment costs associated with removal of oil as the LLC.

Specific Organics

A variety of processes have been developed to remove total organics as well as specific compounds from industrial wastes. Performance of these processes varies considerably, depending on the compounds present and the sophistication of the process. As pretreatment, six categories are considered in the section:

1. Biological processes
2. Ozone, peroxide and other oxidants
3. Ozone-ultraviolet
4. Ultraviolet and radiation
5. Other advanced processes
6. In-plant control

Biological Processes

Technology evolved for municipal waste treatment has been used to provide microbial decomposition of industrial wastes. These facilities are usually tanks with aeration devices, recycling capabilities, equalization and clarification. The method of operation involves biomass generation, sludge separation and effluent discharge. Combining the sludge with the effluent, the organic loss is on the order of 50%. The behavior of specific priority pollutants, such as substituted aromatics, phenolics and long-chain hydrocarbons is not well established in such biological processes. In fact, a study of textile dyes (Little *et al.* 1974) indicated that little or no decomposition of these compounds occurred. Thus, as a pretreatment to reduce specific dyes, liquid-areation systems are not highly effective.

What role might such aeration facilities have in a pretreatment-land application system? One benefit would exist for an industrial waste in which the total oxygen demand were the LLC. That is, for soil and waste conditions in which the land size is dictated by the area needed for an adequate flux of oxygen to satisfy aerobic microbial processes, the pretreatment reduction in waste oxygen demand reduces the land area needed. Discussions and calculations for this aggregate assimilative capacity are given in Chapter 7. With aggregate organics as the LLC, liquid tank aeration to reduce the total oxygen demand must be evaluated on a cost basis in comparison to savings in the overall land system (Chapter 12).

The extended aeration process recycles sludge biomass to reduce the total sludge organic matter. This would be an advantageous process where the total waste (effluent plus sludge) is to be land applied and where oxygen demand is the LLC. Processes such as activated sludge and UNOX normally produce greater biomass and result in less overall loss of carbon.

In industrial wastes for which a specific compound, such as phenol, cresol or toluene is the LLC, the tank, liquid aeration pretreatment is generally not very useful. Soil microbial populations, diversity and evolutionary capacity are much greater than those used in liquid aeration systems. Thus, when comparing investment and operational costs, it is usually much less expensive to satisfy oxygen demand (as the LLC) in a plant-soil system than in a conventional liquid aeration biological process.

One exception to the non-use of conventional biological unit processes for industrial wastes prior to land application is with certain complex and changing wastes. It has been found from analysis for Consent Decree priority pollutants that industrial raw waste contains numerous compounds, while the effluent from conventional biological treatment contains few of these compounds. In one instance, the number of organic priority pollutants was reduced from 26 to 8, following an extended aeration process. This decrease in complexity reduces the intensive data needs for development of plant-soil assimilative capacities for these compounds. As the data base for land treatment criteria of specific organics is expanded with new systems and studies, the complexity of waste will become less of a factor in design of land-based systems and the advantages of conventional aeration reduced.

Ozone, Peroxide and Other Oxidants

Chemical oxidation of specific organics and total carbon is a viable pretreatment where these constitutents are determined to be the LLC. The oxidation process follows the pathway of parent compounds, intermediates and then stable end products. Rate of oxidation is determined by the compound present and the strength of the oxidant. For the purposes of this chapter, the discussion will center on ozone, with some comparison to the other less active oxidizing agents.

From a mechanistic viewpoint, ozone is a three-oxygen molecule and is a highly reactive allotrope. In aqueous wastes, the ozone can create other reactive species from water (hydroperoxide and hydroxy). Ozone also spans carbon-carbon double bonds, forming ozonides that rupture to yield shorter carbon chain fragments. This process continues until 2-4 carbon acids, CO_2 and H_2O are formed. With evolution of CO_2, the short chain compounds are the remaining carbon. This reaction sequence is depicted in Figure 13.7, in

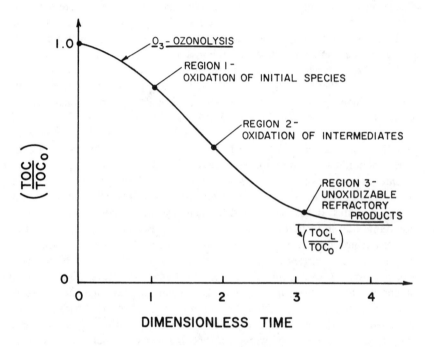

Figure 13.7 Ozonolysis stages of organic compound (Prengle, 1977).

which there is disappearance of parent or intial compounds, formation and oxidation of intermediates, and then formation of short-chain acids. A wide variety of compounds are known to react readily with ozone. Table 13.3 is a listing of reactive and very slowly reactive classes of industrial waste constituents. There is a certain amount of compatibility between ozonolysis and the plant-soil receiver in that many of the compounds that do not oxidize further with ozone are readily degradable in a land treatment system. Thus, only partial oxidation, often of the parent compound, is needed to convert a specific LLC organic to an easily assimilated species.

Table 13.3 Organic Constituent Categories and Relative Ozonation Reactivity

Substrates that React With Ozone
    Alkenes (olefins, rubbers, unsaturated polymers)
    Alcohols (but not degraded)
    Alkynes (acetylines)
    Aldehydes
    Phenols
    Amines
    Aromatics in General
    Epoxides

Substrates that React Very Slowly With Ozones
    Alkanes (polyethylene, polypropylene, etc.)
    Saturated Halocarbons (Cl, Br, I, F, *i.e.,* CH_2Cl_2, CH_3Cl, etc.)
    Ethers (polyethylene gylcol, etc.)
    Some Waxes, Fats and Oils
    Most Greases
    Fused Aromatics React but not Degraded
    Carboxylic Acids

Ozonation equipment consists of a generator of ozone gas, agitated vessel with waste liquid input, and exit parts for gas and liquid (Figure 13.8). The ozone pilot plant is typical of batch reactors, while the full-scale unit demonstrates the continuous reactor arrangement. The characteristic feature of a gas-liquid reaction, such as the ozonation of industrial wastewaters, is that one or more of the reactants must pass through an interface between the phases to react. The overall relationship between mass transfer and the reaction kinetics must be studied to identify the controlling regime. Once the proper regime is identified, an appropriate reactor may be specified. Reactor types range from a bubble column for use with a slow reaction-controlling compound to a packed column for mass transfer-controlled reactions.

Yocum *et al.* (1977) has analyzed the relative controlling steps (mass transfer vs reactions kinetics) based on analyses of the ozone concentration in both the gas and liquid phase. This procedure is based on a combination of Fick's law of diffusion and a material balance for each component in both phases over the entire reactor system. This can be expressed by the following equation:

$$\frac{dC_j}{dt} = k_L a\,(C_{je} - C_j) - r_j \qquad (13.1)$$

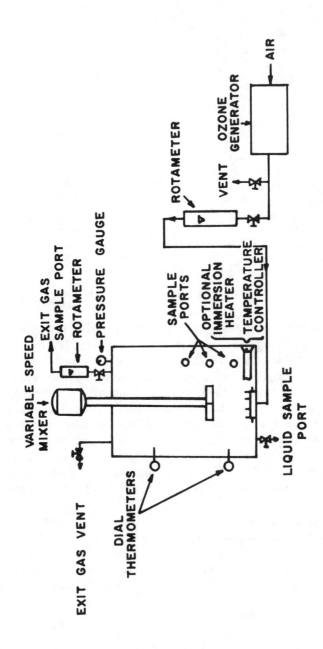

BATCH CONTACTOR

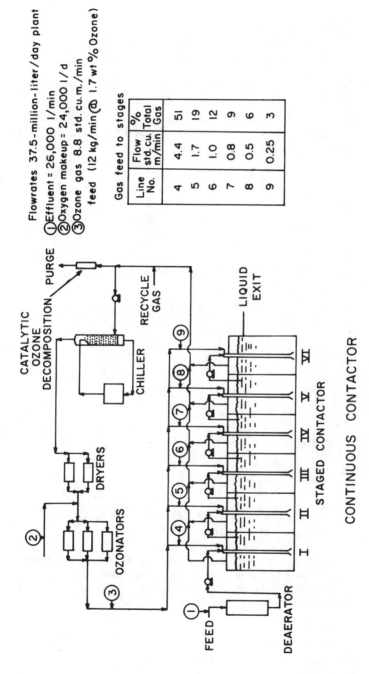

Flowrates 37.5-million-liter/day plant
① Effluent = 26,000 l/min
② Oxygen makeup = 24,000 l/d
③ Ozone gas 8.8 std. cu. m./min
    feed (12 kg/min @ 1.7 wt% Ozone)

Gas feed to stages

Line No.	Flow std. cu. m/min	% Total Gas
4	4.4	51
5	1.7	19
6	1.0	12
7	0.8	9
8	0.5	6
9	0.25	3

Figure 13.8   Batch and continuous contactors for ozonation.

where C_j =   concentration of organic species j

K_L =   liquid phase mass transfer coefficient

a   =   interfacial area

C_{je} =   phase equilibrium concentration of organic species j

r_j =   oxidation or decomposition reaction rate for organic species j

The macroscopic view is illustrated in Figure 13.9 with a plot of the ozone concentration in the liquid phase as a function of time. In the absence of a liquid reactant, the ozone is adsorbed exponentially to the solubility limit. This concentration profile of ozone in the liquid phase can be found by equating r_j to zero and solving Equation 13.1 to get

$$C_B = C_{Be} (1 - e^{-\phi t}) \tag{13.2}$$

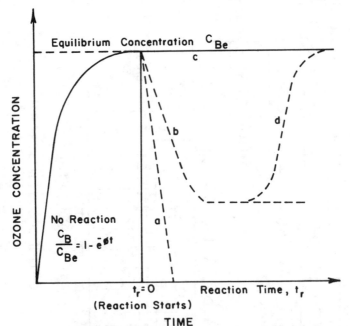

Figure 13.9 Ozone reaction stages (Yocum, 1977).

where ϕ = expression relating volumetric mass transfer coefficient to the ozone decomposition reaction rate coefficient

C_B = ozone concentration in liquid phase

The ozone concentration profile is shown for the three most frequent gas-liquid reaction regimes. After an injection of liquid reactant at time equals zero, the ozone concentration will immediately drop to zero if the reaction is instantaneous, resulting in a truly mass transfer limited system (a). A quasi-equilibrium concentration is reached if mass transfer is controlling with a slow reaction (b), and the ozone concentration will remain at the solubility limit if the system is truly reaction rate controlled (c). Such batch-scale tests are needed for a particular industrial waste in order to select the most cost-effective reactor.

While ozone is frequently used for oxidation of industrial wastes, other oxidants are also feasible. The nature of the LLC compound that was to be removed by pretreatment dictates the oxidizing power necessary. The relative strength of oxidants was developed by Prengle and Mauk (1977) and is given in Table 13.4. Bench-scale tests would be essential to select the most cost-effective oxidant, depending on the industrial waste.

Table 13.4   Relative Reactivity of Pretreatment Oxidants

Species	Oxidation Potential Volts	Relative[a] Oxidation Power
Fluorine (F_2)	3.06	2.25
Hydroxyl Radical (OH)	2.80	2.05
Atomic Oxygen (O)	2.42	1.78
Ozone (O_3)	2.07	1.52
Hydrogen Peroxide (H_2O_2)	1.77	1.30
Perhydroxyl Radicals (HO_2)	1.70	1.25
Hypochlorous Acid (HOCl)	1.49	1.10
Chlorine (Cl_2)	1.36	1.00

[a]Based on Cl_2 as a reference.

The use of ozone as a pretreatment prior to land application of an industrial waste is often different from conventional ozonation. As a pretreatment, the goal is selective removal of the specific organic controlling the land receiver size (the LLC). Thus, only partial oxidation is needed in which the parent compound is removed. Generally, the formation of intermediates represents compounds more easily degraded, so these do not become the LLC. A recent study by Yocum et al. (1977) demonstrated this effect of ozonolysis. The total carbon of a polyethylene glycol wastewater was only slightly reduced following ozone pretreatment (Table 13.5). The biochemical

Table 13.5   Pretreatment Partial Ozonation of Wastewater Containing Polyethylene Glycol to Remove Recalcitrant Compounds

	TOC	BOD_5	BOD_5/TOC
Raw Waste	830	93.1	0.11
Ozonated	626	613.5	0.98

oxygen demand (BOD) was increased dramatically, indicating the formation of readily degradable intermediates from the recalcitrant initial wastewater compounds. The need for only partial ozone contact to fragment particular compounds, implies that for a given volume of industrial waste, the ozone reactor volume and O_3 generation rate are much lower. Thus, the pretreatment costs are reduced when compared to the usual objective of complete oxidation to the minimum possible effluent TOC level.

A number of successful systems using ozone have been demonstrated, although none has been used in the specific mode described above for pretreatment prior to land application. A partial list of ozone-treated wastes includes toluene, diisocyanate, ethylene dichloride, phenol, cyanide-containing wastes, polyethylene glycol and styrene production effluent.

Another industrial use for ozone would be for certain conditions requiring disinfection. Ozone does not generate chlorinated by-products that may be of environmental concern. Storage of chlorine or sodium hypochloride is not needed. Contact times for ozone are much shorter than with chlorine, so reactor size is greatly reduced. Thus, ozone can be used to disinfect certain pathogenic industrial wastes and can offset the need for extensive land treatment buffer zones.

It is very difficult to obtain accurate cost evaluations, specifically for partial oxidation likely to be used in pretreatment. The most logical approach is to conduct bench-scale tests to determine the necessary reaction conditions to remove the specific LLC. Then, allow the several suppliers of commercial ozone equipment to prepare cost estimates to remove 10%, 30%, 50%, 70% and 90% of the LLC, for the particular industrial waste conditions under investigation.

From a qualitative view, some comparative cost studies are available. Ozone is about twice as expensive as chlorine usage and is directly comparable in cost to sodium hypochloride when used to disinfect a secondary treated effluent (*Chemical Week,* 1971). However, ozone has a lot of additional environmental advantages over chlorination, so the comparison is not direct. When used as an advanced wastewater treatment process, ozone was slightly less expensive on an annual basis than activated carbon (*Chemical Engineering,* 1970).

Ozone-ultraviolet

The addition of ultraviolet radiation to the use of ozone with industrial waste results in a considerable expansion in the ability to remove specific organic constituents. These compounds would, of course, have to be the LLC determined from Chapter 2. Ultraviolet radiation is an additional energy source that (1) produces a large number of free radicals in an aqueous medium (similar to the free radicals from ozone addition), and (2) produces excited states of the organic compounds. This latter mechanism is a photolysis, which fragments the organic species producing intermediates. The UV wavelengths most commonly used are in the 180-400 nm range.

The presence of UV substantially increases the decomposition effect of ozone on specific organic compounds. With UV, the more stable short-chain and/or alkane species not removed by ozone are oxidized completely (Figure 13.10). Another example of the additional benefits of UV was given by

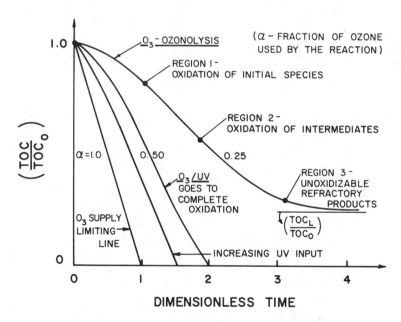

Figure 13.10   Enhanced organic compound degradation utilizing ultraviolet energy in addition to ozone (Prengle, 1977).

Prengle and Mauk (1977). Malathion in a waste stream was unaffected by ozone (Figure 13.11). With the addition of UV radiation a substantial decomposition occurred. More radiation per unit of Malathion further increased the breakdown. The reaction pathway is presented in Figure 13.12.

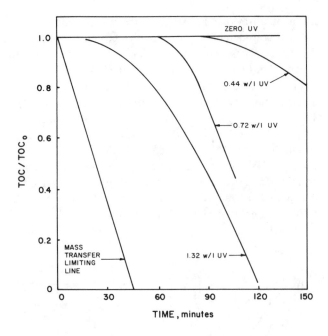

Figure 13.11   Increased Malathion decomposition with various ultraviolet doses (Prengle, 1977).

The enhanced breakdown of organics with the UV-ozone processes is caused by a number of factors. Less ozone is lost by autodecomposition or in the exit gas stream because the UV increases the reaction kinetic rates so that the mass transfer of ozone itself controls the overall decomposition. For specific compound pretreatment desired prior to land application, ultraviolet radiation has more potential. Examination of the adsorption spectrum of the compound to be removed allows the use of narrow wavelength bands, which provide maximum adsorption. In this manner this compound receives a greater dose and is more readily placed in excited or free radical state leading to decomposition. This selection can reduce the pretreatment costs by making more efficient use of the oxidation inputs.

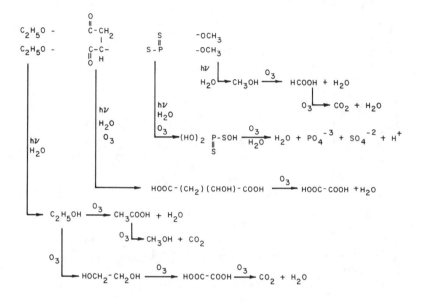

Figure 13.12 Malation reaction mechanisms (Prengle, 1977).

As a consequence of numerous observations (Prengle and Mauk 1977), the following molecular structure groups are determined to be vulnerable to enhanced attack by ozone-UV photooxidation:

1. exposed halogen atoms, *e.g.*, chloride ion appears rapidly, electronegative atoms distort charge distribution;
2. unsaturated resonant carbon ring structures;
3. readily accessible multibond carbon atoms; on the other hand, shielded multibonds, sulfur and phosphorus are much less vulnerable; and
4. alcohol and ether linkages.

A number of specific compounds and the respective wavelength for enhanced adsorption are given in Table 13.6.

As with ozonation, the use of ozone-UV is often different as a pretreatment prior to land application than for the conventional waste treatment usage. Where land is the receiver system, the objective is to remove specific compounds that control the land area requirements. Thus ozone-UV must be designed for this task. An example of this difference can be seen in Figure 13.13. Assuming pentachlorophenol (PCP) was the LLC of the given industrial waste, the loss of PCP occurred between 0 and 19 minutes, with corresponding chlorine liberation. In this time frame the pretreatment would be

Table 13.6 Example Compounds for Ozone-UV Pretreatment, with Theoretical Ozone Requirement (b), Average Specific Absorbance ($<\bar{A}>$), and UV Absorption Range ($\Delta\lambda$) (Prengel and Mauk 1977)

Compound (M)	Formula MW	$b(\dfrac{mgO_3}{mgM})$	$<\bar{A}>\ (\dfrac{cm^2}{m.mol})$	$\Delta\lambda$ (nm)
1) Malathion	$(CH_3O)_2(PS)SCH\ COOC_2H_5$ $\vert$ $CH_2COOC_2H_5$ 330	4.65	1.86 x 10³	200-270
2) Baygon	$CH_3NH(CO)OC_6H_4OCH(CH_3)_2$ 198	6.08	4.23 x 10³ 0.561 x 10³	180-240 240-290
3) Vapam	$CH_3NH(CS)SNa \cdot 2H_2O$ 129	3.63	1.42 x 10³ 0.078 x 10³	180-290 210-300
4) DDT	$ClC_6H_4(CHCCl_3)C_6H_4Cl$ 354.5	4.40	153 x 10³	190-280
5) Pentachlorophenol	C_6Cl_5OH 266.5	2.07	25 x 10³	200-335
6) o-Dichlorobenzene	$C_6H_4Cl_2$ 147	4.57	0.5 x 10³	180-285
7) Polychlorinated Biphenyls (PCB)	$C_6H_aCl_5C_6H_cCl_d$ 321	m 5.13	58 x 10³	180-250
8) Phenol	C_6H_5OH 94	7.15	0.6 x 10³	180-290

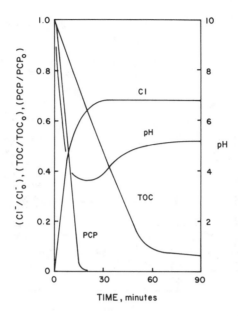

Figure 13.13   Specific compound decomposition using ozone-ultraviolet pretreatment (Prengle, 1977).

halted, depending on the percentage of original PCP desired to be removed. However, to produce an effluent for stream discharge, the process would have to continue until the desired levels of TOC were achieved. Thus, shorter reaction times, less ozone and UV, and lower costs are needed for this pretreatment objective.

Full-scale installations of ozone-UV equipment have been established for more than 100 industries (Prengle and Mauk 1977). A typical process schematic is given in Figure 13.14. Industrial wastes treated with ozone-UV include cyanide and cyanide complexes, Malathion, polyvinyl chloride, polychlorinated aromatics, polychlorinated biphenyls and numerous pesticides. Since these waste streams were oxidized to very low organic levels, it can be concluded that partial ozonation will remove the specific compounds likely to be the LLC; hence, ozone-UV is a viable pretreatment unit process for specific organics.

Since use as a partial oxidant has not been implemented prior to land treatment, the pretreatment economics are not available. Cost values have been developed for high-level removal of specific organics. Figure 13.15 presents the investment and operation/maintenance costs for polychlorinated

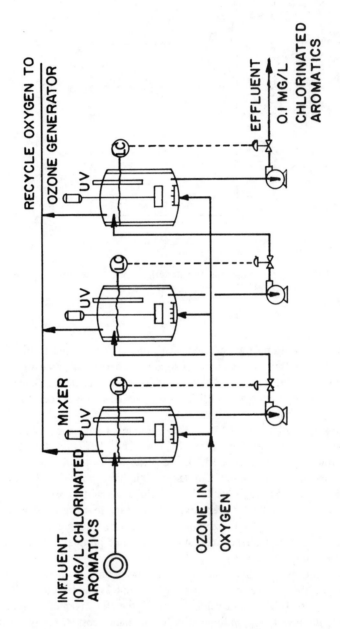

Figure 13.14 Process schematic for ozone-ultraviolet pretreatment (Prengle, 1977).

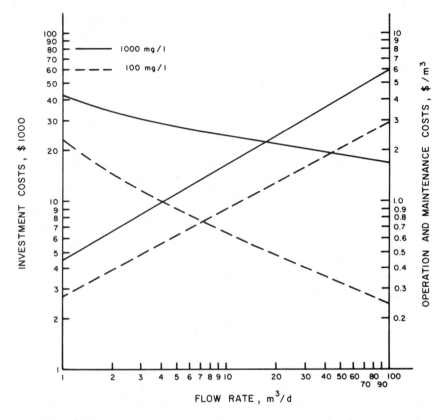

Figure 13.15 Investment and O & M costs for removal of polychlorinated aromatic by ozone-ultraviolet process (Prengle, 1977).

aromatic removal. These would represent upper limit economics. On a relative basis, the ultraviolet portion is far less expensive than the ozone part. As with ozonation, equipment manufacturers could be relied on to develop full scale economic estimates.

Ultraviolet and Other Radiation

Radiation can be used to decompose specific organics. An advantage of this pretreatment is that it performs well in nonpolar media, as well as in aqueous solutions. However, the efficiency of destruction is greatly improved by the presence of water.

The mechanism for radiation breakdown of compounds is the formation of unstable radicals, which decompose to fragments of the original compound.

In a nonpolar situation, the organic itself is the critical excited species. Thus, selective removal based on absorption spectra can be performed. With water present, hydroxyl and hydrogen radicals are also formed—the Compton effect (Allen 1961). These are particularly reactive (Table 13.4) and degrade specific compounds present. The decreased efficiency in removal of 3, 4, 2′ trichlorobiphenyl from hexane, when compared to water, is shown in Table 13.7. Over 1,000 times more radiation is needed without water than in the nonpolar hexane.

Table 13.7   Decomposition of 3,4,2′ Trichlorobiphenyl in Polar and NonPolar Media by Radiation Pretreatment (Merrill *et al.* 1977)

	System	Concentration (mg/l)	Dose in kr for 50% Degradation
1.	Hexane	4.0	1350
2.	Hexanol	4.3	1000
3.	Hexanol + 6% H_2O	4.1	800
4.	0.5% Na Stearate	0.7	350
5.	0.1% Na Stearate	0.7	250
	Water	0.07	<1

Radiation treatment of wastes has been demonstrated for municipal sludge (Merrill *et al.* 1977). This complex mixture of lipid phase, aqueous phase and specific compounds is similar to industrial waste situations. In aqueous mixtures, the destruction to various levels of a fixed initial concentration were studied for surfactant (linear alkylate sulfonate); 3, 4, 2′ trichlorobiphenyl; 4 chlorobiphenyl; and monuron (urea type herbicide) (Merrill *et al.* 1977, Rohrer and Woodbridge 1975). Realistic economic evaluations of this pretreatment process are not presently available on a transferable basis.

Other Advanced Processes

Incineration is one possible pretreatment process for industrial wastes, in which a complex mixture of specific organics are the LLC. In this instance, incineration is used to remove organics to a 99[+]% level. The remaining material is predominantly (although not exclusively) inorganics. Land treatment analysis, stage I, would then be performed again and the new LLC established. The land area would then be established for the assimilation of the incineration residue.

On a first estimate it would appear that the energy and operational costs for incineration would exceed the partial oxidation described for ozone or ozone-UV processes. However, in certain instances, incineration of a particular industrial waste may be cost-effective prior to land application.

With certain organic constituents as the LLC, separation and reuse is an effective pretreatment alternative. Solvents and constituents of reaction media are examples of organics that might be separated by distillation, liquid-liquid extraction, centrifugation or other processes. An example of solvent extraction for industrial waste has been reported by Rucker and King (1977). The separation product generally will not be as pure as the original chemicals, but may still be in the usable range. Recycling the separated material back to the manufacturing process for one or more times results in a corresponding reduction of LLC per unit mass of product manufactured or processed. Thus, for a given size manufacturing or processing facility, the land area for this LLC situation is reduced. The economics of such recovery and reuse options would represent the pretreatment cost, which would be balanced against the corresponding savings in the land system (Chapter 12) to determine the degree of recovery feasible.

Microwave plasma is an innovative process for decomposing reclacitrant organics. In this process, an ionized gas is produced by microwave-induced electrons, rather than thermal or radiative sources, as the major contributors of energy for promotion of the chemical reactions. The reactions are initiated by collisions between the reactant molecules and electrons in the plasma. Organic free radicals are generally considered as the primary intermediates in this process. The ionized gas, or plasma, is also derived from the carrier gas, which serves to transport the molecules into the plasma zone. Since the plasma decomposition mechanism involves electronic rather than thermal energy, the microwave applicator power coupling equipment can be maintained at low temperatures, that is, barely hot to the touch. The materials of construction associated with furnace or incinerator devices will therefore be generally unnecessary, and maintenance and repair expenses should be low or nonexistent.

This process has decomposed in excess of 99% of such materials as Kepone, Malathion, polychlorinated biphenyl, phenylmecuric acetate and methyl bromide (Bailin and Hertzler 1977). By controlling the waste feed rate, less or more decomposition can be achieved. This process is not particularly selective, so removal of a specific organic cannot be readily achieved. Such a process would, however, be effective in reducing the diversity of a really complex industrial waste or sludge.

In-plant Control

All industrial waste constituent categories determined to be the LLC in a land treatment system can be removed by means of in-plant options. For specific organics, the in-plant options include:

1. alteration of process variables, and
2. substitution of chemicals.

Before either option is considered in detail, the stage I LLC analysis must be performed to identify whether a particular compound or class of organic compounds is in fact the LLC. If not, then pretreatment or in-plant removal of organics is economically contraindicative.

If a specific organic(s) is identified as the LLC, the designer must attempt to determine the origin of the compound. It is a reactant or process input, an intermediate or by-product, or the final reaction product? Different courses of action are required for these three cases. If the LLC is a manufacturing process input, then process changes to ensure greater conversion to product will reduce the critical constituent. An intermediate or by-product as the LLC can generally be modified by altering the process conditions (temperature, residence time, solvent, etc.) to provide less of this material. For an organic LLC that is the actual manufacturing product, better separation of product from waste stream can reduce this parameter in the material to be land applied. These proposed solutions should be considered as possible choices, with technical input from process engineers and designers to establish the best option.

Substitution is also an effective in-plant control technique. The designer must establish the purpose for a particular chemical in the manufacturing sequence and then determine what alternative chemical can be used. From the possible substituents, selection is made of compounds with greater plant-soil assimilative capacity. For example, in one process, ethylene dichloride was being used. From the stage I analysis, this compound was determined to be the LLC. On investigation, the ethylene dichloride was used as a solvent, for which toluene could be used instead. Toluene was more readily assimilated in a plant-soil system, thus substituting solvents had no extra cost but resulted in a savings by reducing the required land area.

The diversity of in-plant alternatives is nearly as broad as the number of manufacturing plants in existence. The purpose of this section is to continue to focus on this option for reducing the LLC. Selection of in-plant techniques to remove a specific organic constituent must be determined as an economic trade-off between the land system and the cost of such in-plant change.

Metals, Anions and Salts

These industrial waste constituents have two "pretreatment" options to achieve a net removal, in an environmentally acceptable form:

1. concentration of constituent(s) with reuse, and
2. in-plant source control.

While these pretreatment options cannot be designed for uniform perform-ance with respect to all parameters included in these constituents categories, detailed discussions for each would be prohibitive. The discussion instead will highlight the alternatives generally available.

Concentration for Reuse

Electrolytic recovery can be designed to extract selected metals or to recover a broad spectrum of ionic species. The former process would be aimed at the specific metal defined to be the LLC (EPA 1973). On the other hand, recovery of a broad spectrum of species would then allow reversal to create plating or other solutions usable in the original manufacturing proces-ses. Considerable sophistication of electrolytic recovery processes has been developed recently, as municipal surcharge for metals has been enforced.

Ion exchange resins are available to remove specific or a broad spectrum of cationic or anionic species present in industrial wastes. A typical unit is illustrated in Figure 13.16. Having loaded the exchange complex, a backwash solution is used to strip the waste constituent. Further concentration can occur in the regenerant solution by means of electrolytic processes. End uses are thus as a specific metal or ion, or as an ionic solution for use in a manufacturing process. However, if no end use can be achieved, then the concentration process will only have achieved the formation of two streams for land treatment.

A third, frequently used concentration technique for these constituent categories is evaporative recovery. Either vacuum evaporators or atmospheric evaporators can be used (EPA 1973). Removal of water concentrates the waste stream until a usage can be successful at various levels. For example, when the concentration of the original process solution bath is achieved, then substitution into the manufacturing process is feasible (Figure 13.17). An ultimate use must exist for the concentrated solution.

In-plant Source Control

Metals as the LLC are sometimes present in the manufacturing raw materials as trace contaminants. In a recent industrial land treatment design, cadmium was determined as the LLC, yet it was not known to be inten-tionally added to the process. Further investigation revealed the cadmium was present in a salt solution purchased by the plant. Substitution of another supplier reduced the cadmium such that it was not the LLC. The in-plant source control program had reduced the LLC for the small difference of cost between suppliers of the salt solution.

For ions such as sodium, an imbalance present in the waste can be cor-rected by substituting calcium and magnesium for sodium. However, if both

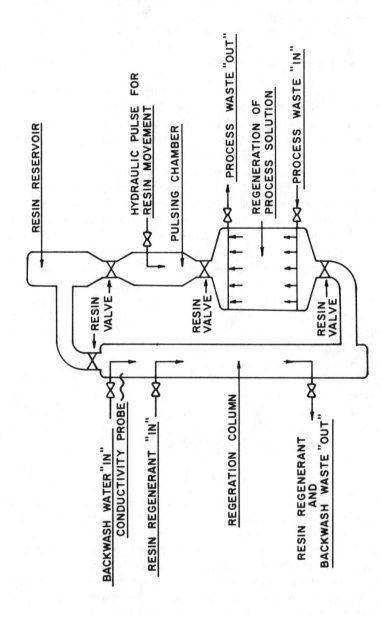

Figure 13.16 Ion exchange and regeneration process schematic for recovery of metal solutions (EPA, 1973).

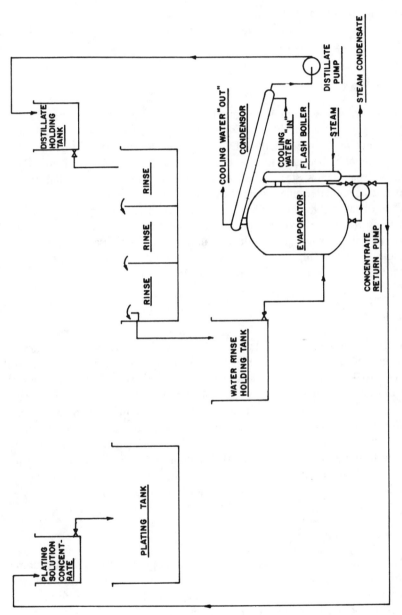

Figure 13.17 Evaporative concentration of metal plating solutions (EPA, 1973).

sodium imbalance and salinity exist, then substitution of one ion for others can not produce an acceptable waste for land treatment. Dilution or absolute reduction in-process of salts are the only viable "pretreatment" alternatives. The discussion of other in-plant options for these constituents is similar to that presented previously for phosphorus, sulfur and others.

Acids and Bases

Neutralization is the most commonly used pretreatment for acids or bases. Selection of a reactive addition must be done carefully to avoid adding species that might become the LLC. Organic acids and bases for neutralization are superior because they are readily degraded when applied to land. For acids and bases in waste, neutralization is required regardless of what other constituents are determined to be the LLC.

Concentration of acidic or basic waste streams is feasible in certain special situations where an industrial use can be found for the resultant material. Also, in-plant changes to conduct manufacturing processes as closely as possible to neutral pH, or in as little acid/base as possible, has a beneficial effect of minimizing the need for neutralization.

Nitrogen

Pretreatment removal of nitrogen involves the conversion of organic nitrogenous compounds and ammonium to an environmentally acceptable or industrially useful form. Most frequently, pretreatment for nitrogen as the LLC of an industrial waste is a nitrification-denitrification sequence, leading to N_2 gas. Another possible pretreatment is ammonium ion exchange until enough material is retained that backwashing yields a commercially acceptable ammonium solution. Other nitrogen pretreatment processes utilized in the municipal or industrial field do not give a net nitrogen removal from liquid and solids (sludge settling or nitrification) or evolve an unacceptable material to the atmosphere (ammonia volatilization).

Nitrification-Denitrification

The biochemical pathways required for the pretreatment nitrification-denitrification sequence are the same as those presented for the soil system in Chapter 11. The two-stage process is first aerobic, for conversion to nitrate, and then anaerobic, for conversion to N_2 gas.

In full-scale industrial systems the nitrification stage invariably requires the oxidation of all the biologically available carbonaceous material present in the waste. Thus, the oxygenation requirements are for the total oxygen demand. It is possible to achieve nitrification with many varied process flowcharts. These fit in two broad categories: (1) combined carbonaceous and nitrogenous oxidation, or (2) separated organic carbon oxidation vessel from the nitrification vessel. The relative advantages and disadvantages of these approaches have been described in detail elsewhere (EPA 1975).

Equipment to nitrify industrial wastes are designed to support either attached growth processes or suspended growth processes. In the first category are trickling filters, packed-bed vessels and rotating biodisc units. In the second category are activated sludge, oxidation ditch and extended aeration processes. Detailed discussion of these nitrification options is presented in summary form as Table 13.8.

Denitrification of a nitrified effluent must usually be accompanied by addition of a carbon source to provide substrate for biological denitrification and maintain anaerobic conditions. The carbon source may be some convenient biologically available compound added to the nitrified waste. Alternatively, it may be a part of the waste itself if careful design is used balancing nitrate formation with sufficient remaining carbon for denitrification arrangements (EPA 1975), and only a relative comparison is presented here (Table 13.9).

Since the amount of net nitrogen removal is to be determined by the relative pretreatment vs land system costs, no single level of removal can be fixed as a standard. Control of the percentage of nitrogen removed is probably best done by limiting the amount of nitrate formation. In this manner energy and expense are limited to the minimum, with nitrogen removal ranging from 10-95%.

Ion Exchange for Ammonium Concentration

Industrial wastewater, when passed through clinoptilolite (a zeolite), allows ammonium ions present to displace calcium, magnesium and sodium, and be absorbed. The common process flowchart for ammonium recovery is given in Figure 13.18. To obtain a usable ammonium product, the ion exchange column must be regenerated or stripped of ammonium. Stream regeneration of the column produces a solution of about 1% ammonia, which could be used in certain industrial processes. A gas-liquid countercurrent tower with recycle of the sodium chloride regenerant solution has been shown to yield up to 50% ammonium concentration (EPA 1974). The costs of selective ammonium exchange are estimated to be in the range of $0.04-0.06/1,000 liter.

Table 13.8 Evaluation of Alternative Nitrification Processes as a Part of Nitrogen
Constituent Pretreatment

System Type	Advantages	Disadvantages
Combined Carbon Oxidation-Nitrification		
Suspended growth	Combined treatment of carbon and ammonia in in a single stage. Very low effluent ammonia possible. Inventory control of mixed liquor stable due to high BOD_5/TKN ratio.	No protection against toxicants. Only moderate stability of operation. Stability linked to operation of secondary clarifier for biomass return. Large reactors required in cold weather.
Attached growth	Combined treatment of ammonia in a single stage. Stability not linked to secondary clarifier as organisms on media.	No protection against toxicants. Only moderate stability of operation. Effluent ammonia normally 1-3 mg/l (except RBD). Cold weather operation impractical in most cases.
Separate Stage Nitrification		
Suspended growth	Good protection against most toxicants. Stable operation. Very low effluent ammonia possible.	Sludge inventory requires careful control when there is a low BOD_5/TKN ratio. Stability of operation linked to operation of secondary clarifier for biomass return. Greater number of unit processes required than for combined carbon oxidation-nitrification.
Attached growth	Good protection against most toxicants. Stable operation. Less sensitive to low temperatures. Stability not linked to secondary clarifier as organisms on media.	Effluent ammonia normally 1-3 mg/l. Greater number of unit processes required than for combined carbon oxidation-nitrification.

Table 13.9 Evaluation of Alternative Denitrification Processes as a Part of Nitrogen Constituent Pretreatment

System Type	Advantages	Disadvantages
Suspended growth using methanol following a nitrification stage	Denitrification rapid, small structures required. Demonstrated stability of operation. Few limitations in treatment sequence options. Excess methanol oxidation step can be easily incorporated. Each process in the system can be separately optimized. High degree of nitrogen removal possible.	Methanol required. Stability of operation linked to clarifier for biomass return. Greater number of unit processes required for nitrification-denitrification than in combined systems.
Attached growth (column) using methanol following a nitrification stage.	Denitrification rapid, small structures required. Demonstrated stability of operation. Stability not linked to clarifier as organisms on media. Few limitations in treatment sequence options. High degree of nitrogen removal possible. Each process in the system can be separately optimized.	Methanol required. Excess methanol oxidation process not easily incorporated. Greater number of unit processes required for nitrification-denitrification than in combined system.
Combined carbon oxidation-nitrification-denitrification is suspended growth reactor using endogenous carbon source	No methanol required. Lesser number of unit processes required.	Denitrification rates very low; very large structures required. Lower nitrogen removal than in methanol-based system. Stability of operation linked to clarifier for biomass return. Treatment sequence options limited when both N and P removal required. No protection provided for nitrifiers against toxicants. Difficult to optimize nitrification and denitrification separately.

Table 13.9 Continued.

System Type	Advantages	Disadvantages
Combined carbon oxidation-nitrification-denitrification in suspended growth reactor using wastewater carbon source	No methanol required. Lesser number of unit processes required.	Denitrification rates low; large structures required. Lower nitrogen removal than in methanol-based system. Stability of operation linked to clarifier for biomass return. Tendency for development of sludge bulking. Treatment sequence options limited when both N and P removal required. No protection provided for nitrifiers against toxicants. Difficult to optimize nitrification and denitrification separately.

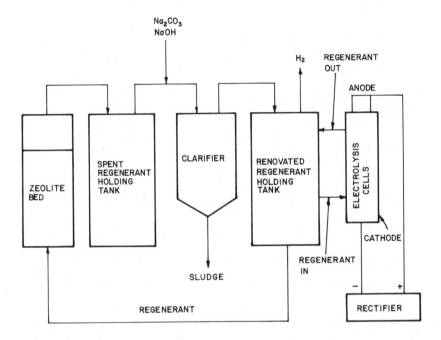

Figure 13.18 Ion exchange removal and concentration of regenerant solution as nitrogen constituent pretreatment (EPA, 1974).

In-plant Source Control

The principles used in evaluating in-plant options for reducing nitrogen levels are:

1. determine source or nitrogen;
2. establish capacity to reduce the amount of N appearing in the waste stream, and
3. examine the possibility of substituting for nitrogen in the manufacturing or processing facility a greater plant-soil assimilative capacity.

Since nitrogen is one industrial constituent that can be pretreated to yield a net loss in an environmentally acceptable form (N_2), the detailed discussion of in-plant options is presented earlier for conservative elements such as phosphorus and sulfur.

REFERENCES

Allen, A. O. *The Radiation Chemistry of Water and Aqueous Solutions* (New York: D. Van Nostrand Co., 1961).

Allison, R. C., C. B. Cobb and H. M. McCastlain. "Optimization of Water Use in Plywood Manufacturing," *Proc. 2nd Nat. Conf. Complete Watereuse,* AIChE, New York (1975), pp. 1176-1180.

Bailin, L. J., and B. L. Hertzler, "Development of Microwave Plasma Detoxification Process for Hazardous Wastes. Phase I," EPA-600/2-77-030, Cincinnati, OH (1977).

Fleischman, M. "Reuse of Wastewater Effluent as Cooling Tower Makeup Water," *Proc. 2nd Conf. on Complete Wateruse,* AIChE, New York (1975), pp. 501-514.

Little, L. W., J. C. Lamb III, M. A. Chillingworth and W. B. Durkin. "Acute Toxicity of Selected Commercial Dyer to the Fathead Minnow and Evaluation of Biological Treatment for Reduction of Toxicity," *Proc. 29th Purdue Ind. Waste Conf.* 145:524-534 (1974).

Merrill, E. W., D. R. Mabry, R. B. Schulz, W. D. Coleman, J. G. Trump and K. A. Wright. "Destruction of Trace Toxic Compounds in Water and Sludge by Ionizing Radiation," *Water-77* 245-250 (1977).

Prengle, H. W., Jr., and C. E. Mauk. "New Technology: Ozone/UV chemical Oxidation Wastewater Process for Metal Complexes, Organic Species and Disinfection," *Water* 228-244 (1977).

Quigley, R. E., and E. L. Hoffman. "Flotation of Oily Wastes," *Proc. 21st Purdue Ind. Waste Conf.* 527-533 (1966).

Rohrer, D. M., and D. D. Woodbridge. "Reduction in Surfactants by Irradiation," *Bull. Environ. Contam. Toxicol.* 13:31-36 (1975).

Ruker, N. L., and C. J. King. "Solvent Extraction for Treatment of Wastewaters from Acetic-Acid Manufacture," AIChE Symp. Series, *Water 1977,* 178(74):204-209 (1977).

Sayers, R. H., and R. J. Langlais. "Removal and Recovery of Sulfide from Tannery Wastewater," EPA-600/2-77-031, Cincinnati, OH (1977).

Seng, W. C. "Recovery of Fatty Materials from Edible Oil Refinery Effluents," EPA-660&2-73-015, Washington, DC (1973).

Sittig, M. *Pollutant Removal Handbook,* Noyes Data Corporation, Park Ridge, NJ (1973).

Sittig, M. *How to Remove Pollutants and Toxic Materials from Air and Water. A Practical Guide,* Noyes Data Corporation, Park Ridge, NJ (1977).

Springer, A. M., and D. W. Marshall. "Progress with Water Recycling in Nonintegrated Fine Paper and Board Manufacturing Operations," *Proc. 2nd Nat. Conf. Complete Watereuse,* AIChE. New York (1975), pp. 1155-1163.

U.S. Environmental Protection Agency. "Upgrading Metal-Finishing Facilities to Reduce Pollution," EPA 625/3-73-002 (1973).

U.S. Environmental Protection Agency. "Physical-Chemical Nitrogen Removal Wastewater Treatment," EPA 625/4-74-008, Technology Transfer, Washington, DC (1974).

U. S. Environmental Protection Agency. "Process Design Manual for Nitrogen Control," Technology Transfer, Washington, D.C. (1975).

Waggener, J. E., Jr., and J. C. North. "Ammonia Removal and Water Recycle in a Rayon Production Plant," *Proc. 2nd Nat. Conf. Complete Watereuse,* New York (1975), pp. 1200-1203.

Wallace, A. T., G. A. Rohlich and J. R. Villemonte. "The Effect of Inlet Conditions on Oil-Water Separations at SOHIO's Toledo Refinery," *Proc. 20th Purdue Ind. Waste Conf.* (1965), pp. 618-625.

Yocum, F. H., J. H. Mayes and W. A. Myers. "Pretreatment of Industrial Wastes with Ozone," *Water-1977* 217-227 (1977).

CHAPTER 14

TOTAL SYSTEM DESIGN—STAGE IV

INTRODUCTION

In this final phase of design for a total system involving land treatment of industrial waste, several items are available to the designer. The specific waste and site characteristics have been established. Land limiting constituent(s) (LLC) and total land area for the raw waste have been established (stage I). An in-depth analysis of all required components of the land system has been performed (stage II), with a resulting cost curve over a range of possible land area sizes (representing different percentages of the complete raw waste LLC). The complete range of possible pretreatment or in-plant source control options has been reviewed (stage III). Any processes that remove the LLC without an environmentally acceptable use for it are rejected. Approximate economics are developed over a range of removal of the LLC. Care has been taken to continually iterate the stage I analysis so that, as the initial LLC is removed, it is determined at what level some other constituent also becomes limiting. Beyond this level, the pretreatment selection and economics represent removal of all LLC(s).

Available Information from Previous Stages

Stage IV integrates the previous results to achieve an initial, least-cost pretreatment-land application system for the industrial waste under investigation (Figure 14.1). Within the overall methodology developed in this book, stage IV is the final phase in specifying the balance between land treatment of the limiting waste constituents and the pretreatment net removal of those constituents from the industrial waste. The balance between pretreatment vs land application is an economic matter, not a technical feasibility decision. That is, in stage I, the designer has established the assimilative criteria for all relevant waste constituents, so the land area of

637

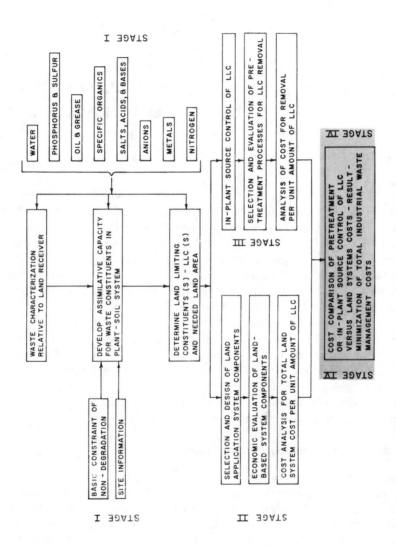

Figure 14.1 Relationship of Chapter 14 to overall design methodology for industrial pretreatment-land application systems.

from 100% to 0% of the raw waste is established. The economic decisions then dictate whether it is more effective to apply 100%, 0% or any intermediate percentage of the land limiting constituents to the plant-soil system, with the remainder being removed via pretreatment.

Type of Economic Balance that can be Used

Economic analyses performed in stage IV can involve various cost factors. The designer must establish which is the most relevant for the particular industrial case. The two most common cost analysis factors used are:

1. investment expenditure, or
2. total annual cost (amortized investment plus annual operation/maintenance expenses).

In the latter cost, the designer must select a depreciation interest rate and lifetime relevant to that particular industrial facility. With either investment or total annual cost, the economic curves for pretreatment (from stage III) and for the complete land system (from stage II) are plotted vs the percentage of the total raw waste that reaches the land treatment system (Figure 14.2). The sum of these two curves is the total system cost. The designer would then choose the relative levels of pretreatment and land application associated with the minimum total system cost (Figure 14.2)

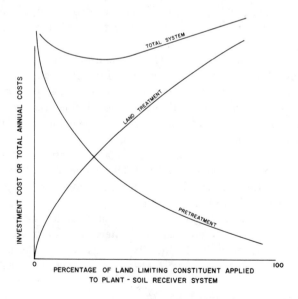

Figure 14.2. Representative economic balance between pretreatment and land treatment to minimize total system costs.

Selection of this minimum total cost point is thus the result of stage IV and represents the specification of the recommended total system. From this point, detailed drawings, specifications and bid documents can be prepared prior to implementation of the industrial land treatment system.

The remainder of Chapter 14 is devoted to four detailed examples of the overall methodology for design of pretreatment-land application of industrial wastes. Through these case studies, the principles developed in stages I-IV are illustrated using full-scale system design. The four case studies are:

1. industry A—pharmaceutical manufacturing (antibiotic synthesis)
2. industry B—organic and inorganic chemicals manufacture
3. industryC—poultry processing
4. industry D—electroplating

The authors wish to specifically acknowledge Soil Systems, Inc.* for the permission to utilize the diverse range of examples cited in Chapter 14.

INDUSTRY A—ANTIBIOTIC SYNTHESIS

Land treatment was to be considered as one waste management alternative in the construction of a new manufacturing facility. Waste generation was to occur in two streams: (1) a solids-laden (10-18%) slurry, and (2) a predominantly liquid discharge from cleaning and cooling operations. Land treatment was based on the combined waste to permit completion of the stage I analysis.

A complete analysis of the waste parameters relevant for a plant-soil receiver is given in Table 14.1, and on an individual stream basis, in Table 14.2. The sanitary wastes (about 35 people) were handled by a septic tank system. This separation of domestic waste was valuable because it removed the excessive buffer zone and pretreatment restrictions associated with domestic waste land application.

Two possible sites were investigated initially to estimate the relative advantages for land treatment (Table 14.3). The second site had less exchange capacity and less free drainage. In-depth testing to establish the assimilative capacity was then performed on site 1 only.

Water or hydraulic loading for these soils depends on the permeability of the various soil levels, the distances of interflow liquid travel to outlet and the climatic variables in the vicinity of this site. In the preliminary design stages it was recognized that the waste was a concentrated industrial

---

*525 Webb Industrial Dr., Marietta, GA 30062.

Table 14.1 Total Waste Load Characterization—Industry A

Parameter	Total Waste Generation	
	(kg/wk)	(kg/yr)
COD	6,100	320,000
N	350	18,200
NH_4-N	60	2,850
P	100	6,500
Cu	0.15	7.5
Zn	0.30	15.4
Ni	0.30	7.5
K	35	1,790
Water	5.6 ha·cm	291 ha·cm
Hg	< 0.008	< 0.41
Cd	< 0.008	< 0.41
Pb	< 0.065	<3.4

material, the soils were all in the well-drained category and the terrain contained reasonable slope. At site 1 detailed borings were taken to 5 m, and the simulation over available weather records established a conservative hydraulic application for the worst period in the annual cycle of 1.25 cm/wk.

Nitrogen (N) was to be assimilated in the land application system as a major crop nutrient. Other processes, such as ammonia volatilization, or organic nitrogen mineralization and nitrate leaching, were accounted for along with plant uptake. The industry A waste contained about 15% of the total nitrogen in the ammonium form (Table 14.1). Waste application would be by surface application; therefore, the research with other wastes had indicated that essentially all the NH_3 was expected to be lost. The low clay content of the sandy loam soils at this site also favored the enhanced loss of ammonia (Ryan and Keeney, 1975). Therefore, the net nitrogen for field assimilation is the total N minus the NH_4-N.

The available nitrogen assimilation is related to the crop uptake rates. To provide for maximum annual capacity for crop growth and nutrient uptake, a combination of grasses was used. The combination of areas for warm and cool season species was established based on months of growth, nitrogen uptake per season and site-specific soil and vegetation characteristics, which controlled growth of the respective species. Utilizing techniques in Chapter 11, the effective annual nitrogen assimilation at site 1 was 660 kg/ha/yr.

The assimilative capacity for metals, copper (Cu), zinc (Zn), lead (Pb), cadmium (Cd) and nickel (Ni) were taken as an accumulation process, then spread over a projected lifetime for this site. Annual monitoring of crops

Table 14.2 Waste Characterization for Separate Waste Streams–Industry A

Parameter	Solids (mg/l)	Liquid (mg/1)	Solid (mass, kg/wk)	Liquid (mass, kg/wk)
COD	51,000	550	5,580	560
N	2,900	35	315	35
NH_3-N	450	5	55	5
P	1,050	12.5	110	13
Ca	1,650	25	180	25
K	290	3	32	3
Na	245	25	180	25
Mg	82	1	9	1
Cu	1.25	0.01	0.14	0.01
Zn	2.5	0.025	0.27	0.025
Ni	1.25	0.01	0.14	0.01
Cd	< 0.07	0	< 0.008	0
Pb	< 0.6	< 0.006	< 0.065	< 0.006
H_2O	54,650	$0.5 \times 10^6 1$	–	–
Solids	14%	–	–	–

allows readjustment to possibly extend the site life according to the crop response. The site lifetime was set at 50 years. The permissible accumulative levels were those taken from Chapter 10. The sodium absorption ratio (SAR) was calculated as a measure of the balance of sodium and other cations:

$$SAR = \frac{Na}{[(Ca + Mg)/2]^{\frac{1}{2}}}$$

where all concentrations were in meq/l. From the clay analysis at site 1 it was determined that if SAR was kept less than 12-15, no sodium imbalance problems would occur. The waste, Table 14.1, has an SAR of about 1.3, so was easily assimilated in the plant-soil system.

Potassium (K) was evaluated for assimilation in crop growth, so application rates are for crop uptake. With forage hay crop fertilization, a ratio of 4:1 (N:K) was recommended. This waste had a N:K ratio of about 8.6:1 and therefore was deficient in potassium. Hence, the waste K was satisfactorily assimilated by this plant-soil system. Supplementary K would have to be added at this site.

Organics are assimilated under aerobic conditions maintained in the plant-soil system by the diffusion of oxygen from the atmosphere. The nature of the industry A waste is biological matter, so the degradation rate

Table 14.3 Soil Properties at Alternative Land Treatment Sites—Industry A

Site	Soil	Phosphorus Adsorption Maximum (mg P/100 g)	Cation Exchange Capacity (meq/100 g)	Bulk Density (g/cm³)	Acidity (meq/100 cm³)	Organic Matter (g/cm³)	pH	Soil Concentration, kg/ha·20 cm					
								P	K	Mg	Ca	S	Mn
1. Appling, cm													
	0-5	5.7	10.2	1.26	0.4	0.6	6.8	26	235	>240	1,020	28	30
	15-30	12.5	16	–	–	–	–	–	–	–	–	–	–
2. Madison, cm													
	0-15	8.7	6.0	1.31	1.2	0.6	5.6	39	225	178	620	–	15
	15-30	14.1	6.2	–	–	–	–	–	–	–	–	–	–

in the soil system will be rapid if aerobic conditions are maintained. There-fore, the organics assimilative capacity is determined by the supply of oxygen to exceed the waste oxygen demand in the soil. The soil porosity was measured as 0.2, and the hydrologic simulation gave a reaeration time of 3 days per week. No specific nonbiological organic species were identified for separate assimilative calculations.

The assimilative capacity for phosphorus is based primarily on the phosphorus fixation capacity and crop uptake. Phosphorus sorption maximum of 12.5 mg P/100 g soil (Table 14.3) was measured at site 1. Thus, a value of 58,000 kg/ha was used as the phosphorus fixation capacity plus crop uptake, which over a 50-year site life would be 1,160 kg/ha/yr.

Having established the waste generation and developed the plant-soil assimilative capacity on a constituent basis, the ratio of these two values was the area necessary for the environmentally acceptable land application of each constituents (Table 14.4). These values are determined for the soils at site 1 and for an operational period of 50 years. The design is also based on a two crop system of areas determined earlier: (1) fescue grass and (2) Coastal Bermudagrass. Such an arrangement will allow year-round application.

Table 14.4 Determination of Land Limiting Constituent (LLC) at Site I for Waste from Industry A

	Waste Generation (kg/yr)	Assimilative Capacity (kg/ha/yr)	Land Area (ha)	Acres
P	6,500	1,200	5.4	13.3
K	1,790	165	11	27
Cu	7.5	5	1.5	3.7
Zn	15.4	10	1.5	3.8
Pb	< 3.4	20	0.17	0.42
Ni	7.5	2	3.8	9.3
Cd	< 0.41	0.5	0.85	2.1
H_2O	5.6 ha cm/wk	1.25 cm/wk	4.5	11
COD	320,000	70,000	4.5	11.2
N (net)	15,400	600	23	56

Table 14.4 shows the limiting constituent for site 1 and waste from industry A was the nitrogen content. Total land area to be used for waste application will be 23 ha (56 ac), with about 36% being a warm season species and 64% being a cool season forage. Based on N as the LLC, the vegetative cover must be kept operational to ensure a satisfactory system operation.

Having established the LLC and the total land area requirements, the land system was then developed. In a preliminary study three alternatives were considered:

1. direct irrigation of both solids and liquids;
2. combination of solids and liquid in a storage pond with irrigation of the effluent; and
3. separate handling of waste streams with liquid irrigation and solids spreading

Approximate cost analysis identified the third alternative as the least expensive, and subsequent land system design was based on it (Figure 14.3).

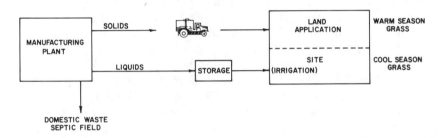

Figure 14.3. Land systems schematic for Industry A.

Land System Design

The solids stream was to be stored for time periods in a covered tank (see Chapter 12). Thus, relevant storage increments were periods of non-application, crop cycle, and repair and emergency allowance. With special flotation spreader equipment, the solids spreading was permissible nearly all year round, except during freezing and heavy rainfall conditions. Pump repair took 1 week while crop harvesting restriction was 10 days. The resulting solids storage volume was for 27 days.

The liquid stream must be stored in a separate pond because of the large effluent volume. The relevant pond volume elements from Chapter 12 were: (1) volume necessary to keep odor production at a reasonable minimum, (2) an emergency 24-hour storm with probability of occurrence equal to 25 years, and (3) a period of nonirrigation. The retention time derived from these data was 16.4 weeks, thus a volume of 10,000 m^3 was designed.

The solids produced were applied to the land by means of spreader truck. Utilizing process storage pumps, the solids (14% dry matter) were transferred to the truck at the plant site. Solids application would be 5 times per growing season on the Coastal Bermudagrass and 7 times per annual growing season for the fescue areas. The spreader truck will contain a 6,000-l tank designed for sludge applications. Top hatch loading was used to fill the spreader tank. Flotation tires were used so that nearly all soil conditions could be tolerated. Ground pressures were about 90-95 kPa (13-14 lb/in.^{2}) beneath the flotation tires.

A traveling gun sprinkler irrigation system was designed for the 3-ha spray field (Figure 14.4), shown by diagonal lines and identified as F-1 (Fescue irrigated). Three travel lanes were established 67 m apart. The direction of travel was as indicated by arrows and the pie-shaped circle showing the area of coverage of the sprinkler. The traveler system was installed, operated and maintained according to the manufacturer's recommended procedures. The traveling sprinkler moved at a speed of 0.8 m/min over a design length of run of 335 m and irrigated 2.2 ha in 7 hours operating time, applying 2.5 cm at a precipitation rate of 0.75 cm/hr.

To complete the harvesting of the 23 ha, a judicious combination of labor and machine automation is needed. In the Coastal Bermudagrass area (7 ha), the design nitrogen application will result in 16-18 metric ton/ha/yr of dry matter, which will be removed in 4-5 cuttings during the growth period (May-September). This yield will mean 115-130 metric ton/yr of dry matter, which, for a protein content of 15%, is 17-20 metric ton/yr of protein.

For the fescue areas (16 ha) the nitrogen application is projected to produce about 5.5-7 metric ton/ha/yr of dry matter. There will be approximately two cuttings in the winter-spring and two cuttings in the fall-winter periods. Total dry matter yield would be 90-110 metric ton/yr, which, with a protein content of 12%, is 11-13 metric ton/yr of protein. Both crops were harvested in a similar manner.

The major emphasis of the monitoring program was to assess the plant-soil assimilative capacity relative to the LLC and verify the performance relative to the groundwater quality. To achieve the first monitoring objective, samples must be taken of the soil and grass for analysis of NPK levels as well as the basic agricultural trace nutrients. Sampling areas are (1) Coastal Bermudagrass—solids spreading, (2) fescue—solids spreading, and (3) fescue—irrigation. In August and February, soil samples in the two fescue and the Coastal Bermuda area were collected for analyses. Yield measurements were to be made throughout the growing season and a forage quality measurement in April of each year. Three monitoring wells were to be used to measure groundwater quality.

A schematic of the total system and the field layout is shown in Figure 14.4. The land system components were developed with the larger elements described above. Site I was purchased with sufficient land for the manufacturing plant. The cost for the entire land system was calculated

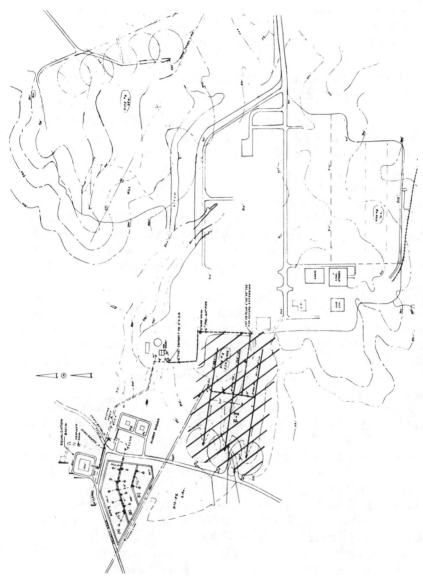

Figure 14.4 Field layout of Industry A land application system.

at $245,000, including land ($2,500/ha). For this system there was no feasible pretreatment option. In-plant, all the nitrogen came in proportion to the product sold, so reduction would have limited manufacturing. Nitrification-denitrification was too expensive because the land system was relatively inexpensive. Therefore, no further optimization calculations were made, and land treatment of the entire waste was implemented.

INDUSTRY B—ORGANIC AND INORGANIC CHEMICAL MANUFACTURE

The implementation of BPT, toxics regulation and the future requirements of BAT caused industry B to reevaluate its waste management philosophy. Compliance with fish bioassay, priority pollutant analyses and nutrient loading criteria were immediate problems. Land treatment was selected as a viable alternative leading to a nondischarge system. Hence, a program of design and analysis was instituted.

Waste characterization was the first step undertaken. Parameters selected were those relevant to land as the ultimate receiver systems (Table 14.5). A number of specific organic compounds were included as major reactant or product species. The entire plant waste was considered for land treatment.

An initial soil and hydrological investigation was undertaken to select the best site from six possible locations (Table 14.6). With implications from the waste characterization phase, site 1 was identified for in-depth site testing. Drilling, horizon analyses for treatment capability and areal compositing were performed on site 1 (Figure 14.5). These results were used to determine the constituent assimilative capacity as a part of the stage 1 analyses.

Assimilative data were not available for four of the specific organics identified in Table 14.5. A pilot-scale testing program was established to determine the critical phytotoxic level and the rate of degradation. These data were obtained and utilized in the LLC analysis (Table 14.7). Utilizing a grass receiver system, 20 ha of land were needed based on pentachlorobenzene as the LLC.

The stage II design of the land system for industry B was presented previously as Table 12.20. The cost analysis was performed to cover a range of pentachlorobenzene levels (Figure 12.38).

For this specific organic as the LLC there were several pretreatment or in-plant source control options. The pentachlorobenzene was found to be a reaction intermediate. Product research had determined that good quality could be maintained at 65°C in reaction vessels. Higher temperatures did not improve the product, so energy savings suggested the use of a 65°C

Table 14.5 Detailed Waste Characterization–Industry B.

Parameter	Concentration (mg/l)[a]	Mass Generated (kg/yr)	(kg/wk)
BOD$_5$	35	5,690	110
COD	552	89,800	1,730
TOC	130	21,200	410
TS	2,670	434,100	8,350
TDS	2,624	426,600	8,200
TSS	46	7,500	144
TKN	17.3	2,800	54
NO$_3$-N	0.3	49	0.94
NH$_4$-N	–	–	–
TN	17.6	2,860	55
TP	51.6	8,400	160
Conductivity	1.75 (mmhos/cm)	–	–
pH	7.2 (S.U.)	–	–
Na	45	7,320	140
K	10	1,625	31
Ca	0.96	156	3.0
Mg	0.97	158	3.0
SAR (no units)	7.7	–	–
Cu	< 0.004	< 0.65	< 0.01
Zn	0.05	8.1	0.16
Cd	< 0.001	< 0.16	< 0.003
Pb	< 0.01	< 1.6	< 0.03
Surfactant (nonionic)	33.4	5,400	104
Pentachlorobenzene	3.3	530	10.3
o-cresol	56.3	9,100	175
Pyruvic Acid	130	21,000	404
Succinic Acid	46	7,500	144
Water	–	163 x 10^6 l/yr	3.13 x 10^6 l/wk

[a]Unless otherwise noted.

temperature. However, pilot tests revealed that at 75°C, the pentachlorobenzene was reduced 15%, while at 85%, the reduction was 22%. Presumably the higher temperature caused more complete second-stage reaction, thus removing more intermediate. Greater heat input at the process source could reduce the LLC.

Further removal of pentachlorobenzene could be achieved by ozone addition. Pilot-scale studies revealed that 26% removal could be achieved using 5 times the rated flow in an ozonation reactor normally designed for complete detoxification and carbon removal. Addition of ultraviolet under the same flow conditions increased the removal to 32%. The

Table 14.6 Soil Characteristics of Alternative Sites—Industry B

Soil Characteristic	Site 1 (Spray Field)	Site 2 (Pinefield-spray)	Site 3 (Old Blg-West side)	Site 4 (Old Parking)	Site 5	Site 6 (Along E-W:00)
Mechanical						
Sand, %	49.7	64.3	43.6	59.4	50.7	46.7
Silt, %	13.9	12.9	22.9	26.6	17.4	19.7
Clay, %	36.4	22.8	33.5	14.0	31.9	33.6
Textural Class	Sandy clay loam	Sandy loam	Clay loam	Sandy loam	Sandy clay loam	Sandy clay loam
Organic matter, %	0.5	2.1	0.7	0.2	1.6	0.2
N, %	0.03	0.12	0.04	0.01	0.09	0.01
Exchange Acidity meq/100 g soil	8.2	7.3	6.4	8.6	10.7	10.3
pH (1:1, soil water)	4.6	4.7	6.3	4.7	4.9	4.5
Exchangeable Ions (meq/100 g soil)						
Calcium	0.44	1.4	3.5	0.4	0.98	0.07
Magnesium	0.27	0.3	0.3	0.1	0.32	0.1
Potassium	0.1	0.1	0.15	0.1	0.1	0.15
Sodium	0.13	0.0	0.0	0.03	<0.005	<0.005
Cation Exchange Capacity (meq/100 g soil)	9.14	9.10	10.35	9.23	12.11	10.63
	9.14	9.10	10.35	9.23	12.11	10.63
Base Saturation, %	10.3	19.8	38.2	6.8	11.60	3.1
Extractable Ions, ppm						
Aluminum	53.1	28.8	8.8	46.9	20.6	87.5
Manganese	1.25	0.96	0.12	7.15	4.60	1.85
Zinc, copper cadmium, chromium nickel, lead	All less than 1 ppm					

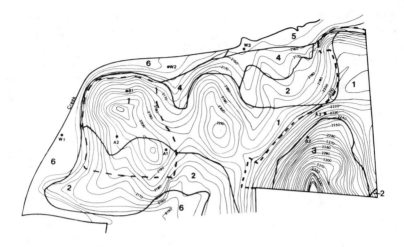

Figure 14.5 Field site layout including soil testing locations–Industry B.

Table 14.7 Stage I Assimilative Capacity Analysis–Industry B

Parameter	Waste Generation (kg/yr)	Assimilative Capacity (kg/ha/yr)	Land Area Required (ha)	Land Area Required (ac)
Total N	2,800	675	4.1	10.1
Total P	8,400	640	13.1	32.2
Zn	8.1	7.5	1.1	2.7
Cu	<0.65	3.8 3.8	0.17	0.42
Pb	<1.6	15 15	0.11	0.27
Cd	< 0.16	0.15	1.1	4.2
COD	89,900	152,000	0.59	1.4
Surfactant (nonionic)	6,400	360	15	36.9
Pentachlorobenzene	530	27	19.6	48.2
o-cresol	9,100	45,600	0.20	0.49
Pyruvic ACid	21,000	10,500	2.0	4.9
Succinic acid	7,500	12,500	0.60	1.5
Water	163×10^6 l/yr	238 cm/yr	6.8	16.7

probable pretreatment and in-plant costs were determined in conjunction with various equipment manufacturers (Figure 14.6). Investment costs was selected as the criterion for evaluating the land system and the pretreatment options.

Combining the various cost curves the total system investment was determined (Figure 14.6). A minimum total cost exists for which the relative proportions are 24% pretreatment removal and 76% land treatment of the LLC pentachlorobenzene. At this level pentachlorobenzene is still the LLC. In this instance a process temperature above 85°C and a small, high-rate ozonator involved nearly the same cost, so the process temperature change was selected.

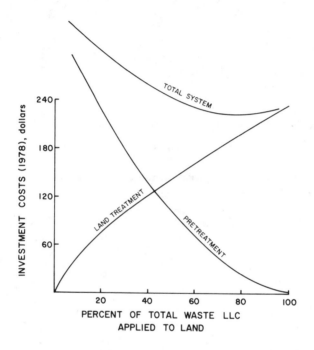

Figure 14.6 Total system investment costs—Industry B.

INDUSTRY C—POULTRY PROCESSING

An existing processing plant was in danger of noncompliance with stream standards and 1977 BPT requirements. Rather than continue with escalating restrictions on stream discharge, management purchased nearby land to achieve a zero-discharge system. Plant production was 70,000 birds/day on a 5 day/wk basis. The entire waste was in one effluent stream, which had received solids, blood and grease. This pretreatment was economically self-sustaining through sale of recovered material to renderer. This pretreatment was not the result of the land treatment system analysis, so the effluent from the recovery processes was taken as the raw waste.

The first portion of stage I was to obtain a representative waste characterization on a mass basis, *i.e.*, determine the mass of each constituent generated per month for an average production of 70,000 birds/day. A flume flow recorder and flow proportional sampler (N-CON Sentinel VI) were installed to obtain a 24-hour composite sample. This procedure was repeated 8 times to reliably determine constituent generation. Results are given in Table 14.8.

Table 14.8 Waste Load Generated—Industry C

Parameter	Concentration (mg/l)	Mass Generation (kg/day)
COD	1,200	4,500
TKN	49	185
Oil and Grease	246	920
TP	3	11
Cd	<0.02	<0.075
Cu	<0.02	<0.075
Zn	0.25	0.94
Pb	<0.1	<0.38
Na	30	112
K	5	19
Ca	0.5	1.9
Mg	1.5	56
pH	7.4 (S.U.)	—
SAR	4.8	—
H_2O	—	3.75×10^6 l/day

The second portion of stage I was to evaluate the critical land treatment characteristics. Experience with three previous poultry processing systems emphasized the need to establish accurately the hydrologic assimilative capacity. A high-intensity boring grid was established with undisturbed soil samples, analyzed by saturated falling head permeability technique. Various soil chemical and vegetation capacity tests were performed to establish the full range of site capabilities. The assimilative capacities for industry C waste constituents were then developed.

An LLC analysis was the completion step of Stage I at the designated site. Matching the constituent waste generation to the site assimilative capacity, it was determined that water or hydraulic loading was the LLC (Table 14.9). Total raw waste land treatment required 98.7 ha.

Table 4.9   Stage I, LLC Analysis for Selected Site and Waste from Industry C

Parameter	Waste Generation (kg/day)	Assimilative Capacity (kg/day/ha)	Area Required (ha)
Water	3.75×10^6 l/day	0.38 cm/day	98.7
Nitrogen	185	2.9	64
Oil and Grease	246	150	1.6
Total Phosphorus	11	0.58	18.9
Cadmium	<0.075	0.0035	<21.4
Copper	<0.075	0.086	< 0.9
Zinc	0.94	0.172	5.5
Lead	<0.38	0.36	< 1.04
Chemical Oxygen Demand	4,500	300	15

Stage II in the overall design process was next implemented. The land system components needed at this site were:

1. application system,
2. storage,
3. farming equipment,
4. monitoring, and
5. transmission to the site.

Selection of a solid set spray irrigation was made to accommodate the surface soil infiltration. Buried PVC with 1-m risers and rotating impact sprinklers were designed for the site. The storage pond was specified to accommodate the periods described in Chapter 12. The results of this analysis were similar to those presented in Table 12.10 with 3.5 months of storage. Monitoring the land treatment site was developed to maintain viable plant-soil performance and to test for impact on surrounding water bodies from land treatment. In tier 1 were nitrogen, water and phosphorus, while cadmium, sodium and oxygen demand were measured at the tier 2 level. Transmission line to the site was buried PVC for a distance of 1.2 km. The site was established with a double cropping system of Coastal Bermudagrass, with a cool season over-seed of annual rye grass. Because of the size of the land system, two sets of farm equipment were needed to ensure reliable harvesting. For the complete raw waste, the land system as designed was estimated to cost $280,000. The land system design and cost estimate was repeated for 10%, 20%, 30% and 35% reduction in the LLC. Below this level nitrogen also became the LLC, thus requiring nitrogen removal to further reduce the land area.

Stage III in the overall design methodology was to evaluate and cost account methods for reducing the LLC—water. In-plant source control held the most promise for lowering water usage. A list of 15 modifications was developed (Table 14.10), along with investment cost and water reduction amounts for each (Figure 14.7). These water reductions were arranged in

Table 14.10 Water Flow Modifications to Reduce Water Use at Poultry Processing Plant

1. Eviscerating Trough, Handwashers
2. Reduction in Scalder Flow
3. Modification to Trim Table
4. Feather Removal Changes
5. Blood Removal and Lung System
6. Plant Cleanup
7. Final Bird Washers
8. Gizzard Splitters
9. Side Pan Wash of Eviscerating Tough
10. Dry Solids Removal—Yard and Handling Area
11. Measure Ice to Chillers
12. Whole Bird Washers
13. Heart, Liver and Gizzard Handling System
14. Improved By-product Handling Equipment
15. Chiller Water to Scalder

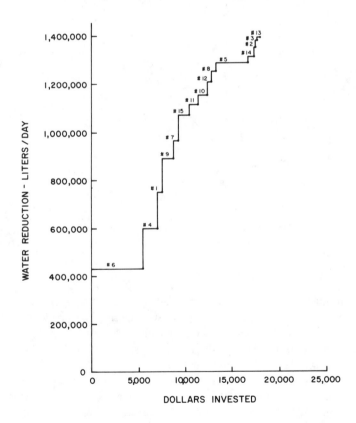

Figure 14.7 In-plant source control of water as LLC—Industry C.

order of decreasing effectiveness to evaluate the cost of reducing water (the LLC) usage from 100% of the present raw waste to 65% of that amount.

The final stage for industry C was to determine the least-cost total system of pretreatment-land application. Results of stage II and III were plotted over the range of 0-35% reduction of water usage (Figure 14.8). In this instance the minimum cost corresponded to the maximum water controlled in-plant. That is, any investment to reduce water usage was more than paid for by savings in the ultimate land receiver system. Thus, an implementation plan was developed to reduce water by 35%, so that land treatment was controlled by water and nitrogen.

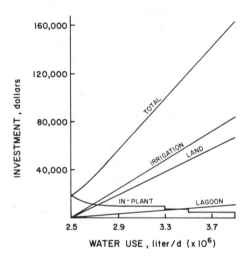

Figure 14.8 Total system investment cost—Industry C.

INDUSTRY D—ELECTROPLATING

Continually increasing surcharges led industry D to evaluate alternative solutions for its electroplating bath wastes. During the previous two years, three nearby tracts of land became available and were bought by industry D at $6,500-9,000/ha. At the suggestion of state regulatory personnel they began a complete land treatment design.

At the electroplating industry location, two sites were identified within their property (Figure 14.9). Complete soil testing was performed at each site, with particular emphasis on soil chemistry. These results are given in Table 14.11. The field testing involved considerable professional input, since

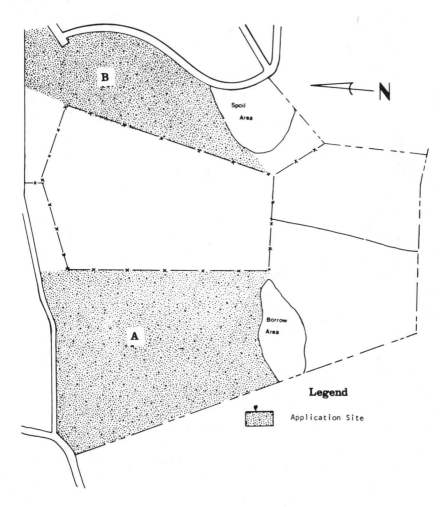

Figure 14.9 Site selection for soil testing—Industry D.

careful compositing, testing site selection and detailed field logging were required. Considerable field equipment, including test vehicles, was required, as in every one of the previous examples.

Waste characterization was completed at this plant. The several waste streams had varying flowrates as well as some changes with product mix. Hence, a six-month program of flow proportional sampling was performed to determine the average and range of parameter concentration. These results are given in Table 14.12.

Table 14.11 Site Evaluation—Industry D

	Site 1						Site 2					
	A Horizon		B Horizon		C Horizon		A Horizon		B Horizon		C Horizon	
Property	Mean	Range	Mean	Range	Mean	Range	Mean	Range	Mean	Range	Mean	Range
Depth, in.	2.5	0-5	27	5-48	54	48-60	4	0-8	28	8-48	66	48-84
Sand, %	55	40-65	40	25-55	55	45-60	35	25-45	35	25-45	50	40-60
Silt or Clay, %	45	35-60	60	45-75	45	40-55	65	55-75	65	55-75	50	40-60
Texture[a]	Loam to fine sandy loam		Sandy clay loam to clay loam		Sandy clay loam to fine sandy loam		Silt loam to loam		Clay loam to loam		Clay loam	
Available Water, in.	0.8	0.6-1.0	7.2	5.2-8.6	1.6	1.3-1.8	1.4	1.1-1.8	6.6	5.6-7.6	3.2	2.2-4.3
pH	0.8	4.5-5.5	5.0	4.5-5.5	5.0	4.5-5.5	5.0	4.5-5.5	5.0	4.5-5.5	5.2	4.5-6.0
Permeability, in./hr	4.0	2.0-6.0	1.3	0.6-2.0	4.0	2.0-6.0	3.3	0.6-6.0	1.3	0.6-2.0	1.3	0.6-2.0
Cation Exchange Capacity (CEC),[b] meq/100 g[d]	17.7	15.3-19.3	9.3	5.3-14.2			12.9		8.5			
Calcium (Ca),[c] meq/100 g	0.13	0.08-0.22	0.06	0.05-0.07			0.75		0.09			
Magnesium (Mg),[c] meq/100 g	0.10	0.07-0.14	0.05	0.02-0.07			0.33		0.07			
Potassium (k),[c] meq/100 g	0.22	0.19-0.27	0.14	0.13-0.15			0.36		0.25			
Sodium (na),[c] meq/100 g	0.25	0.16-0.34	0.18	0.19-0.29			0.35		0.21			
Base Saturation, %	3.9	3.5-4.7	4.8	3.9-5.8			13.9		7.3			

[a] SCS classification.
[b] Cation exchange capacity.
[c] Exchangeable.
[d] meq/100 g—milliequivalents per 100 grams of soil.

Table 14.12 Average Flow-Weighted Wastewater Characteristics—Industry D

Constituent	Concentration (mg/l)
Total Suspended Solids	22.2
Copper	0.94
Nickel	0.29
Total Chromium	0.62
Iron	0.85
Lead	0.08
Fluoride	1.74
Phosphate	3.14
Total Cyanide	0.09
BOD	13.6
Fecal Coliform	0 per ml
pH	7.2 (S.U.)
Water	1.5×10^6 l/wk

The mass assimilative capacities (kg/ha/yr) were determined utilizing measured soil properties. A system lifetime of 100 years was selected. With a second site available, the total lifetime of nearly 200 years was felt to be an acceptable criterion for design. That is, management wanted a conservatively designed system to ensure satisfactory long-term performance. Data from Chapters 4-11 were used to establish the assimilative capacity, and, hence, the LLC (Table 14.13). Chromium was identified as the LLC, and a land area of 36.2 ha was needed for complete raw waste land treatment.

The land system design and economic analysis were then undertaken. A spray irrigation system was selected as the application method because of the relatively dilute nature of the waste. The terrain permitted use of a traveler, which was then included in the design. Storage was essential at this

Table 14.13 Stage I, LLC Analysis—Industry D

Parameter	Generation Rate (kg/wk)	Assimilative Capacity (kg/ha/wk)	Area Required (ha)
Copper	1.4	0.048	29
Nickel	0.44	0.019	23
Total Chromium	2.8	0.077	36.2
Iron	1.28	0.48	2.7
Lead	0.12	0.19	0.6
Fluoride	2.6	0.08	32.5
Phosphate	4.7	23	0.2
Total Cyanide	0.14	0.0093	15
Cadmium	0.014	0.002	7.0
Water	1.5×10^6 l/wk	2 cm/wk	7.5

site and volume elements of nonirrigation, repair and emergency, single 25-year, 24-hour rainfall and freeboard were included. A total of 7.5 weeks equivalent of waste volume were used as the storage pond criterion. Utilizing the site topography, the pond was constructed with dirt equalling about 20% of the storage volume, thus saving considerable money.

The selected site was 0.9 km from the electroplating plant. Transmission of waste to the storage pond was by means of buried PVC. The storage pond also served as a sump, with a pump and pumphouse located at the pond. No

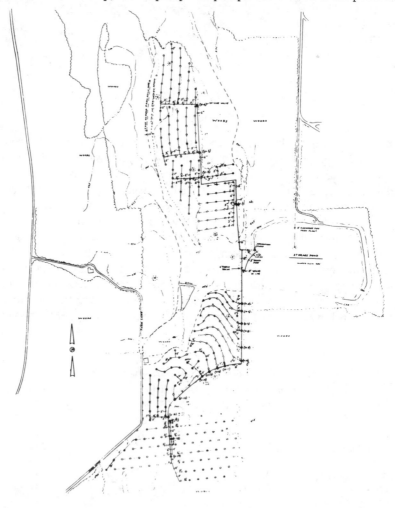

Figure 14.10 Overall system layout—Industry D.

diversion structures at the irrigation site were necessary. Monitoring involved 3 chemicals as tier 1 and 2 chemicals as tier 2. Annual samples for tier 1 were taken of soil, plants and groundwater.

No sanitary wastes were present and so a 10-m buffer zone was included. This site already was established in bromegrass. Farm equipment specified at the site were tractor, mower, rake, baler, wagons and hay barn. All of the land system components cost $312,000 as initial investment (including land used).

No pretreatment alternatives were investigated because of the LLC involved. Subsequently, construction plans and documents were prepared to allow land treatment system construction. The overall land system schematic is given in Figure 14.10.

REFERENCES

Ryan, J. A., and D. R. Keeney. "Ammonia Volatilization from Surface-applied Wastewater Sludge," *J. Water Poll. Control Fed.* 47:386-393 (1975).

CONVERSION TABLES

Multiply	by	to Obtain
Acres	43,560	Square feet
Acres	4,047	Square meters
Acres	1.562×10^{-3}	Square miles
Acres	4840	Square yards
Acres	0.4042	Hectares
Acre-feet	43,560	Cubic feet
Acre-feet	3.259×10^5	Gallons
Acre-inch	2.75×10^4	Gallons
Acre-inch	1.03	Hectare-centimeters
Angstrom units	3.937×10^{-9}	Inches
Atmospheres	76.0	Centimeters of mercury
Atmospheres	29.92	Inches of mercury
Atmospheres	33.90	Feet of water
Atmospheres	10,333	Kilograms/square foot
Atmospheres	14.70	Pounds/square inch
Atmospheres	1.058	Tons/square foot
Barrels (British, dry)	5.780	Cubic feet
Barrels (British, dry)	0.1637	Cubic meters
Barrels (British, dry)	36	Gallons (British)
Bars	0.9869	Atmospheres
Bushels	1.244	Cubic feet
Bushels	2150	Cubic inches
Bushels	0.03524	Cubic meters
Bushels	4	Pecks
Bushels	64	Pints (dry)
Bushels	32	Quarts (dry)
Centimeters	0.0328083	Feet (U.S.)
Centimeters	0.3937	Inches
Centimeters	0.01	Meters
Centimeters	10	Millimeters

Multiply	by	to Obtain
Centimeters of mercury	0.01316	Atmospheres
Cord-feet	4 feet x 4 feet x 1 foot	Cubic feet
Cords	8 feet x 4 feet x 4 feet	Cubic feet
Cubic centimeters	3.531×10^{-5}	Cubic feet
Cubic centimeters	6.102×10^{-2}	Cubic inches
Cubic centimeters	10^{-6}	Cubic meters
Cubic centimeters	1.308×10^{-6}	Cubic yards
Cubic centimeters	2.642×10^{-4}	Gallons
Cubic centimeters	10^{-3}	Liters
Cubic centimeters	2.113×10^{-3}	Pints (liquid)
Cubic centimeters	1.057×10^{-3}	Quarts (liquid)
Cubic centimeters	0.033814	Ounces (U.S. fluid)
Cubic feet	2.832×10^{4}	Cubic centimeters
Cubic feet	1728	Cubic inches
Cubic feet	0.02832	Cubic meters
Cubic feet	0.03704	Cubic yards
Cubic feet	7.481	Gallons
Cubic feet	28.32	Liters
Cubic feet	59.84	Pints (liquid)
Cubic feet	29.92	Quarts (liquid)
Cubic feet of water (60° F)	62.37	Pounds
Cubic inches	16.39	Cubic centimeters
Cubic meters	10^{6}	Cubic centimeters
Cubic meters	35.31	Cubic feet
Cubic meters	61,023	Cubic inches
Cubic meters	1.308	Cubic yards
Cubic meters	264.2	Gallons
Cubic meters	10^{3}	Liters
Cubic meters	2113	Pints (liquid)
Cubic meters	1057	Quarts (liquid)
Cubic meters	8.1074×10^{-4}	Acre-feet
Cubic meters	8.387	Barrels (U.S., liquid)
Cubic yards	7.646×10^{5}	Cubic centimeters
Cubic yards	27	Cubic feet
Cubic yards	0.7646	Cubic meters
Days	1440	Minutes
Days	86,400	Seconds
Feet	30.48	Centimeters
Feet	12	Inches
Feet	0.3048	Meters

Multiply	by	to Obtain
Feet	1/3	Yards
Feet (U.S.)	1.893939×10^{-4}	Miles (Statute)
Gallons (British)	4516.086	Cubic centimeters
Gallons (British)	1.20094	Gallons (U.S.)
Gallons British	10	Pounds (avordupois) of of water at $62°F$
Gallons (U.S.)	128	Ounces (U.S. Fluid)
Gallons	3785	Cubic centimeters
Gallons	0.1337	Cubic feet
Gallons	231	Cubic inches
Gallons	3.785×10^{-3}	Cubic meters
Gallons	4.951×10^{-3}	Cubic yards
Gallons	3.785	Liters
Gallons	8	Pints (liquid)
Gallons	4	Quarts (liquid)
Gallons/minute	2.228×10^{-3}	Cubic feet/second
Gallons/minute	0.06308	Liters/second
Grams	10^{-3}	Kilograms
Grams	10^3	Milligrams
Grams	0.03527	Ounces
Grams	0.03215	Ounces (troy)
Grams	0.07093	Poundals
Grams	2.205×10^{-3}	Pounds
Grams/cubic meters	0.43700	Grains/cubic foot
Grams/centimeter	5.600×10^{-3}	Pounds/inch
Grams/cubic centimeter	62.43	Pounds/cubic foot
Grams/cubic centimeter	0.03613	Pounds/cubic inch
Grams/cubic centimeter	3.405×10^{-7}	Pounds/mil foot
Grams/cubic centimeter	8.34	Pounds/gallon
Grams/liter	58.417	Grains/gallon (U.S.)
Grams/liter	9.99973×10^{-4}	Grams/cubic centimeter
Grams/liter	1000	Parts/million (ppm)
Grams/liter	0.06243	Pounds/cubic foot
Hectares	10^{-4}	Square meters
Hectares	2.46	Acres
Horsepower	42.44	British thermal units/ minute
Horsepower	33,000	Foot-pounds/minute
Horsepower	550	Foot-pounds/second
Horsepower	1.014	Horsepower (metric)
Horsepower	0.7457	Kilowatts
Horsepower	745.7	Watts

Multiply	by	to Obtain
Horsepower (metric)	0.98632	Horsepower
Horsepower-hours	2547	British thermal units
Horsepower-hours	1.98×10^6	Foot-pounds
Horsepower-hours	2.684×10^6	Joules
Horsepower-hours	641.7	Kilogram-calories
Horsepower-hours	2.737×10^5	Kilogram-meters
Horsepower-hours	0.7457	Kilowatt-hours
Hours	60	Minutes
Hours	3600	Seconds
Inches	2.540	Centimeters
Inches of Mercury	0.03342	Atmospheres
Kilograms	10^3	Grams
Kilograms	2.2046	Pounds
Kilograms	1.102×10^{-3}	Tons (short)
Kilograms/cubic meter	10^{-3}	Grams/cubic meter
Kilograms/cubic meter	0.06243	Pounds/cubic foot
Kilograms/cubic meter	3.613×10^{-5}	Pounds/cubic inch
Kilograms/hectare	0.894	Pounds/acre
Kilometers/hour	54.68	Feet/minute
Kilometers/hour	0.9113	Feet/second
Kilowatts	56.92	British thermal Units/minute
Kilowatts	4.425×10^4	Foot-pounds/minute
Kilowatts	737.6	Foot-pounds/second
Kilowatts	1.341	Horsepower
Kilowatts	14.34	Kilogram-calories/minute
Kilowatts	10^3	Watts
Kilowatt-hours	3415	British thermal units
Kilowatt-hours	2.655×10^6	Foot-pounds
Kilowatt-hours	1.341	Horsepower hours
Liters	10^3	Cubic centimeters
Liters	0.03531	Cubic feet
Liters	61.02	Cubic inches
Liters	10^{-3}	Cubic meters
Liters	1.308×10^{-3}	Cubic yards
Liters	0.2642	Gallons
Liters	2.113	Pints (liquid)
Liters	1.057	Quarts (liquid)
Liters/minute	5.885×10^{-4}	Cubic feet/second
Liters/minute	4.403×10^{-3}	Gallons/second
Meters	100	Centimeters
Meters	3.2808	Feet

Multiply	by	to Obtain
Meters	39.37	Inches
Meters	10^{-3}	Kilometers
Meters	10^3	Millimeters
Meters	1.0936	Yards
Meters	10^{10}	Angstrom units
Meters	6.2137×10^4	Miles
Micrograms	10^{-6}	Grams
Microliters	10^{-6}	Liters
Microns	10^{-6}	Meters
Miles	5280	Feet
Miles	1.6093	Kilometers
Miles	1760	Yards
Milligrams	10^{-3}	Grams
Milliliters	10^{-3}	Liters
Millimeters	0.1	Centimeters
Millimeters	0.03937	Inches
Millimeters	39.37	Mils
Months	30.42	Days
Months	730	Hours
Months	43,800	Minutes
Months	2.628×10^6	Seconds
Ounces	28.35	Grams
Ounces	0.0625	Pounds
Ounces (fluid)	1.805	Cubic inches
Ounces (fluid)	0.02957	Liters
Parts/million	0.0584	Grains/U.S. gallon
Parts/million	0.7016	Grains/Imperial gallon
Parts/million	8.345	Pounds/million gallons
Pascals	10^{-5}	Atmospheres
Pounds	7000	Grains
Pounds	453.6	Grams
Pounds	16	Ounces
Pounds	32.17	Poundals
Pounds/cubic foot	0.01602	Grams/cubic centimeter
Pounds/cubic foot	16.02	Kilograms/cubic meter
Pounds/cubic foot	5.787×10^{-4}	Pounds/cubic inch
Pounds/cubic foot	5.456×10^{-9}	Pounds/mil foot
Pounds/cubic inch	27.68	Grams/cubic centimeter
Pounds/cubic inch	2.768×10^4	Kilograms/cubic meter
Pounds/cubic inch	1728	Pounds/cubic foot
Pounds/cubic inch	9.425×10^{-6}	Pounds/mil foot
Quarts (dry)	67.20	Cubic inches

Multiply	by	to Obtain
Quarts (liquid)	57.75	Cubic inches
Quarts (U.S. liquid)	0.033420	Cubic feet
Quarts (U.S. liquid)	32	Ounces (U.S. fluid)
Quarts (U.S. liquid)	0.832674	Quarts (British)
Square centimeters	1.076×10^{-3}	Square feet
Square centimeters	0.1550	Square inches
Square centimeters	10^{-6}	Square meters
Square feet	2.296×10^{-5}	Acres
Square feet	929.0	Square centimeters
Square feet	144	Square inches
Square feet	0 09290	Square meters
Square feet	3.587×10^{-8}	Square miles
Square feet	1/9	Square yards
Square inches	6.452	Square centimeters
Square inches	6.944×10^{-3}	Square feet
Square meters	2.471×10^{-4}	Acres
Square meters	10.764	Square feet
Square meters	3.861×10^{-7}	Square miles
Square meters	1.196	Square yards
Square miles	640	Acres
Square miles	27.88×10^{6}	Square feet
Square miles	2.590	Square kilometers
Square miles	3.098×10^{6}	Square yards
Square yards	9	Square feet
Square yards	0.8361	Square meters
Square yards	3.228×10^{-7}	Square miles
Temperature (°C) + 273	1	Absolute temperature (°C)
Temperature (°C) + 17.8	1.8	Temperature (°F)
Temperature (°F) + 460	1	Absolute temperature (°F)
Temperature (°F) - 32	5/9	Temperature (°C)
Tons (long)	1016	Kilograms
Tons (long)	2240	Pounds
Tons (metric)	10^{3}	Kilograms
Tons (metric)	2205	Pounds
Tons (short)	907.2	Kilograms
Tons (short)	2000	Pounds
Weeks	168	Hours
Weeks	10,080	Minutes

Multiply	by	to Obtain
Weeks	604,800	Seconds
Yards	91.44	Centimeters
Yards	3	Feet
Yards	36	Inches
Yards	0.9144	Meters
Years (common)	365	Days
Years (commons)	8760	Hours